Developments in Plant and Soil Sciences

Volume 6

Series ISBN 90-247-2405-8

NITROGEN CYCLING IN ECOSYSTEMS OF LATIN AMERICA
AND THE CARIBBEAN

United Nations Environment
Program (UNEP)

Nitrogen Cycling in Ecosystems of Latin America and the Caribbean

edited by

G.P. ROBERTSON

R. HERRERA

T. ROSSWALL

Reprinted from *Plant and Soil* Vol. 67 (1982)

1982

MARTINUS NIJHOFF / DR W. JUNK PUBLISHERS
THE HAGUE / BOSTON / LONDON

Proceedings of a regional workshop arranged by the SCOPE / UNEP International Nitrogen Unit of the Royal Swedish Academy of Sciences under UNEP contract FP / 1303-78-01 (1330) at the Centro Internacional de Agricultura Tropical (CIAT), Cali, Colombia, 16-21 March, 1981; a meeting sponsored by SCOPE, UNEP, MAB and COSTED.

Distributors

for the United States and Canada

Kluwer Boston, Inc.
190 Old Derby Street
Hingham, MA 02043
USA

for all other countries

Kluwer Academic Publishers Group
Distribution Center
P.O.Box 322
3300 AH Dordrecht
The Netherlands

Library of Congress Cataloging in Publication Data
Main entry under title:

Nitrogen cycling in Latin American and Caribbean
    escosystems.

    (Developments in plant and soil sciences ; v. 6)
    "Reprinted from Plant and soil, vol. 67."
    Papers from a workshop held at Ciat, Cali, Colombia,
in March 1981, which was arranged by the SCOPE/UNEP
International Nitrogen Unit.
    Includes bibliographies and index.
    1. Nitrogen cycle--Latin America--Congresses.
2. Nitrogen cycle--Caribbean area--Congresses.
3. Agricultural ecology--Latin America--Congresses.
4. Agricultural ecology--Caribbean area--Congresses.
I. Robertson, G. P. II. Herrera, R. III. Rosswall, T.
(Thomas) IV. SCOPE/UNEP International Nitrogen Unit.
V. Series.
QH106.5.N57 1982   631.4'17   82-12631
ISBN 90-247-2719-7

ISBN-13: 978-94-009-7641-2   e-ISBN-13: 978-94-009-7639-9
DOI: 10.1007/978-94-009-7639-9

Preface

The large and rapidly expanding body of literature related to nitrogen cycling in both managed and native terrestrial ecosystems reflects the importance accorded to the behaviour of this vital and often limiting nutrient. Research at the organism, ecosystem and landscape levels commonly addresses questions concerning nitrogen acquisition, internal cycling and retention. Goals for this research include increased agricultural productivity and a better understanding of human impact on local, regional and global nitrogen cycles.

Nitrogen cycle research in tropical regions has a long and distinguished history. Research on different aspects of nitrogen cycling in ecosystems of the tropics has been carried out in many regions. In relatively few instances has there, however, been a focus on the biogeochemical cycles at the ecosystem level. The meeting resulting in this volume was an attempt to bring together existing information on nitrogen cycling in ecosystems of Latin America and the Caribbean and discuss this in an ecosystem context.

The papers represent the proceedings of a workshop on Nitrogen Cycling in Ecosystems of Latin America and the Caribbean, the third workshop on nitrogen cycling within particular regions organized by the SCOPE/UNEP International Nitrogen Unit of the Royal Swedish Academy of Sciences, Stockholm. The purpose of the workshop was fivefold: 1) to emphasize the importance of the nitrogen cycle in the different ecosystems of the region, 2) to provide a forum for scientists from the region to present papers describing ongoing nitrogen-cycle research, 3) to compile available data into coherent nitrogen budgets for the region's main ecosystems, and 5) to define nitrogen-cycle research priorities for the region. Previous workshops have been held in West Africa[1] and in Southeast Asia[2]. The three workshops have been supported by UNEP under contract FP/1303–78.01(1330).

The present workshop was held 16–21 March, 1981, at CIAT (Centro Internacional de Agricultura Tropical) in Cali, Colombia. Three days of symposia and contributed paper sessions were followed by two days of workgroup discussions organized around major ecosystems of the region. These included shifting cultivation and traditional agroecosystems, sugarcane, cereal and grain crops, coffee and cacao plantations, savannas and shrublands, forests, and wetlands and aquatic systems. Workgroups were charged with building informal nitrogen budgets of the respective systems and thereby summarizing the current state of knowledge regarding nitrogen cycles in each system. They were also asked to discuss research priorities, which were later reviewed by the plenary session. These priority rankings will, we hope, be useful for efficiently focusing increasingly scarce research resources on important but little-understood nitrogen-cycle processes.

The volume contains most of the papers presented at the meeting and the work group reports. Three additional papers by scientists from the region unable to participate in the meeting are also included. A number of papers were originally presented in Spanish or Portuguese. In order to ensure as large an audience as possible for these reports we decided to publish all papers in English with a Spanish summary.

Co-sponsors of the meeting apart from SCOPE and UNEP included the Man and the Biosphere (MAB) programme of Unesco, and the Committee on Science and Technology in Developing Countries (COSTED). We are greatly indebted to all the sponsoring organizations for their interest and support. The organizers also extend particular thanks to CIAT Director General J. L. Nickel and his hospitable staff, and also to the simultaneous translators.

We are also indebted to Britta Myrvik, Gudrun Sunnerstrand and Peter Wigren for artwork revisions and to Dina Söderström and Gun Martinsson for typing the final manuscript.

Despite minor difficulties with communicating in three languages, we think most participants will agree that the workshop was a success and that its major objectives were well-met.

East Lansing and Uppsala, April, 1982

G. P. Robertson, R. Herrera and T. Rosswall

References

1 Rosswall T (Ed.) 1980 Nitrogen Cycling in West African Ecosystems. Stockholm: Royal Swedish Academy of Sciences, 450 p.
2 Wetselaar R, Simpson J R and Rosswall T (Eds.) 1981 Nitrogen Cycling in South-East Asian Wet Monsoonal Ecosystems. Canberra: Australian Academy of Sciences. 216 p.

Contents

Plant and Soil 67, 1–13 (1982).   0032-079X/82/0671-0001$01.95.

Plant assimilation and nitrogen cycling

Asimilación de nitrógeno por las plantas y el ciclo de este elemento

A. A. FRANCO

Empresa Brasileira de Pesquisa Agropecuária (EMBRAPA), SNLCS-PFBN-km 47, Seropédica, Rio de Janeiro 23460, Brazil

and D. N. MUNNS

Department of Land, Air and Water Resources, University of California, Davis, California 95616, USA

Key words N-assimilation   N-cycling   N-fertilization N_2-fixation   Nitrogenase   N-mineralization   N-reductase   pH changes

Abstract   Nitrogen, an abundant and yet limiting nutrient for crop and food production, enters the plant as nitrate or ammonium, or as dinitrogen through biological fixation by procaryotes associated with the plant. Nitrogen incorporation into the soil-plant-animal system is ultimately restricted by rates of biological and industrial fixation. Biological fixation conserves fossil fuel, but fertilization is preferred in most present agriculture. Nitrogen-metabolism research has the practical objectives of allowing more efficient N-fertilizer utilization by plants, including those that fix N_2 but benefit from fertilizer-N supplements.

Nitrogen accumulation by harvested crops results in changes in soil acidity, with the direction of change depending on the N-source. There is little information on long-term effects of crop N-nutrition on acidity, and acidity is a critical factor that affects agricultural productivity in many tropical soils. Thus, plant control of pH and the acid/base balance in the soil as a consequence of nitrogen uptake and assimilation are important areas of future research.

Resumen   El nitrógeno, abundante pero sin embargo limitante para los cultivos, entra en las plantas en forma de nitrato o amonio o es incorporado al sistema a través de fijación biológica. La incorporación del nitrógeno al sistema suelo-planta-animal está limitado por las tasas de fijación biológica e industrial. La primera ahorra energía fosil pero la segunda fuente es la predominante en la agricultura moderna. La investigación del metabolismo de nitrógeno tiene objetivos prácticos tales como el permitir un uso mas eficiente de los fertilizantes nitrogenados por los cultivos, incluyendo aquellos que puedan fijar N_2 pero se benefician de suplementos de fertilizantes nitrogenados.

La acumulación de nitrogeno en los cultivos trae como consecuencia cambios de acidez en el suelo cuya dirección depende de la forma de nitrógeno utilizada. Aun existe poca información sobre los efectos a largo plazo de la fertilización nitrogenada sobre la acidez del suelo, factor que es determinante de la productividad de muchos suelos en los trópicos. Asi pues, el control de pH por las plantas y el balance de acidez en el suelo son areas de interés para futuras investigaciones.

Introduction

Nitrogen is abundant but is the nutrient that most often limits crop and food production. A crop can accumulate up to $800\,kg$ N ha^{-1} yr^{-1}; most non-fertilized tropical soils under agriculture deliver less than $50\,kg$ N ha^{-1} yr^{-1} (Sanchez[53]).

The availability of nitrogen in soil is limited by rates of organic matter decomposition unless availability is enhanced by biological N_2-fixation or additions of fertilizer-N. Optimizing these three sources for crop production is desirable for economic and ecological reasons.

In this paper we discuss the turnover and availability of nitrogen in the soil and plant uptake and nitrogen assimilation, including the special case of legumes and the balance of pH as a consequence of N uptake. pH balance is of special importance because of its impact on the already acid soils found in most of the tropics.

Nitrogen in the soil

Forms of nitrogen

Inorganic nitrogen in soil has three main sources: soil organic matter, atmospheric N_2, and N-fertilizer. During decomposition of organic matter in most agricultural soils, excess NH_4^+ not utilized by microbes is released, and subsequently usually oxidized mostly by autotrophic bacteria to NO_2^- and then NO_3^-. Nitrite does not usually accumulate except temporarily in the special situation where pH is above 7 and excess NH_4OH (or urea hydrolysing to NH_4OH) together inhibit the NO_2^- oxidizers.

When atmospheric N_2 is fixed, the first form of combined nitrogen to appear is NH_4^+, and most of it is immediately assimilated into organic forms so that very little is exuded to soil. Both organic and exuded NH_4^+ and fertilizer-N follow the same path as N derived from organic matter.

Total nitrogen in soil varies with soil organic matter content; soil organic matter usually contains ca. 5% N. In surface mineral soils, values of 0.03% N (Vertisol from Sudan) to 0.69% N (Oxisol from Brazil) have been reported[53]. Levels of nitrate to 0.006% NO_3–N (60 mg NO_3–N kg dry soil^{-1}) have been noted by Chapman[15], while ammonium is usually much less except in waterlogged soils after the addition of NH_4^+–N, urea, or nitrification inhibitors.

Factors affecting nitrogen release

In the absence of fertilizer N, the major source of fixed nitrogen the soil supplies to the plant is from soil organic matter undergoing decomposition.

As illustrated in Fig. 1, the availability of nitrogen to plants depends on the amount and type of organic matter present, and on the presence of microbial populations and conditions favoring their activity. In general, microbial activity is favored under those conditions that are optimal for plant growth, though microbes generally have a wider tolerance range than plants. The direction of N-transformation processes is dictated by the C:N ratio in the soil. A carbon limitation results in net N-mineralization; a nitrogen limitation (for example after the addition of excess carbon in the form of organic material with less than 1.3–

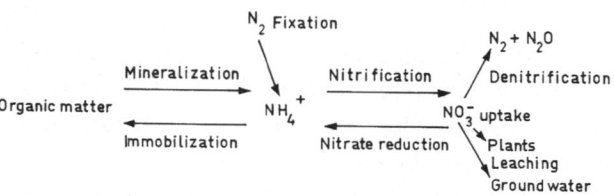

Fig. 1.   Principle forms and pathways for soil nitrogen (from Broadbent[7]).

1.5% N) results in the net immobilization of nitrogen until the C:N ratio is lowered to 20–30, when net N-mineralization is reinitiated[8].

During periods of excess carbon, the free-living N_2-fixers are also favored, and fixation of considerable amounts of nitrogen may occur. However, the availability of nitrogen to plants during this period is restricted by competition with the large population of heterotrophs in the soil.

Under aerobic conditions in most agricultural soils, nitrification is faster than nitrate reduction, mineralization faster than immobilization, and nitrification NH_4-limited[7]. As a consequence, there is often a continual conversion of organic nitrogen to nitrate, with little or no accumulation of ammonium. Excess water in the soil will inhibit part of the soil aerobic microbial population, particularly fungi and actinomycetes. Also, aerobic metabolism is more efficient for cell synthesis than is anaerobic metabolism. Thus, as saturation is approached, immobilization decreases faster than mineralization, and mineral-N accumulates[7]. In rice fields fertilized with high rates of ammonium, nitrification and denitrification proceed simultaneously: nitrification in the oxidized layer at the soil surface and above, and denitrification below, where anaerobiosis is dominant[48]. This is reflected in the better response of rice to deeper application of ammonium or urea than to broadcast application[10].

Nitrogen mineralization proceeds slowly in soils too dry for crops to grow[61]. There is also an increase in N-mineralization in re-wetted soil relative to soil kept moist[4], and this may account for the initial flush of available nitrogen in pot experiments or when rain or irrigation follows dry weather when plant growth has been restricted. The effects of soil water on microbial activity depend to some extent on temperature. In one set of incubation studies[13], for example, rates of mineralization dropped by a factor of 3 as soil moisture potential dropped from the optimum 0.5 bar (25% H_2O) to 2 bar (18% H_2O), but the effect of water was most marked at the highest and most favorable temperature, 30°C. In general, low temperature slows down microbial activity, and maximum activity usually occurs at *ca.* 40°C. The period of N-immobilization is subsequently shorter at higher temperatures, although the total amount of nitrogen immobilized is not greatly affected[7].

The surface horizon normally contains nearly all of the mineralizable-N in a soil profile. However, there are exceptions, for example deep undifferentiated alluvial soils in which more than half of the nitrogen released may come from depths below 20 cm[13].

Turnover of nitrogen in the soil

Rates of nitrogen immobilization and mineralization have been quantified by using labeled nitrogen. Mineralization of recently-immobilized nitrogen is slow, sometimes requiring years or even decades[11,56]. But as recently-introduced nitrogen is immobilized, other organic-N is mineralized. The equation $N_c = At^m$, where N_c = crop uptake of N, t = time, and A and m are constant, predicted reasonably well the release to crops of tagged and untagged nitrogen from the soil organic fraction in different soils[7]. The calculated turnover time was 5–6 years for added-N and 20–30 years for humus-N. These values may be lower for tropical soils. Nevertheless, nitrogen turns over very slowly in both temperate and tropical agricultural soils.

Nitrogen-uptake efficiency

Nitrogen uptake efficiency can be defined as the fraction of added fertilizer recovered in the plant. Broadbent and Carlton[9] applied ^{15}N-depleted fertilizer to corn and found that in their most responsive site the percent of fertilizer-N appearing in the crop declined from 61% and 67% at applied rates of 112 and 224 kg N ha^{-1}, respectively, to only 35% at 560 kg N ha^{-1}. They also observed that over five years the maximum utilization of soil and fertilizer-derived nitrogen was at rates just sufficient to produce maximum yield. Only 3 to 8% of fertilizer-N was recovered in the following crop. The overall efficiency of nitrogen recovery was higher than the 50% (or less) that is generally reported in the literature[1,8].

Estimation and prediction of N-availability

Soil nitrogen mineralized under optimal temperature and moisture conditions has long been suggested as a basis for predicting the amount of soil-N mineralized in the field[58,62]. However, the modifying effects of soil temperature[64] and water content[61] under field conditions must also be considered[39]. The buried polyethylene bag technique has been used to study N-mineralization under field conditions[58] and Westerman and Crothers[68] obtained a good correlation ($r = 0.98$) between such mineralization estimates and N-uptake by corn and potato. They estimated nitrogen uptake by using the difference between soil $NO_3^- $–N in buried bags and that in the plant root zone plus that in the plant.

Several rapid laboratory techniques correlate well with the N-supplying capacity of soil[13,17,24,60,63], but none have the simplicity required by the routine soil testing laboratory. Fox and Piekielek[25] obtained a significant correlation ($r = 0.865$) between ultraviolet absorption (at 260 nm) by a 0.01 M NaHCO$_3$ soil extract and the soil's capacity to supply nitrogen to corn in field experiments, as indicated by the total amount of nitrogen taken from the soil by the crop. They used an equation proposed by Stanford[59] to evaluate how much fertilizer-N to apply once the N-supplying power of the soil was determined:

$$Nf = \frac{Np - Ns}{E},$$

where Nf = N-fertilizer recommendation,
Np = plant N-requirement,
Ns = N supplied by the soil,
and E = N-fertilizer efficiency.

Fox and Piekielek[25] estimated Ns by regressing UV absorbance versus soil N-supplying capacity.

Where total soil nitrogen is low, as is usually the case in tropical soils under agriculture, estimating N-fertilizer requirements based on crop N-requirements seems adequate.

Plant N-uptake and assimilation

Kinetics

Plants differ markedly in their N-uptake strategies: legumes can grow exclusively on N_2, the majority of plants utilize both nitrate and ammonium, and some plants lack the ability to absorb or reduce nitrate[31].

Nitrate assimilation involves absorption (uptake) and subsequent reduction to ammonium. In some situations nitrate accumulates before reduction. Typically, NO_3^--uptake rates are initially exponential and then linear[14, 32, 38]. The exponential curve indicates an inducible transport system. In fact, studies with inhibitors of RNA-synthesis and protein synthesis[38, 54] indicate an active nitrate-transport system that can be induced by nitrate in the external medium.

There is convincing evidence of separate enzyme systems for uptake and reduction. For example, tungstate and vanadate inhibited nitrate reductase activity in both tobacco cells[34] and barley seedlings[51], but did not affect NO_3^--uptake, and Neurospora mutants lacking a viable nitrate reductase system still developed a normal transport system[54]. Further, Hallmark and Huffaker[32] found that nitrate reduction was more affected by increased temperature than was uptake.

Nitrate reduction occurs in two steps, the first mediated by nitrate reductase (NR-ase) and the second by nitrite reductase (Ni-ase), both nitrate inducible[6, 37]:

$$NO_3^- + 2H^+ + 2e^- \xrightarrow{\text{NR ase}} NO_2^- + H_2O \tag{1}$$

$$NO_2^- + 8H^+ + 6e^- \xrightarrow{\text{Ni ase}} NH_4^+ + 2H_2O \tag{2}$$

There are indications that NR-ase is in cytoplasm and Ni-ase is in chloroplasts (in leaves) or other organelles in roots[55]. Once ammonium has been produced it is assimilated via glutamine synthetase and glutamate synthase under normal, low

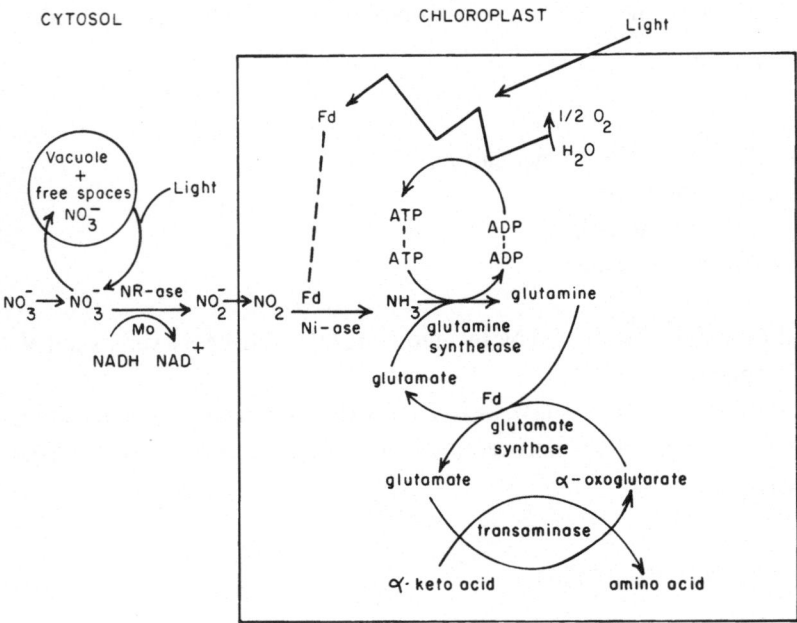

Fig. 2.   Proposed route of nitrate into amino acids in leaves (after Lea and Miflin[40] and Aslam *et al.*[2]).

NH_4^+ concentrations (Fig. 2), and possibly *via* glutamate dehydrogenase under high NH_4^+ concentrations.

Photosynthetic CO_2-fixation may enhance nitrate reduction with a reductant (NADH) supplied from primary products. The proposed triose phosphate shuttle between chloroplasts and cytoplasm (R. C. Huffaker, personal communication) could accomplish this. Nitrate uptake decreases in the absence of light[32], and light can regulate the efflux of nitrate from storage pools to metabolic pools[2]. In fact, high rates of fertilization combined with low light intensities may allow plant-nitrate to accumulate to toxic levels[28,66].

Transport

Assimilated nitrogen is transported from root to shoot and then attached to carbon skeletons that vary widely with plant species, the plant's age, its nitrogen source, and environmental stress[46,47]. But the skeletons have a N:C ratio higher than 0.4[5]. The amides glutamine and asparagine and amino acids closely related to them are dominant carriers in some species, while in others, alkaloids, ureides or certain unusual nonprotein amino acids are the main nitrogen compounds transported to the shoot[46]. Allantoic acid and allantoin (4N:4C) have been found in some legumes growing on N_2[36,41,65]; this is apparently the most energetically efficient transport system (N:C ratio is 1). However, plants that have this transport system do not differ significantly from other N_2-fixing plants in their overall efficiency of N_2-assimilation.

Nitrate reductase and nitrogenase activities as indicators of plant nitrogen nutrition

In wheat and corn, a good correlation between the amount of reduced nitrogen supplied to the plant (estimated by the *in vivo* or *in vitro* NR-ase assay) and the actual amount accumulated by the plant has been demonstrated[12, 18, 21]. This supports the hypothesis that nitrate reduction is the rate-limiting step in the assimilation of nitrate to reduced N. Although tissue-slice NR-ase activity differs from actual whole-plant nitrate reduction and most probably overestimates actual assimilation[23], it does give an estimate of the relative amount of NR-ase present in the cells[32], and is valid for studies comparing relative rates of NO_3^--assimilation under different environmental treatments over the plant cycle, and for studies concerning the effects of heredity on reduction[37].

Nitrogenase is the enzyme responsible for the transformation of N_2 to ammonia in nitrogen fixation, and its activity as measured by acetylene reduction assays may also be used to compare treatments and follow seasonal patterns.

Felker and Bandurski[22] proposed a plant ideotype for minimum-energy-input agriculture; the plant had growth characteristics that resulted in minimal soil nutrient loss, little or no need for irrigation, the ability to fix nitrogen, and a high yield of high-quality protein. Species of the genus Prosopis whose roots reach depths of 60 to 80 meters and that produce up to 20,000 kg of pods per hectare per year matched their ideal most closely. Most of our crops differ greatly from this, but the possibility of using both biological nitrogen fixation and nitrogen fertilizer may be one way to attain high yields at relatively low cost. The biological potential of this approach is implied by studies of the relationship between nitrate assimilation and N_2 fixation in crop plants throughout the growing season. The seasonal patterns of nitrate uptake and reduction and the patterns of N_2 fixation in soybean[26, 33, 67] and Phaseolus bean[27] indicate that the processes of nitrate assimilation and N_2 fixation are successive events, each contributing nitrogen at defined stages of plant development. In soybean, nitrate reduction appears to be more important at the preflowering stages, while maximal nitrogenase activities have been observed after the decline of NR-ase activity. Foliar application of fertilizers containing phosphorus, potassium, sulphur and reduced-N (ammonium or urea) to soybeans at pod filling stage have shown positive yield responses in a few cases, but generally have not increased and in some cases have even decreased soybean seed yield[45].

In Phaseolus, the opposite has been observed – maximum nitrate reductase activity after flowering with maximum nitrogenase before flowering – indicating that nitrogen fertilization at flowering might complement fixation. Recent results have indicated that detopping Phaseolus to delay flowering can prevent the decline of nitrogen fixation[3]. However, the general applicability of these results should be tested with bean cultivars with different times of onset and decline of nitrogenase activity[30].

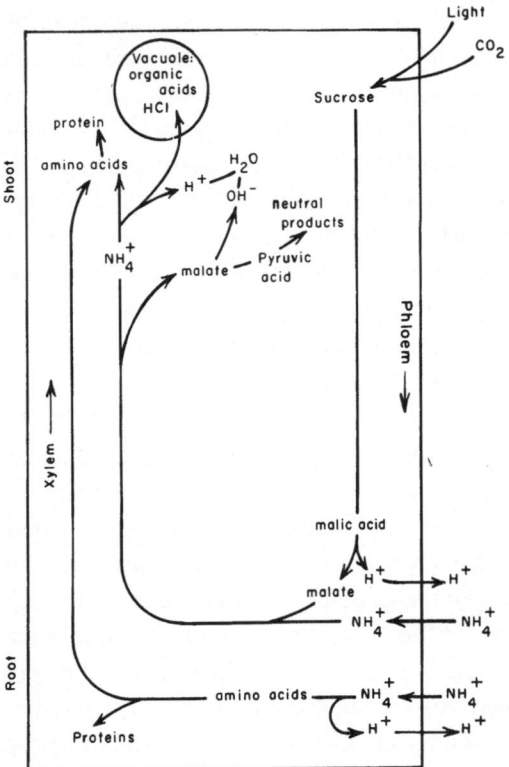

Fig. 3.   Fate of H^+ during NH_4^+ assimilation by plants (after Raven and Smith[52]).

Energy relations

In symbiotic N_2-fixing systems there are energy costs associated with nodule formation, nitrogenase activity, hydrogen loss, and transport of fixed-N. The carbon cost of N_2-reduction has been recently estimated by Philips[49] to be 2.57 g C per g N, while the cost of the entire system apparently varies from 0.3 to 20 g C per g N.

Even though there are no great differences in energy consumption for plants growing on nitrate nitrogen *vs* N_2 [29,42,43], the total energetic cost of industrially-fixed nitrogen is roughly twice as much as that of biological fixation, since approximately the same amount of energy is spent to fix N_2 industrially as is used in biological N_2-fixation[19] and then again to reduce nitrate in the plant. However, since most current crops prefer fixed nitrogen and active N_2-fixation is restricted to a few species, complementation of biological N_2-fixation with industrial fixation seems an attractive option.

pH balance and the nitrogen cycle

Plant control of pH

The effect of nitrogen on soil pH depends on the form of nitrogen being put into or taken out of the plant-soil system, since the complete unaltered cycle is a

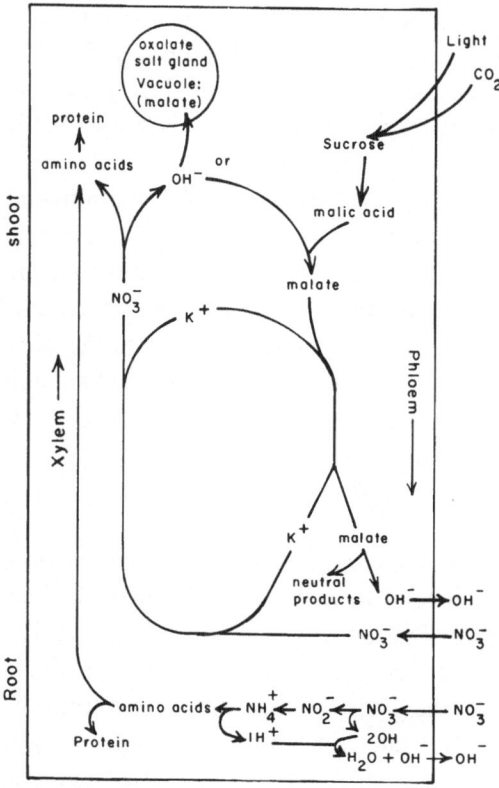

Fig. 4. Fate of OH^- during NO_3^- assimilation by plants (after Raven and Smith[52]).

neutral process[35]. Nitrogen taken up by plants may be in several forms, but nitrate, ammonium and atmospheric nitrogen account for most nitrogen taken up in natural ecosystems and in agriculture. This topic has been well covered by Raven and Smith[52]. They point out that assimilation of ammonium in plant cell cytoplasm produces at least one H^+ per NH_4^+ assimilated, that N_2 fixation generates 0.1 to 0.2 H^+ per N assimilated, and that NO_3^- assimilation produces almost one OH^- per assimilated NO_3^-. Any H^+ or OH^- produced in excess of that required to maintain cytoplasmic pH must be neutralized or removed from the metabolic stream. There is evidence that most NH_4^+ assimilation is in the roots[69], and that the excess H^+ generated is actively exuded to the soil solution[57], partly in exchange for cations. In some situations NH_4^+ may be transported as the malate salt and assimilated in the shoot. In this case malic acid is the intermediary for H^+ transport from the shoot to the root where the H^+ is exchanged with the external soil solution. Some acidity from NH_4^+ assimilation may be stored in the vacuole, although such neutralization is not often quantitatively significant (Fig. 3).

When NO_3^- is assimilated in the shoot, the excess OH^- produced may be temporarily neutralized by its storage as osmotically neutral products in shoot tissue. This involves either the synthesis of oxalic acid and its precipitation as the inert Ca or Mg salt, or the synthesis and storage of Ca and Mg carbonates in specialized salt glands[52]. However, a large proportion of the excess OH^- is exuded to the soil via roots. Corn may excrete to soil 50% of the OH^- generated during assimilation of nitrate[16, 20]; this may be sufficient OH^- to increase soil pH (Fig. 4).

In general, plant effects on soil pH depend on the relative uptake of cations and anions. If the plant takes up more anions than cations, as is normal where nitrate is the main source of N, it will maintain electrical neutrality by exporting hydroxyl or bicarbonate, raising the soil pH. If the plant takes up more cations than anions, as is normal if ammonium or N_2 are the main sources of N, the plant will export protons and acidify the soil. It is possible to estimate net acidification by subtracting total estimated removal of major anions (NO_3^-, Cl^-, PO_4^{-2} and SO_4^{-2}) from removal of major cations (Ca, Mg, K and NH_4^+)[44, 50].

Acid-base balance in the soil

Helyar[35] discussed in detail the effect of nitrogen cycling on soil acidification. Denitrification and nitrate absorption tend to make the soil more alkaline, whereas ammonium absorption and nitrification tend to make it more acid. The complete cycling of nitrogen is, however, a neutral process. Essentially, changes in soil acidity or alkalinity depend on the gains or losses of nitrogen from the system. Maximum acidification would occur with ammonium as a nitrogen source, with high rates of removal of metal cation nutrients in harvested material, and with high rates of nitrate leaching. (During nitrification, protons are produced, and cumulatively add to the soil's acidity when they displace exchangeable metal cations that are subsequently removed by uptake or by leaching with nitrate). Nitrate leaching may be especially marked where excessive nitrogen fertilizer is applied, and where drought or cold are followed by rain that leaches accumulated nitrate before it can be taken up by actively-growing plants. When whole shoots are harvested by grazing animals, a large proportion of bases are returned as animal droppings and urine, lessening the effects on soil pH. The use of green manure is acidifying only if nitrification is followed by nitrate leaching[35].

Nyatsanga and Pierre[44] attempted to quantify the amount of $CaCO_3$ necessary to bring the cropped soil pH to the pH of uncropped soil. The fixation of 270 kg N in a clay loam soil produced acidity equivalent to 600 kg of $CaCO_3$. Helyar[35] estimated that 250 kg of lime per hectare were necessary to prevent the pH of a podzolic soil from dropping when plants fixed 70 kg N per hectare. Even so, it would be more economical to apply 250 kg of lime than to apply the 140 kg of fertilizer-N that would be required to supply 70 kg N to the plant without affecting pH (assuming 50% fertilizer efficiency).

Soils that support tropical rainforest vegetation may be under little acidifying influence because nitrogen recycling is rapid and the complete nitrogen cycle is neutral. However, after clearing, these soils may be rapidly acidified due to nitrification coupled with low rates of absorption of nitrate by the plants and consequent leaching of bases as they accompany leached nitrate.

References

1   Allison F E 1955 The enigma of soil nitrogen balance sheets. Adv. Agron. 7, 213–250.
2   Aslam M, Oaks A and Huffaker R C 1976 Effect of light and glucose on the induction of nitrate reductase and on the distribution of nitrate in etiolated barley leaves. Plant Physiol. 58, 588–591.
3   Baird L M 1980 Morphogenesis of effective and ineffective root nodules in *Phaseolus vulgaris* L. Ph.D. Thesis, University of California, Davis, California. 67 p.
4   Birch H F 1960 Nitrification in soils after different periods of dryness. Plant and Soil 12, 81–96.
5   Bollard E G 1960 Transport in the xylem. Annu. Rev. Plant Physiol. 11, 141–166.
6   Breteler H, Cate C H T and Nissen P 1979 Time-course of nitrate uptake and nitrate reductase activity in nitrogen-depleted dwarf bean. Physiol. Plant. 47, 49–55.
7   Broadbent F E 1968 Turnover of nitrogen in soil organic matter. Pontificiae Academiae Scientarum Scripta Varia 32, 61–88.
8   Broadbent F E 1973 Sources and sinks of nitrate in soils. *In* Proc. of the 1st Annual Trace Contaminants Conference, Oak Ridge National Laboratory, National Science Foundation, Washington, D.C., pp 108–119.
9   Broadbent F E and Carlton A B 1978 Field trials with isotopically labeled nitrogen fertilizer. *In* Nitrogen in the Environment, Vol. 1, pp 1–41. Nielsen D N and MacDonald J G (Eds). Academic Press, New York.
10 Broadbent F E and Mikkelsen D S 1968 Influence of placement on uptake of tagged nitrogen by rice. Agric. J. 60, 674–677.
11 Broadbent F E and Nakashima T 1967 Reversion of fertilizer nitrogen in soils. Soil Sci. Soc. Am. Proc. 31, 648–652.
12 Brunetti N and Hageman R H 1976 Comparison of *in vivo* and *in vitro* assays of nitrate reductase in wheat (*Triticum aestivum* L.) seedlings. Plant Physiol. 58, 583–587.
13 Cassman K G and Munns D N 1980 Nitrogen mineralization as affected by soil moisture, temperature, and depth. Soil Sci. Soc. Am. J. 44, 1233–1237.
14 Chantarotwong W, Huffaker R C, Miller B L and Granstedt R C 1976 *In vivo* nitrate reduction in relation to nitrate uptake, nitrate content, and *in vivo* nitrate reductase activity in intact barley seedlings. Plant Physiol. 57, 519–522.
15 Chapman H D (Ed.) 1965 Diagnostic Criteria for Plants and Soils. Homer D. Chapman, Riverside, California. 319 p.
16 Coic Y 1971 Influence du métabolisme de nitrate dans les racines sur l'état nutritional de la plante. *In* Recent Advances in Plant Nutrition Vol. I, pp 217–227. Samish R M (Ed.). Gordon and Breach, New York.
17 Dahnke W C and Vasey E H 1973 Testing soil for nitrogen. In Soil Testing and Plant Analysis, revised edition, pp 97–114. Walsh L M and Beaton J D (Eds). Soil Sci. Soc. Am. Madison, Wis.
18 Deckard E L, Lambert R J and Hageman R H 1973 Nitrate reductase activity in corn leaves as related to yields of grain and grain protein. Crop Sci. 13, 343–350.
19 Delwiche C C 1970 The nitrogen cycle. Sci. Am. 223, 137–146.
20 Dijkshoorn W 1971 Partition of ionic constituents between organs. *In* Recent Advances in Plant Nutrition, Vol. 2. pp 447–476. Samish R M (Ed.). Gordon and Breach, New York.
21 Eilrich G L and Hageman R H 1973 Nitrate reductase activity and its relationship to accumulation of vegetative and grain nitrogen in wheat (*Triticum aestivum* L.). Crop Sci. 13, 59–66.
22 Felker P and Bandurski R S 1979 Uses and potential uses of leguminous trees for minimal energy input agriculture. Econ. Bot. 33, 172–183.

23   Fillipe G M, Dale J E and Marriott C 1975 The effect of irradiance on uptake and assimilation of nitrate by young barley seedlings. Ann. Bot. 39, 43–55.

24   Fox R H and Piekielek W P 1978 Field testing of several nitrogen availability indexes. Soil Sci. Soc. Am. J. 42, 747–750.

25   Fox R H and Piekielek W P 1978 A rapid method for estimating the nitrogen-supplying capability of a soil. Soil Sci. Soc. Am. J. 42, 751–753.

26   Franco A A, Fonseca O O M and Marriel I E 1978 Efeito do nitrogênio mineral na actividade da nitrogenase e nitrato-reductase, durante o ciclo da soja no campo. Rev. Bras. Ci. Solo 2, 110–114. (In Portuguese, English summary.)

27   Franco A A, Pereira J C and Neyra C A 1979 Seasonal patterns of nitrate reductase and nitrogenase activities in *Phaseolus vulgaris* L. Plant Physiol. 63, 421–424.

28   George J R, Rhykerd C L and Noller C H 1971 Effect of light intensity, temperature, nitrogen and stage of growth on nitrate accumulation and dry matter production of a sorghum × Sudan Grass hybrid. Agron. J. 63, 413–415.

29   Gibson A H 1966 The carbohydrate requirements for symbiotic nitrogen fixation: a 'whole plant' growth analysis approach. Aust. J. Biol. Sci. 19, 499–515.

30   Graham P H and Rosas J C 1977 Growth and development of indeterminate bush and climbing cultivars of *Phaseolus vulgaris* L. inoculated with Rhizobium. J. Agric. Sci. Cambr. 88, 503–508.

31   Greidamus T, Peterson A, Schrader L E and Dana M N 1972 Essentiality of ammonium for cranberry nutrition. J. Am. Soc. Hortic. Sci. 97, 272–277.

32   Hallmark W B and Huffaker R C 1978 The influence of ambient nitrate, temperature, and light on nitrate assimilation in Sudan grass seedlings. Physiol. Plant. 44, 147–152.

33   Harper J E and Hageman R H 1972 Canopy and seasonal profiles of nitrate reductase in soybeans (*Glycine max* L. Merr.) Plant Physiol. 49, 146–154.

34   Heimer Y M and Filner P 1971 Regulation of the nitrate assimilation pathway in cultured tobacco cells. III. The nitrate uptake system. Biochem. Biophys. Acta 230, 362–372.

35   Helyar K R 1976 Nitrogen cycling and soil acidification. J. Aust. Inst. Agric. Sci. 42, 217–221.

36   Herridge D F, Atkins C A, Pate J S and Rainbird R M 1978 Allantoin and allantoic acid in the nitrogen economy of the cowpea (*Vigna unguiculata* [L.] Walp.). Plant Physiol. 62, 495–498.

37   Jackson W A 1978 Nitrate acquisition and assimilation by higher plants: processes in root systems. *In* Nitrogen in the Environment, Vol. 2. Soil-Plant-Nitrogen Relationships, pp 45–88. Nielsen D R and MacDonald J G (Eds). Academic Press, New York.

38   Jackson W A, Flesher D and Hageman R H 1973 Nitrate uptake by dark-grown corn seedlings. Some characteristics of apparent induction. Plant Physiol. 51, 120–127.

39   Kafkafi U, Bar-Yosef B and Hadas A 1978 Fertilization decision model. A synthesis of soil and plant parameters in a computerized program. Soil Sci. 125, 261–268.

40   Lea P L and Miflin B J 1974 Alternative route for nitrogen assimilation in higher plants. Nature London 251, 614–616.

41   McClure P R and Israel D W 1979 Transport of nitrogen in the xylem of soybean plants. Plant Physiol. 64, 411–416.

42   McCree K J and Silsbury J H 1978 Growth and maintenance requirements of subterranean clover. Crop Sci. 18, 13–18.

43   Minchin F R and Pate J S 1973 The carbon balance of a legume and the functional economy of its root nodules. J. Exp. Bot. 24, 259–271.

44   Nyatsanga T and Pierre W H 1973 Effect of nitrogen fixation by legumes on soil acidity. Agron. J. 65, 936–940.

45   Parker M B and Baswell F C 1980 Foliage injury, nutrient intake, and yield of soybeans as influenced by foliar fertilization. Agric. J. 72, 110–113.

46   Pate J S 1973 Uptake, assimilation and transport of nitrogen compounds by plants. Soil Biol. Biochem. 5, 109–119.

47 Pate J S 1980 Transport and partitioning of nitrogenous solutes. Annu. Rev. Plant Physiol. 31, 313–340.

48 Persall W H 1950 The investigation of wet soils and its agricultural implications. Emp. J. Exp. Agric. 18, 289–298.

49 Phillips D A 1980 Efficiency of symbiotic nitrogen fixation in legumes. Annu. Rev. Plant Physiol. 31, 29–49.

50 Pierre W H and Banwart W L 1973 The excess-base and excess-base/nitrogen ratio of various crop species and plant parts. Agron. J. 64, 91–96.

51 Rao K P and Rains D W 1976 Nitrate absorption by barley. I. Kinetics and energetics. Plant Physiol. 57, 55–58.

52 Raven J A and Smith F A 1976 Nitrogen assimilation and transport in vascular land plants in relation to intracellular pH regulation. New Phytol. 76, 415–431.

53 Sanchez P A 1976 Properties and Management of Soils in the Tropics. John Wiley and Sons, New York. 618 p.

54 Schloemer R H and Garrett R H 1974 Nitrate transport in *Neurospora crassa.* J. Bacteriol. 118, 259–269.

55 Schrader L E, Beevers L and Hageman R H 1967 Differential effects of chloramphenicol on the induction of nitrate and nitrate reductase in green leaf tissue. Biochem. Biophys. Res. Commun. 26, 14–17.

56 Shields J A, Paul E A, Lowe W E and Parkinson D 1973 Turnover of microbial tissue in soil under field conditions. Soil Biol. Biochem. 5, 753–764.

57 Smith F A and Raven J A 1979 Intracellular pH and its regulation. Annu. Rev. Plant Physiol. 30, 289–311.

58 Smith S J, Young L B and Miller G E 1977 Evaluation of soil nitrogen mineralization potentials under modified field conditions. Soil Sci. Soc. Am. J. 41, 74–76.

59 Stanford G 1973 Rationale for optimum nitrogen fertilization in corn production. J. Environ. Qual. 2, 159–166.

60 Stanford G 1978 Evaluation of ammonium release by alkaline permanganate extraction as an index of soil nitrogen availability. Soil Sci. 126, 244–253.

61 Stanford G and Epstein E 1974 Nitrogen mineralization-water relations in soil. Soil Sci. Soc. Am. Proc. 38, 103–107.

62 Stanford G and Smith S J 1972 Nitrogen mineralization potentials of soils. Soil Sci. Soc. Am. Proc. 36, 465–472.

63 Stanford G and Smith J 1978 Oxidative release of potentially mineralizable soil nitrogen by acid permanganate extraction. Soil Sci. 126, 210–218.

64 Stanford G, Frere M G and Schwaninger D H 1973 Temperature coefficient of soil nitrogen mineralization. Soil Sci. 115, 321–323.

65 Streeter J G 1979 Allantoin and allantoic acid in tissues and stem exudates from field-grown soybean plants. Plant Physiol. 63, 478–480.

66 Sumner D C, Martin W E and Echegaray H S 1965 Dry matter and protein yields and nitrate content of piper Sudan grass (*Sorghum sudanense* [Piper] Stapf.) in response to nitrogen fertilization. Agron. J. 57, 351–354.

67 Thibodeau P S and Jaworsky E G 1975 Patterns of nitrogen utilization in the soybean. Planta 127, 133–147.

68 Westerman D T and Crothers S E 1980 Measuring soil nitrogen mineralization under field conditions. Agr. J. 72, 1009–1012.

69 Yoneyama T and Kumazawa K 1974 A kinetic study of the assimilation of ^{15}N-labelled ammonium in rice seedlings. Plant Cell Physiol. 15, 655–659.

Plant and Soil 67, 15–34 (1982).   0032-079X/82/0671-0015$03.00.
© 1982 *Martinus Nijhoff/Dr W. Junk Publishers, The Hague.*

Microbiological regulation of the biogeochemical nitrogen cycle

Regulación microbiana del ciclo biogeoquímico del nitrógeno

T. ROSSWALL
Department of Microbiology, Swedish University of Agricultural Sciences, S-750 07 Uppsala, Sweden

Key words Acetylene Denitrification Immobilization Mineralization Microbial processes N-cycling N_2-fixation Nitrification Nitrate reduction Oxygen.

Abstract Most nitrogen transformations in soil are carried out by micro-organisms. An understanding of the microbiological processes is thus necessary in order for us to devise management practices in agricultural ecosystems, which will optimize plant root uptake of nitrogen and minimize nitrogen losses from the systems. Some aspects of the individual microbiological processes in the nitrogen cycle are discussed and their importance for an efficient management of agroecosystems.

In soil various groups of organisms compete for available inorganic nitrogen and quantitative data are needed on the uptake kinetics for these various groups in order to be able to assess their competitive ability under different conditions.

The influence of abiotic factors such as oxygen concentration, inorganic nitrogen concentration and pH is discussed in relation to the different processes.

The importance of acetylene as a tool in nitrogen cycling studies is discussed briefly.

Resumen La mayoría de las transformaciones del nitrógeno en el suelo ocurren a través de los micro-organismos. Se requiere asi un conocimiento de los procesos microbiológicos con el fin de desarrollar las prácticas de manejo de los sistemas agrícolas que optimicen la absorción de nitrógeno por las raíces y que minimicen las pérdidas de nitrógeno de los sistemas. Se discuten algunos aspectos de ciertos procesos microbiológicos en el ciclo de nitrógeno como su importancia para el manejo eficiente de agroecosistemas.

Varios grupos de microorganismos compiten por el nitrógeno disponible y se requieren datos cuantitativos sobre la cinética de absorción de estos grupos de manera de estimar su capacidad de competir bajo diferentes condiciones.

La influencia de los factores abióticos tales como la concentración de oxígeno, la concentración de nitrógeno inorgánico y el pH se discuten en relación a los diferentes procesos.

Se discute también la importancia del acetileno como herramienta para estudiar el ciclo de nitrógeno.

Introduction

Understanding the major biogeochemical cycles is of paramount importance for our possibilities to sustain an ever growing population in a fragile environment. The global environment has developed over several billion years, but the geologically recent expansion of living organisms has drastically altered primordial biogeochemical cycles. In recent decades man has, himself, through various activities begun to affect global biogeochemical cycles, and it is possible that human influence may be harmful to the global environment and may change the initial equilibrium conditions.

The biogeochemical cycles and energy flow comprise the life sustaining system on which the biosphere depends. When human influence changes the biogeochemical cycles, both quantitatively and qualitatively, the fundamental basis for life is threatened. An assessment of the possible effects man has on the global ecosystem can only be made if we have a thorough understanding of pre-industrial and present biogeochemical cycles.

Nitrogen is one of the major nutrients for all living organisms, and one of the most important factors limiting crop yield. Since prehistoric times man has known that N_2-fixing legumes and the addition of manure are beneficial to agricultural production. The Maya, Aztec and Inca cultures, for example, were dependent on high productivities of food and fodder crops to sustain the large human and animal populations. They developed intricate irrigation systems and also intercropped non-legume crops such as maize with nitrogen-fixing legumes. Linneaus, in his travels in Sweden in the mid-eighteenth century, noted that also Swedish farmers considered legumes in a crop rotation to increase soil fertility.

In the first part of the nineteenth century, von Liebig showed that plants require soluble inorganic nitrogen. Later it was realized that the beneficial effects of legumes on subsequent crops were due to their ability to fix molecular nitrogen. In the light of these discoveries, man began to intensify agricultural production through the manufacture and application of commercial N-fertilizers. The production of commercial fertilizers continues to increase more rapidly than most other commodities. It is estimated that by the end of this century the production of nitrogen fertilizers will equal the amounts of nitrogen fixed biologically in all terrestrial ecosystems[50].

The biogeochemical nitrogen cycle (Fig. 1) is unusually complex as nitrogen can occur in many valance states. The processes regulating its cycling are governed by a number of factors, the qualitatively most important of which is the redox potential, as certain processes occur only aerobically while others only anaerobically.

In all ecosystems, microbial processes play a paramount role in the cycling of nitrogen. Micro-organisms are the sole or major group of organisms responsible for such vital processes as nitrogen fixation, nitrification and denitrification (Table 1). In this paper I discuss some of the important processes of the biogeochemical nitrogen cycle, the regulatory role that micro-organisms play in these processes, the effect of some major abiotic factors on the micro-organisms and some practical implications for agriculture and forestry.

Biological nitrogen fixation

Biological nitrogen fixation has aroused considerable interest in recent years, and a large number of scientific meetings have been devoted solely to this topic, e.g.[5, 15, 22, 23, 35, 36]. The effects of some abiotic factors on nitrogen fixation are

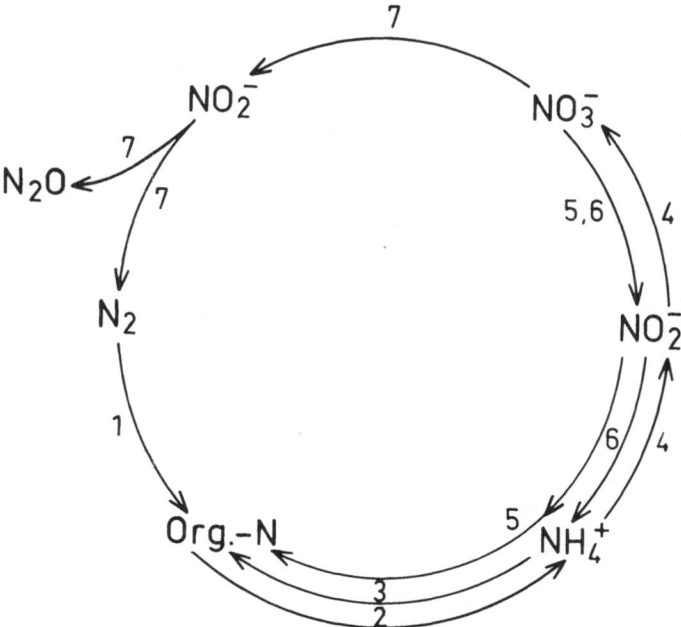

Fig. 1. Schematic view of the biogeochemical nitrogen cycle (Söderlund and Rosswall[49]). 1, nitrogen fixation; 2, mineralization; 3, immobilization; 4, nitrification; 5, nitrate assimilation; 6, dissimilatory nitrate reduction; 7, denitrification.

listed in Table 2. It is not possible to cover this subject in any detail here, and I will focus my attention on two points only.

The first is related to the biochemical nature of the nitrogenfixation enzyme system and how this imposes certain constraints on nitrogen-fixing organisms. The enzymes responsible for the fixation of atmospheric nitrogen, which is globally abundant but in a form inaccessible to most forms of life, are highly sensitive to oxygen. This places an evolutionary constraint on organisms which have developed the nitrogen-fixing capacity in that they must have evolved a suitable system for protecting these enzymes from molecular oxygen. An understanding of how organisms have developed these protective measures will also lead to an understanding of the organisms' specific ecological niches.

As discussed by Gallon[21], there are a number of mechanisms by which N-fixers achieve O_2-protection (Fig. 2). Anaerobic bacteria which fix nitrogen have not needed to evolve such mechanisms, because they live in environments devoid of oxygen (Fig. 2a). Examples of such organisms are photosynthetic bacteria (*Rhodospirillum* spp.) and certain other species, such as *Clostridium*. These forms may have some importance in, for example, the anaerobic soils of rice paddies, but in general their contributions to the nitrogen economy of ecosystems are small. Other organisms, in particular the cyanobacteria, have evolved protective barriers (Fig. 2b) such as the thick walls of the heterocysts. Such organisms may be very important to the N-balance of many systems, as they can live in aerobic habitats. Cyanobacteria, for example, can occur both free-living and in different

Table 1. Important non-industrial processes in the biogeochemical nitrogen cycle

Process	Name	Abiotic	Biotic*		
			M	A	P
$N_{org} \rightarrow NH_4^+$	Mineralization, ammonification	+	+	+	+
$NH_4^+ \rightarrow N_{org}$	Immobilization, assimilation	−	+	−	+
$NH_4^+ \rightarrow NH_3$	Volatilization	+	−	−	+
$NH_4^+ \rightarrow NO_2^-$	Nitrification	−	+	−	−
$NO_2^- \rightarrow NO_3^-$	Nitrification	−	+	−	−
$NO_2^- \rightarrow N_2O$	Chemo-denitrification, nitrifier denitrification	+	+	−	−
$NO_3^- \rightarrow NO_2^-$	Nitrate reduction, nitrate respiration	−	+	−	−
$NO_3^- \rightarrow NO, N_2O, N_2$	Denitrification	−	+	−	−
$NO_3^- \rightarrow NH_4^+$	Dissimilatory nitrate reduction	−	+	−	−
$NO_3^- \rightarrow N_{org}$	Nitrate assimilation, immobilization	−	+	−	+
$N_2 \rightarrow N_{org}$	Nitrogen fixation	−	+	−	(+)
$N_2 \rightarrow NO_x$	Nitrogen fixation	+	−	−	−
$NH_3 \rightarrow NO_x$		+	−	−	−
$N_2O \rightarrow NO_x$		+	−	−	−

* M = Micro-organisms, A = Animals, P = Plants. The + sign signifies that a process occurs, a − sign that it does not. Nitrogen fixation occurs in plants but only in association with bacteria, which is denoted by (+)

symbiotic forms (lichens and Azolla) and may contribute significantly to the nitrogen economy of certain ecosystems, examples being the lichen-dominated heathlands in tundra areas and rice paddies with Azolla. The use of Azolla for introducing combined nitrogen to lowland rice fields shows great promise and the efficiency of nitrogen additions to such systems through Azolla can certainly be raised by management practices. Protective barriers are also important in legumes, in which the root nodules restrict oxygen diffusion to the bacteroids of

Table 2. Relative effects of various factors in different nitrogen transformations*

Factor	$N_2 \rightarrow$	$N_{org} \rightarrow$	$NH_4^+ \rightarrow$	$NO_2^- \rightarrow$	NO_3^-
O_2	(− − −)	+ +	+ + +	+ + +	
NH_4^+	− − −	+	+ + +	0	
Low pH	− −	− −	− −	− − −	
NO_3^-	−	(+ +)	0	0	
C_2H_2	− −	0?	− − −	0	

* + signs signify different degrees of stimulation, − signs inhibition and ? = unknown

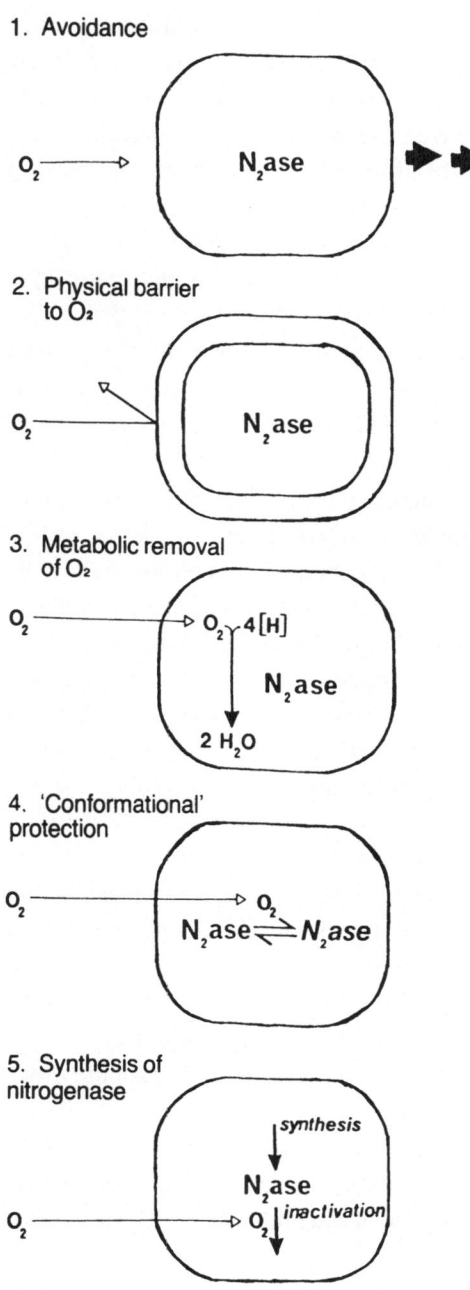

Fig. 2. Protective mechanisms by micro-organisms for protecting nitrogenase enzymes from inactivation by oxygen (Gallon[21]).

Rhizobium. The legume symbiosis is, of course, the most important nitrogen-fixing system available to man at present.

The metabolic removal of O_2 is a third O_2-protection mechanism, and is manifested in *Azotobacter* spp., where high respiratory activity removes oxygen from the cells (Fig. 2c). Azotobacter probably has the highest respiratory activity of any known microorganisms. Nitrogen-fixing bacteria may also contain an uptake hydrogenase, which recycles the hydrogen formed during nitrogen fixation, using oxygen as a terminal electron acceptor and thus generating ATP and consuming oxygen. The N-contribution of free-living nitrogen-fixing bacteria to most ecosystems is generally low, and proper management to increase this fixation is difficult. A conformational protection of the sensitive nitrogenase system (Fig. 2d) may occur in, for example, Azotobacter, whereby active nitrogen fixation can continue even in the presence of certain amounts of oxygen.

Finally, certain micro-organisms seem to possess enzymes with a very rapid turnover, so that constant resynthesis of new enzymes counterbalances the O_2-inactivation of other older enzymes (Fig. 2e). This has been suggested to occur in the cyanobacterium *Anabaena flos-aquae*.

My second point concerning N_2-fixation is that for the future, it will probably be rewarding to look for presently underexploited species of plants with a nitrogen-fixing symbiont. The stem nodules on Sesbania[17] and on *Aeschynomene indica*[63], both having Rhizobium as the N_2-fixing organism, are particularly interesting. Aeschynomene N_2-fixation is inhibited by inorganic nitrogen in the soil[62], but Sesbania seems to be able to fix nitrogen even at $3\,mM\ NH_4NO_3$ nitrogen in the growth medium of hydroponically grown plants[17], a concentration which is usually inhibitory to nodulation and nitrogen fixation. Inoculation is performed simply by spraying an appropriate Rhizobium strain on the above-ground parts of the Sesbania[17].

The use of nitrogen-fixing tree species for fuel wood, or for other uses, is also of great interest. These species include both legumes and non-legumes. The Frankia symbiosis with *Alnus* spp. and *Casuarina* spp. is especially important as a potential source of fuel wood in wood-poor areas such as the W. African savannas. There is a great potential for the utilization of such species, which until recently have been accorded very little interest. Their nitrogen-fixing capacity is largely unknown, but in savanna ecosystems, for example, they are probably very important for the overall nitrogen economy[47].

The recent claim that eucaryotic green alga can fix nitrogen[61] shows that we still have not discovered the full extent of the presence of nitrogen-fixing organisms in the biosphere, and this is an area of research which will continue to receive wide interest.

Mineralization

The mineralization of nitrogen is mainly carried out by micro-organisms, and through this process organically bound nitrogen, which is the major form in which nitrogen occurs in terrestrial systems, is liberated as ammonium nitrogen (process 2 in Fig. 3). The subsequent fate of ammonium nitrogen is dependent on a number of biotic and abiotic factors. As a consequence of the large number of

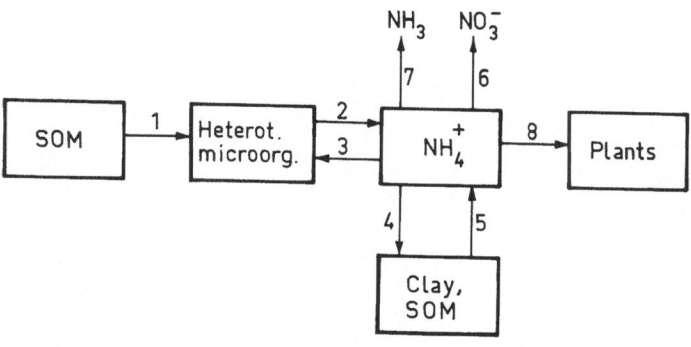

Fig. 3. Possible fates of ammonium-nitrogen in soil. Numbers refer to processes discussed in text. SOM = soil organic matter excluding living micro-organisms and plant roots.

processes competing for available ammonium-N (Fig. 3), its concentration in soils with a vegetation cover is usually very low or less than 5 mg kg^{-1} soil[12]. A low concentration of ammonium-N in the soil is, however, not an indication of low mineralization rates, as it can, for example, indicate rapid nitrification or plant uptake.

Micro-organisms need nitrogen for growth, and whether nitrogen is mineralized or immobilized (process 3 in Fig. 3) by micro-organisms is dependent on the carbon-nitrogen ratio of the substrate as compared to that of the decomposer organisms. The substrate is used both for synthesis of new biomass and for energy production. In aerobic processes, carbon dioxide is the end product of energy production and a fraction of the substrate-C is thus lost. If the substrate has a low C/N ratio, nitrogen will be in excess and ammonium-N will be liberated. The proportion of carbon in the substrate that ends up as cell biomass depends on the assimilatory efficiency of the organisms. The efficiency of soil micro-organisms for utilizing the carbon substrate for biomass production is poorly known, but based on available data an assimilatory efficiency of 40% seems realistic[25]. The C/N ratio of microbial biomass varies. While the content of carbon is usually 50% or somewhat lower, the nitrogen content will vary considerably, depending on growth conditions. Most literature data indicate a nitrogen content of 8–12%, but these estimates are usually based on determinations of N-contents of micro-organisms grown on laboratory culture media. In general, owing to the limiting conditions for microbial growth in soil, micro-organisms grow at N-starvation levels, and a realistic value for the nitrogen content of microbial biomass in soil is probably around 4 percent[4,43].

In Fig. 4, the substrate C/N ratio is plotted against the assimilation efficiency of micro-organisms with different nitrogen contents. From these three parameters it is possible to determine under what conditions net nitrogen mineralization will occur. The assimilation efficiency is dependent on the quality of the substrate, however, and it is not possible to judge if net mineralization will occur based only on knowledge of the C/N ratio of the substrate.

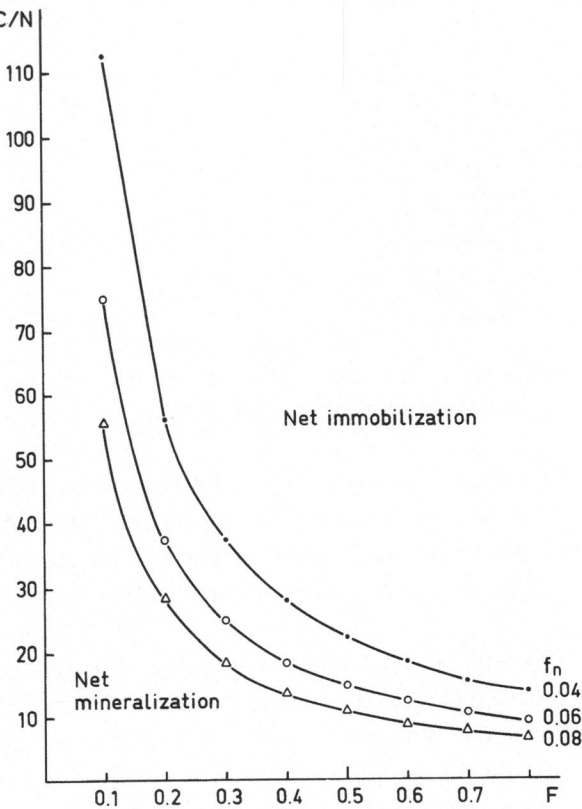

Fig. 4. Dependence of net mineralization *versus* immobilization on the C/N ratio of the substrate, the assimilatory efficiency of the decomposers (F), and the nitrogen content of microbial biomass (f_n). 0.04 and 0.08 (f_n) are equivalent to C/N ratios of *ca.* 12 and 6 (from Rosswall[44]).

In many cases it has been observed that there is a net increase in total nitrogen in plant litter during the first stages of decomposition (scenarios A and B, Fig. 5), especially if the C/N ratio of the substrate is high[8]. Such an accumulation can occur through translocation of nitrogen by fungi from the surrounding medium, by nitrogen fixation stimulated by the availability of carbon sources, by inputs from dry and wet deposition, and by the migration of small animals into the substrate. Thus, it is not possible to calculate nitrogen mineralization rates from plant litter by assuming that the relative rates of weight loss and nitrogen net-mineralization are the same. The critical C/N ratio for mineralization of nitrogen in litter which exhibits an initial accumulation phase can vary between 26 and 167[8]. Lignin in litter seems to play an important role in litter decomposition by regulating the rate of weight loss and consequently also of nitrogen mineralization[7] (Fig. 6). Ammonium-N inhibits ligninolytic activity in some wood-decomposing fungi[18,19]. This may be an important factor in controlling rates of decomposition in fertilized ecosystems and may partly account for increased soil organic matter contents in certain agro-ecosystems receiving

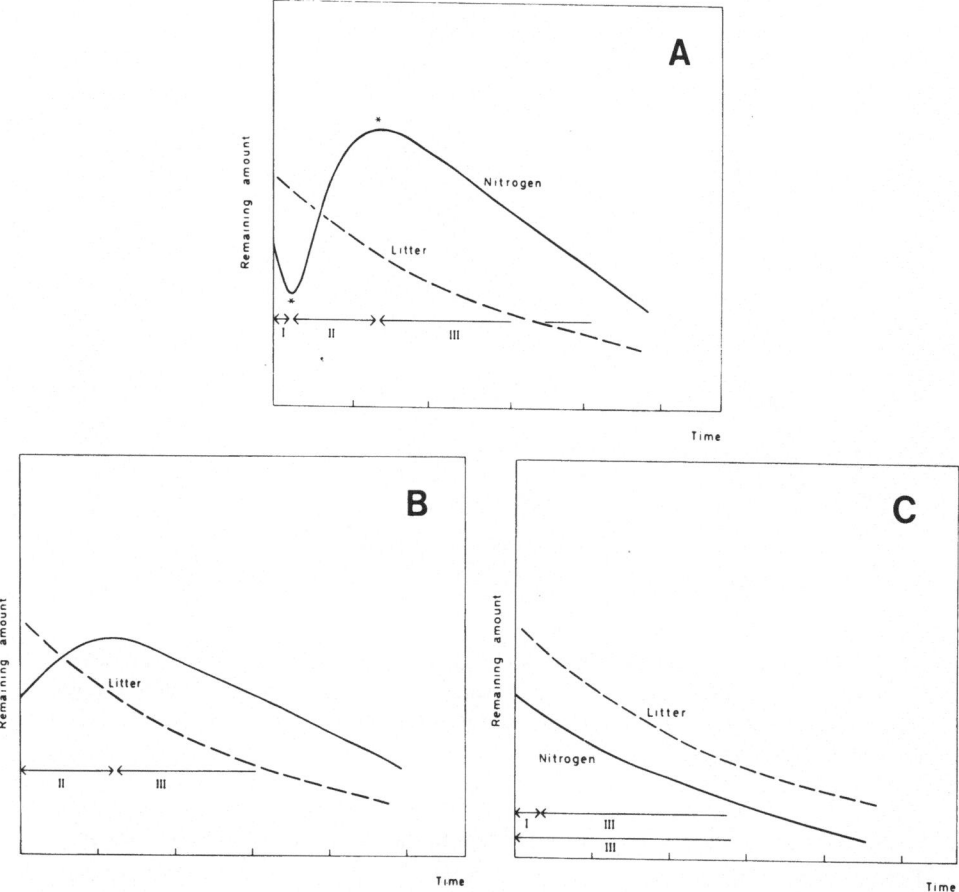

Fig. 5. Schematic illustration of the different phases of nitrogen gains and losses during decomposition of needle litter: I – leaching, II – accumulation and III – mineralization. A–C refers to different cases depending on substrate characteristics (Berg and Staaf[8]).

nitrogen fertilizers as observed by Persson[39] and others. Berg and Staaf[7] have suggested that ammonium-N plays an important role in regulating decomposition of pine needle-litter lignin.

Soil animals also have a major role in regulating nitrogen mineralization, as reviewed by Anderson *et al.*[1], and animals can be of direct importance by excreting ammonium-N. Protozoa and nematodes are probably especially important groups of soil animals in this respect. These groups have a fairly low biomass C/N ratio and are mainly microbial feeders. Ammonium-N will thus be liberated during the process of grazing on micro-organisms. Field studies have shown that peaks in bacterial biomass are often followed by peaks in numbers of protozoa[14], which consume the bacteria with liberation of inorganic nitrogen. This pattern is influenced by the presence of plant roots, which increases the populations of both bacteria and protozoa in microcosm experiments[14]. The data of Clarholm[14] indicate that nitrogen mineralization through the grazing of

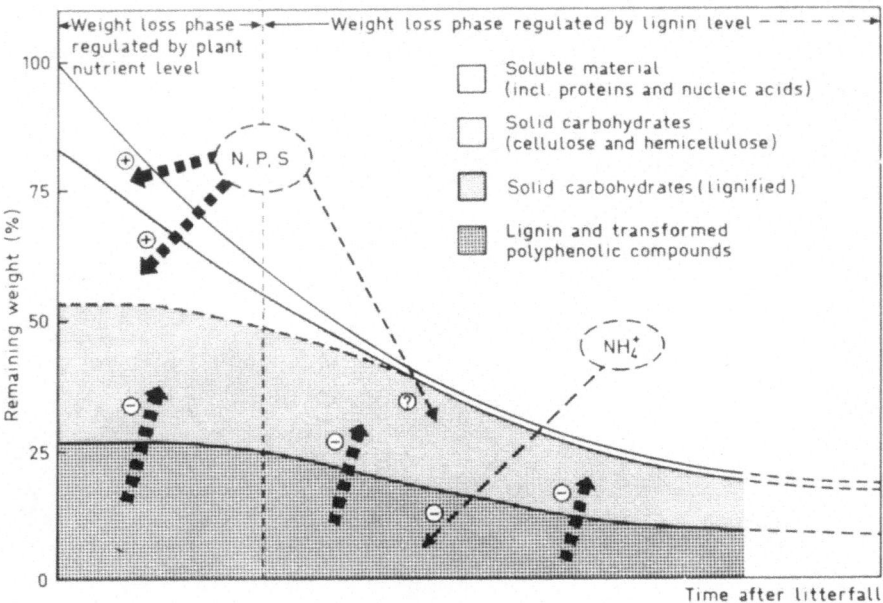

Fig. 6. Schematic diagram of the weight loss rates of some organic chemical components in Scots pine needle litter, and how these rates are influenced by increasing concentrations of nitrogen, phosphorus, sulphur and lignin. The possible influence of ammonium-N on lignin decomposition is also indicated (Berg and Staaf[7]).

bacteria by protozoa was three times higher in the presence of roots than without. The grazing on microorganisms is important also in the context of the nitrogen economy of agro-ecosystems, since it has been estimated that nematodes may mineralize 14–124 kg N ha^{-1} yr^{-1} under field conditions[1]. It is probable that protozoa play a similar role.

Immobilization

Immobilization of nitrogen can occur through both biotic and abiotic processes. Ammonium-N is efficiently immobilized by clays (process 4 in Fig. 3) both in an exchangeable and non-exchangeable (fixed) form[37]. The exchangeably bound ammonium-N is in dynamic equilibrium with ions in the liquid phase and is thus available for biological immobilization (process 3 in Fig. 3). It is generally believed that the non-exchangeable ammonium, which is fixed in the lattice of certain clay minerals, has a low biological availability. Several investigators have, however, shown that the fixed ammonium-N is available to micro-organisms and plant roots, and Kudeyarov[31] has estimated that in most soils 30–60% of the fixed ammonium is available for biological uptake. Ammonium-N can also be bound to soil organic matter in forms which makes it only slowly available for biological uptake.

When micro-organisms and plant roots compete for ammonium-N (processes 3, 6 and 8 in Fig. 3) in nitrogen limited systems, the heterotrophic micro-organisms have an advantage as a result of their higher substrate affinity (Table 3). If a substrate with a high C/N ratio is added to soil, for example by ploughing-under straw or by other ways of returning agricultural residues to the field, rapid immobilization by micro-organisms of soil inorganic nitrogen will occur. Plant roots are, however, probably more efficient competitors for NH_4^+ than are nitrifying bacteria, which oxidize ammonium-N to nitrite-N and nitrate-N (process 6 in Fig. 3). There is thus a sequence of reactions leading to ammonium-N depletion in the soil by plant roots and different types of micro-organisms.

Nitrification

Through nitrification ammonium-N is oxidized to nitrite-N and nitrate-N, mainly by autotrophic nitrifying bacteria of the genera Nitrosomonas and Nitrobacter. Recent observations suggest, however, that Nitrosomonas is not always the most abundant NH_4^+-oxidizer in soil. Walker[59] reported that both Nitrosolobus and Nitrosospira are common in soils and that Nitrosolobus is the dominant NH_4^+-oxidizer in agricultural soils.

Nitrification is a key process for determining the fate of nitrogen in an ecosystem. Nitrate is more mobile than ammonium and thus more readily lost through leaching, and is also available for reduction to nitrous oxide (N_2O) or dinitrogen (N_2) by denitrifying bacteria. From a agro-ecosystem management viewpoint, nitrification can be regarded as a negative factor[56].

Autotrophic nitrification is a strictly aerobic process (Table 2) and it is often inhibited at low pH values, even though nitrification has been shown to occur at pH values as low as 4[10]. Nitrite oxidation is more sensitive to low pH values than is ammonia oxidation[2], and this may lead to nitrite accumulation under acid conditions. Nitrite may, however, also accumulate at high pH in alkaline soils, especially after addition of ammonium fertilizers[48]. Available phosphorus may

Table 3. Michaelis constants (K_m) for different soil processes affecting ammonium- and nitrate-N

	K_m (mg N l^{-1})	Reference
Ammonium-N		
Plant root uptake	0.28–2.4	38
Immobilization	0.2	33
Nitrification	8.0–18.0	3,52
Nitrate-N		
Plant root uptake	0.35– 8.4	38
Immobilization	1.0	33
Denitrification	3.2	28
Dissimilatory reduction to ammonium-N	?	

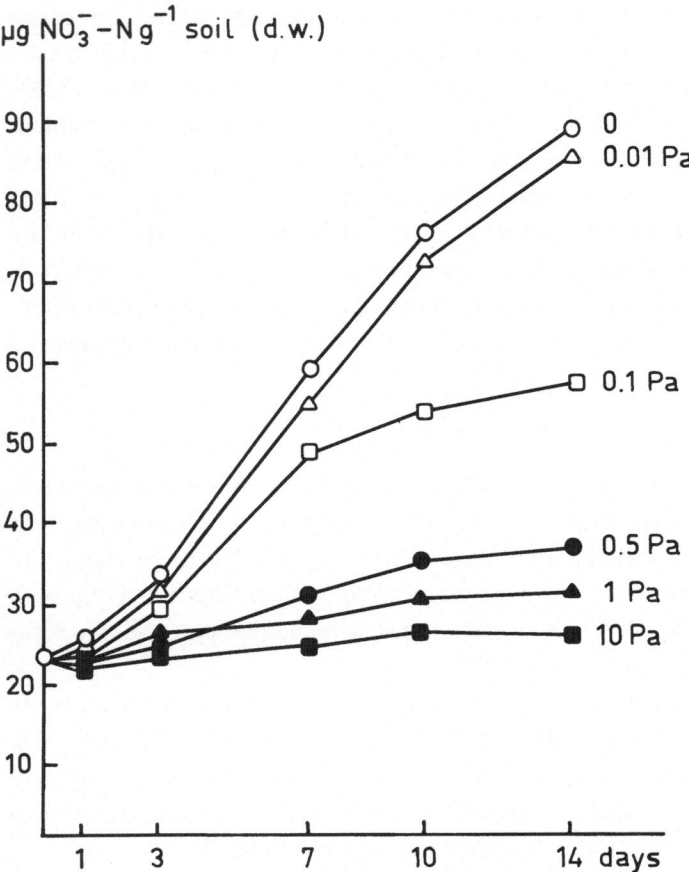

Fig. 7. Production of NO_3^--N at different partial pressures (Pa) of C_2H_2 as a function of time. Standard error $\leq \pm 2$ percent (Berg et al.[9]).

be an important factor for controlling nitrification rates in some ecosystems[40], and nitrite may accumulate under phosphorus deficiency[55], but it seems probable that the main factor influencing the nitrification rate is the concentration of available NH_4^+-N[42].

Special attention has been devoted to the role of nitrification in regulating nitrogen cycling in forest ecosystems. In general, aggrading forests exhibit a comparatively closed nitrogen cycle with small inputs and losses. However, after clear-felling nitrogen may be lost from such sites at appreciable rates, especially if the nutrient conditions are good[53,57,58]. In undisturbed sites, heterotrophic micro-organisms and plant roots together with their mycorrhizal symbionts efficiently scavenge the soil for available nitrogen and little is left for the nitrifier population with its low competitive ability. When the plant roots and mycorrhizas are removed by clear-cutting, excess nitrogen mineralization will increase ammonium-N concentrations to levels at which efficient nitrification can occur.

Nitrate reduction

Nitrate can be used by most bacteria and many fungi as a source of nitrogen for growth. In general, however, ammonium-N is preferred. The process of nitrogen immobilization is, however, only one of three processes by which micro-organisms use nitrate. In the other two cases nitrate is used as a terminal electron acceptor under anaerobic conditions. These processes can result in the following three sets of end-products: 1) nitrite (in the case of nitrate respiration *sensu stricto*), 2) nitric oxide (NO), nitrous oxide (N_2O) and dimolecular nitrogen (N_2) (*via* denitrification), and 3) ammonia (*via* dissimilatory nitrate reduction to ammonia). These processes, discussed below, are much less understood than most other processes in the biogeochemical nitrogen cycle; the dissimilatory reduction to ammonia in particular has only recently received attention as a mechanism which may be important in specific ecosystems.

Assimilatory nitrate reduction

The assimilatory reduction of nitrate to ammonium is believed to occur in only two steps[32]:

$$NO_3^- \xrightarrow{2e^-} NO_2^- \xrightarrow{6e^-} NH_4^+$$

The assimilatory and the dissimilatory nitrate reductases differ in that the former are not membrane-bound[64], and function only under anaerobic conditions (Table 4). As eight electrons are needed for the reduction of nitrate to ammonia, the process requires a source of energy (Table 5). From an energetic viewpoint, ammonium-N should be the preferred nitrogen source for microbial assimilation. Assimilatory nitrate-nitrite-reductases are fairly common in bacteria[24], but comparatively few fungi seem to be able to utilize nitrate as a nitrogen source[16].

The fact that many plants, especially agronomic species, generally take up nitrate may be the result of the better competitive ability of heterotrophic micro-organisms for ammonium-N, leaving excess mineral nitrogen as nitrate available

Table 4. Relative effects of various factors on nitrate assimilation and dissimilatory nitrate reduction*

Factor	$NO_3^- \rightarrow NH_4^+$	$NO_3^- \rightarrow N_{org}$
O_2	– – –	+ +
NO_3^-	+ +	+ +
NH_4^+	0	– – –
Low pH	?	?
C_2H_2	–?	0?

* See Table 2 legend for explanation of symbols

for plant root uptake. The fact that nitrate is produced despite the fact that roots seem to be more efficient than nitrifiers in using NH_4^+–N as discussed earlier is probably a result of the ubiquitous nature of the nitrifiers in soils outside of the rhizosphere, where they can oxidize NH_4^+–N to NO_3^-–N, which, due to its high mobility, can diffuse to the plant roots for subsequent uptake.

Denitrification

Denitrification is the process whereby nitrate and nitrite are reduced to gaseous forms of nitrogen (NO, N_2O and N_2). The use of nitrate as a terminal electron acceptor generates nearly as much energy as aerobic respiration and much more than the common fermentative pathways (Table 5). Denitrification is repressed by oxygen, as nitrate is used as an alternate electron acceptor in place of the preferred oxygen. The process is not repressed by the presence of ammonium (Table 6). As yet, only bacteria have been shown to possess this metabolic pathway. Generally organic substances are oxidized, although there are certain species which can grow autotrophically on H_2 and CO_2 or reduced sulphur

Table 5. Free energy changes ($\Delta G_0'$) in inorganic nitrogen metabolism reactions (from Roswall[44])

	$\Delta G_0'$ (kJ mol^{-1})
Nitrate respiration	
Escherichia coli	
$NO_3^- + [H_2] \rightarrow NO_2^- + H_2O$	− 161
Denitrification	
Pseudomonas aeruginosa	
$2NO_3^- + 2H^+ + 5[H_2] \rightarrow N_2(g) + 6H_2O$	−1121
Other possible reactions	
$N_2O(g) + [H_2] \rightarrow N_2(g) + H_2O$	− 340
$NO_2^- + 1/2[H_2] + H^+ \rightarrow NO(g) + H_2O$	− 76
$2NO(g) + [H_2] \rightarrow N_2O(g) + H_2O$	− 306
$2NO_2^- + 2H^+ + 2[H_2] \rightarrow N_2O(g) + 3H_2O$	− 459
Assimilatory nitrate reduction	
$NO_3^- + 2H^+ + H_2O \rightarrow NH_4^+ + 2O_2$	+ 348
Nitrate fermentation	
Clostridium perfringens	
$NO_3^- + 2H^+ + [4H_2] \rightarrow NH_4^+ + 3H_2O$	− 591
Nitrification	
Nitrosomonas	
$NH_4^+ + 1/2O_2 \rightarrow NH_2OH + H^+$	+ 15
$NH_2OH + O_2 \rightarrow NO_2^- + H_2O + H^+$	− 289
Nitrobacter	
$NO_2^- + 1/2O_2 \rightarrow NO_3^-$	− 77

compounds[29]. It has further been shown that a Hyphomicrobium strain can use nitrate as an electron acceptor even in the presence of oxygen, using methylated amines as an oxidizable substrate[34].

Unlike the nitrate reductase of the assimilatory pathway, the nitrate reductase in denitrification is membrane-bound. A larger number of organisms can reduce nitrate for energy production than can reduce nitrite. If the organism can only use the first reduction step, there must, however, be a mechanism for removing the nitrite, since this is toxic even at low concentrations.

In general, nitric oxide (NO) is rarely a quantitatively important free intermediate in the denitrification pathway, and the main end-products are nitrous oxide (N_2O) and dimolecular nitrogen (N_2). Oxygen inhibits denitrification, but the last step – the reduction of N_2O – seems to be most sensitive to O_2 (Table 6). If O_2 concentrations are increased, the relative amounts of N_2O to N_2 increase. The production of N_2O is also regulated by pH, larger amounts being formed at low pH values[20]. High concentrations of NO_3^- also seem to inhibit N_2O reduction, whereas ammonium-N does not affect production of either N_2O or N_2. In certain instances organisms lack nitrous oxide reductase and N_2O is the sole endproduct[41]. There do not seem to be any bacteria which form NO as the sole end-product during denitrification.

Denitrifying bacteria are cosmopolitan in soil even if conditions prevent them from using their capacity to utilize nitrate instead of oxygen as an electron acceptor. One example is afforded by the tundra mire at Stordalen, Sweden[46]. The low pH in the peat of this mire prevents nitrification; nitrifying bacteria or nitrate have never been observed. In spite of this, 20% of the bacteria isolated from the peat were able to reduce nitrate in pure culture[45]. However, it was impossible to induce denitrification in soil incubations even after additions of nitrate and an easily decomposable carbon source unless the pH was raised from four to seven[28].

Dissimilatory nitrate reduction to ammonium

Dissimilatory reduction of nitrate to ammonium should be an ideal electron acceptor system in anaerobic habitats; and in systems which are more or less permanently anoxic, such as sediments[30,51], and rumen[27], this process seems to

Table 6. Relative effect of various factors on denitrification* (modified from Knowles[29])

Factor	NO_3^-	→	NO_2^-	→	NO	→	N_2O	→	N_2
O_2	$--$		$--$?		$--$?		$---$		
NO_3^-	$++$		$++$		$(-)++$		$(-)++$		
NH_4^+	0		0		0		0		
Low pH	$-$		$-$		$-$		$--$		
C_2H_2	0		0		0		$---$		

* See Table 2 legend for explanation of symbols

be quantitatively important. The process has been shown to occur in soil[13,54], but its quantitative importance is uncertain. In an elegant double-labelling experiment, Tiedje et al.[54] determined the relative rates of certain nitrogen transformations in soil (Table 7). The addition of an energy source (glucose) stimulated the dissimilatory reduction to ammonium, but the rate was still only 4% of the denitrification rate. Dissimilatory reduction is a better electron sink than denitrification and if it were possible to stimulate the dissimilatory reduction to ammonium, this would have considerable importance for maintaining the fertility of many soils in which large losses of nitrogen occur as a result of denitrification.

The effects of acetylene on nitrogen-cycle processes

Acetylene (C_2H_2) affects many of the nitrogen transformations (Tables 2, 4 and 6). The acetylene reduction technique for the determination of rates of nitrogen fixation is well known; the nitrogenase enzyme system not only reduces dinitrogen to ammonia but also acetylene to ethylene. This method has received wide acceptance for determining rates of nitrogen fixation despite many difficulties associated with it, and the correct $C_2H_4 : N_2$ conversion factor must always be determined using other techniques. The theoretical factor is 3, but this is generally not valid; it is usually higher, largely as a result of different amounts of molecular hydrogen being formed.

Acetylene has also found wide use in attempts to determine denitrification rates, since the gas quantitatively inhibits the reduction of N_2O to N_2 at certain concentrations[6,65]. This technique has been developed for determining denitrification rates in sediments[6] and soils[28,66].

Acetylene also affects nitrification by inhibiting the oxidation of ammonium to hydroxylamine[26,60]. This process is much more sensitive to the presence of C_2H_2 than either denitrification or nitrogen fixation, and is completely inhibited by 0.5 Pa ($5 \cdot 10^{-6}$ atm)[9]. The effects of various concentrations of acetylene on rates of nitrification are shown in Fig. 7. This effect is important especially when determining denitrification rates by the use of the acetylene inhibition method since, in view of the fact that no nitrate is formed by nitrification, the rate of nitrate reduction will decline because of substrate limitation. The effects of

Table 7. Calculated rates ($\mu g\ N\ g^{-1}$ dry soil day^{-1}) for selected nitrogen transformations in the presence (1 mg C g soil) or absence of glucose (Tiedje et al.[54])

	Denitrification	Dissimilatory reduction	Mineralization	Immobilization
No glucose	15	0.3–0.6	1–3	0–0.6
Plus glucose	25	1.0	8	2

acetylene on nitrification in soil can persist up to one week after removal of the acetylene[9].

Acetylene will prove to be a powerful tool in nitrogen cycling studies and hopefully its mode of action will soon be determined so that we can use the technique as efficiently as possible.

Acknowledgements The paper has been prepared under the auspices of the SCOPE/UNEP International Nitrogen Unit and the research project 'Ecology of Arable Land. The Role of Organisms in Nitrogen Cycling'. This support is gratefully acknowledged. Dr. G. P. Robertson has made valuable comments on a draft of this manuscript. The excellent typing of Gun Martinsson is appreciated.

References

1   Anderson R V, Coleman D C and Cole C V 1981 Effects of saprotrophic grazing on net mineralization. *In* Clark F E and Rosswall T (Eds). Terrestrial Nitrogen Cycles. Processes, Ecosystem Strategies and Management Impacts. Ecol. Bull. Stockholm 33, 201–216.

2   Anthonisen A C, Loehr R C, Prakasam T B S and Srinath E G 1976 Inhibition of nitrification by ammonia and nitrous acid. J. Water Pollut. Contr. Fed. 48, 835–852.

3   Ardakani M S, Rehbock J T and McLaren A D 1974 Oxidation of ammonium to nitrate in a soil column. Soil Sci. Soc. Am. Proc. 38, 96–99.

4   Ausmus B S, Edwards N T and Witkamp M 1976 Microbial immobilization of carbon, nitrogen, phosphorus and potassium: Implications for forest ecosystem processes. *In* Anderson J M and MacFadyen A (Eds). The Role of Terrestrial and Aquatic Organisms in Decomposition Processes, pp 397–421. London: Blackwell.

5   Ayanaba A and Dart P J (Eds) 1977 Biological Nitrogen Fixation in Farming Systems of the Tropics. Chichester-New York-Brisbane-Toronto: John Wiley and Sons. 377 p.

6   Balderstone W L, Sherr B and Payne W J 1976 Blockage by acetylene of nitrous oxide reduction in *Pseudomonas perfectomarinus*. Appl. Environ. Microbiol. 31, 504–508.

7   Berg B and Staaf H 1980 Decomposition rate and chemical changes of Scots pine needle litter. II. Influence of chemical composition. *In* Persson T (Ed.). Structure and Function of Northern Coniferous Forests – An Ecosystem Study. Ecol. Bull. Stockholm 32, 373–390.

8   Berg B and Staaf H 1981 Leaching, accumulation and release of nitrogen in decomposing forest litter. *In* Clark F E and Rosswall T (Eds). Terrestrial Nitrogen Cycles. Processes, Ecosystem Strategies and Management Impacts. Ecol. Bull. Stockholm 33, 163–178.

9   Berg P, Klemedtsson L and Rosswall T 1982 Inhibitory effect of low partial pressures of acetylene on nitrification. Soil Biol. Biochem. 14, 301–303.

10   Bhuiya Z H and Walker, N 1977 Autotrophic nitrifying bacteria in acid tea soils from Bangladesh and Sri Lanka. J. Appl. Bacteriol. 42, 253–257.

11   Bolin B, Crutzen P J, Vitousek P M, Woodmansee R G, Goldberg E D and Cook R B 1983 Interactions of biogeochemical cycles. *In* Bolin B and Cook R B (Eds). The Biogeochemical Cycles and Their Interactions. SCOPE Report. Chichester-New York-Brisbane-Toronto: John Wiley and Sons. (*In press*).

12   Bowen G D and Smith S E 1981 The effects of mycorrhizae on nitrogen uptake by plants. *In* Clark F E and Rosswall T (Eds). Terrestrial Nitrogen Cycles. Processes, Ecosystem Strategies and Management Impacts. Ecol. Bull. Stockholm 33, 237–247.

13   Caskey W H and Tiedje J M 1979 Evidence for clostridia as agents of dissimilatory reduction of nitrate to ammonium in soils. Soil Sci. Soc. Am. J. 43, 931–936.

14   Clarholm M 1981 Protozoan grazing of bacteria in soil – impact and importance. Microb. Ecol. 7, 343–350.

15   Döbereiner J, Burris R H and Hollaender A (Eds) 1978 Limitations and Potentials for
     Biological Nitrogen Fixation in the Tropics. New York-London: Plenum Press. 398 p.

16   Downey R J 1978 Control of fungal nitrate reduction. In Schlessinger, D. (Ed.). Microbiology –
     1978, pp 320–323. Washington D.C.: American Society for Microbiology.

17   Dreyfus B L and Dommergues Y R 1981 Stem nodules on the tropical legume, Sesbania
     rostrata. In Gibson A H and Newton E (Eds). Current Perspectives in Nitrogen Fixation, p. 471.
     Amsterdam-New York-Oxford: Elsevier/North Holland Biomedical Press.

18   Fenn P and Kirk T K 1981 Relationship of nitrogen to the onset and supression of ligninolytic
     activity and secondary metabolism in Phanerochaete chrysosporium. Arch. Microbiol. 130, 59–
     65.

19   Fenn P, Choi S and Kirk T K 1981 Ligninolytic activity of Phanerochaete chrysosporium:
     Physiology of supression by NH_4^+ and 1-glutamate. Arch. Microbiol. 130, 66–71.

20   Focht D D 1974 The effect of temperature, pH and aeration on the production of nitrous oxide
     and gaseous nitrogen – a zero-order kinetic model. Soil Sci. 118, 173–179.

21   Gallon J R 1981 The oxygen sensitivity of nitrogenase: a problem for biochemists and
     microrganisms. Trends Biochem. Sci. 6, 19–23.

22   Gibson A H and Newton W E (Eds) 1981 Current Perspectives in Nitrogen Fixation. Proc. 4th
     Intl. Symp. Nitrogen Fixation. Amsterdam-New York-Oxford: Elsevier/North Holland
     Biomedical Press. 534 p.

23   Granhall U (Ed.) 1978 Environmental Role of Nitrogen-fixing Blue-green Algae and
     Asymbiotic Bacteria. Ecol. Bull. Stockholm 26, 391 p.

24   Hall J 1978 Nitrate-reducing bacteria. In Schlessinger D (Ed.). Microbiology – 1978, pp 296–
     298. Washington D.C.: American Society for Microbiology.

25   Heal O W and MacLean S F 1975 Comparative productivity in ecosystems – secondary
     productivity. In Van Dobben W H and Lowe-McConnell R H (Eds). Unifying Concepts in
     Ecology, pp 89–108. The Hague: Dr. W. Junk and Wageningen: Pudoc.

26   Hynes R K and Knowles R 1978 Inhibition by acetylene of ammonium oxidation in
     Nitrosomonas europea. FEMS Microbiol. Lett. 4, 319–321.

27   Kaspar H F and Tiedje J M 1981 Dissimilatory reduction of nitrate and nitrite in the bovine
     rumen: nitrous oxide production and effect of acetylene. Appl. Environ. Microbiol. 41, 705–709.

28   Klemedtsson L, Svensson B H, Lindberg T and Rosswall T 1977 The use of acetylene
     inhibition of nitrous oxide reductase in quantifying denitrification in soils. Swed. J. Agric. Sci. 7,
     179–185.

29   Knowles R 1981 Denitrification. In Clark F E and Rosswall T (Eds). Terrestrial Nitrogen
     Cycles. Processes, Ecosystem Strategies and Management Impacts. Ecol. Bull. Stockholm 33,
     315–329.

30   Koike I and Hattori A 1978 Denitrification and ammonia formation in anaerobic coastal
     sediments. Appl. Environ. Microbiol. 35, 278–282.

31   Kudeyarov V N 1981 Mobility of fixed ammonium in soil. In Clark F E and Rosswall T (Eds).
     Terrestrial Nitrogen Cycles. Processes, Ecosystem Strategies and Management Impacts. Ecol.
     Bull. Stockholm 33, 281–290.

32   Losada M, Guerrero M G and Vega J M 1981 The assimilatory reduction of nitrogen. In Bothe
     H and Trebst A (Eds.) Biology of Inorganic Nitrogen and Sulphur, pp 30–63. Berlin-Heidelberg-
     New York: Springer-Verlag.

33   McGill W B, Hunt H W, Woodmansee R G and Reuss J O 1981 Phoenix, a model of the
     dynamics of carbon and nitrogen in grassland soils. In Clark F E and Rosswall T (Eds).
     Terrestrial Nitrogen Cycles. Processes, Ecosystem Strategies and Management Impacts. Ecol.
     Bull. Stockholm 33, 49–115.

34   Meiberg J B M, Bruinenberg P M and Harder W 1980 Effect of dissolved oxygen tension in the
     metabolism of methylated amines in Hyphomicrobium X in the absence and presence of nitrate:
     evidence for 'aerobic' denitrification. J. Gen. Microbiol. 120, 453–463.

35 Newton W, Postgate J R and Rodríguez-Barrueco C (Eds) 1977 Recent Developments in Nitrogen Fixation. Proc. 2nd Intl. Symp. London-New York-San Francisco: Academic Press. 622 p.

36 Newton W E and Orme-Johnson W H (Eds) 1980 Nitrogen Fixation, Vol. 1 and 2. Baltimore: University Park Press. 394 and 325 p.

37 Nömmik H 1981 Fixation and biological availability of ammonium in soil clay minerals. In Clark F E and Rosswall T (Eds). Terrestrial Nitrogen Cycles. Processes, Ecosystem Strategies and Management Impacts. Ecol. Bull. Stockholm 33, 273–279.

38 Nye P H and Tinker P B 1977 Solute Movement in the Soil-Root System. Oxford-London-Edinburgh-Melbourne: Blackwell Scientific Publications. 342 p.

39 Persson J 1981 Rapporteur's comments: Nitrogen fertilizer effect on nitrogen cycle processes. In Clark F E and Rosswall T (Eds). Terrestrial Nitrogen Cycles. Processes, Ecosystem Strategies and Management Impacts. Ecol. Bull. Stockholm 33, 562–564.

40 Purchase B S 1974 The influence of phosphate deficiency on nitrification. Plant and Soil 41, 541–547.

41 Renner E D and Becker G L 1970 Production on nitric oxide and nitrous oxide during denitrification by Corynebacterium nephridii. J. Bacteriol. 101, 821–826.

42 Robertson G P and Vitousek P M 1981 Nitrification potentials in primary and secondary succession. Ecology 62, 376–386.

43 Rosswall T 1976 The internal nitrogen cycle between vegetation, microorganism and soil. In Svensson B H and Söderlund R (Eds). Nitrogen, Phosphorus and Sulphur – Global Cycles. Ecol. Bull. Stockholm 22, 157–167.

44 Rosswall T 1981 The biogeochemical nitrogen cycle. In Likens G E (Ed.). Some Perspectives of the Major Biogeochemical Cycles, pp 25–49. SCOPE Report 17. Chichester: John Wiley and Sons, Ltd.

45 Rosswall T and Clarholm M 1974 Characteristics of tundra bacterial populations and a comparison with populations from forest and grassland soils. In Holding A J, Heal O W, MacLean, S F and Flanagan P W (Eds). Soil Organisms and Decomposition in Tundra, pp 93–108. Stockholm: Tundra Biome Steering Committee.

46 Rosswall T and Granhall U 1980 Nitrogen cycling in a subarctic ombrotrophic mire. In Sonesson M (Ed.). Ecology of a Subarctic Mire. Ecol. Bull. Stockholm 30, 209–234.

47 Rosswall T and Vitousek P M (Rapporteurs) 1980 Research priorities and future co-operation. In Rosswall T (Ed.). Nitrogen Cycling in West African Ecosystems, pp 421–428. Stockholm: SCOPE/UNEP International Nitrogen Unit, The Royal Swedish Academy of Sciences.

48 Scarsbrook C E 1965 Nitrogen availability. In Bartholomew, W V and Clark F E (Eds). Soil Nitrogen. Agronomy 10, 481–502. Madison, Wisc.: American Society of Agronomy.

49 Söderlund R and Rosswall T 1982 The nitrogen cycles. In Hutzinger O (Ed.) Environmental Chemistry. Berlin-Heidelberg-New York: Springer Verlag (in press).

50 Söderlund R and Svensson B H 1976 The global nitrogen cycle. In Svensson B H and Söderlund R (Eds). Nitrogen, Phosphorus and Sulphur – Global Cycles. Ecol. Bull. Stockholm 22, 23–73.

51 Sørensen J 1978 Capacity for denitrification and reduction of nitrate to ammonia in a coastal marine sediment. Appl. Environ. Microbiol. 35, 301–305.

52 Starr J L, Broadbent F E and Nielsen D R 1974 Nitrogen transformations during continuous leaching. Soil Sci. Soc. Am. Proc. 39, 284–289.

53 Tamm C O, Holmen H, Popovic B and Wiklander G 1974 Leaching of plant nutrients from soils as a consequence of forestry operations. Ambio 3, 211–221.

54 Tiedje J M, Sørensen J and Chang Y-Y L 1981 Assimilatory and dissimilatory nitrate reduction: perspectives and methodology for simultaneous measurement of several nitrogen cycle processes. In Clark F E and Rosswall T (Eds). Terrestrial Nitrogen Cycles. Processes, Ecosystem Strategies and Management Impacts. Ecol. Bull. Stockholm 33, 331–342.

55   Verstraete W 1981 Nitrification. *In* Clark F E and Rosswall T (Eds). Terrestrial Nitrogen
     Cycles. Processes, Ecosystem Strategies and Management Impacts. Ecol. Bull. Stockholm 33,
     303–314.
56   Verstraete W 1981 Nitrification in agricultural systems: Call for control. *In* Clark F E and
     Rosswall T (Eds). Terrestrial Nitrogen Cycles. Processes, Ecosystem Strategies and
     Management Impacts. Ecol. Bull. Stockholm 33, 565–572.
57   Vitousek P M 1980 Nitrogen losses from disturbed ecosystems – Ecological considerations. *In*
     Rosswall T (Ed.). Nitrogen Cycling in West African Ecosystems, pp 39–53. Stockholm:
     SCOPE/UNEP International Nitrogen Unit, The Royal Swedish Academy of Sciences.
58   Vitousek P M 1981 Clear-cutting and the nitrogen cycle. *In* Clark F E and Rosswall T (Eds).
     Terrestrial Nitrogen Cycles. Processes, Ecosystem Strategies and Management Impacts. Ecol.
     Bull. Stockholm 33, 631–642.
59   Walker N 1978 On the diversity of nitrifiers in nature. *In* Schlessinger, D. (Ed) Microbiology –
     1978, pp 346–347. Washington D.C.: American Society for Microbiology.
60   Walter H M, Keeney D R and Fillery I R 1979 Inhibition of nitrification by acetylene. Soil Sci.
     Soc. Am. J. 43, 195–196.
61   Yamada T and Sakaguchi K 1980 Nitrogen fixation associated with hotspring green algae.
     Arch. Microbiol 124, 161–167.
62   Yatazawa M and Susilo H 1980 Development of upper stem nodules in *Aeschynomene indica*
     under experimental conditions. Soil Sci. Plant Nutr. 26, 317–319.
63   Yatazawa M and Yoshida S 1979 Stem nodules of *Aeschynomene indica* and their capacity of
     nitrogen fixation. Physiol. Plant. 45, 293–295.
64   Yordy D M and Ruoff K L 1981 Dissimilatory nitrate reduction to ammonia. In Delwiche, C.
     C. (Ed.). Denitrification, Nitrification and Atmospheric Nitrous Oxide, pp 171–190. New York-
     Chichester-Brisbane-Toronto: John Wiley.
65   Yoshinari T and Knowles R 1976 Acetylene inhibition of nitrous oxide reduction by
     denitrifying bacteria. Biochem. Biophys. Res. Commun. 69, 705–710.
66   Yoshinari T, Hynes R and Knowles R 1977 Acetylene inhibition of nitrous oxide reduction and
     measurement of denitrification and nitrogen fixation in soil. Soil Biol. Biochem. 9, 177–183.

Plant and Soil 67, 35–43 (1982). 0032-079X/82/0671-0035$01.35.
© 1982 *Martinus Nijhoff/Dr W. Junk Publishers, The Hague.*

Simulation of nitrogen in agro-ecosystems: Criteria for model selection and use

Simulación de nitrógeno en agro-ecosistemas: Criterios para selección y uso del modelo

P. S. C. RAO, R. E. JESSUP and A. G. HORNSBY*

Soil Science Department, University of Florida, Gainesville, Florida 32611, USA

Key words Agro-ecosystems Mathematical models N-cycling Simulation modeling.

Abstract Available simulation models for describing nitrogen behavior in agro-ecosystems vary in two characteristics: (*i*) *conceptual completeness* in terms of the number of processes considered, and (*ii*) the *level of detail* at which each process is modeled. These model characteristics are determined by both the objectives that the model is designed to meet and the current state-of-the-art understanding of the various processes included in the model. The levels of conceptual completeness and detail in a model govern the potential applications for which the model may be used. Applications of models may be research-oriented, management-oriented, or planning-oriented. A model suitable for a given application should have an appropriate level of completeness and detail to accomplish the stated objective.

Criteria to aid in the selection and evaluation of nitrogen simulation models for a particular application include: i) the availability of computational facilities, ii) the spatial and temporal scales of application, iii) the intended use of the simulations, iv) the availability of model input data, and v) the confidence regions associated with the model output.

Resumen Los modelos de simulación del comportamiento en agro-ecosistemas difieren en dos características: (i) *entereza o minuciosidad conceptual*, en términos del número de procesos que considera, y (ii) *el nivel de detalle* en el cual cada proceso es modelado. Estas características del modelo son determinadas por los objetivos para los cuales el modelo es diseñado y por el grado actual del conocimiento de los procesos incluídos en el modelo. Los niveles de entereza conceptual y de detalle en un modelo determinan las aplicaciones potenciales para las cuales dicho modelo puede ser utilizado. Las aplicaciones pueden ser orientadas a la investigación, manejo o planificación. Un modelo adecuado para una aplicación determinada debería tener un nivel apropiado de entereza y detalle para lograr el objetivo establecido.

Los criterios de ayuda en la selección y evaluación de los modelos de simulación del nitrógeno para una aplicación particular incluyen: (i) la disponibilidad de facilidades computacionales, (ii) las escalas espacial y temporal de aplicación, (iii) el uso deseado de las simulaciones, (iv) la disponibilidad de datos de entrada al modelo, y (v) los ámbitos de confianza asociadas con las salidas del modelo.

Introduction

The primary objective of traditional agronomic research on nitrogen has been the measurement of crop yield response to fertilizer nitrogen. With an increasing

---

* Permanent address: R. S. Kerr Environmental Research Lab., U.S. Environmental Protection Agency, Ada, Oklahoma

concern for nitrate pollution of surface and groundwater resources, recent research emphasis has shifted towards a quantitative description of nitrogen behavior in agro-ecosystems, especially on the quantities of nitrogen leached beyond the crop root zone. This type of research has focused on investigating the factors and processes that govern nitrogen dynamics in the root zone; these in turn, determine nitrogen availability and subsequent crop yield response. The common goal of both agronomists seeking to increase crop yields and environmental scientists interested in minimizing nitrate pollution of groundwater is to maximize fertilizer-use efficiency. This goal can be achieved by management practices based on a mechanistic understanding of nitrogen cycling in agro-ecosystems.

The fate of nitrogen in agro-ecosystems is governed by a variety of complex and inter-related processes. Various inorganic nitrogen species (NH_4^+, NO_3^-, NO_2^-, N_2O, NO, and N_2) as well as organic forms may exist simultaneously. These nitrogen species undergo both reversible and irreversible transformations caused by chemical and microbiological processes. The water-soluble nitrogen species (principally NH_4^+ and NO_3^-) are leached along with soil-water following irrigation and rainfall. NH_4^+ and NO_3^- are taken up by plant roots, at rates dependent on the plant species, its growth stage, meteorological conditions, and soil physical properties. These and other processes pictured schematically in Fig. 1 are dynamic in nature and occur simultaneously.

Various simulation models have been developed during the past decade for describing nitrogen dynamics in soil-water-crop-atmosphere systems. Excellent reviews of these nitrogen models have been presented [22,23,24]. A number of books and conference proceedings dealing with nitrogen in the environment have recently been published [1,8,9,13,14,15,17,21].

Because of the availability of these excellent books and review articles, the objective of this paper will be limited to a discussion of various factors that govern the development and use of nitrogen simulation models. Certain criteria are suggested to aid in selection and use of the available nitrogen models.

Overview of nitrogen simulation models

Using systems analysis concepts, a simulation model for nitrogen should consist of various submodels, each describing a given subsystem or process. An example of nitrogen model development using such a block building technique is shown in Fig. 2. Each of the blocks pictured may consist of other, smaller blocks and the submodels may receive inputs from other submodels. An ideal simulation model for nitrogen would include all of the processes dictating nitrogen dynamics in the system to be modeled. Furthermore, each process (*i.e.* submodel) should be described at a level of detail that represents the current state-of-the-art in understanding. But such an idealized approach to model development for complex and heterogeneous biological systems such as agro-ecosystems is

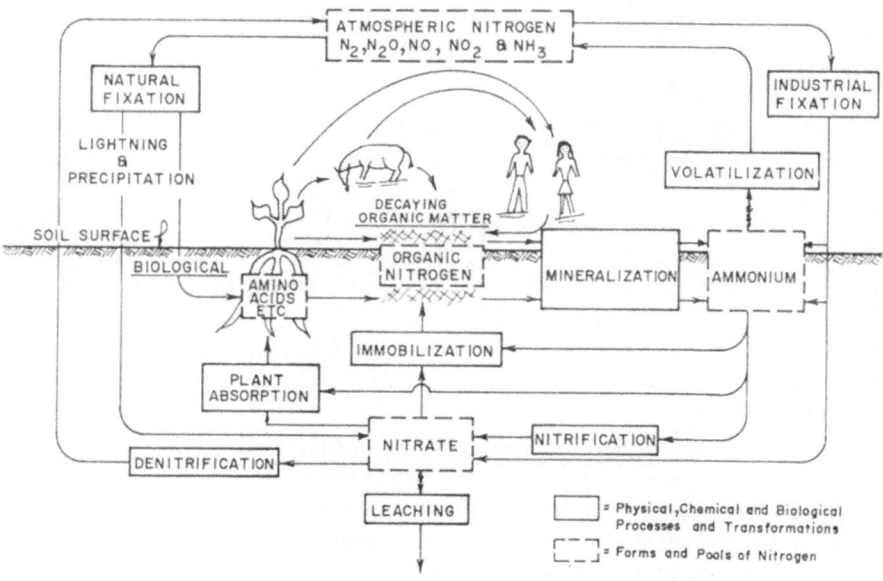

Fig. 1. Schematic representation of the nitrogen cycle in agro-ecosystems. Adapted from Hornsby[11].

impractical. In most instances not all the system processes are well understood and the coupling among them may be unknown. In this sense all simulation models may be considered to be incomplete. Each modeler introduces various approximations and simplifications in the development of a specific model for a particular application.

Simulation models vary in two principal characteristics:

(i) *conceptual completeness* in terms of the number of system processes included in the model, and

(ii) the *level of detail* or complexity at which each system process is modeled.

A given model may include all known system processes but may not describe all of them at a high level of detail. Conversely, in a given model some of the processes may be described at a high level of detail but the model may not necessarily include all the possible processes.

When system processes are initially unknown and a model consists of empirical functions derived by inductive reasoning, the modeling approach is considered to be 'black-box' modeling. On the other hand, when complete quantitative descriptions of the system processes are available and the model is deduced from established laws, a 'white-box' label is used. Thus, simulation models may be considered to have various 'shades of gray'[12]. The conceptual completeness and the level of detail in a given simulation model are determined, in part, by the following factors:

(i)   current state-of-the-art in understanding the system to be modeled,

AGRO—ECOSYSTEM MODEL

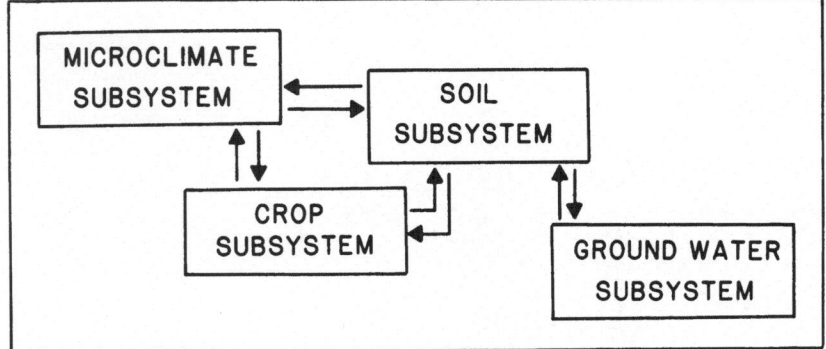

SOIL SUBSYSTEM MODEL

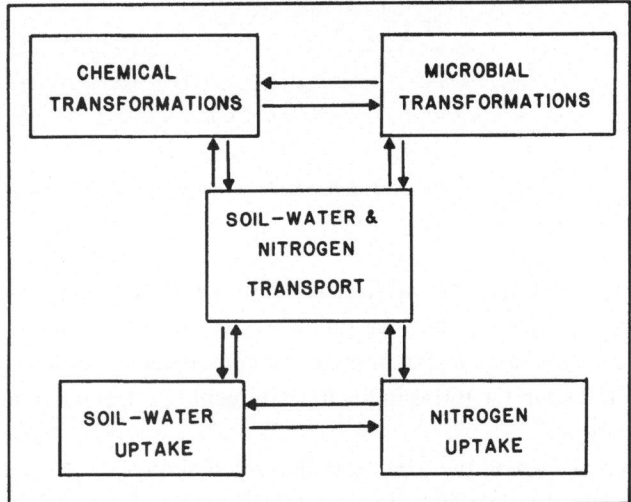

MICROBIAL TRANSFORMATIONS MODEL

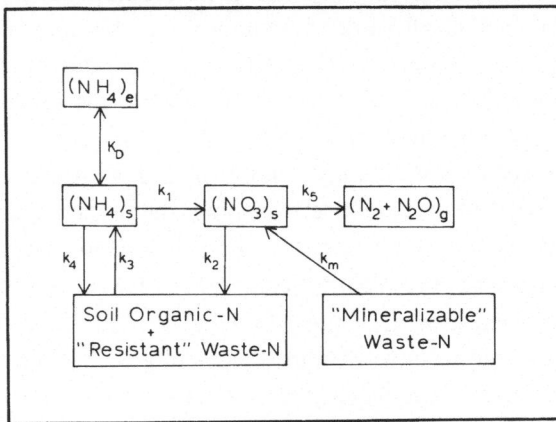

Fig. 2. An example of model development using systems analysis concepts. Nitrification of mineralized waste-N is assumed to be instantaneous.

(ii) the modeler's conceptualization of the system processes,

(iii) the modeling approach and the error bounds allowable in approximations required to solve the problem, and

(iv) the spatial and temporal scales of intended model applications.

The modeler's conceptualization of the system is usually determined by his scientific expertise and background. For example, hydrologists and soil physicists developing nitrogen models tend to focus primary attention on a detailed mechanistic description of soil-water and solute transport phenomena, while biological processes such as plant uptake and microbial transformations are treated as sink/source terms in the transport equations[4,5,25,31]. Biological scientists on the other hand generally emphasize microbial transformations and crop-related processes, with secondary importance given to physical processes (e.g.[19,28]).

Applications of nitrogen simulation models

Simulation models for describing nitrogen dynamics, as is the case for other models, have been developed for two principal reasons. First, models present a logical synthesis of existing knowledge and provide quantitative descriptions of the systems processes. Models can thus reveal gaps in our understanding and provide directions for future research efforts. Second, a properly validated and verified simulation model can be used to predict system responses to a given perturbation (e.g. change in irrigation practices or increased fertilizer input). This can reduce the need for extensive experimentation to study a system's behavior under various treatments.

One of the common uses of nitrogen models has been to better understand fundamental scientific principles or phenomena. An example is a recently-developed model that describes mineralization of soil organic nitrogen[28]. These types of models describe a process in as much detail as possible and quantify corresponding rates and transfer coefficients. Such single-process models can be linked together to examine interactions among basic processes which are not mutually independent. For example, models developed to describe over-all nitrogen behavior in the crop root zone attempt to couple soil-water and solute-transport models with those for transformations and plant uptake[4,10,20,25]. These models are very complex and require a large number of inputs; further, spatial and temporal scales of resolution for such models usually are of the order of a few centimeters and minutes. In principle, these models can be used to describe nitrogen dynamics on a field or watershed scale. However, such applications are often impractical because computer costs are prohibitively expensive and the model input parameters as well as measured data for verifying model output are generally unavailable.

For management-oriented applications several simplifying assumptions can be incorporated into comprehensive models. Simplifications usually involve a reduction in the level of detail and the elimination of certain 'minor' processes. The sacrifice in detailed specification of individual processes is justified by the finding that sufficiently accurate descriptions of nitrogen dynamics at field-scale are often provided by models that use only limited amounts of input data[3,6,18]. Such models may be useful for designing crop production systems that optimize use efficiency and minimize groundwater pollution. For applications such as irrigation and fertilizer application scheduling, the model may treat fields of 10 to 100 hectares and time scales of a few days to a few weeks.

Simulation models have also been developed to assist regional planning and feasibility studies. For such purposes, long-term rather than short-term effects of agricultural management practices need to be evaluated. As the transient changes caused by individual processes are of little interest, they are aggregated over space and time in order to estimate the total contributions of various sources and sinks. Mass-balance steady-state models proposed by Hornsby[11], Fried et al.[7], and Tanji et al.[26,27] may be placed in this category. The spatial and temporal scale of these models is considerably larger than those for the comprehensive models. For example, irrigation or reclamation projects may have spatial dimensions ranging from 1,000 to 100,000 hectares and temporal scales varying from a few months to several decades.

Criteria for nitrogen model selection and use

Currently there are about twenty different simulation models available for simulating nitrogen dynamics in agricultural ecosystems (see ref.[9,23]). Many of these models are site-specific and because of the complexity of agro-ecosystems, most are very complex. Without sufficient documentation, another modeler, let alone a prospective user, may not fully comprehend a given model's intricacies. Even if adequate documentation were available for all models, the potential user is still faced with the dilemma of selecting a model suitable for his application. Some criteria which may aid in the selection and evaluation of nitrogen models presently available include:

(i)   the intended use of the model simulations,
(ii)   the spatial and temporal scales of application,
(iii) the availability of computational facilities,
(iv)   the availability of model input data, and
(v)   the confidence regions associated with model output.

The intended use of the model simulations must focus on two model attributes: the type and the scale of the model output. Usually outputs from models consist of the total amounts (mass) and fluxes (mass per time per unit area) associated with various sinks and sources. As indicated earlier, each model is coded with characteristic spatial and temporal scales which define the lower limit at which they may be applied. Thus, a model specifically designed for management or planning purposes would probably not be useful for most research applications.

Simulation models are formulated as a set of mathematical operations, usually partial differential equations, that quantify various system processes. These operations are performed using numerical methods with the aid of digital computers. A simulation model can therefore be thought of in two distinct terms: as a compilation of mathematical operations for describing the system, and as a computer program (or coding) to perform these operations. One criterion in model selection must then be the availability of adequate computational facilities (hardware, software, and personnel). A user with access to limited facilities, *e.g.* microcomputer, may not be able to use comprehensive simulation models that require large amounts of computer memory. Even if the user has access to sophisticated computers, more often than not he may need to introduce various minor modifications into the computer coding because of slight variations in high level computer languages from one computer system to another. It may also be desirable to make certain modifications in the input-output routines of the computer program to accomodate the user's particular needs. Sometimes the program may need to be entirely rewritten in a new language (*e.g.* from CSMP to FORTRAN IV) in order to run it on a given computer. Thus, the services of a competent computer programmer may be an essential element in the selection and use of a model.

The level of confidence a user can place on the model output must also be evaluated when selecting a simulation model. At the simplest level this evaluation may involve checks to ensure that there are no errors in the computer program or that calculations are being made within the range of conditions over which the program and the model were tested. More importantly, however, the user must also evaluate the appropriateness and validity of the assumptions and simplifications made in developing the model. For example, the assumption in several models of complete soil-water displacement may not be valid for well-structured soils.

The confidence to be placed on the model output also depends upon the uncertainties associated with the input parameters. As we have discussed earlier, with increasing complexity (*i.e.* level of detail and conceptual completeness) the input requirements of a model increase. The use of comprehensive models at a field-scale or watershed-scale is limited by the unavailability of the necessary input data. Furthermore, the physical, chemical and biological parameters for characterizing the ecosystem vary considerably, both spatially and temporally, even in a single field[2,16,29]. Such heterogeneity will be even greater for watersheds or river basins. Therefore, unless a large number of samples are taken, the estimates for these parameters can be in considerable error[30]. With the exception of a single case[25], all simulation models for nitrogen consider the input parameters to be deterministic and do not account for their stochastic nature. Errors in input parameters can propagate through a model and, depending upon the model's sensitivity to a given parameter, can result in rather large variations in model output. Spatial and temporal heterogeneities are also reflected in variations in the measured data which are to be used for model verification. Thus, in selecting and using a nitrogen simulation model for field-scale applications, the user must be cognizant of the limitations imposed by uncertainties in model input parameters and the associated confidence limits on the model output. A good rule-of-thumb might be that a model should not be expected to simulate a system's behavior any better than one's ability to measure it.

References

1   Beek J and Frissel M J 1973 Simulation of Nitrogen Behaviour in Soils. Wageningen: Pudoc. 67 p.
2   Biggar J W and Nielsen D R 1976 Spatial variability of the leaching characteristics of soils. Water Resour. Res. 12, 78–84.
3   Davidson J M and Rao P S C 1978 Use of mathematical relationships to describe the behaviour of nitrogen in the crop root zone. *In* Pratt, P. F. (Ed.). Management of Nitrogen in Irrigated Agriculture: Proceedings of a National Conference, pp 291–319. Riverside: Univ. of Calif. Press.
4   Davidson J M, Graetz D A, Rao P S C and Selim H M 1978 Simulation of Nitrogen Movement, Transformations, and Uptake in the Plant Root Zone. Washington D.C.: EPA – 600/3-78-029. 116 p.
5   Donigian A S Jr. and Crawford N H 1976 Modeling Pesticides and Nutrients on Agricultural Lands. Washington D.C.: EPA – 600/3-76-043. 332 p.
6   Duffy J, Chung C, Boast C and Franklin M 1975 A simulation model of biophysiochemical transformations of nitrogen in tile-drained corn belt soils. J. Environ. Qual. 4, 447–486.
7   Fried M, Tanji K K, and van de Pol R M 1976 Simplified long term concepts for evaluating leaching of nitrogen from agricultural land. J. Environ. Qual. 5, 197–200.
8   Frissel M J (Ed.) 1978 Cycling of Mineral Nutrients in Agricultural Ecosystems. New York: Elsevier Scientific. 356 p.
9   Frissel M J and van Veen J A (Eds) 1981 Simulation of Nitrogen Behaviour of Soil-Plant Systems. Wageningen: Pudoc. 277 p.
10 Hagin J and Amberger A 1974 Contribution of Fertilizers and Manures to the N- and P-Loads of Waters: A Computer Simulation. A Final Report to Deutsche Forschungsgemeinschaft from Technion Foundation, Israel. 123 p.
11 Hornsby A G 1973 Prediction Modeling for Salinity Control in Irrigation Return Flows. Washington D.C. EPA – R2-73-168. 55 p.
12 Karplus W J 1976 The future of mathematical models of water resource systems. *In* Vansteenkiste G C (Ed.). System Simulation in Water Resources, pp 11–18. Amsterdam: North-Holland Publishing Co.
13 Law J P and Skogerboe G V (Eds) 1977 Irrigation Return Flow Quality Management: Proceedings of National Conference. Ft. Collins: Colorado State University Press. 451 p.
14 Nielsen D R and MacDonald J G (Eds) 1978 Nitrogen in the Environment. Vol. 1. New York: Academic Press. 526 p.
15 Nielsen D R and MacDonald J G (Eds) 1978 Nitrogen in the Environment. Vol. 2. New York: Academic Press. 528 p.
16 Nielsen D R, Biggar J W and Erh K T 1973 Spatial variability of field-measured soil-water

properties. Hilgardia 42, 215–259.

17 Pratt P F (Ed.) 1978 Management of Nitrogen in Irrigated Agriculture: Proc. of National Conference. Sacramento: Univ. of California Press. 442 p.

18 Rao P S C, Davidson J M and Jessup R E 1981 Simulation of nitrogen behaviour in the root zone of cropped land areas receiving organic wastes. *In* Frissel M J and van Veen J A (Eds). Simulation of Nitrogen Behaviour of Soil-Plant Systems, pp 81–95. Wageningen: Pudoc.

19 Reuss J O and Innis G S 1977 A grassland nitrogen flow simulation model. Ecology 58, 379–388.

20 Shaffer M J, Ribbens R W and Huntley C W 1977 Detailed Return Flow Salinity and Nutrient Simulation Model. Vol. V. Prediction of Mineral Quality of Irrigation Return Flow. Washington D.C.: EPA – 600/2-77-179. 243 p.

21 Stevenson F J (Ed.) 1981 Nitrogen in Agricultural Soils. Agronomy Monograph No. 22. Madison: American Society of Agronomy.

22 Tanji K K 1980 Problems in modeling nonpoint sources of nitrogen in agricultural systems. *In* Overcash M R and Davidson J M (Eds). Environmental Impact of Nonpoint Source Pollution. pp 165–183. Ann Arbor: Ann Arbor Sci. Publ.

23 Tanji K K 1981 Modeling of the soil nitrogen cycle. *In* Stevenson F J (Ed.). Nitrogen in Agricultural Soils. Agronomy Monograph No. 22. Madison: American Society of Agronomy.

24 Tanji K K and Gupta S K 1978 Computer simulation modeling of nitrogen in irrigated croplands. *In* Nielsen D R and MacDonald J G (Eds). Nitrogen in the Environment, Vol. 1, 526 p. New York: Academic Press.

25 Tanji K K, Mehran M and Gupta S K 1981 Water and nitrogen fluxes in the root zone of irrigated maize. *In* Frissel M J and van Veen J A (Eds). Simulation of Nitrogen Behaviour of Soil-Plant Systems, pp 51–67. Wageningen: Pudoc.

26 Tanji K K, Fried M and van de Pol R M 1977 A steady-state conceptual nitrogen model for estimating nitrogen emissions from cropped lands. J. Environ. Qual. 6, 155–159.

27 Tanji K K, Broadbent F E, Mehran M and Fried M 1979 An extended version of a conceptual model for evaluating annual nitrogen leaching losses from croplands. J. Environ. Qual. 8, 114–120.

28 van Veen J A and Frissel M J 1981 Simulation model of the behaviour of nitrogen in soil. *In* Frissel M J and van Veen J A (Eds). Simulation of Nitrogen Behaviour of Soil-Plant Systems, pp 126–144. Wageningen: Pudoc.

29 Warrick A W and Nielsen D R 1980 Spatial variability of soil physical properties in the field. *In* Hillel D (Ed.). Applications of Soil Physics, pp 319–344. New York: Academic Press.

30 Warrick A W, Mullen G J and Nielsen D R 1977 Predictions of soil-water flux based upon field-measured soil-water properties. Soil Sci. Soc. Am. Proc. 41, 14–19.

31 Watts D G and Hanks R J 1978 A soil-water nitrogen model for irrigated corn on sandy soils. Soil Sci. Soc. Am. J. 42, 492–499.

Plant and Soil 67, 45–59 (1982).   0032-079X/82/0671-0045$02.25.

Nitrogen supply and crop yield: The global scene

La provisión de nitrógeno a los cultivos y su rendimiento: Una visión general

D. J. GREENWOOD

National Vegetable Research Station, Wellesbourne, Warwick, UK

Key words   Cereals Crop response Legumes N-fertilixation. N_2-fixation Nitrogen losses N-supply.

Abstract   Agricultural yields are limited by acute deficiencies of at least one major nutrient in those parts of the world where most people live. Crop responses to fertilizer are invariably considerable and average yields per ha of cereals (the main component of man's food) in the major countries are nearly proportional to the amounts of $N + P_2O_5 + K_2O$ applied as fertilizer. Often responses to nitrogen fertilizer are restricted by shortage of some other nutrient, but in West Europe where the soils are well endowed with phosphorus, potassium and sulphur average yields of wheat per country are almost directly proportional to the level of N-fertilizer applied.

Much N-fertilizer is wasted because of difficulties in forecasting levels and methods of application for different conditions. Predictions based on simple statistical interpretation of the results of field trials have proved to be unsatisfactory. The new mechanistic modelling approaches that take far greater account of existing principles about key processes have been more successful.

Nitrogen recycling is small in existing agriculture and there is much scope for improvement. Biological fixation provides much nitrogen for world agriculture. Under the right conditions legumes can fix at least 300 kg N ha^{-1} yr^{-1}, which is more than sufficient for maximum growth. A major drawback of legumes, however, is that grain yields are inherently much lower than those of cereals.

Sufficient N-fertilizer to grow all the food required for mankind can be synthesised from only 2% of the present world consumption of fossil fuel. Despite massive increases in oil prices, the cost of nitrogen fertilizer relative to that of food has remained virtually unchanged. It is still very profitable to apply nitrogen fertilizer in most parts of the world.

Serious problems in the future are likely to result from essential resources (energy and minerals) being very unevenly distributed in relation to where they are needed to grow food.

Resumen   Los rendimientos de los cultivos están limitados en la agricultura por severas deficiencias de por lo menos uno de los nutrimentos principales en aquellas regiones del mundo donde vive una major proporción de la población. La respuesta de los cultivos a la fertilización es en general apreciable tanto que los rendimientos promedios de los cereales, la principal fuente alimenticia del hombre, son casi proporcionales al las cantidades de $N + P_2O_5 + K$ aplicadas en los principales paises productores. Frecuentemente la respuesta a los fertilizantes nitrogenados se ven limitadas por deficiencias en otro nutrimento pero en Europa Occidental, donde los suelos están bien dotados de fósforo, potasio y azufre, los rendimientos promedio de trigo son casi directamente proporcionales a los niveles de N aplicado.

Debido a las dificultades para pronosticar los niveles y métodos de aplicación apropiados para condiciones diferentes, gran parte del nitrógeno aplicado no es aprovechado. Las predicciones basadas en interpretaciones estadísticas de ensayos de campo han dado resultados poco satisfactorios mientras que los nuevos métodos basados en modelos mecanísticos que prestan mayor atención a los principios que controlan procesos claves, han dado mejores resultados.

Muy poca proporción del nitrógeno aplicado en la agricultura actual es reutilizado y existen grandes oportunidades de obtener mejoras en este respecto. Bajo condiciones adecuadas las leguminosas pueden fijar cantidades importantes de nitrógeno (*ca.* 300 kg ha^{-1} año^{-1}) que son mas que suficiente para obtener crecimientos máximos, sin embargo las leguminosas tienen la gran desventaja de rendir inherentemente menos que los cereales. La cantidad de abono nitrogenado necesario para producir suficiente alimento para la humanidad, puede ser sintetizado utilizando solo el 2% del consumo mundial de combustible fósil. A pesar de grandes aumentos en los precios del petroleo, el precio del abono nitrogenado se ha mantenido casi constante en relación al de los alimentos. Aun sigue siendo muy ventajoso aplicar abonos nitrogenados en la agricultura de la mayor parte del mundo.

Algunos problemas graves en el futuro podrían presentarse como resultado de la mala distribución de recursos esenciales tales como energía y minerales en relación a los sitios de mayor demanda por la agricultura.

Introduction

Crop yield is influenced, often decisively, by the extent to which the plants' requirements for nitrogen can be met. This varies greatly from crop to crop and from soil to soil and from one climate to another. Nevertheless, there seem to be some principles governing the more important aspects of nitrogen nutrition that apply almost everywhere. The object of this paper is to identify these principles and then to discuss the opportunities for relieving nitrogen stress and thus increasing yields.

Crop demand for nitrogen

The demand for nitrogen of every well-grown crop is the product of the dry weight of the crop and the fraction of N in it.

When supplies of nutrients and water are adequate, the growth patterns of many crops in Western Europe are similar during the main growing period (May to September). The total dry weight first increases nearly exponentially until crop cover is almost complete, then it increases linearly until senescence. Examples of growth during this linear phase are given in Fig. 1 for four widely-different C3 crops. The rate of dry matter increase for these crops is remarkably constant, about 200 kg ha^{-1} day^{-1}, despite differences in weather conditions. Of the 150 days in Western Europe when rapid growth can occur, there are fewer than 100 days of complete crop cover. In consequence, the maximum weight of dry matter is approximately $20 \cdot 10^3$ kg ha^{-1}. About half the total dry matter of many crops can be eaten, so maximum yields should be about $10 \cdot 10^3$ kg ha^{-1} of edible material. This has often been obtained for cereal grain and storage roots.

The per cent N in the dry matter of C3 crops grown with the optimum levels of nutrients appears to be almost entirely determined by the total weight of dry matter at harvest. Fig. 2 shows that the average per cent N in the total dry matter (excluding fibrous roots) at harvest of 22 different agricultural and horticultural crops fell closely about the same curve when plotted against dry weight of the crops per unit area. The per cent N fell from about 4 when the dry weight was 1×10^3 kg ha^{-1} to about 1.1 when the weight was 20×10^3 kg ha^{-1}. By

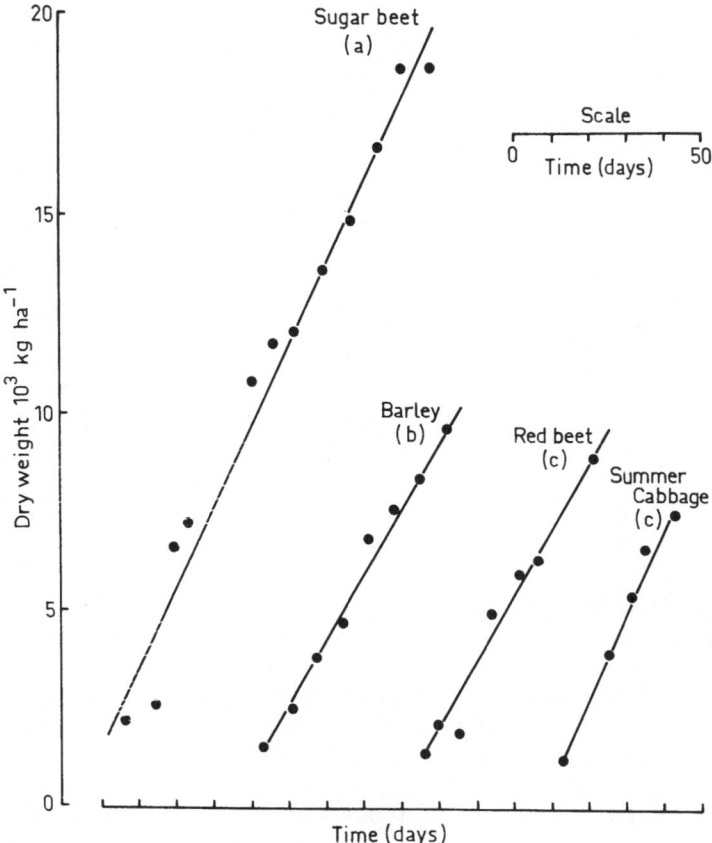

Fig. 1.   Dry weight (10^3 kg ha^{-1}) accumulation for above ground parts of various crop plants. The horizontal scale indicates the number of days between any point of the graph and the time corresponding to the first measurement. Subscripts indicate the source of the data: (a) Sibma[38], (b) Biscoe et al.[4], (c) Greenwood et al.[20].

comparison with the effect of plant weight on per cent N, the effect of species was negligible. It also follows from Fig. 2 that if high yields (20×10^3 kg ha^{-1} of total dry matter) are to be obtained, the crop must absorb more than 200 kg N ha^{-1}.

In warmer climates potential yields and thus removals of nitrogen are often larger because the higher temperatures permit C4 plants with their more efficient photosynthetic pathways to be grown in place of C3 plants, and because growth can take place on more days during the year than in temperate areas. Annual dry matter growth of $30 \cdot 10^3$ kg ha^{-1} and more have been obtained for corn, sugarcane, and various grasses[30]. Thus, in many situations uptake in excess of 300 kg N ha^{-1} yr^{-1} might be needed for maximum yields.

Soils' ability to supply nitrogen

Rainfall and decomposition of organic matter in soil provides some inorganic nitrogen for crops. Cooke[9] estimated that this process on average supplies 40 kg

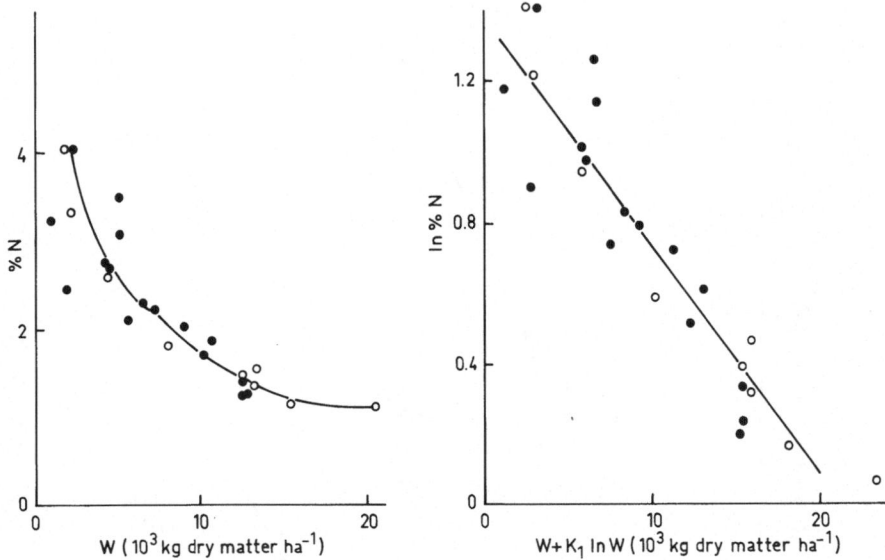

Fig. 2. Relation between average % N and total dry weights (W) of 22 different crop species at harvest after being grown with adequate levels of fertilizer-N. Values are for entire plants excluding fibrous roots. Some crops (●) were grown on plots that received the optimum levels of N-fertilizer in fertilizer experiments (data derived from Greenwood et al.[21]); others (o) were crops that attained exceptionally high yields and for which data were kindly provided by A. E. Johnston of Rothamsted Experimental Station, U.K. K_2 is a constant equal to $10^3 \, kg \cdot ha^{-1}$.

N ha^{-1} yr^{-1} in United Kingdom (UK) soils that have been continuously cropped with cereals. Continuously cropped soils nearer the equator can generally supply less nitrogen because they usually have much smaller reserves of organic matter. Thus, in general an unfertilized arable soil can provide no more than about one fifth of the nitrogen needed for high yields. The ability of such soils to supply phosphate and potassium also often falls far short of crop requirement[18]. In the absence of fertilizer, yields seldom exceed one-tenth of the potential maximum[13,45].

Fertilizers

Applications of one or more major nutrients increase yields considerably throughout the world. FAO carried out over 100,000 fertilizer trials in 40 developing countries using indigenous crop varieties and cultural practices[43]. On average, yields were increased by 72% with the best treatment and the mean value for the cost ratio (increased value of crop divided by the cost of fertilizer) was 4.8. This figure represents a very high return on investment compared with other investments in agriculture.

The average yield of cereals (man's main food; Evans[12]) in each of the largest countries of the world is almost proportional to the rate of fertilizer application (Fig. 3); the regression removes 75 per cent of the total variance in yield (n = 20,

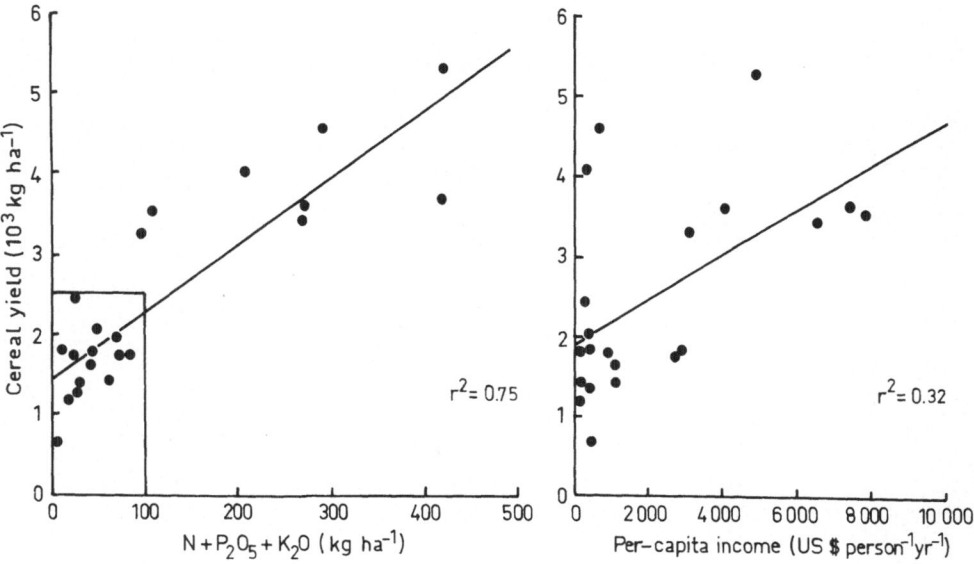

Fig. 3. Average yield of cereals (10^3 kg ha^{-1}) in 1976 for countries with a population of more than 35×10^6 people plotted against (a) total application of N + P_2O_5 + K_2O (in kg ha^{-1}) to arable and permanent crops, and (b) per capita income in U.S. $ person^{-1} yr^{-1}. The total population of countries within the rectangle is 2.51×10^9 and the population of countries outside is 630×10^6. Data were derived from FAO[14,15] and Kurian[26].

$p < 0.001$). Such a high correlation is notable in view of the wide diversity of crops, cultural practices, and soil and weather conditions. Of course, countries which apply high levels of fertilizer may also tend to adopt other practices that increase yields, so that in reality less than 75 per cent of the variance can be accounted for by differences in rates of fertilizer application. Even so, no other factor, as far as I am aware, correlates so well with yield. Japan and South Korea have among the highest agricultural yields and China and India amongst the lowest. Fig. 3 shows yields are very small in those countries inhabited by 2.5×10^9 people, which is most of the world's population.

It therefore seems reasonable to conclude that crop growth in a very large proportion of soils in the world is severely limited by lack of nutrients, and that unless steps are taken to rectify these deficiencies, there can be no appreciable increase in yield.

Applications of fertilizers containing only one major nutrient do not give much improvement in yield unless the levels of the others in soil are sufficient to satisfy crop demand for them. The nutrient that is most severely deficient varies from continent to continent and from site to site within a continent[35,43]. It is often phosphate, especially in South America; less frequently it is potassium. But few good crops of non-legumes are obtained anywhere without N-fertilizer.

The importance of N-fertilizer for attaining high yields is well illustrated by considering yields in relation to fertilizer practice in almost any area of the world. In Western Europe, for example, where the soils are generally well endowed with

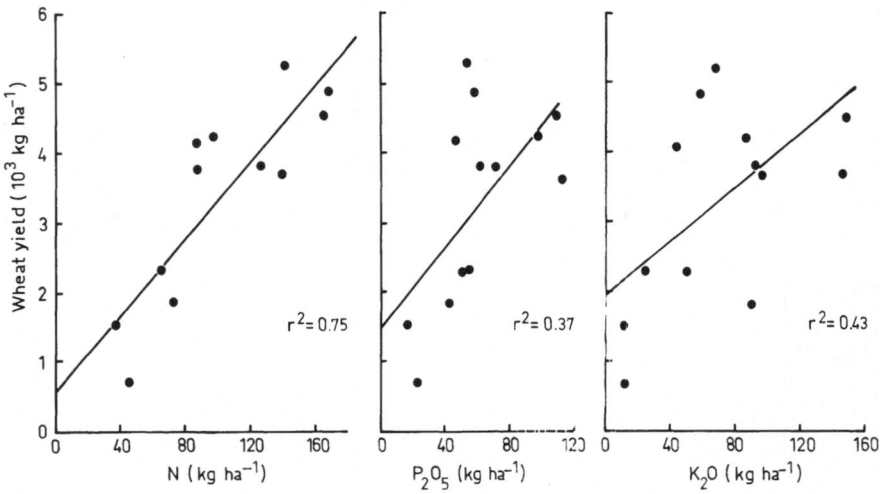

Fig. 4. Average yield of wheat in 1977 plotted against average rates of N, P_2O_5, and K_2O applied to arable and permanent crops in all countries of Western Europe where average N-applications were less than 200 kg N ha^{-1} yr^{-1} (derived from FAO[14,15]).

phosphorus, potassium, and sulphur, the average yield of wheat grain for each country increases almost linearly with the rates of N-fertilizer applied to arable and permanent crops, at least until this exceeds 160 kg of N ha^{-1} (Fig. 4). N-application rates remove 75% of the variance in wheat yield, whereas K_2O or P_2O_5 fertilizer removes less than 43%.

Globally, nitrogen fertilizer experiments with wheat generally show that when there are no deficiencies of phosphate or potassium, an extra 10–25 kg wheat grain ha^{-1} are produced for every kg applied-N ha^{-1} over a wide range of fertilizer rates[35,43]. The corresponding value: cost ratios of 3–8 emphasise that it is generally very profitable to apply nitrogen fertilizer.

Recovery of nitrogen by crops

Nevertheless, as wheat grain contains about 2% N, responses of 10–25 kg grain per kg N indicate that on average only about 35% of the nitrogen applied as fertilizer is recovered by the grain; less than half the fertilizer-N is taken up by the crop. Lower recoveries are found for many other crops in different parts of the world[41]. These low recoveries may be caused by nitrogen being lost from soil by volatilization, denitrification and leaching. More than 70% of nitrogen in urea applied to the soil surface can be volatilized as ammonia[25]. Ammonium ions are also nitrified to nitrate, which then can move to anaerobic micro-sites where it is denitrified to form nitrogen gas[17]; these processes occur in most soils and have resulted in losses of at least 3 kg N ha^{-1} day^{-1} in some agricultural soils[1]. Leaching following nitrification can also be serious; about half the fertilizer nitrogen applied to crops has been found in the drainage water of some catchment areas[40].

Losses of nitrogen by each of these processes is exacerbated if more fertilizer-N is applied than is required by the crop. Thus, in Britain, any nitrate remaining in soil after harvest of the crops in autumn is often completely leached out of the rooting zone by the winter rains, so that none remains for the next crop. It is therefore essential to forecast responses to N-fertilizer reasonably well if the fertilizer is to be used efficiently.

Responses depend upon several factors, but chiefly on the crop's demand for nitrogen, on how soil physical conditions influence root proliferation, and on the distributions of roots, nitrate, and water relative to one another. It is not surprising, therefore, that very large differences among response curves of yield against level of N-fertilizer are always found in nationwide trials[35].

Unfortunately, however, most methods for forecasting nitrogen-fertilizer recommendations for different soil and weather conditions are unsatisfactory. In some UK fertilizer experiments, the measured optimum levels on particular sites differed by a factor of two from those that were recommended for those sites by current procedures (e.g.[3]).

In my view, work on prediction has relied too heavily on carrying out field experiments and attempting to correlate the response with a measurable feature of the soil or weather conditions. Such a simple statistical approach does not permit sufficient input of existing scientific knowledge, and it is difficult to see how this approach can provide a satisfactory means of predicting the effects of the many interacting factors that influence response. In fact very few – if any – predictions made in this way have been tested by independent experiments.

We need a more mechanistic approach that takes greater account of existing principles of soil science and plant nutrition. Work along these lines led to a practical system of advice in the Netherlands, West Germany and France. It is based on estimating the potential demand of the crop for nitrogen, the effective depth of rooting, the amount of nitrate in the soil to that depth at the start of the growing season, and the amount that is released by breakdown of the soil organic matter. So far the procedure has given satisfactory predictions of nitrogen-fertilizer requirements (e.g.[24]).

Recently it has become clear that some of the key processes influencing crop response to N-fertilizer can be defined by simple equations. For instance, leaching in unstructured soils can be defined in terms of rainfall, the water holding capacity of the soil, and the initial nitrate distribution down the soil profile[7]. Other simple models include equations for the release of nitrogen from soil[31] and for nitrogen uptake by crops[8]. Models such as these have been combined into dynamic models for forecasting the day to day increases in growth of crops, and have correctly predicted the effects of weather conditions on the shapes of the response curves to nitrogen fertilizer[2]. These new modelling approaches offer considerable opportunities for improving the reliability of N-fertilizer predictions and thus of reducing waste.

Retention

The fact that heavy applications of N-fertilizer are needed to sustain high yields (Fig. 4) implies that retention of nitrogen in agricultural systems may be very small. Some support for this view is provided by the results of experiments. Passing a continuous stream of air over grass cuttings resulted in a 27% loss of their total-N to the atmosphere within 15 days[36]. A compost heap of straw and poultry manure lost more than a third of its initial nitrogen to the atmosphere during eight weeks of storage under conditions which minimized N-loss[42]. Far less nitrogen was recovered by crops that received nitrogen as farmyard manure than from those that received N as mineral fertilizer even when account was taken of long term benefits (Table 1). In addition, animals excrete much of their nitrogen as urea which is quickly hydrolysed to ammonia upon reaching soil. More than half of the nitrogen in animal excreta may be volatilized as ammonia gas and lost to agriculture[11]. In many parts of the world very little of the nitrogen in human excreta is returned to land. This is partly because of economic pressures and partly because sewage purification is primarily aimed at preventing spread of disease. The net result is that in Britain, of the 1×10^9 kg of nitrogen applied as fertilizer, only 0.18×10^9 kg is consumed as food by the population and only about one tenth of the nitrogen that enters the sewage system is returned to the land[11]. There is therefore considerable scope for adopting more efficient methods of recycling nitrogen and thereby for reducing the need for N-fertilizer.

Biological nitrogen fixation

Applications of N-fertilizer suppress biological N-fixation. Nevertheless, many crops receive very little fertilizer (Fig. 3) and about two-thirds of the nitrogen utilized in world agriculture is fixed biologically[22]. Though many organisms fix nitrogen, probably most is fixed by legumes. They can fix far more than the 200 kg N ha^{-1} required for high yields of dry matter (Table 2). Fixation, however, is often low because of the difficulty of introducing efficient strains of Rhizobium that (i) are compatible with agronomically good cultivars and (ii) have the ability to compete successfully with native Rhizobium in forming nodules.

Legume grains contain about the same amount of metabolizable energy per unit weight as cereal grain[34], but both average[15] and maximum yields of legume grains (dry matter and metabolizable energy) are much lower. For example, the highest grain yield I have found recorded for any legume is $5.6 \cdot 10^3$ kg ha^{-1} for soybean[37], compared with $15.8 \cdot 10^3$ kg ha^{-1} for wheat[12]. The key to the difference appears to be that soybean grains contain three times more protein and therefore nitrogen than do cereal grains[39]. 5.6 kg of soybean contains 340 g of N, and 15.8 kg of wheat grain contains 350 g N; these values are remarkably similar and close to the maximum N-contents of any annual crop. If soybean were to yield

Table 1. Recovery of nitrogen added as organic manures and inorganic fertilizer by crops. Applications of FYM (farmyard manure) had been made for many years to establish a new equilibrium before the measurements were made (from Johnston[23])

Crop	Annual application of N (kg/ha)		Recovery of added N by crop (Per cent)
	Fertilizer	FYM	
Spring barley	0	265	23
Spring barley	144	265	28
Potatoes	0	262	13
Potatoes	216	262	26

Table 2. Fixation of nitrogen by legumes (from Nutman[32])

Crop	N-fixation (kg N ha^{-1} yr^{-1})
Lucerne	56–463
Clover	45–673
Soybean	1–168
Cowpea	73–354
Pigeon pea	168–280
Lentil	88–114

15.8×10^3 kg ha^{-1} they would have to take up 966 kg N, an extraordinarily high rate of N-uptake.

Since the maintenance of protein necessitates metabolism of photosynthate, there should be an upper limit on the total amount of protein a plant can contain and yet grow. If there is sufficient protein there may be no photosynthate left over for growth of new structural material[19].

Legumes *versus* cereal grains

A far higher population could be supported from a given area of land if most people continued to satisfy the bulk of their food requirements by direct consumption of cereals[12] rather than by consumption of legume grain. Man requires a minimum of about 10.9 MJ of metabolizable energy and 40 g of protein (containing 6.7 g N) per day[27]. If these requirements were to be met by a single grain, it would contain about 1% N (40 g protein contains 6.7 g N and 640 g food dry matter contain about 10.9 MJ energy). Legume grain contains 4–6% N and cereal grain 1–2% N[39].

Cereal grains contain all the essential amino acids needed by man, but the amounts of lysine are generally lower than in an ideal protein. Lysine is, however, contained in ample quantities in other edible plants, and indeed is extracted commercially from sugar beet waste[16].

It is therefore yields of metabolizable energy rather than of protein that limit the population that can be supported by any system of cropping. As yields of grain and of metabolizable energy from cereals are at least double those from legumes, it follows that a given area of land if cropped with cereals could support about twice the population it could support if cropped with legumes.

Nitrogen nutrition of crops in the future

It is worth considering the supply of N-fertilizer in relation to actual need on a global scale. For this purpose I assume that cereal grain contains 1.5% N and likewise that all other crop yields contain 1.5% N, which is roughly true when yields are high (Fig. 2). If all the nitrogen in the crops were derived from fertilizer at an efficiency of only 25%, then it would require only 64×10^9 kg fertilizer-N yr^{-1} to grow enough food for the present global population (Table 3). The energy required to manufacture this fertilizer is equivalent to 128×10^9 kg of oil (Table 3), or about 2% of the world's annual fossil fuel consumption (Table 4). Alternatively, the world's requirements for N-fertilizer could be met indefinitely by using one-third of existing hydro-electric power (Table 4). There can therefore be no doubt that shortages of energy on a world scale, at least, are not serious constraints to the production and use of N-fertilizer.

Current reserves of fossil fuels that could be extracted economically at today's prices are sufficient to meet the world's energy requirements for about 100 years at present levels of consumption (Table 4). Resources of all fossil fuels that could be extracted with current technology if the price were high enough could meet global requirements for about 1000 years. There are, however, problems. One is that much of the world's industry and infrastructure is geared to the use of only one of the fossil fuels, oil, which is being consumed faster than new reserves are being discovered[6]. Another is that energy reserves are confined to only a few countries, and these have a natural tendency to conserve them for their descendants. A third is that 'present consumption level' omits future consumption by regions that are currently underdeveloped.

Oil prices have risen dramatically in recent years and have had an impact on the cost of every commodity. From 1965 to 1979 oil prices increased by almost a factor of ten, whereas those of N-fertilizer and wheat doubled (Table 5). But the cost of N-fertilizer has not risen relative to that of wheat (Table 5), or for that matter relative to that of rice or maize[33]. It is therefore still very profitable to buy fertilizer-N and to use it to grow grain.

Considerations such as these suggest that the application of N-fertilizer is the most effective way of meeting the nitrogen required for food production in much

Table 3. Nitrogen needed to grow sufficient food for the world's population

Per capita energy requirement per day	12 MJ d^{-1}
Weight of food dry matter (FDM) to provide 12 MJ of energy assuming 17 KJ per g food (Paul and Southgate[34])	0.71 kg
Protein contained in 1 kg FDM assuming 1.5% N (protein = 16% N)	94 g
Weight of protein absorbed per person per day (0.71 kg FDM × 94 g kg FDM^{-1})	67 g day^{-1}
Weight of nitrogen in protein consumed (67 g protein × 0.16 g N g protein^{-1})	11 g day^{-1}
Annual nitrogen uptake needed to feed world population (365 d × 11 g N d^{-1} person^{-1} × 4 × 10^9 people).	16 × 10^9 kg yr^{-1}
N fertilizer requirement if recovery of fertilizer-N is 25%	64 × 10^9 kg yr^{-1}
Since 2 kg of oil-equivalent are required to synthesise 1 kg of fertilizer-N* the total oil equivalent required to synthesise 64 × 10^9 kg of fertilizer N is	128 × 10^9 kg yr^{-1}

* 1 kg of oil produces about 45 MJ (White[44]) and about 90 MJ are required to synthesise 1 kg of fertilizer-N (Lewis and Tatchell[29]).

Table 4. World consumption and reserves of energy. Reserves are those that can be extracted economically at current world prices (from British Petroleum Ltd.[5]). Energy needed to produce all nitrogen fertilizer is 128 × 10^9 kg oil equivalent year^{-1} (see Table 3).

Energy	kg of oil equivalent	
	Consumption 1979	Reserves 1979
Oil	3.12 × 10^{12}	87.4 × 10^{12}
Natural gas	1.30 × 10^{12}	66.3 × 10^{12}
Coal	1.98 × 10^{12}	372.2 × 10^{12}
Hydro-electric power	0.41 × 10^{12}	
Nuclear energy	0.16 × 10^{12}	
Total	6.96 × 10^{12}	525.9 × 10^{12}

of the world. It is therefore encouraging that many countries have put a high priority on increasing their production and use of N-fertilizer: over the period 1965–1979 there has been a massive increase in N-fertilizer consumption throughout the world; in the developing countries consumption has increased from 6 × 10^9 to 16 × 10^9 kg N[43]. There has also been a large increase in food production, but unfortunately there has been an almost equal increase in population, so that many people are no better off than initially[18].

Table 5. Changes in prices of oil, urea, and wheat from 1965 to 1979

Year	Oil[a] U.S. $ per barrel	Urea[b] U.S. $ per 10^3 kg	Wheat[c] U.S. $ per 1×10^3 kg	Ratio of price per kg of urea-N[d] divided by price per kg of wheat[d]
1965	1.80	95	65	3.2
1972	2.48	75	62	2.7
1975	12.37	140	170	1.8
1979	14.34	170	140	2.7

[a] Saudi Arabian Light Oil Port Price F.O.B.,
[b] F.O.B. Western Europe estimated from von Peter[43],
[c] International wheat price (Saine: Home-grown Cereal Authority, London),
[d] Urea-N, not urea as in column 3; urea contains 45% N.

Furthermore, although fertilizer-N is profitable, it is also expensive, and much of the world is inhabited by people who do not have any reserves of natural fossil fuel energy, nor have credit to buy fertilizer or the technology and fuel needed to make it. Their only practical way of improving the nitrogen nutrition of crops is by more efficient recycling of nitrogen and by greater biological fixation of nitrogen. Undoubtedly some progress can be made without much difficulty, but the technical problems of larger improvements are considerable.

What is even more serious, however, is that there can be no benefits from improved nitrogen nutrition of crops if the soil is seriously deficient in phosphate, potassium, sulphur or trace elements. Such deficiencies are widespread, as discussed previously, and the only way of relieving them within a reasonable period is by applying fertilizer. All too often, however, this has to be bought at the expense of precious foreign currency because the reserves of minerals rich in these nutrients are confined to few countries[18].

Indeed, I believe that the most serious threat of widespread starvation arises from a failure to appreciate that yields in the countries supporting most of the world's population cannot be increased rapidly unless more nutrients are applied as fertilizer in ways that enable plants to absorb them.

Conclusions

1. Crops must absorb at least 200 kg N ha^{-1} if high yields are to be obtained. Highly productive arable cropping relies on heavy applications of fertilizer.

2. Less than half the nitrogen applied as fertilizer is usually recovered by the crop; the remainder is usually wasted. New mechanistic modelling approaches offer much promise for saving fertilizer by providing more reliable ways of predicting effective fertilizer levels and methods of application.

3. Recycling of nitrogen is generally very poor in agricultural systems and there is considerable scope for improvement.

4. Huge quantities of nitrogen can be fixed by biological means and there is scope for increasing biological fixation in agriculture.

5. A major drawback of legumes is that yields of grain are much lower than those of cereals.

6. Only 2% of present annual global consumption of fossil fuel energy is needed to provide sufficient N-fertilizer to grow enough food for the present global population.

7. The sharp rise in oil prices has not increased the price of N-fertilizer relative to that of food. It remains extremely profitable to apply optimum levels of N-fertilizer to soils well-endowed with other essential nutrients.

8. Deficiencies of phosphorus, potassium or sulphur in soil are often so acute that there can be little benefit from improved N-nutrition of crops without fertilizers containing these other nutrients as well.

9. Problems in the future are likely to result from the fact that the main reserves of non-nitrogen fertilizer nutrients as well as capital and energy required for N-fertilizer are very unevenly distributed in relation to where they are needed to grow food.

References

1 Arnold P W 1954 Losses of nitrous oxide from soil. J. Soil Sci. 5, 116–128.

2 Barnes A, Greenwood D J and Cleaver T J 1976 A dynamic model for the effects of potassium and nitrogen fertilizers on the growth and nutrient uptake of crops. J. Agric. Sci. Camb. 86, 225–244.

3 Batey T 1976 Some effects of nitrogen fertilizer on winter wheat. J. Sci. Food Agric. 27, 287–297.

4 Biscoe P V, Scott R K and Monteith J L 1975 Barley and its environment. III. Carbon budget of the stand. J. Appl. Ecol. 12, 269–293.

5 British Petroleum Limited 1979 B.P. Statistical Review of the World Oil Industry. British Petroleum Limited, London.

6 British Petroleum Limited 1979 Oil Crisis... Again? A Brief by the Policy Review Unit. British Petroleum Limited, London.

7 Burns I G 1977 Nitrate movement in soil and its agricultural significance. Outl. Agric. 9, 144–148.

8 Burns I G 1980 A simple model for predicting the effects of leaching of fertilizer nitrate during the growing season on the fertilizer need of crops. J. Soil Sci. 31, 175–185.

9 Cooke G W 1975 The energy costs of the nitrogen fertilizers used in Britain, the returns received and some savings that are possible. J. Sci. Food Agric. 26, 1065–1069.

10 Cooke G W 1976 A review of the effects of agriculture on the chemical composition and quality of surface and underground waters. In Ministry of Agriculture Fisheries and Food, Technical Bulletin 32, Agriculture and Water Quality, pp 5–57. Her Majesty's Stationery Office, London.

11 Cooke G W 1977 Waste of fertilizers. Phil. Trans. R. Soc. London B 281, 231–241.

12 Evans L T 1975 Crops and world food supply. In Crop Physiology – Some Case Histories. pp 1–22. Evans L T (Ed.). Cambridge University Press, Cambridge.

13   FAO 1974 Fertilizers, the first decade: A summary of results achieved between 1961 and 1971.
     FAO, Rome.

14   FAO 1978 Fertilizer Yearbook 1977. FAO, Rome.

15   FAO 1978 Production Yearbook 1977. Vol. 31. FAO, Rome.

16   Fowden L 1980 Amino acids: production by plants and the requirements by man. In Food
     Chains and Human Nutrition. pp 135–155. Blaxter K (Ed.). Applied Science Publishers,
     London.

17   Greenwood D J 1962 Nitrification and nitrate dissimilation in soil. II. Effect of oxygen
     concentration. Plant and Soil 17, 378–391.

18   Greenwood D J 1980 Fertilizer use and food production: World scene. Fert. Res. 2, 33–51.

19   Greenwood D J and Barnes A 1978 A theoretical model for the decline in the protein content in
     plants during growth. J. Agric. Sci. Camb. 91, 461–466.

20   Greenwood D J, Cleaver T J, Loquens S M H and Niendorf K B 1977 Relationship between
     plant weight and growing period for vegetable crops in the United Kingdom. Ann. Bot. 41, 987–
     997.

21   Greenwood D J, Cleaver T J, Turner M K, Hunt J, Niendorf K B and Loquens S M
     H 1980 Comparisons of the effects of nitrogen fertilizer on the yield, nitrogen content and
     quality of 21 different vegetable and agricultural crops. J. Agric. Sci. Camb. 95, 471–485.

22   Hall D O 1980 World production of organic matter. In Food chains and human nutrition, pp
     51–92. Blaxter K (Ed.). Applied Science Publishers, London.

23   Johnston A E 1976 Additions and removals of nitrogen and phosphorus in long term
     experiments at Rothamsted and Woburn and the effect of the residues on total soil nitrogen and
     phosphorus. In Ministry of Agriculture Fisheries and Food, Technical Bulletin 32, Agriculture
     and Water Quality, pp 111–144. Her Majesty's Stationery Office, London.

24   Jungk A and Wehrmann J 1978 Determination of nitrogen fertilizer requirements by soil and
     plant analysis. Plant Nutrition 1978. Proc. 8th Int. Colloq. Plant Analysis and Fertilizer
     Problems, Auckland, New Zealand. DSIR Information Series No. 134, Vol 1, pp 209–224.
     Government Printer, Wellington, New Zealand.

25   Kresge C B and Satchwell D P 1960 Gaseous loss of ammonia from nitrogen fertilizer applied
     to soils. Agron. J. 52, 104–107.

26   Kurian G T 1978 The Book of World Rankings. Macmillan, London.

27   Lapedes D N 1977 McGraw Hill Encyclopedia of Food and Agriculture, p 3. McGraw Hill,
     New York.

28   Lauer D A, Bouldin D R and Klausner S D 1976 Ammonia volatilization of dairy manure
     spread on the soil surface. J. Environ. Qual. 5, 134–144.

29   Lewis D A and Tatchell J A 1978 The role of fertilizer energy in agricultural production.
     Phosph. Agric. 74, 1–13.

30   Loomis R S and Gerakis P A 1975 Productivity of agricultural ecosystems. In Photosynthesis
     and Productivity in Different Environments, pp 145–172. Cooper J P (Ed.). Cambridge
     University Press, Cambridge.

31   Mary B and Remy J C 1979 Essai d'appréciation de la capacité de minéralisation de l'azote des
     sols de grande culture. 1. Signification des cinetiques de mineralisation de la matière organique
     humifée. Ann. Agron. 30, 513–527.

32   Nutman P S 1976 IBP field experiments on nitrogen fixation by nodulated legumes. In
     Symbiotic Nitrogen Fixation in Plants, IBP Synthesis Ser. Vol. 7, pp 211–217. Nutman P S (Ed.).
     Cambridge University Press, Cambridge.

33   OECD 1980 Instability of agricultural markets. OECD Observer No. 102, 19–23.

34   Paul A A and Southgate D A T 1978 McCance and Widdowson's The Composition of Foods,
     p 9. Her Majesty's Stationery Office, London.

35   Richards I R 1979 Response of tropical crops to fertilizer under farmers conditions – analysis of
     results of the FAO fertilizer programme. Phosph. Agric. 76, 147–156.

36 Salt P D 1965 An apparatus for measuring losses of ammonia from decomposing plant materials. Chem. Ind. p 461.

37 Shibbles R M, Anderson I C and Gibson A H 1975 Soybean. *In* Crop Physiology – Some Case Histories, pp 151–189. Evans L T (Ed.). Cambridge University Press, Cambridge.

38 Sibma L 1968 Growth of closed green crop surfaces in the Netherlands. Neth. J. Agric. Sci. 16, 211–216.

39 Sinclair T R and Wit C T de 1975 Photosynthate and nitrogen requirements for seed production by various crops. Science 189, 565–567.

40 Stewart W D P, May E and Tuckwell S B 1976 Nitrogen and phosphorus from agricultural land and urbanization and their fate in shallow fresh water lochs. *In* Ministry of Agriculture Fisheries and Food, Technical Bulletin 32, Agriculture and Water Quality, pp 111–114. Her Majesty's Stationery Office, London.

41 Terman G L 1979 Volatization losses of nitrogen as ammonia from surface-applied fertilizers, organic amendments and crop residues. Adv. Agron. 31, 189–223.

42 Tinsley J and Nowakowski T Z 1969 The composition and manurial value of poultry excreta, straw-droppings compost and deep litter. II. Experimental studies on composts. J. Sci. Food Agric. 10, 150–167.

43 von Peter A 1980 Fertilizer requirements in developing countries. Proc. Fert. Soc. 188, 1–58.

44 White D J 1977 Prospects for greater efficiency in the use of different energy sources. Phil. Trans. Roy. Soc. Lond. B. 281, 261–275.

45 Wit C T de 1968 Plant Production. *In* Agricultural Sciences and the World Food Supply. Miscellaneous Papers 3(1968), pp 42–44. Landbouwhogeschool Wageningen. Veenman & Zonen, Wageningen.

Plant and Soil 67, 61–71 (1982).   0032-079X/82/0671-0073$01.65.   SU-05
© 1982 *Martinus Nijhoff/Dr W. Junk Publishers, The Hague.*

The role of the atmosphere in nitrogen cycling

El papel de la atmosfera en el ciclo de nitrógeno

E. SANHUEZA

Instituto Venezolano de Investigaciones Científicas (IVIC), Fotoquímica y Contaminación Atmosférica, Apartado 1827, Caracas 1010-A, Venezuela

Key words   Ammonia   Atmosphere   N-cycling   Nitric acid   Nitric oxide   Nitrogen dioxide   Nitrous oxide   Tropospheric-N

Abstract   In this work an analysis of the sources, atmospheric concentration, chemical reactions and sinks of the principal atmospheric nitrogen compounds is made. Atmospheric emissions of N_2O and NH_3 are almost entirely due to biological activity on the continents and in the oceans. The combustion of fossil fuels and biomass is the principal source of NO_x. The only relevant chemical transformations in the troposphere are the oxidation of NO_x to NO_3^- and the formation of ammonium salts. Only 10% of the NH_3 emitted is oxidized. Washout of NH_4^+ and NO_3^- by rainfall is the principal mechanism for removing nitrogen compounds from the atmosphere. Part of the N_2O enters the stratosphere and part must be removed in the biosphere by processes not yet established.

NO_x produced in the atmosphere by the burning of fossil fuels and biomass and by lightning represents between 30 and 40% of the total nitrogen fixed. A complete nitrogen balance for the troposphere is presented.

Since the photochemical oxidation of NO_x is rapid and atmospheric transport is relatively slow with respect to the cycling of water in the troposphere, nitrogen compounds return to the earth's surface close to where they were emitted. Fixed-nitrogen inputs to the continents and oceans due to biological and industrial fixation are slightly greater than those due to rain water. However, since rain falls everywhere, input from this source is only important on soils not subject to intensive agriculture.

Resumen   En el presente trabajo se hace un análisis de las fuentes, concentraciones, reacciones químicas y depósitos de los principales compuestos nitrogenados atmosféricos. Las emisiones a la atmósfera de N_2O y NH_3 provienen casi exclusivamente de procesos biológicos en continentes y océanos. Los NO_x son producidos principalmente en la quema de combustibles fósiles y de biomasa. La oxidación de los NO_x a NO_3^- y la formación de sales de amonio son las únicas transformaciones químicas relevantes en la tropósfera. Sólo el 10% del NH_3 emitido es oxidado. La remoción de la atmósfera de los compuestos nitrogenados se produce fundamentalmente por lavado por lluvia de NH_4^+ y NO_3^-. Parte del N_2O pasa a la estratósfera y parte debe ser removido en la biósfera a través de un proceso no establecido.

En la atmósfera ocurren importantes procesos de fijación de nitrógeno y el NO_x producido en la quema de combustibles fósiles y de biomasa y en las descargas eléctricas representa entre 30 y 40% del total del nitrógeno fijado. Se plantea un balance del nitrógeno troposférico.

Teniendo en consideración que la oxidación fotoquímica del NO_x es rápida y que el transporte atmosférico es relativamente lento con respecto al ciclo troposférico del agua, se encuentra que los compuestos nitrogenados vuelven a la superficie de la tierra en lugares cercanos a donde fueron emitidos. Se establece que la entrada de nitrógeno fijo a los continentes y océanos por fijación biológica e industrial es escasamente superior a la producida en el agua de lluvia. Sin embargo, debido a que la lluvia cae en todas partes, el aporte de este nitrógeno fijo en el agua de lluvia es sólo significativo en los suelos en donde no se practica una agricultura intensiva.

Introduction

The atmosphere plays an essential role in nitrogen cycling. Fig. 1 illustrates the most important interactions that occur between the troposphere and the other parts of the earth (continents, oceans and stratosphere) and that play a part in the transformation or distribution of nitrogen compounds. Since there is a possibility that increased biological nitrogen fixation and fertilization may affect the integrity of the stratospheric ozone layer by increasing N_2O emissions[1,9,21,23], increased attention has been paid in recent years to the various aspects of the earth's nitrogen cycles.

This paper examines the qualitative and quantitative changes of nitrogen compounds in the troposphere as related to the global nitrogen cycle.

Tropospheric nitrogen compounds

Gaseous molecular nitrogen represents 78% of the atmosphere, and its total mass is approximately 4×10^{18} kg. This value represents ca. 18% of the total amount of nitrogen that exists on and in the earth[4]. A great portion of the remaining nitrogen is located in deep reservoirs within the earth's crust, and does not play a part in nitrogen cycling. Furthermore, within the mobile 18% only a small portion is not molecular nitrogen. The main atmospheric N-compounds include N_2, NH_3, N_2O, NO_x, HNO_3, NH_4^+ and NO_3^-.

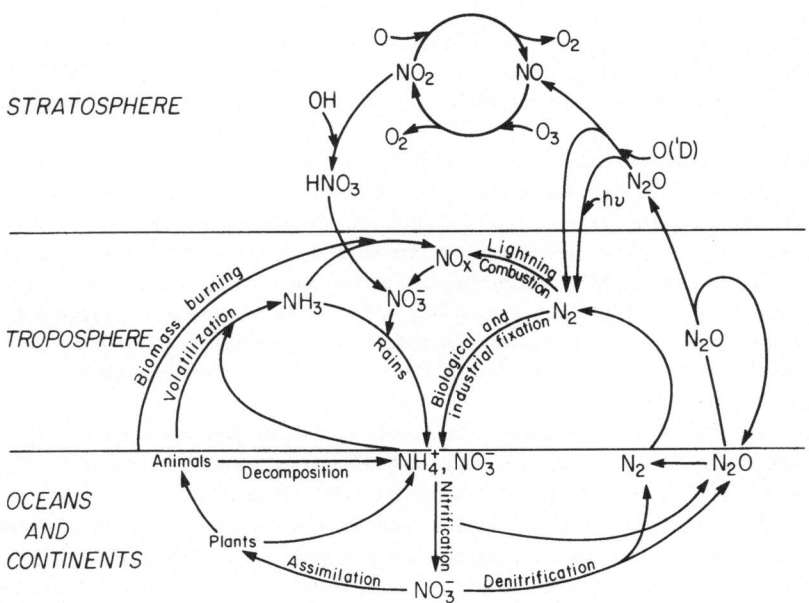

Fig. 1.   Nitrogen cycling in the atmosphere.

Ammonia (NH_3)

Ammonia emitted into the atmosphere is produced naturally from decomposing organic matter, from NH_3 volatilized from animal and human urine and other excreta and from coal. A global emission of $27–50 \times 10^9$ kg N yr^{-1} was estimated to be lost by this pathway by Söderlund and Svensson[27]. Combining estimates made by Söderlund and Svensson[27] and Dawson[12] yields a total ammonia flux into the atmosphere of $73–100 \times 10^9$ kg N yr^{-1} (Crutzen[10]). This estimate includes emissions from unfarmed lands.

NH_3 diffuses towards the upper atmospheric layers where it is removed through chemical reactions and by rain. It mainly reacts with nitric and sulphuric acids to produce ammonium salts. These salts are eventually rainwashed towards the earth's surface. Reaction with OH radicals (the most important radical in the troposphere) is also a sink of NH_3 in the atmosphere:

$$NH_3 + OH \rightarrow NH_2 + H_2O \qquad (1)$$

Nevertheless, only approximately 10% of the NH_3 emitted is consumed in this manner because of the low reaction rate constant (with the concentration of the OH radical 10^6 molecules cm^{-3}).

If the atmospheric cycle of water takes approximately one week[19] and if NH_3 and ammonium salts are removed by rain processes, then at least 90% of the emitted NH_3 will return to the surface in a period of 7 days.

NH_3 is the only gaseous atmospheric constituent with basic characteristics. Thus its role in the atmosphere is a vital one, especially because of problems related to the formation of acid rain.

Nitrous oxide (N_2O)

Nitrous oxide is the most abundant oxide of nitrogen in the troposphere, with an average concentration of 0.3 ppm. Emissions of this compound to the atmosphere have been frequently discussed in recent years[1]. The role of the oceans is a matter of strong disagreement and estimates range from an overall net production of 135×10^9 kg N per year[16] to an overall net removal of 40×10^9 kg N per year[22]. Söderlund and Svensson[27] estimated that emissions ranged from 20 to 80×10^9 kg N yr^{-1}, including continents as well as oceans. Hahn and Junge[17] suggested 32 and 45×10^9 kg N yr^{-1} for continents (soils and continental water bodies) and oceans, respectively.

More recently, Cohen and Gordon[7] have concluded that oceans represent an N_2O source equal to or less than 10×10^9 kg N yr^{-1}. These authors suggest that N_2O is produced by the oxidizing regeneration of nitrates and not by denitrification, which occurs mainly in soils. Bremner and Blackmer[3] and Elkins *et al.*[13] have also suggested that N_2O is also released during nitrification in soils.

Hahn and Junge[17] estimate that emissions due to burning fossil fuels amount to 2.5×10^9 kg N yr^{-1}. Crutzen *et al.*[11] indicate that approximately 8×10^9 kg N yr^{-1} is emitted by burning nutrient-rich biomass.

N_2O in the troposphere is practically inert, and it is estimated that only $\sim 2 \times 10^9 \, kg \, N \, yr^{-1}$ are removed through different tropospheric processes[17]. Its transfer to the stratosphere is the only important mode of removal currently assessed. It is estimated that $13 \times 10^9 \, kg \, N_2O{-}N \, yr^{-1}$ are diffused into the stratosphere[17]. This means that the final destination of 90% of the N_2O emissions is unknown, though probably the N_2O is degraded in the soil[2,6] or the emissions from soils and oceans are grossly overestimated. According to Blackmer and Bremner[2] the degradation is a microbial process that involves the reduction of N_2O to N_2, and is accelerated by anaerobic conditions and retarded by the presence of nitrates. Seasonal changes, soil humidity, and temperature, as well as many other factors, may play a role in the regulation of this process[6]. However, the actual importance of this N_2O sink has not yet been measured at a global level, and its significance is still uncertain.

Nitric oxide (NO) and nitrogen dioxide (NO$_2$)

Anthropogenic emissions of NO_x (NO plus NO_2) originate from burning fossil fuels and been estimated by Pratt et al.[25] to reach $20 \times 10^9 \, kg \, N \, yr^{-1}$. Natural emissions from the soil have been estimated at $10 \times 10^9 \, kg \, NO_x{-}N \, yr^{-1}$ (ref.[15]). Estimates of emissions resulting from atmospheric electric discharges range from 8 to $40 \times 10^9 \, kg \, N \, yr^{-1}$ (ref.[5,24]). Quantities produced by NH_3 oxidation are below $8 \times 10^9 \, kg \, N$ per year[10]. Crutzen et al.[10] have recently suggested that a total of $50 \times 10^9 \, kg \, NO_x{-}N$ are produced each year, mainly in the tropics during the dry season when considerable biomass is burned.

Measurements of the atmospheric levels of NO_x to date show large variations[24]. In general, they range between 0.1 and 0.3 ppb NO_x[8,24] with similar concentrations of NO and NO_2. From the large variations in concentrations observed in 'non-contaminated' atmospheres (0.1 ppb–4 ppb NO_x) and from the high values found in cities (over 100 ppb NO_x), it is clear that the tropospheric residence time of NO_x is extremely short compared with the time scale for global mixing. From other arguments (e.g., reaction (2) below) it has been estimated to be ca. 1 day.

NO_x is very important in atmosopheric chemistry, especially because of the roles it plays in the chemistry of radicals and in the production and destruction of tropospheric and stratospheric ozone[10]. NO_x disappears from the troposphere by oxidation to nitric acid:

$$NO_2 + OH \rightarrow HNO_3$$

Nitric acid (HNO$_3$)

The most extensive study of the levels of nitric acid in the clean atmosphere has been conducted by Hueber and Lazrus[18], who observed values between 0.2 and 0.8 ppb over the U.S. and Canada at medium latitudes. The levels observed over the Pacific Ocean were considerably lower, ranging between less than 0.03 and 0.15 ppb.

Table 1. Production, removal and subsequent balance of the tropospheric nitrogen compounds. Values in parentheses are tropospheric transformations; consequently the number of tropospheric nitrogen atoms remains constant. Values for denitrification and N_2-fixation are from Hahn and Junge[17]

Compound	Sources/Sinks	10^9 kg N yr^{-1}
Production		
N_2	Denitrification	
	Continental	125
	Oceanic	115
NH_3	Biological decay	73–100
N_2O	Denitrification	
	Continental	32
	Oceanic	10–45
	Nitrification	?
	Industrial processes	(2.5)
	Burning biomass	8
NO_x	Combustion	(20)
	Biological (in soils)	10
	Lightning	(8–40)
	NH_3 oxidation	(<8)
HNO_3	NO_x oxidation	(50–100)
	Diffusion from stratosphere	1–2
Net input of N		424–487
Removal		
N_2	Biological fixation	
	Continental	180
	Oceanic	85
	Industrial fixation	40
	NO_x formation	(20–60)
NH_3, NH_4^+	Deposition (wet and dry)	73–100
	Oxidation to NO_x	(<6.3)
N_2O	Diffusion to stratosphere	13
	Chemical reactions	(2)
	Unknown	57
NO_x	Oxidation to HNO_3	(50–100)
HNO_3, NO_3^-	Deposition (wet and dry)	50–100
Net output of N		465–575

If practically all NO_x is oxidized to HNO_3, HNO_3 production by this process ranges between 50 and 100×10^9 kg N yr^{-1}. The contributions of stratospheric HNO_3 are *ca.* 1–2 kg N per year[27]. There is no evidence for surface emissions of HNO_3.

HNO_3 is removed by its neutralization to nitrate salts as discussed earlier and by rain. HNO_3, together with H_2SO_4, is one of the main constituents of acid rain. In areas with little contamination by SO_2 emissions, HNO_3 would be the main cause of an acid pH in rain water. The half-life of HNO_3 in the atmosphere is approximately one week.

Balance of tropospheric nitrogen

Table 1 summarizes the balance of tropospheric nitrogen based on the preceding information. The balance considers all significant nitrogen species in the troposphere. In some cases, the removal of one N-compound implies the production of another that remains in the troposphere, and in these cases it should be remembered that the number of atoms of tropospheric nitrogen remains unchanged.

The role of the atmosphere in the global nitrogen cycling

The atmosphere supplies N_2 both for biotic fixation and for fixation in the atmosphere itself. Yung and McElroy[28] have estimated that in the prebiotic low-oxygen atmosphere the fixation of sufficient nitrogen for the development of life was carried out through reactions triggered by electric discharges:

$$H_2O \quad \rightleftharpoons OH + H, \tag{3}$$

$$2OH \quad \rightleftharpoons H_2O + O \tag{4}$$

$$O + N_2 \rightleftharpoons NO + N, \text{ and} \tag{5}$$

$$N + OH \rightleftharpoons NO + H, \text{ with} \tag{6}$$

NO later oxidized to precipitable products, mainly HNO_3.

Nitrogen fixed by electric discharge today apparently represents approximately 5 to 10% of total N-fixation (biological plus anthropogenic). Electrical fixation mainly occurs at low and medium latitudes[24] and therefore deposition of nitrogen from this source ought to be latitude-dependent.

Nitrogen fixation from burning fossil fuels (due to N_2 and O_2 reactions in heated air) also represents about 5 to 10% of the total. This fixation occurs mainly in heavily populated or industrialized areas, so that there is a notable difference between the northern and southern hemispheres (5–10 times more of this fixation for the Northern Hemisphere). Further, even within the northern hemisphere there is a latitude dependence, since emissions occur mostly at middle latitudes (northern U.S. and central Europe).

Nitrogen may also be fixed in the atmosphere by burning biomass, as recently suggested by Crutzen et al.[11] If these authors are correct, this source of nitrogen would be very significant (ca. 10–20% of the total) and, as they point out, would occur mainly in tropical zones during the dry season, creating regional and seasonal patterns of fixed nitrogen in the troposphere. Experimental work to confirm this hypothesis is needed.

The regional deposition of fixed nitrogen, whether fixed directly in the atmosphere or having been emitted from the earth's surface, will depend on atmospheric dispersion processes and rain patterns of the region. Although there are no detailed calculations available for the distribution and transport of tropospheric trace components, parameters used in box models provide a reasonable estimate of mixing velocities in the troposphere. Based on these parameters[20], the vertical mixing time in the troposphere is approximately one month. Longitudinal mixing time ranges from 15 to 60 days, and latitudinal mixing within each hemisphere takes about 30 days. Transportation across the equator is much slower, taking more than one year. Exchange with the stratosphere is even slower, taking up to two years.

Table 2 shows the residence time of water as a function of latitude. Residence time (τ) was obtained from the formula:

$$\tau_{H_2O} = \frac{\text{Precipitable water (g cm}^{-2})}{\text{Precipitation (g cm}^{-2}\text{ yr}^{-1})}$$

Residence times of atmospheric soluble compounds (e.g. NH_4^+, NO_3^-) should be similar to those of water, and we may infer then that their residence time also depends on latitude. Residence times in subtropical areas (10–30° latitude) are approximately twice as long as those calculated at medium latitudes (40–60°). The mean 'average' residence time is approximately 10 days[19].

Since NO_x is rapidly oxidized to HNO_3 (within ~1 day), since water-soluble nitrogen compounds (NH_3, NH_4^+, HNO_3 and NO_3^-) have a residence time of less than 10 days, and since dispersion phenomena are relatively slow, fixed-nitrogen compounds in the atmosphere (except N_2O) are deposited on the earth's

Table 2. Average residence time of water vapor in the atmosphere as a function of latitude. Values in parentheses are extrapolated. All values are from Junge[19]

	Latitude (degrees)								
	0–10	10–20	20–30	30–40	40–50	50–60	60–70	70–80	80–90
Average precipitable water (g cm^{-2})	4.1	3.5	2.7	2.1	1.6	1.3	1.0	(0.7)	(0.45)
Average precipitation (g cm^{-2} year^{-1})	186	114	82	89	91	77	42	19	11
Residence time (days):	8.1	11.2	12.0	8.7	6.4	6.2	8.7	(13.4)	(15.0)

surface relatively close to their source point. Long-distance latitudinal transport is of little importance.

Regional transport, however, can be important. For example, in subtropical regions evaporation losses of fixed nitrogen to the atmosphere during the 3–4 month dry season and its further transport to other latitudes may be significant. Dawson[12] developed a model for the production and emission of ammonia based on the following assumptions:

i) Soils are sources of atmospheric NH_3 and the processes can be adequately described by a microbiological and physico-chemical model that is not highly sophisticated.

ii) Soil ammonium concentrations are derived from relative decomposition and nitrification rates. Henry's Law provides equilibrium concentrations of NH_3 in the soil, and flow to the atmosphere can be estimated by a simplified diffusion equation.

iii) Local biomass and variations of temperature, humidity and soil pH provide the parameters required to obtain a latitude dependence.

NH_3 emissions predicted by Dawson[12] follow a strong latitude dependence (Fig. 2) with most NH_3 emitted at intermediate latitudes. Fig. 3 shows the latitudinal distribution of NH_4^+ in rainwater[14]. Values obtained in the Caracas, Mauritius and Buenos Aires stations are not included, since these stations reported extremely high values, and it is suspected that they were influenced by local sources[12]. The strong agreement between predicted NH_3 emissions and NH_4^+ deposition in rain supports the previous suggestion that nitrogen compounds return to the earth relatively close to their emission point. Rainwater over oceans has less than half of the NH_4^+ found in rainwater over continents[19].

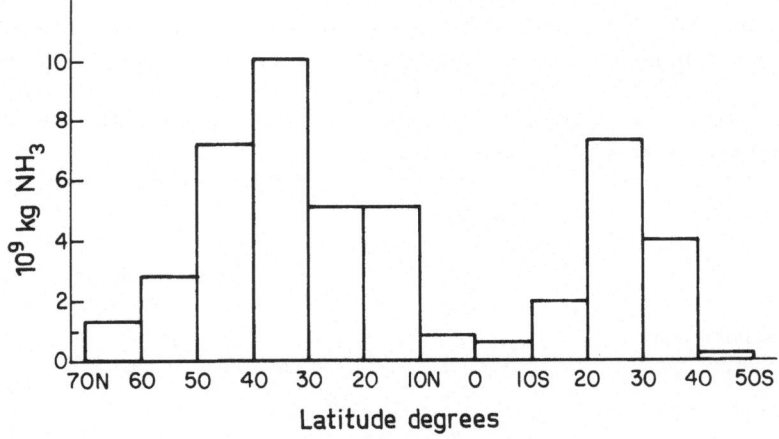

Fig. 2.   Calculated NH_3 emissions (not NH_3–N) as a function of latitude (Dawson[12]).

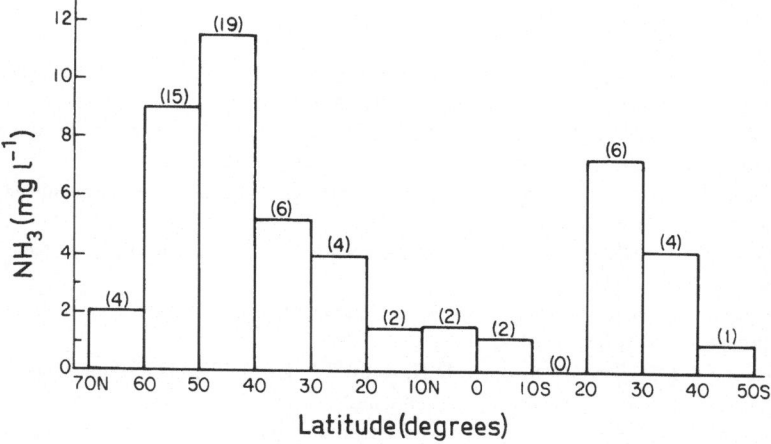

Fig. 3. Latitudinal distribution of NH_3 (not NH_3–N) in rain water (Eriksson[14]). The numbers in parentheses are the numbers of stations at that latitude. Note that y-axis units are different from those in Fig. 2.

Data for $NO_3{}^-$ in rainwater are scarce, but Junge[19] estimated that its latitude dependence should be more or less similar to that of $NH_4{}^+$. However, if electric discharges and burning biomass (which occur mainly at low latitudes) supply a significant fraction of fixed nitrogen, the differences between low and intermediate latitudes should be smaller than those observed for $NH_4{}^+$. Possible effects of burning biomass on ammonia emissions are as yet unknown.

Total fixed-nitrogen inputs to the soil from biological and industrial fixation processes ($\sim 305 \times 10^9$ kg N yr^{-1}) are slightly larger than inputs from rainwater ($\sim 240 \times 10^9$ kg N yr^{-1}). However, since rain falls on non-agricultural land, its significance for intensive agriculture is less relevant. For example, nitrogen fixation by leguminous crops can reach 340 kg ha^{-1} yr^{-1}, while rainwater inputs are about 5 kg ha^{-1} yr^{-1} at medium latitudes. Consequently, in fertilized soils most of the nitrogen emitted to the atmosphere is lost from the system since after atmospheric redistribution much will be deposited in non-fertile areas and in the ocean. The degree of loss for each region will depend mainly on rain patterns.

In addition to the role played by the atmosphere in the terrestrial and marine nitrogen cycles, these cycles also play an important role in atmospheric chemistry[10]. Changes in the flow of nitrogen compounds to the atmosphere can alter the existing frail physico-chemical balances. The increase of N_2O emissions is perhaps one of the most pressing problems in this regard, since it could endanger the integrity of the stratospheric ozone layer. Bermejo and Sanhueza[1] have discussed this issue in detail.

References

1   Bermejo H A and Sanhueza E 1980 The nitrogenated fertilizers and the stratosphere's ozone shield. Interciencia 5, 231–238. (In Spanish, English summary.)

2   Blackmer A M and Bremner J M 1976 Potential of soil as a sink of atmospheric nitrous oxide. Geophys. Res. Lett. 3, 739–742.

3   Bremner J M and Blackmer A M 1978 Nitrous oxide: emission from soils during nitrification of fertilizer-nitrogen. Science 199, 295–296.

4   Campbell I M 1977 Energy and the Atmosphere: A Physical-Chemical Approach. New York: J. Wiley. 198 p.

5   Chameides W. Stedman D, Dickerson R, Rusch D and Cicerone R 1977 NO_x production in lightning. J. Atmos. Sci. 34, 143–149.

6   Cicerone R J, Shetler J D, Stedman D H, Kelly T J and Liu S C 1978 Atmospheric N_2O: measurements to determine its sources, sinks, and variations. J. Geophys. Res. 83, 3042–3050.

7   Cohen Y and Gordon L I 1979 Nitrous oxide production in the ocean. J. Geophys. Res. 84, 347–353.

8   Cox R A 1977 Some measurements of ground-level NO, NO_2, and O_3 concentrations at an unpolluted maritime site. Tellus 29, 356–362.

9   Crutzen P J 1976 Upper limits on atmospheric ozone reductions following increased application of fixed nitrogen to the soil. Geophys. Res. Lett. 3, 169–172.

10 Crutzen P J 1979 The role of NO and NO_2 in the chemistry of the troposphere and stratosphere. Annu. Rev. Earth Planet. Sci. 7, 443–472.

11 Crutzen P J, Leroy E H, Krasnec J P, Pollock W H and Seiler W 1979 Biomass burning as a source of the atmospheric gases CO, H_2, N_2O, NO, CH_3Cl and COS. Nature London 282, 253–256.

12 Dawson G A 1977 Atmospheric ammonia flux from undisturbed land. J. Geophys. Res. 82C, 3125–3133.

13 Elkins J W, Wofsy S C, McElroy M B, Kolb C E and Kaplan W A 1978 Aquatic sources and sinks for nitrous oxide. Nature London 275, 602–606.

14 Eriksson E 1952 Composition of atmospheric precipitation. I. Nitrogen compounds. Tellus 4, 215–232.

15 Galbally I E and Roy C R 1978 Loss of fixed nitrogen from soils by nitric oxide exhalation. Nature London 275, 734–735.

16 Hahn J 1974 The North Atlantic Ocean as a source of atmospheric N_2O. Tellus 26, 160–168.

17 Hahn J and Junge C 1977 Atmospheric nitrous oxide: a critical review. Z. Naturforsch. 32a, 190–214.

18 Huebert B J and Lazrus A L 1978 Global tropospheric measurements of nitric acid vapor and particulate nitrate. Geophys. Res. Lett. 5, 577–580.

19 Junge C E 1963 Air Chemistry and Radioactivity, p 10. New York. Academic Press. 375 p.

20 Levy H 1974 Photochemistry of the troposphere. – Adv. Photochem. 9, 368–524.

21 Liu S C, Cicerone R J, Donahue T M and Chameides W L 1977 Sources and sinks of atmospheric N_2O and the possible ozone reduction due to industrial-fixed nitrogen fertilizers. Tellus 29, 251–263.

22 McElroy M B 1977 Chemical process in the solar system: A kinetic perspective. Phys. Chem. Ser. 2, 9–15. London: Butterworths.

23 McElroy M B, Wofsy S C and Yung Y L 1977 The nitrogen cycle: perturbations due to man and their impact on stratospheric N_2O and O_3. Trans. Roy. Soc. London 277B, 159–181.

24 Noxon J F 1978 Tropospheric NO_2. J. Geophys. Res. 83C, 3051–3057.

25 Pratt P F, Barber J C, Corrin M L, Goering J, Hauck R D, Johnston H S, Klute A, Knowles R, Nelson D W, Pickett R C and Stephens E R 1977 Effect of increased nitrogen fixation on stratospheric ozone. Climate Change 1, 109–135.

26 Robinson E and Robbins R C 1970 Gaseous nitrogen compound pollutants from urban and natural sources. J. Air Poll. Contr. Assoc. 20, 303–306.

27 Söderlund R and Svensson B H 1976 The global nitrogen cycle. *In* Svensson B H and Söderlund R (Eds.). Nitrogen, Phosphorus and Sulphur – Global Cycles. SCOPE Report 7. Ecol. Bull. Stockholm 22, 23–77.

28 Yung Y L and McElroy M B 1979 Fixation of nitrogen in the prebiotic atmosphere. Science 203, 1002–1004.

Plant and Soil 67, 73–79 (1982). 0032-079X/82/0671-0073$01.05.

Regional nitrogen budgets: Approaches and problems

Los balances regionales de nitrógeno: Enfoques y problemas

G. P. ROBERTSON

Department of Microbiology, Swedish University of Agricultural Sciences, S-750 07 Uppsala, Sweden*

Key words Models N-budgets N-inputs N-Losses Tropics

Abstract Regional nitrogen budgets are useful for assessing what is known about nitrogen cycling in important ecosystems of a region, for placing the various regional fluxes and pools into perspective, and for providing insight into the processes that regulate both regional and global nitrogen cycling. Existing regional budgets have been used both to study groundwater nitrate pollution and to help identify local ecosystems that are important on a land-use basis but that are poorly described biogeochemically. Avoidable problems common to many budgets include inappropriate compartment components, inadequate documentation, and unjustified certainty. Though imprecise, large-scale nutrient budgets at our present stage of understanding offer to researchers and system managers important advantages that would otherwise not be available.

Resumen Los balances regionales de nitrógeno son útiles para estimar el estado del conocimiento sobre el ciclo de nitrógeno en los ecosistemas mas importantes de la región, para enfocar los flujos y reservas regionales en perspectiva y para adentrarse en los procesos que regulan tanto los ciclos regionales como los globales. Los balances regionales existentes se han utilizado para estudiar la contaminación de aguas freáticas con nitratos y para identificar aquellos ecosistemas localmente importantes desde el punto de vista del uso de la tierra pero que son poco conocidos biogeoquímicamente. Algunos problemas que son comunes a muchos balances son: la selección de compartimientos inapropiados, documentación inadecuada y la certeza injustificada. Aun cuando sean imprecisos, los balances en gran escala, en el estado actual del conocimiento, aportan a investigadores y a quienes manjean los sistemas, algunas ventajas importantes que no estarían disponibles de otro modo.

Introduction

Ecosystem nutrient budgets for individual watersheds have become an increasingly common way of examining nutrient cycling in both agricultural and natural systems. The value of this approach, in which all significant pools, sources and sinks of a nutrient as well as turnover rates between the various compartments are quantified, needs little elaboration: nutrient budgets provide a convenient and biologically meaningful context within which to organize what is known about a system's biogeochemical cycles, help to force nutrient pools and fluxes into perspective, and can lend considerable insight into processes that

---

* Present address: Departments of Crop & Soil Science and of Microbiology & Public Health, Michigan State University, East Lansing, MI 48823, USA.

regulate nutrient cycling. They thus help to guide system management decisions and direct the course of future research.

A logical outgrowth of individual-watershed studies has been an expansion of system boundaries to include wider geographic areas. Global budgets represent the far end of this spectrum. Large-scale compilations are useful for the same reason smaller-scale studies are valuable, but because the scale is different, the specific questions addressed and the approaches taken in their construction differ.

The major difference between well-constructed watershed and landscape budgets is the massive aggregation present in the latter. Aggregation simplifies and thus makes large syntheses manageable by allowing many smaller pools and processes to be lumped into single black-box compartments. In general, aggregation expands with geographic size. The degree of simplification is a function of both our ignorance regarding certain processes and the time and effort available for close resolution, and it is made at the expense of both detailed insight and precision.

Erosion, for example, is commonly a victim of aggregation in large-scale budgets. Although erosion may be extremely important on a local level, in landscape budgets it is effectively ignored except for losses from the system via river transport. This is because soil washed from one local system largely ends up in another within the same region, and such a transfer can be considered internal and consequently ignored, in much the same way that NH_3 volatilized from beneath an unfertilized forest or corn canopy and subsequently trapped by the vegetation is considered an internal transfer in watershed studies. Insight into the local importance of erosion is thus restricted because values for regional erosion losses are usually based upon very little hard information. Data for sediment loads and nutrient contents of rivers draining most regions are unavailable, and rough extrapolation from other systems or time periods becomes necessary. Söderlund and Svensson[10], for example, estimated the global transfer of organic-N from terrestrial to marine systems via rivers by extrapolating values from Sweden, U.S. and Amazon rivers to total non-desert land areas. Even though such a value may be correct to only a factor of five, still it contributes significantly to our understanding of the global importance of this process and to our understanding of how it intereacts with others.

Taken to the extremes, of course, extrapolation yields negligible benefit. The global budget presented in Fig. 1, for example, offers very little insight into nitrogen cycle processes. More detailed global budgets (e.g.[9, 10]), however, can help to gauge the relative importance of various N-pools and processes (e.g. anthropogenic vs natural sources of fixed nitrogen and nitrogen oxides), and to draw attention to how little is known about others (e.g. nitrogen fixation in the open ocean and natural sources and sinks of nitrous oxide). Information of this nature is vital if the impact of human activity on processes in the biosphere is to be adequately assessed.

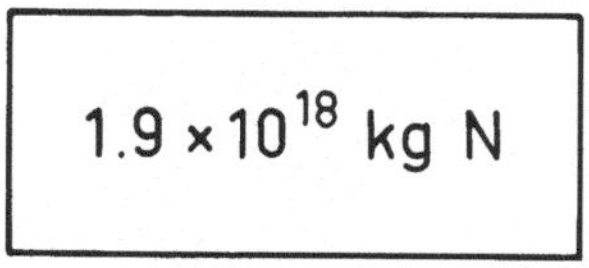

$$1.9 \times 10^{18} \text{ kg N}$$

Fig. 1. A one-compartment global nitrogen model. Pool size is from Söderlund and Svensson[10].

Regional budgets

General approaches

Nutrient budgets for landscape units between the watershed and global extremes are as yet rare. Only five such budgets, all for nitrogen, presently exist in the open literature: two riverbasin studies for parts of sourthern California[1,6], two U.S. state-wide models, one for Wisconsin and one for peninsular-Florida[7], and a budget for West Africa south of the Sahara[8].

At its most detailed, a regional budget is simply a compilation and synthesis of individual-watershed budgets for the important systems of a region. Since local budgets are presently lacking for many significant system types within most regions, a major potential use for regional budgets at present is to help identify those local systems that are important on a land-use basis but that are not well-described biogeochemically. Regional syntheses at this early stage are also useful for addressing specific management problems; for example, four of the existing regional nitrogen budgets were designed to address potential groundwater pollution by nitrates. Other budgets might focus on agricultural productivity or crop residue management, and all such budgets are useful for building more detailed and presumably more precise global budgets.

As for nutrient budgets of any geographic size, the two most important decisions to be made when compiling a regional budget are first, geographic boundaries, and second, compartment components. Both choices depend on the questions to be addressed with the budget. Regional studies developed in the 1970's to better understand groundwater nitrate pollution illustrate the importance of appropriate choices. The general model used for the two California river basin studies[1,6] incorporated seven broad compartments to emphasize hydrologic fluxes (Fig. 2). Compartments included pools of atmospheric nitrogen, land-surface nitrogen, surface-water nitrogen, soil nitrogen, substrata nitrogen, and groundwater nitrogen. The model presents a logical conceptual framework for studying the movement of nitrate through the system, but because it concentrates terrestrial activity (both human and natural) into one massive compartment, the model offers little quantitative insight into ways to better focus management efforts to reduce nitrate percolation at its sources. It further fails to quantify external sources and sinks of nitrogen that are not waterborne. In agricultural areas, both nitrogen gained in fertilizer and nitrogen gained and lost *via* crop import (*e.g.* animal feed) and export are important fluxes for the entire system, and to assume that these inputs and outputs balance is unrealistic[7].

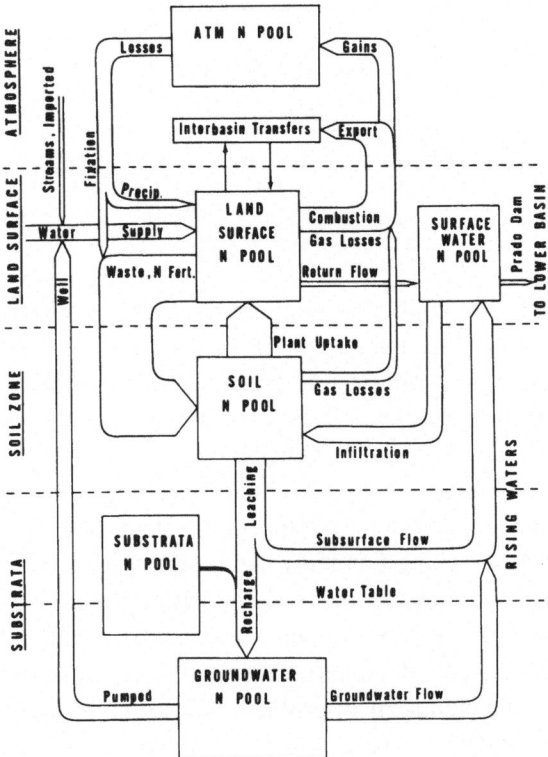

Fig. 2.   Nitrogen pools and fluxes quantified in the California Santa Ana and San Joaquin Valley river basin studies[1].

Alternative hydrologic models might include these external fluxes and divide the land-surface pool into a number of separate compartments. The peninsular-Florida model[7], for example, divided the land-surface pool into animal, human, and harvested organic-N pools (Fig. 3) and further divided each of these into subcompartments where available information permitted (e.g. Fig. 4). Such detail can help managers accurately pinpoint nitrate hotspots in the system and suggest causal relationships as well as indicate those areas for which important information is lacking.

Common to both regional nutrient budgets and to budgets for other landscape units are at least two important problems of interpretation that are often readily avoidable. The first concerns documentation, and occurs when values are assigned to compartments and fluxes that leave the careful reader wondering 1) how the values were derived, 2) what assumptions were involved in their derivation, and 3) what scientific justifications exist for the assumptions. Careful documentation can lead to a cumbersome final product, but it is difficult to justify leaving such documentation out of a budget that is inherently unverifiable.

The second important problem is unjustified certainty. This occurs when specific values are assigned to fluxes or pools for which a wide range of values

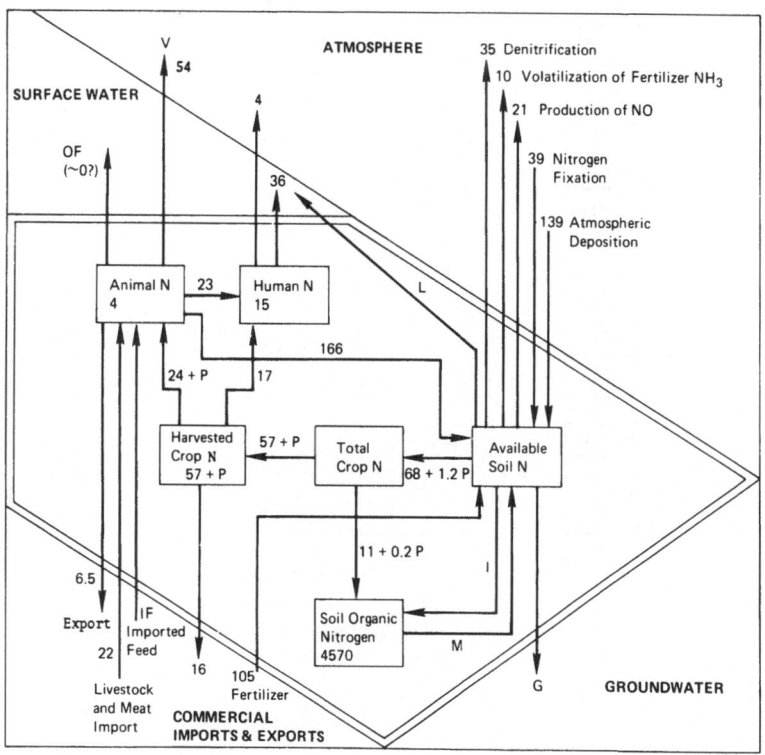

Fig. 3.   Nitrogen pools and fluxes in peninsular-Florida[7]. Flux values are 10^6 kg N yr^{-1}; pool sizes are 10^6 kg N. P = pasture nitrogen, V = volatilized NH$_3$–N, I = microbial immobilization of fertilizer-N, L = nitrogen leached to surface water, G = nitrogen leached to groundwater, IF = imported nitrogen in feeds, OF = overland flow, and M = mineralization.

would be as accurate. This is particularly the case for basic processes such as NH$_3$ volatilization and denitrification which are still poorly quantified at even the watershed level. It is also the case for relatively well-understood processes such as nitrogen fixation by legumes, however, for which a high degree of uncertainty stems from uncertainty associated with actual land-use practices (*e.g.*, fertilization and planting densities). A more appropriate approach in both cases is to assign ranges of likely values to pools and fluxes. There seems little to be gained by implying certainty where none exists, and such a practice can cause considerable damage if values so-reported are actually used as a basis for system management decisions.

Approaches for tropical regions

Nutrient budgets for tropical regions present several methodological problems not usually encountered with temperate budgets. First among these are ecosystem heterogeneity and the very uneven availability of basic information regarding land use in the region and nitrogen cycling in particular important ecosystems.

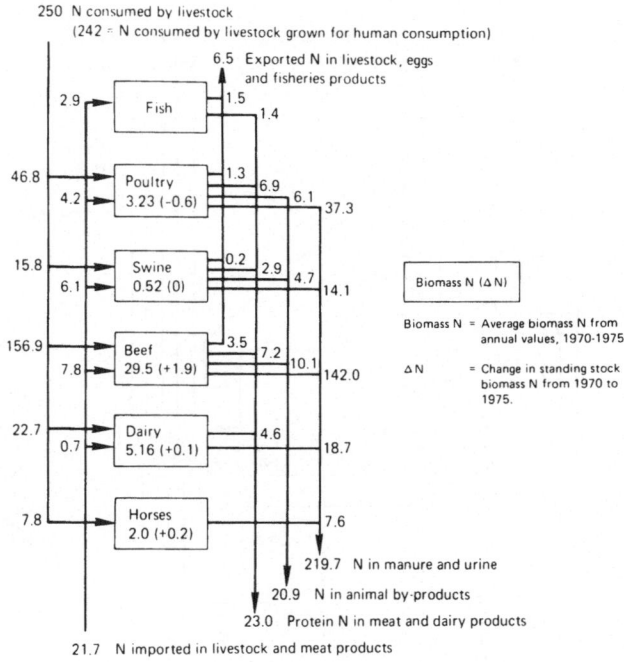

Fig. 4.   Sub-compartment model of the animal-N pool in the peninsular-Florida model[7]. All values are 10^6 kg N yr^{-1}.

System heterogeneity results from a high diversity of both crops and cropping systems, in addition to often significant areas of undisturbed natural systems in various states of ecological succession. Regional values for rates such as nitrogen uptake and nitrogen fixation by crops, for example, are difficult to estimate in agricultural systems in which a large variety of crops are grown in polycultures of varying compositions and densities. The usual approach for temperate-system budgets is to calculate fluxes per unit area of monoculture and then extrapolate to regional rates based on known areas under a particular crop. In multiple-crop systems, however, the density of a given crop species may vary by an order of magnitude or more from one field to another even within the same watershed. Without data for densities and without N-uptake or N-fixation data for a range of densities, the only practical way to arrive at reasonable estimates for these fluxes is to back-calculate from total regional yield (available from FAO statistics) to total residue biomass by using a yield:residue ratio, and thereby calculate total crop biomass. The mean nitrogen content of a crop can then be used to estimate regional crop-N uptake, and then regional-N fixation by estimating from ^{15}N studies the percentage of the crop's N-uptake that represents crop-fixed nitrogen. This procedure is fairly trivial for crops with a well-defined and documented harvest (yield:residue) index, but unfortunately such indices are not often available. Indirectly calculating the index from scattered literature sources adds to the budget's imprecision; future budgets will benefit significantly from agronomic reports that include explicit harvest indices as well as nitrogen content data for all major crop components.

The lack of both land-use documentation and information on nitrogen cycling in particular systems further complicate tropical large-scale nitrogen budgets. Although nitrogen budgets of varying completeness have existed for citrus and oil palm since before 1940 (e.g. [3, 4]) and for many non-cultivated systems such as lowland rainforest since 1960 (e.g. [2, 5]) even simple budgets for some important crops and native systems have yet to appear in the open literature. The nitrogen budgets presented in this volume for coffee and cacao for example, appear to be among the very first published.

These and other difficulties result in regional budgets for both tropical and non-tropical regions that contain a large degree of uncertainty, and this is reflected in the very wide range of values that are at least implicitly assigned to most compartments and fluxes. The resulting inelegance is unfortunate, and for system managers hoping to use such budgets for policy decisions, probably frustrating. For formal first approximations, however, such uncertainty is unavoidable.

Successively better and more precise budgets, constructed perhaps with information from research stimulated by earlier budgets that have focused attention on neglected systems and processes, should make regional budgets at this stage of understanding well worthwhile.

References

1   Ayers R S and Branson R L (Eds) 1973 Nitrates in the upper Santa Ana River basin in relation to groundwater pollution. California Agric. Exptl. Sta. Bull. No. 861. Riverside, California: University of California. 59 p.

2   Bartholomew W V, Meyer J and Laudelout H 1953 Mineral nutrient immobilization under forest and grass fallow in the Yangambi (Belgian Congo) region. I.N.E.A.C. Ser. Sci. Brussels 57, 1–27.

3   Cameron S H and Appleman D 1933 The distribution of total nitrogen in the orange tree. Am. Soc. Hortic. Sci. 30, 341–348.

4   Georgi C D V 1931 The removal of plant nutrients in oil palm cultivation. Malayan Agric. J. 19, 484–489.

5   Greenland D J and Kowal J M L 1960 Nutrient content of the moist tropical forest of Ghana. Plant and Soil 12, 154–174.

6   Miller R J and Smith R B 1976 Nitrogen balance in the southern San Joaquin Valley. J. Environ. Qual. 5, 274–278.

7   NAS 1978 Nitrates: an Environmental Assessment. Washington, D.C.: National Academy of Sciences. 721 p.

8   Robertson G P and Rosswall T 1982 A regional nitrogen budget for West Africa south of the Sahara. (In preparation).

9   Rosswall T, Melillo J M, Ayyad M, Fogg, G E, Kudeyarov V N, Richey J F, Robertson G P and Singh K D 1982 The global nitrogen cycle revised. (In preparation).

10 Söderlund R and Svensson B H 1976 The global nitrogen cycle. In Svensson B H, Söderlund R (Eds). Nitrogen, Phosphorus, and Sulfur. Global Cycles. SCOPE Report 7. Ecol. Bull. Stockholm 22, 23–73.

Plant and Soil 67, 81–90 (1982). 0032-079X/82/0671-0081$01.50.   SU-07
© 1982 *Martinus Nijhoff/Dr W. Junk Publishers, The Hague.*

Some priority research areas in nitrogen studies

Algunas areas prioritarias de investigación en los estudios de nitrógeno

A. J. HOLDING
Department of Agricultural and Food Bacteriology, The Queen's University of Belfast, Belfast BT9 5PX, Northern Ireland, UK

Key words   Fertilizers N-budgets N_2-fixation N-losses N-resources Research priorities Soil processes

Abstract   The more efficient management of the world's nitrogen resources, leading to greater plant productivity, depends on research which (a) improves the economic efficiency of nitrogen inputs into ecosystems, (b) makes more effective use of the nitrogen within the plant environment, and (c) reduces nitrogen losses, particularly from the root region. Ecosystem nitrogen budget studies can emphasize gaps in knowledge of nitrogen cycle processes.

Resumen   El uso mas eficiente de los recursos nitrogenados del mundo que llevará a una mayor productividad vegetal depende de las investigaciones que (a) aumenten la eficiencia económica del uso de nitrógeno en los ecosistemas, (b) hagan mas efectivo el uso del nitrógeno dentro del ambiente de las plantas y (c) reduzcan las pérdidas de nitrógeno mas allá de la zona radical. Los estudios del balance de nitrógeno pueden servir para identificar lagunas en el conocimiento de los procesos del ciclo de nitrógeno.

Introduction

Methods for increasing the input of nitrogen into plant ecosystems, for improving the efficiency of the utilization of this nitrogen by plants, and for reducing losses of nitrogen are three major research priorities. In all types of terrestrial ecosystems, plant productivity is often regulated by the rates of nitrogen cycle processes. This paper discusses nitrogen cycling in agricultural and natural terrestrial environments, but aquatic, industrial, urban and atmospheric nitrogen cycles will frequently influence these terrestrial cycles. The paper is intended to focus on research priorities and a comprehensive review of work in any area has not been attempted. Emphasis is given to methods for increasing the uptake of nitrogen by plant roots and for reducing losses of nitrogen from the root region. Nitrogen budget studies for small areas, which can be particularly valuable for determining major nitrogen flow rates, also merit attention.

Inputs into the ecosystem

Nitrogen fixation

Comprehensive reviews of nitrogen fixation rates for both symbiotic and free-living systems are available[2,22,35].

Leguminous plant – Rhizobium symbioses generally are capable of fixing nitrogen at an order of magnitude higher than other systems and they continue to receive priority attention. Maximum fixation will only occur if both the legume and Rhizobium components are able to function under optimal conditions. For the plant, general agronomic considerations relating to nutritional factors, climate, moisture and acidity are all of major importance. Nutritional factors have been reviewed by Edwards[13]. Some dramatic increases in nitrogen fixation in tropical soils after applications of lime, phosphatic fertilizers and trace elements[11] emphasize this priority. It should also be recognized that legume symbioses are often more susceptible to certain nutritional deficiencies, for example to deficiencies of boron and molybdenum, than are non-legumes. Potassium deficiency also has been shown to reduce nodule metabolism[30] and the phosphorus requirements of nodulated soybeans are higher than for non-nodulated plants[32]. The adaptation of grain legumes to moisture and elevation has been discussed by Rachie[39].

A comparison of the rates of nitrogen fixation by different legume crops under similar agronomic conditions remains a high priority. Nitrogen fixation by different legumes growing in temperate regions like the U.K. has been compared by Nutman[36] and Sprent et al.[42]. Considerable variation occurs in different climatic and geographical regions. For example with lucerne, amounts of nitrogen fixed annually range from 56 to 463 kg N ha^{-1}, with clovers, from 45 to 673 kg N ha^{-1}, and with field beans, from 45 to 600 kg N ha^{-1}. Fixation rates for forage legumes can be markedly influenced by the grazing animal (discussed later).

The characteristics of rhizobia that are necessary to achieve high fixation rates have been extensively discussed but are poorly understood[38]. The survival of rhizobia in soil is clearly a complex association of physical, chemical and nutritional factors, both in the presence and absence of plant roots. Attachment onto and entry into root cells requires the ability to compete against other rhizobia and other soil organisms for infection sites. The primary agronomic objective is to ensure that rhizobia that are highly effective under field conditions are able to survive in the soil and be highly infective. If these conditions cannot be guaranteed then seed inoculation might help, provided that the inoculated rhizobia survive on the seed until germination takes place and infection is possible. But far more information is required on the relationship between numbers of rhizobia on inoculated seed and subsequent infection levels. In all inoculation trials the cross-contamination of control (uninoculated) treatments can occur, so that where possible, trials should only be undertaken with strains marked by antibiotic resistance or serological characteristics. The level of cross-contamination can then be determined.

Extensive research into the behaviour of rhizobia in soil also needs to be undertaken. The selection of rhizobia tolerant to soil acidity in relation to the growth of mung bean (*Vigna radiata*) demonstrates the value of this type of research[33].

Techniques for the precise measurement of nitrogen fixation in the field continue to be refined. Recent reviews of the use of isotopes of nitrogen[4] and the acetylene reduction technique[24] are noteworthy, but Witty[44] has drawn attention to the possibility of the latter technique's over-estimating nitrogen fixation rates in soil. In all nitrogen fixation studies, any contribution from soil nitrogen compounds also requires estimation.

The scope for increasing the efficiency of nitrogen fixation by legumes and, in fact, nitrogen uptake by all crop plants through selection and breeding procedures is appreciable. With the legume symbiosis the role of the host-plant genetics has been shown both in different cultivars bred for enhanced fixation[25,37] and in the selection by the legume of rhizobia with high effectiveness[34]. In order to obtain optimal nitrogen fixation, both selection and breeding for many factors is required, including the ease of infection, the time and rate of nodule formation and development, and the number and size of nodules. The extensive work on the genetics of rhizobia has so far yielded little information of application to field conditions, but the potential for genetically screening rhizobia for desirable traits for use in inoculants is considerable.

Nitrogen deposition from the atmosphere

Nitrogen-compound deposition from the atmosphere may be attributable to rainfall or to dry deposition. For input in rainfall in tropical regions, Greenland[23] presented figures which varied from $1.5\,kg\,N\,ha^{-1}\,yr^{-1}$ in Northern Australia to approximately $8\,kg\,N\,ha^{-1}\,yr^{-1}$ in Southern Asia and Central America. Batey (pers. comm.) assessed published data from polar regions and concluded that the input from non-anthropic sources is less than $1\,kg\,N\,ha^{-1}\,yr^{-1}$. An average total figure for input by rain in the British Isles is $15\,kg\,N\,ha^{-1}\,yr^{-1}$.

Dry or solid deposition is thought mainly to be associated with the absorption of ammonia and the deposition of aerosol particles containing ammonium salts. The sources of this ammonia are not known but they are assumed to have arisen from many sources including volatilization from the soil surface, particularly from animal faeces, from urine, and from fertilizers. Denmead *et al.*[10] have suggested that there is a closed ammonia-cycle, with the ammonia that originates from terrestrial sources being rapidly readsorbed by the soil or vegetation. Rodgers[40] concluded that dry deposition of the order of 4–$5\,kg\,N\,ha^{-1}\,yr^{-1}$ occurs in the Rothamsted area of England.

Nitrogen fertilizers

A small 'starter dressing' of nitrogen appears to improve the establishment and growth of many legumes prior to the initiation of nitrogen fixation. Legumes vary in their response to higher levels of fertilizer nitrogen. For example, *Phaseolus* beans show a large response to fertilizer nitrogen, but no definite patterns have emerged for *Vicia* spp. and *Pisum* spp. With tropical grain legumes, *Vigna* spp. and *Arachis* spp. in general show little response, and very variable results are reported for *Glycine* spp. In situations where increased yields are obtained with fertilizer nitrogen, the results may be attributable to low maximum nitrogen fixation rates or to the local inhibition of nitrogen fixation by impaired nodule production or function. The precise mechanism of inhibition is not known. Soil inorganic nitrogen may also have an inhibitory effect on nodulation. With some plants ammonium is more inhibitory than nitrate. Symbiotic associations that can continue to fix nitrogen in the presence of high concentrations of ammonia and/or nitrate should be selected; a strain of *Rhizobium phaseoli* has been reported which possesses this property[41].

Increased plant yields resulting from fertilizer nitrogen applications also have important economic implications for cereal and grass production. The application of fertilizer nitrogen may also increase the uptake of indigenous soil nitrogen by the plant[12]. Various arable crops do not differ markedly in their uptake of fertilizer nitrogen, although the distribution in the plant may differ. In the U.K., the mean uptake efficiency into grain is usually between 40–50% but occasionally up to 60%. Economic considerations suggest that farmers can apply fertilizer nitrogen until the 'marginal fertilizer uptake efficiency', which relates the nitrogen content of the crop to the amount of applied fertilizer nitrogen, is as low as 8%. The effective distribution and timing of fertilizer applications to promote greater uptake by the crop justifies further research. The soil types used for arable crops in the U.K. appear to have little effect on the uptake of fertilizer nitrogen except where there might be excessive leaching or denitrification.

The application of fertilizer nitrogen to grassland is often complicated by the grazing animal. Herbage productivity may be measured by shoot dry weight or protein content, animal liveweight gain, output of milk or other measures. Under U.K. conditions, the response of grass pastures to fertilizer nitrogen can increase linearly up to an application of 350 kg N ha^{-1} yr^{-1}, but currently the average application rates are 160 and 96 kg N ha^{-1} yr^{-1} for temporary and permanent grass, respectively. In mixed-grass plus clover swards the decline in legume growth and nitrogen fixation is associated with increased fertilizer nitrogen applications, the decline is offset by increased yield and nitrogen content of the grass crop. For this reason the response to fertilizer nitrogen is usually greater in pure grass swards. Under Northern Ireland conditions, grass alone requires between 150–200 kg N ha^{-1} yr^{-1} to produce yields similar to mixed-clover plus grass pasture[8]. The complexity of the situation with grazing animals was recently demonstrated by Jackson and Williams[27], who showed that compared with cutting experiments, maximum grass yields under a grazing regime are obtained with lower rates of fertilizer nitrogen.

Internal soil processes

Reviews by Allison[1], Campbell[7] and others have stressed the importance of net immobilization or net mineralization of soil nitrogen for plant nutrition. It is generally accepted that under aerobic conditions, a ratio of energy-yielding carbon to utilizable-nitrogen of between 20–30:1 results in little uptake or net release of inorganic nitrogen during decomposition. In ecosystems where there is little accumulation or net decomposition of soil organic nitrogen and a 'steady state' prevails, the considerable movement of inorganic nitrogen in and out of the organic nitrogen pool is often neglected in explanations of nitrogen cycle processes. Where crop rotations lead to a predicted net mineralization of inorganic nitrogen, fertilizer applications can be appropriately amended. The importance of using crop plants and farming systems which can make full use of mineralized nitrogen has been stressed by Bartholomew[3].

Where ammonium is less effective than nitrate as a plant nutrient, nitrification will clearly be beneficial. Otherwise nitrification studies only appear justified in attempts to reduce nitrogen losses through lessened leaching or denitrification of nitrate. Nitrification inhibitors have been extensively investigated[21] and recent work on urease inhibitors to prevent urea hydrolysis emphasizes the continuing interest in studying the retention of nitrogen compounds within the root region.

Nitrogen losses from the ecosystem

Leaching, including drainage

The leaching of nitrogen compounds, both laterally and vertically away from the rooting zone, is recognized as a major factor influencing the nitrogen economy of soils. Climatic, soil and cropping factors can all influence leaching rates. For particular experimental areas, information on leaching is available from lysimeter studies, field drainage monitoring data, and borehole water analyses. In British soils, leaching occurs mainly under arable crops, but significant losses also occur under grass. In the U.K., approximately one third of the rainfall drains from many arable soils, and nitrate equivalent to one third of the fertilizer nitrogen applied often leaches from the root zone. Garwood and Tyson[19], for example, have shown leaching losses at this level from grassland which include a component from the mineralization of soil organic nitrogen. Borehole drilling data confirm the leaching losses demonstrated by lysimeter studies. Typical depth profiles of nitrate concentrations by Young et al.[45] demonstrate the irregular distribution of nitrate down the geological strata. The data emphasize the problems of determining the rate of downward movement of nitrate and predicting future nitrate levels in underground water strata.

While inorganic nitrogen losses have been extensively studied, the leaching of organic nitrogen from the root zone and the mobility and accumulation of this form of nitrogen in aquifers has not received appropriate attention. In the chalk aquifers in England, the inorganic nitrogen content is estimated at 1.52×10^9 kg N and organic nitrogen at 11.1×10^9 kg N. The origin and mobility of this organic nitrogen are not known.

The concentration of nitrate in drainage and surface waters depends on many factors, including the cropping regime. General average figures under United Kingdom conditions are 10 to 25 mg NO_3–N l^{-1} for arable land, 4 to 10 mg NO_3–N l^{-1} for grassland, and less than 1 to 2 mg NO_3–N l^{-1} for rough grazings and unfelled forest areas. In disturbed ecosystems, felled forests for example, considerably greater losses are reported (Wilkinson, B. W. and Saunders, P. J. W., pers. comm.).

Gaseous nitrogen losses

Losses of ammonia, dinitrogen and nitrogen oxides also need to be assessed. Gaseous losses are frequently assumed to be responsible for the differences between nitrogen inputs and readily detected nitrogen losses. In general, measurements of field losses of gaseous nitrogen are imprecise and may even be misleading. The environmental conditions necessary for denitrification to occur under laboratory conditions are well-documented, for example by Bremner and Shaw[6], but experiments with hooded lysimeters using ^{15}N are unsuitable for routine field studies[12]. N_2O evolution from soils has also been used to estimate denitrification losses. However, Firestone et al.[16] and others (see Focht and Verstraete[17]) have clearly demonstrated that the ratio of N_2 to N_2O released from soil varies with several factors. Acetylene has been used to block the reduction of N_2O to N_2[15], leading to the evolution of only N_2O. However, as with the acetylene reduction technique for field estimates of nitrogen fixation, this application of acetylene has limited value in the field. These problems clearly preclude firm estimates being made for gaseous nitrogen losses, but nitrogen balance sheets (e.g.[26,28]) suggest that 30% or more of fertilizer and other nitrogen entering a system may be lost in the gaseous form. Bremner and Blackmer[5] have recently suggested that a proportion of this nitrogen could be lost as N_2O produced during nitrification.

Losses of ammonia from ecosystems may occur from (a) soil or fertilizers applied to soil, (b) decomposing animal faeces, urine and crop residues, and (c) living plants.

The large losses of ammonia which can occur in all soils after nitrogen fertilizer application, but especially in calcareous or limed soils, are well-documented[9]. For example, Volk[43] showed that percentage losses of urea, ammonium sulphate and ammonium nitrate applied to limed grassland were in the ratio 30–40 : 20 : 3. The large losses associated with urea have been reviewed by Gasser[20]. Rapid losses of up to 20% of the nitrogen in urea have been reported for rice fields[31].

Losses from animal wastes, mainly from the volatilization of ammonia, can also be considerable. Lauer *et al.*[29] demonstrated that approximately one-third of the nitrogen applied to soil as dairy cattle waste was lost within a few days after application. Losses of a similar size or even higher have been reported for cattle feedlots and during waste storage as slurry, before field application. These data suggest that more than 50 per cent of the nitrogen in animal wastes can be lost by volatilization. The amount readsorbed onto plants and untreated soil is not known. Ammonia flux-gradient studies, though difficult, deserve further attention.

Net losses of ammonia from growing plants are also difficult to quantify. With certain crops, an analysis of above-ground parts has shown declines in nitrogen content, particularly in senescent leaves, for plants under stress conditions and for plants receiving high levels of fertilizer nitrogen. These losses may be due to internal nitrogen transfer to other plant parts or to leaching onto soil, and are therefore not necessarily a nitrogen loss from the system. Evolution of ammonia from senescing leaves of maize has been clearly demonstrated in pot experiments[14]. The loss was the equivalent of a field rate of $7 \, g \, N \, ha^{-1} \, d^{-1}$. It is far from clear whether canopy recycling of ammonia from ecosystems require further study.

Nitrogen budget studies

Budget studies for catchments can be particularly valuable for assessing the relative importance of a particular input or output of nitrogen, or for determining the magnitude of major gaps in knowledge. An example of a small individual-watershed study in the United Kingdom has been reported by Hood[26], who prepared a balance sheet of the nitrogen added and removed annually from a grassland system over a four-year period. When an estimated total of $377 \, kg \, N \, ha^{-1} \, yr^{-1}$ including $250 \, kg$ N-fertilizer entered the system, only $260 \, kg \, N \, ha^{-1} \, yr^{-1}$ was removed. The balance of $117 \, kg$ N was unaccounted for. Larger catchment studies have been carried out by Foy *et al.*[18], who have shown a positive relationship between increased intensification of agriculture, associated with higher applications of nitrogen fertilizer, and nitrate loadings in rivers. An example of these data is shown in Table 1.

Conclusions

The availability of nitrogen is a major factor controlling plant production throughout the world, covering an ecosystem spectrum from intensively managed agricultural regimes to virgin natural vegetation. Three factors determine whether nitrogen resource management is effectively increasing plant productivity: whether (1) nitrogen is added in a form that is most efficient both in economic terms and in the use of fossil-fuel energy, (2) whether the nitrogen is

Table 1. Relationship between nitrogen fertilizer application and nitrate loading in the Lough Neagh catchment area of Northern Ireland (from Foy et al.[18])

	Year	
	1971	1978
Fertilizer application (kg N ha^{-1} yr^{-1})	51.1	75.8
Mean nitrate concentration in major rivers (mg NO_3–N l^{-1})	1.58	2.71
Total nitrogen loading in major rivers (10^9 kg N yr^{-1})	4.39	10.2

used within the ecosystem at an optimal level, and (3) whether all forms of nitrogen loss from ecosystems are reduced, particularly losses from the root region. A number of research areas require further attention if the unquestioned value of nitrogen as a plant nutrient is to be fully realized in practice.

Acknowledgements The author is Secretary of the Royal Society Study Group on the Nitrogen Cycle (1979–1981). Whilst the views expressed do not necessarily reflect the research priorities of the Study Group, the discussions within the group have been an invaluable help to the author. The author is also indebted to the Royal Society for a Travel Grant to attend the Workshop.

References

1   Allison F E 1973 Soil Organic Matter and its Role in Crop Production. Amsterdam: Elsevier. 637 p.
2   Ayanaba A and Dart P J 1977 Biological Nitrogen Fixation in Farming Systems of the Tropics. Chichester: John Wiley & Sons. 377 p.
3   Bartholomew W V 1977 Soil nitrogen changes in farming systems in the humid tropics. In Ayanaba A and Dart P J (Eds). Biological Nitrogen Fixation in Farming Systems of the Tropics, pp 27–42. Chichester: John Wiley & Sons.
4   Bremner J M 1977 Use of nitrogen-tracer techniques for research on nitrogen fixation. In Ayanaba A and Dart P J (Eds). Biological Nitrogen Fixation in Farming Systems of the Tropics, pp 335–352. Chichester: John Wiley & Sons.
5   Bremner J M and Blackmer A M 1978 Nitrous oxide: emission from soils during nitrification of fertilizer nitrogen. Science 199, 295–296.
6   Bremner J M and Shaw K 1958 Denitrification in soil. J. Agric. Sci. 51, 40–52.
7   Campbell C A 1978 Soil organic carbon, nitrogen and fertility. In Schnitzer M and Khan S U (Eds). Soil Organic Matter, pp 173–271. Amsterdam: Elsevier.
8   Chestnutt D M B and Lowe J 1970 Sect III Sub-Sect B. Agronomy of white clover/grass swards. In Lowe J (Ed.). White Clover Research. Occasional Symposium of the British Grassland Society No. 6, pp 191–213. Maidenhead: British Grassland Society.
9   Cooke G W 1967 Control of Soil Fertility, pp 182–184. London: Crosby-Lockwood.
10   Denmead O T, Freney J R and Simpson J R 1976 A closed ammonia cycle within the plant canopy. Soil Biol. Biochem. 8, 161–164.
11   Döbereiner J 1977 Present and future opportunities to improve the nitrogen nutrition of crops through biological fixation. In Ayanaba A and Dart P J (Eds). Biological Nitrogen Fixation in Farming Systems of the Tropics, pp 3–12. Chichester: John Wiley & Sons.
12   Dowdell R J and Webster C P 1980 A lysimeter study using nitrogen-15 on the uptake of fertilizer nitrogen by perennial ryegrass swards and losses by leaching. J. Soil Sci. 31, 65–75.

13   Edwards D G 1977 Nutritional factors limiting nitrogen fixed by rhizobia. *In* Ayanaba A and Dart P J (Eds.). Biological Nitrogen Fixation in Farming Systems of the Tropics, pp 189–216. Chichester: John Wiley & Sons.

14   Farquhar G D, Wetselaar R and Firth P H 1979 Ammonia volatilization from senescing leaves of maize. Science 203, 1257–1258.

15   Federova R I, Milekhana E I and Ilyukhina N I 1973 Possibility of using the 'gas exchange' method to detect extraterrestrial life, identification of nitrogen fixing organisms. Akad Nauk SSR Isvestia Ser. Bio. 6, 797–806.

16   Firestone M K, Firestone R B and Tiedje J M 1980 Nitrous oxide from soil denitrification: Factors controlling its biological production. Science 208, 749–750.

17   Focht D D and Verstraete W 1977 Biochemical ecology of nitrification and denitrification. Adv. Microb. Ecol. 1, 135–214.

18   Foy R H, Smith R V, Stevens R J and Stewart D A 1980 Factors affecting the nitrogen and phosphorus loadings to Lough Neagh. (Unpublished ms.).

19   Garwood E A and Tyson K C 1977 High loss of nitrogen in drainage from soil under grass following a prolonged period of low rainfall. J. Agric. Sci. 89, 767–768.

20   Gasser J K R 1964 Urea as a fertilizer. Soils Fertil. 27, 175–180.

21   Gasser J K R 1970 Nitrification inhibitors – their occurrence, production, and effects of their use on crop yields and composition. Soils Fertil. 33, 547–554.

22   Granhall U (Ed.) 1978 Environmental Role of Nitrogen-fixing Blue-green Algae and Asymbiotic Bacteria. Ecol. Bull. (Stockholm) No. 26. Stockholm: Swedish National Science Research Council. 391 p.

23   Greenland D J 1977 Contribution of microorganisms to the nitrogen status of tropical soils. *In* Ayanaba A and Dart P J (Eds). Biological Nitrogen Fixation in Farming Systems of the Tropics, pp 13–25. Chichester: John Wiley & Sons.

24   Ham G E 1977 The acetylene-ethylene assay and other measures of nitrogen fixation in field experiments. *In* Ayanaba A and Dart P J (Eds). Biological Nitrogen Fixation in Farming Systems of the Tropics, pp 325–334. Chichester: John Wiley & Sons.

25   Holl F B and La Rue T A 1976 Host genetics and nitrogen fixation. *In* Hill L D (Ed.). World Soybean Research: Fertilization and Nitrogen Fixation, pp 156–163. Danville: Interstate Printers & Publishers.

26   Hood A E M 1976 Nitrogen, grassland and water quality in the United Kingdom. Outl. Agric. 8, 320–327.

27   Jackson M V and Williams T E 1979 Response of grass swards to fertilizer-N under cutting or grazing. J. Agric. Sci. 92, 549–562.

28   Jenkinson D S 1977 The nitrogen economy of the Broadbalk Experiments. Rothamsted Experimental Station Report for 1976 Pt 2, pp 103–109. Harpenden: Lawes Agricultural Trust.

29   Lauer D A, Bouldin D R and Klausner S D 1976 Ammonia volatilization from dairy manure spread on the soil surface. J. Environ. Qual. 5, 134–141.

30   Mengel K, Haghparast M R and Koch K 1974 The effect of potassium on the fixation of molecular nitrogen by roots of *Vicia faba*. Plant Physiol. 54, 535–538.

31   Mikkelson D S, Datta S K de and Obcema W N 1978 Ammonia volatilization losses from flooded rice soils. Soil Sci. Soc. Am. J. 42, 725–730.

32   Mooy C J de and Pesek J 1966 Nodulation responses of soybeans to added phosphorus, potassium, and calcium salts. Agron. J. 58, 275–280.

33   Munns D N, Keyser H H, Fogle V W, Hohenberg J S, Righetti T L, Lauter D L, Larong M G, Clarkin K L and Whiteacre K W 1979 Tolerance of soil acidity in symbioses of mung bean with rhizobia. Agron. J. 71, 256–260.

34   Mytton L R and Felice J de 1977 The effect of mixtures of rhizobium strains on the dry matter production of white clover grown in agar. Ann. Appl. Biol. 87, 83–93.

35 Nutman P S (Ed.) 1976 Symbiotic Nitrogen Fixation in Plants. IBP Synthesis Series, Vol. 7. Cambridge University Press. 584 p.

36 Nutman P S 1976 IBP field experiments on nitrogen fixation by nodulated legumes. *In* Nutman P S (Ed.). Symbiotic Nitrogen Fixation in Plants, IBP Synthesis Series, Vol. 7, pp 211–237. Cambridge: Cambridge University Press.

37 Nutman P S, Mareckova H and Raicheva L 1971 Selection for nitrogen fixation in red clover. Plant and Soil Spec. Vol. 27–53.

38 Obaton M 1977 Effectiveness, saprophytic and competitive ability: three properties of Rhizobium essential for increasing the yield of inoculated legumes. *In* Ayanaba A and Dart P J (Eds). Biological Nitrogen Fixation in Farming Systems of the Tropics, pp 127–133. Chichester: John Wiley & Sons.

39 Rachie K O 1977 The nutritional role of grain legumes in the lowland humid tropics. In Ayanaba A and Dart P J (Eds). Biological Nitrogen Fixation in Farming Systems of the Tropics, pp 45–60. Chichester: John Wiley & Sons.

40 Rodgers G A 1978 Dry deposition of atmospheric ammonia at Rothamsted in 1976 and 1977. J. Agric. Sci. 90, 537–542.

41 Ruschel A P and Reuszer H W 1973 Factors affecting the *Rhizobium phaseoli – Phaseolus vulgaris* symbiosis. Pesquisa Agropecuaria Brasileira, Agronomia 8, 287–292.

42 Sprent J I, Bradford A M and Norton C 1977 Seasonal growth patterns in field beans (*Vicia faba*) as affected by population density, shading and its relationship with soil moisture. J. Agric. Sci. 88, 293–301.

43 Volk G M 1961 Gaseous loss of ammonia from surface-applied nitrogenous fertilizers. J. Agric. Food Chem. 9, 280–283.

44 Witty J F 1979 Acetylene reduction assay can overestimate nitrogen fixation in soil. Soil Biol. Biochem. 11, 209–210.

45 Young C P, Hall E S and Oakes D B 1976 Nitrate in groundwater studies on the chalk near Winchester, Hampshire. Technical Report No. 31, Medmenham, Water Research Centre, 67 p.

Plant and Soil 67, 91–103 (1982). 0032-079X/82/0671-0091$01.95. SU-08

Nitrogen in shifting cultivation systems of Latin America

El nitrógeno en sistemas de agricultura migratoria en America Latina

P. A. SÁNCHEZ

Tropical Soils Program, North Carolina State University, Raleigh, N.C. 27650, USA

Key words Agriculture Forests Land clearing N-cycling N-fertilization Shifting-cultivation Tropics

Abstract Relatively little is known about the dynamics of N in shifting cultivation and related cropping systems in the humid tropics of Latin America. The soils that predominate in 82% of the region, namely Oxisols and Ultisols, have a fairly high total N content. Contrary to conventional wisdom, the bulk of the N in tropical rainforests is present in the soil, and not in the biomass. Losses of N through clearing and burning are about 20–25% of the N existing in the ecosystem. Mechanized land clearing causes larger N losses than the traditional slash and burn method. Ashes can contribute substantial amounts (67–127 kg N ha^{-1}) to the soil, which prevents N deficiency for the first crop sown, but N deficiency is observed from the second crop onwards. The rate of total-N decomposition in the arable layer is high during the first two years after burning, but subsequently reaches a new equilibrium with continuous cultivation.

Continuous production of food crops is feasible in Ultisols and Oxisols of the Amazon with correct agronomic practices. Crops such as maize and rice require N fertilization rates of 80–120 kg N ha^{-1}. The efficiency of applied-N utilization is comparable to that in the temperate zone and varies with planting season and cropping system. Pastures following burning do not cause significant losses of N in the soil, particularly if they consist of properly managed mixtures of grasses and legumes. These observations are based on data collected from only a small number of sites, making generalization difficult. Nitrogen dynamics should be viewed in conjunction with other soil factors such as acidity and the availability of other nutrients.

Resumen Se conoce bastante poco acera de la dinámica del N en sistemas de agricultura migratoria y otras prácticas de cultivo en el trópico húmedo latinoamericano. Los suelos Oxisoles y Ultisoles que predominan en el 82% de la región están bastante bien dotados de N. Contrariamente a lo que comunmente se cree la mayoría del nitrógeno en ecosistemas naturales se encuentra en el suelo y no en la biomasa. Por lo tanto las pérdidas de nitrógeno por la quema son del 20–25% del N existente en el ecosistema. El desmonte mecanizado causa mayores pérdidas de N que el desmonte por tumba y quema tradicional. La ceniza contribuye con cantidades considerables de nitrógeno (67–127 kg N ha^{-1}) al suelo lo cual evita las deficiencias de nitrógeno para el primer cultivo que se siembra. En un Ultisol de la Amazonía la deficiencia de nitrógeno es aguda a partir del segundo cultivo. La tasa de descomposición total para el nitrógeno en la capa arable es alta durante los dos primeros años después de la quema pero se alcanza un nuevo valor de equilibrio posteriormente. La producción contínua de cultivos de ciclo corto es factible en estos suelos mediante prácticas agronómicas correctas. Los cultivos como el maiz y el arroz responden positivamente a dósis de 80–120 kg ha^{-1} de fertilización con N. La eficiencia de utilización del nitrógeno aplicado es comparable a la de zonas templadas y varía con la época de siembra y sistemas de cultivos. El establecimento de pastizales después de la quema no causa pérdidas apreciable de nitrógeno en el suelo, especialmente si se siembran mezclas de gramineas con leguminosas y se manejan bien. Debido a que estas observaciones están basadas en datos recolectados en pocos sitios es por lo tanto difícil generalizar. La fertilización con nitrógeno debe considerarse conjuntamente con el manejo adecuado de otras limitaciones de los suelos tales como la acidez y el estado del suelo con respecto a otros elementos nutritivos.

Introduction

Shifting cultivation is the agricultural system most commonly practised over a vast area of tropical America, the humid tropics, which comprise approximately 46% of the western hemisphere between the Tropics of Cancer and Capricorn. Although known by various names in Latin America (milpa, conuco, roza, chacra, monte and others), shifting cultivation is synonymous with poverty and low productivity per hectare. The system is regarded as ecologically sound in conditions of low population pressure, but ecologically harmful when the shortage of land prevents a sufficiently long fallow interval to regenerate the productivity of the soil. It is important, therefore, to differentiate between traditional shifting cultivation and shifting cultivation in disequilibrium[18]. The former represents a stable situation typical of zones that are not readily accessible, while the latter represents a serious agronomic, social and ecological problem as the agricultural frontier of Latin America advances towards the Amazon and humid tropical areas. Alternative farming systems need to be developed in order to replace shifting cultivation in disequilibrium.

Nitrogen is one of the key elements in ecosystems where shifting cultivation is practised. The aim of this report is to summarize some of the existing knowledge

Table 1. Distribution of soils in the American humid tropics. Adapted from Sánchez and Cochrane[12], with alterations

Soils	Area (10^6 ha)
Oxisols	332
Ultisols	213
Inceptisols:	
Aquepts	42
Andepts	2
Tropepts	17
Entisols:	
Fluvents	6
Psamments	6
Lithics	19
Alfisols	18
Spodosols	10
Vertisols	1
Total	666

of the behaviour of this element in soils, in the biomass after clearing, in continuous food crop production, in pasture systems, and in fallows in the humid tropics of Latin America.

Soils of the American humid tropics

Table 1 shows the distribution of soils in the American humid tropics based on 1 : 5 million maps compiled by Sánchez and Cochrane[12]. It is relevant to note the preponderance of acid and infertile soils classified as Oxisols and Ultisols, which occupy 82% of the area. Poorly drained soils (classified as Aquepts) occupy 6% of the area, well-drained soils of medium to high natural fertility (Andepts, Tropepts, Fluvents, Alfisols and Vertisols) occupy 7% of the region, shallow soils occupy 3%, and extremely infertile sandy soils (Psamments and Spodosols) occupy the remaining 2%.

The total nitrogen content of soils in the humid tropics is not so low as is commonly thought. Table 2 shows a significantly higher reserve of total C and N in the first 100 cm of soils of the humid tropics as compared with those of the

Table 2. Reserves of organic matter (to 1 m depth) in soils of the humid tropics *vs* soils of the humid temperate zone (Sánchez *et al.*[13]) Common superscripts within a column indicate no significant difference ($p < 0.05$)

Region	Number of profiles	Organic-C (kg m^{-2})	Total-N (kg m^{-2})	C/N
Humid tropics	23	10.2[a]	1.12[a]	9.5[a]
Humid temperate	27	7.8[b]	0.78[b]	9.9[a]

Table 3. Nitrogen in virgin-forest biomass and soils in some localities in the humid tropics

Locality	Soil	pH	Forest biomass (kg N ha^{-1})	Soil (0–15 cm) (kg N ha^{-1})	Total eco-system (kg N ha^{-1})	% of system--N in soil	Annual additions to the soil[5] (kg N ha^{-1})
Manaus, Brazil[1]	Orthox	3.8	3294	8906	12200	73	106
Mérida, Venezuela[2]	Tropept	na	1088	4638	5726	81	57
Carare, Colombia[3]	Aquox	3.3	740	1812	2551	71	141
Kade, Ghana[4]	Ustalf	5.2	1017	4336	5353	81	199

[1] Fittkau and Klinge[6].
[2] Fassbender[5].
[3] De las Salas[9].

[4] Greenland and Kowal[7].
[5] *Via* decomposition of litter.
na, not available.

humid temperate zone[13]. When considering other ecosystems, no significant differences in C contents or C/N ratios between tropical soils and soils of the temperate zone were found, but total-N contents at depths of 0–15 and 0–100 cm were higher in tropical soils. It is important to stress that the sub-soil should be included in the calculation of total N reserves, because it usually contains larger amounts of total N than the arable layer, owing to differences in volume.

Nitrogen in rainforest ecosystems

Until a few years ago the bulk of data on nutrient cycling in tropical forests came from Africa. Recently, ecologists and foresters have accumulated valuable information in the humid tropics of America, some of which is summarized in Table 3, in comparison with the classic African data. This table indicates that the amount of N stored in the forest is variable and not directly related to soil type. For example, the biomass N of a very acid Oxisol from Manaus is greater than that produced on soils of high natural fertility such as an Inceptisol from Venezuela or an Alfisol from Ghana. The amount of N accumulated in a poorly drained Oxisol from Colombia, however, was distinctly lower.

In moist tropical forest ecosystems, the bulk of the N is in the soil and not in the live biomass, as Table 3 demonstrates. It is therefore incorrect to state that the bulk of the N in these ecosystems is concentrated in the live biomass and that the soils contribute very little. The same applies in the case of P, but not with K, Ca and Mg in acid soils[11].

The amounts of N added to the soil via litter decomposition are also shown in Table 3. From the agronomic point of view these amounts are similar to N fertilization rates generally applied to crops such as maize on these soils. The soils also receive N via rainwash throughfall, litterfall, root decomposition, and biological N fixation.

Effects of land clearing

The most complete study to date on N dynamics during the process of clearing and burning was carried out on an Inceptisol of volcanic origin in Turrialba, Costa Rica by Ewel et al.[3] Their data, shown in Fig. 1, indicate relatively small N losses after removing the stems suitable for fuelwood, drying the rest over a period of 11 weeks and, finally, burning the remaining material. The data collected include the entire above-ground biomass and the top 3 cm layer of the soil, thereby underestimating the total amount of N in the ecosystem by not taking the subsoil into account. Fig. 1 shows that only 2 and 25% of the total N was lost through felling and burning, respectively. De las Salas and Folster[10] report a loss of 20% of N through felling and burning of a forest on a poorly drained Oxisol in Carare-Opón, Colombia. These studies suggest that felling and burning typical of traditional shifting cultivation causes modest N losses from the ecosystem. This contradicts the widespread views that most of the N is lost when

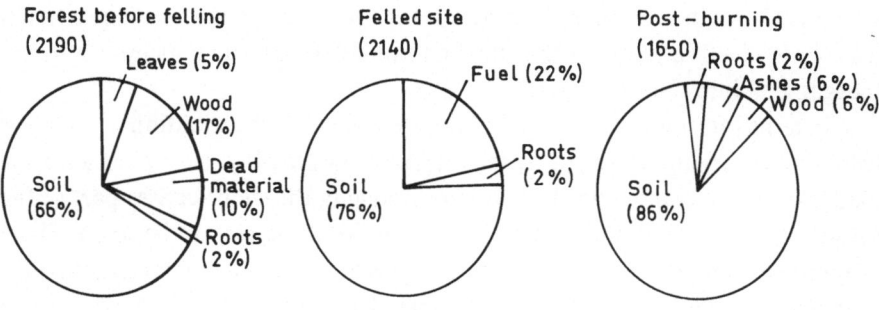

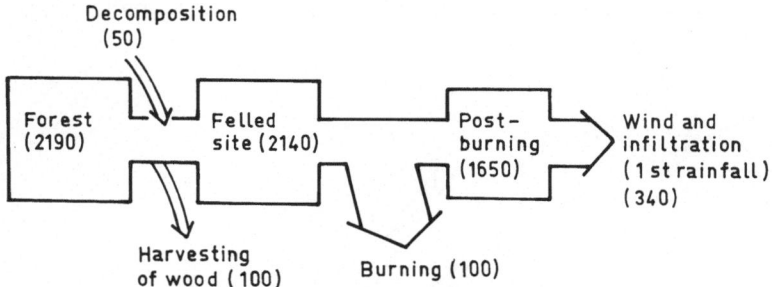

Fig. 1. Dynamics of total nitrogen (biomass + surface soil 0–3 cm) during the clearing and burning of a forest site in Turrialba, Costa Rica (Ewel et al.[3]). Values in parentheses = kg N ha^{-1}.

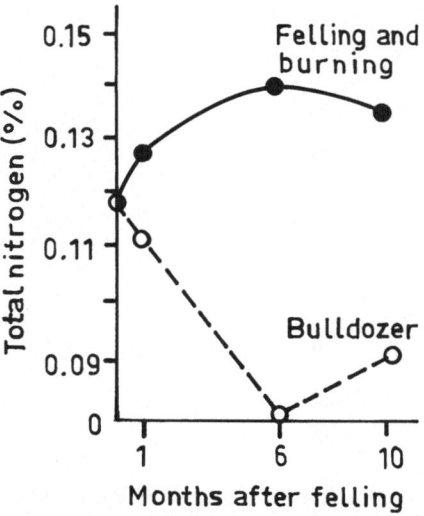

Fig. 2. Effect of land-clearing methods on the total-N content in the first 10 cm of an Ultisol in Yurimaguas, Peru (Seubert et al.[16]).

a tropical forest is cleared and burned. Ewel *et al.*[3] also recorded an additional loss of 15% (340 kg N ha^{-1}) after the first rainfall, which suggests infiltration of inorganic N into the soil (Fig. 1).

When the forest is cleared with bulldozers instead of the traditional system of felling and burning, nitrogen losses increase considerably; this is because the blades of the machines and the action of dragging the stems causes part of the topsoil to be removed and accumulated away from the area to be sown. These changes are illustrated in Fig. 2, from a study of Seubert *et al.*[16] who cleared a 17-year-old secondary forest site on an Ultisol in Yurimaguas, Peru. The increase in total N immediately after burning is not easy to explain but may be due to contamination by plant material that was partially burned or in an advanced stage of decay. Nevertheless the topsoil N status remained higher when the soil was cleared by slashing and burning that when it was cleared by bulldozing.

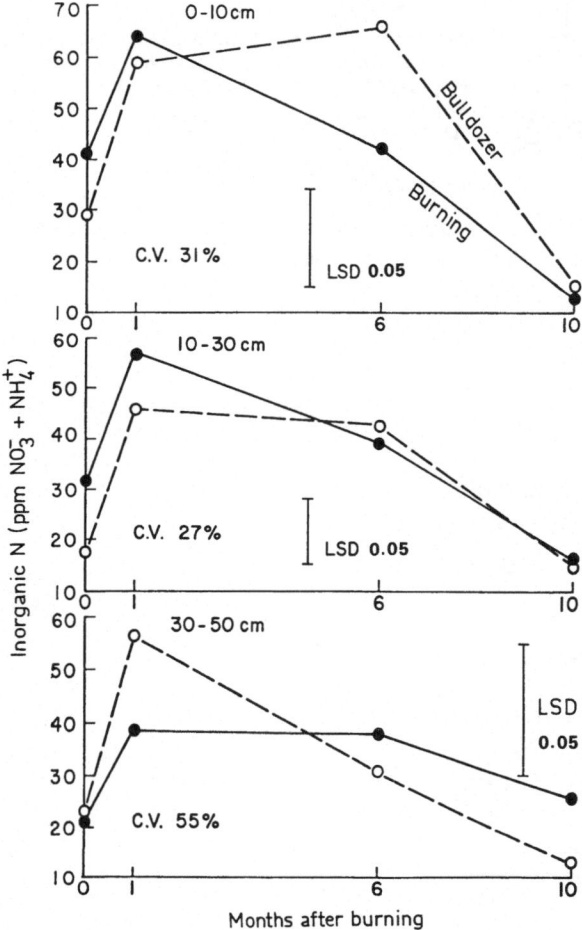

Fig. 3.   Dynamics of inorganic N during the first 10 months after clearing a forest site in Yurimaguas, Peru (North Carolina State University[8]). ppm = mg N·kg soil^{-1}. C.V. = coefficient of variation (standard deviation/n); LSD = least significant difference ($p < 0.05$).

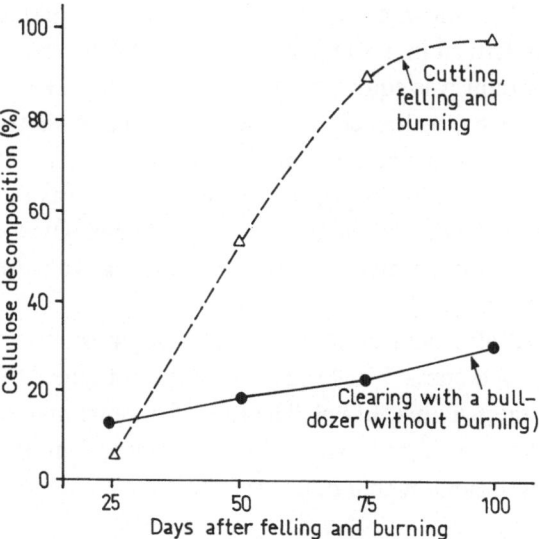

Fig. 4.   Effect of land clearing methods on microbial activity, measured as the rate of cellulolytic decomposition, after clearing a wet forest site in an Ultisol at Barrolândia, Bahia, Brazil. (Adapted from Silva[17]).

Another favourable aspect of burning as compared with mechanized land clearing is the nutrient content of the ashes. Although the effect is more marked in the case of elements that are not volatile (K, Ca, Mg), the ashes also contain appreciable amounts of N that are rapidly incorporated into the soil. Table 4 shows that the quantity of ashes and their N contents vary according to soils and burning conditions. In Yurimaguas, the same forest site yielded different N additions when it was burned in 1972 and in 1980. The 1980 burn was more complete, producing three times the quantity of ashes. Thus, ashes contribute considerable amounts of N to the first crop sown in shifting cultivation systems.

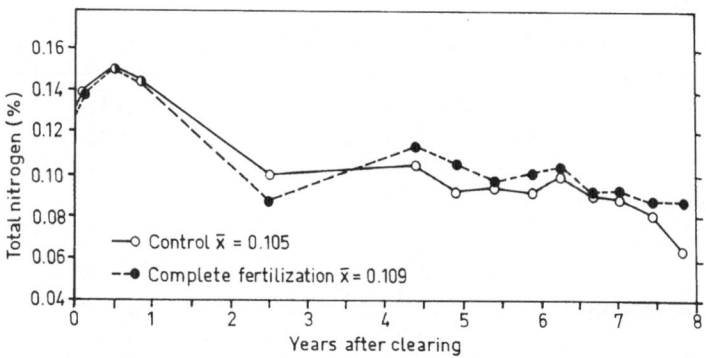

Fig. 5. Dynamics of total N in the first 15 cm of an Ultisol in Yurimaguas, Peru, subjected to continuous cultivation of rice, maize and soybeans for 8 years with and without fertilizing (North Carolina State University[8]).

This is also illustrated in Fig. 3, which shows a considerable increase in inorganic nitrogen in the first 50 cm of an Ultisol in Yurimaguas, Peru, after burning or bulldozer clearing. The total amount was equivalent to an increase of 80 kg N ha^{-1}, which was utilized efficiently by the first crop, *i.e.* rice[16]. Fig. 3 also shows that this effect does not persist beyond the first 6 months after burning: the soil returns to extremely low levels of inorganic nitrogen, and after 10 months the values are similar to what they were before the site was cleared. In Yurimaguas, a severe N deficiency as well as of other elements is reported, starting with the second crop after clearing and burning[11].

The land clearing system used also influences the rate of decomposition of inorganic matter when the crop cycle starts. Fig. 4 shows the results obtained by da Silva[17] in an Ultisol of Southern Bahia, Brazil. The rate of organic matter decomposition was higher when the forest was felled and burned in the traditional way than when it was cleared with a bulldozer.

Food crop production

When a forest is converted into short-cycle crops, soil temperature and the rate of organic matter decomposition increase rapidly, but subsequently reach a new equilibrium. Fig. 5 illustrates the dynamics of N in the first 15 cm of an Ultisol in Yurimaguas during 8 consecutive years after clearing, burning and growing 20 consecutive crops of a rotation of upland rice, maize and soybeans, without fertilization or with an agronomically sound fertilizer treatment (North Carolina State University 1982)[8]. After an initial increase during the first 6 months, there was a steep decline over the first 2 years, with an annual rate of decomposition of approximately 17%. Afterwards, a stable equilibrium can be observed. This figure does not show differences in the N content of the soil attributable to fertilization. It should be stressed that without fertilization, yields were zero from the second year onwards, whereas with fertilization average yields were 2.7 tons

Table 4. Nitrogen content in the ashes after burning moist tropical forest sites

Place and soil	Dry ashes (10^3 kg ha^{-1})	%N	Added N (kg N ha^{-1})
Yurimaguas, Peru (Ultisol, pH 4.0):			
Secondary forest (17 years)[1]	3.97	1.72	67
Secondary forest (25 years)[2]	12.10	1.05	127
Turrialba, Costa Rica (Andept, pH 5.1)[3]	6.70	1.43	96

[1] Seubert *et al.*[16]
[2] North Carolina State University[8]
[3] Ewel *et al.*[3]

Table 5. Balance between added nitrogen and nitrogen removed by 19 consecutive crops in a rice-maize-soybean rotation over a period of 8 years in an Ultisol in Yurimaguas, Peru (North Carolina State University[8])

Component	kg N ha^{-1}
Additions in fertilizers	1480
Removal by crops	1916
Balance	−436

ha^{-1} of dry rice, 2.8 ton ha^{-1} of maize, and 2.3 ton ha^{-1} of soybean, *i.e.* a total of 7.8 tons grain ha^{-1}.yr^{-1} over the last 8 years. This study, which has been described in more detail by Sánchez *et al.*[14] demonstrates that it is agronomically and economically feasible to cultivate these acid Ultisols under a continuous system if the chemical and physical properties of the soil are properly managed, if crop rotations with varieties resistant to the main pests and diseases are used, and if adequate outlets exist for marketing the products.

Table 5 shows the balance between the amount of N added as urea to the 19 crops over the 8-year period and the amounts removed through harvesting. The balance is negative, which probably reflects symbiotic N fixation by soybeans.

After the first harvest the response to N fertilization is strong in crops such as rice and maize. Benites[1] reports increases of 2 to 5 tons ha^{-1} for maize yields when 80 kg N ha^{-1} were applied as urea. The efficiency of utilization of the applied N however, depends on the cropping system and the time of sowing. In Yurimaguas, there are three seasons: a heavy rainy season (January–April), the relatively dry season (May–September) and a moderate rainfall season

Table 6. Efficiency of utilization of the N applied in the form of urea to non-leguminous species in some short-cycle cropping systems in Yurimaguas, Peru (Benites[1]) LSD = least significant difference ($p < 0.05$)

Cropping system	% Apparent recovery of applied N
Maize/rice-cassava/groundnut, cover-crop system	64
Rice-groundnut-maize, in sequence	61
Maize-groundnut-rice, in sequence	35
Maize-maize-maize, in sequence	45
Cassava monoculture	48
LSD	6

(September–December). Table 6 shows the differences in efficiency of utilization of N applied to non-leguminous species for different cropping systems and planting seasons. Recovery of applied N was calculated as the difference in N accumulation by plants fertilized with 80 kg N ha^{-1} minus N uptake by plants receiving 0 kg N ha^{-1}, divided by the rate of N applied and expressed as percentage. This table demonstrates a broad range of nitrogen recovery. The maize/rice-cassava/groundnut intercropped system and the rice-groundnut-maize sequence were the most efficient cropping patterns. Nitrogen efficiency dropped to 35% when maize was planted during the season of heaviest rainfall, which is not recommended.

Except for this latter case, the N utilization efficiencies are similar to those generally obtained in temperate zones, *i.e.* approximately 50% for maize. Thus, the N situation is satisfactory when shifting cultivation is replaced with short-cycle continuous cropping, provided that management practices are adequate and the crop varieties used are suited to the environment.

Pastures

Although little is known about the behaviour of N in pasture-based livestock production systems in the humid tropics, some relevant preliminary data are

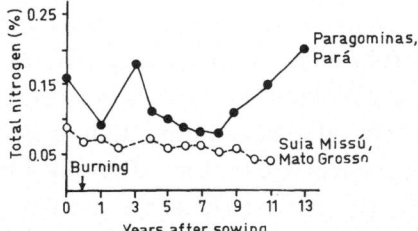

Fig. 6.   Changes in total N (0–15 cm) in two Oxisols (Paragominas, clayey soil and Suia Missú, loamy soil) in the Brazilian Amazon sown to *Panicum maximum* pastures without fertilization. All samples were taken at the same time. (Adapted from Falesi[4] and from Serrão *et al.*[15]).

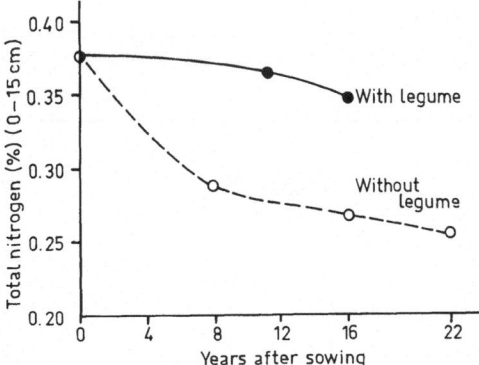

Fig. 7.   Effect of the presence or absence of the legume *Centrosema pubescens* on the N content of the soil in *Panicum maximum* pastures without fertilization but with proper management, as a function of the length of time since clearing, in an Alfisol at South Johnstone, Australia. (Adapted from Bruce[2]).

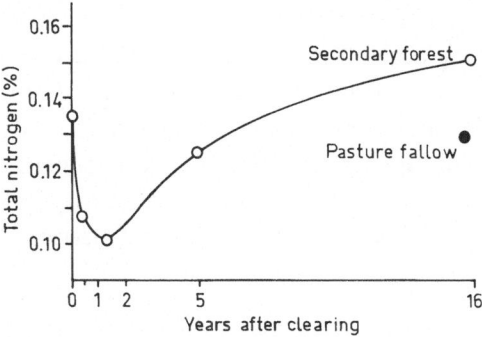

Fig. 8.   N-content of the 0–10 cm soil layer before clearing a virgin forest site, during cultivation, and after its abandonment to a secondary forest or pasture fallow. Samples were taken at the same time from different sites on Aeric Ochraquox soil, in Carare-Opón, Colombia, which has a mean annual rainfall of 3000 mm. (Adapted from de las Salas and Folster[10] and de las Salas[9]).

available. Fig. 6 illustrates the absence of clear trends in the N content of the arable layer of two Oxisols in the Brazilian Amazon, where Falesi[4] sampled soils under *Panicum maximum* pastures of different ages after clearing, at the same time. Although these pastures are severely degraded from the third or fifth year onwards, have not received any fertilization and have not been sown with legumes, it is interesting to observe that there has not been a discernible decline in the reserves of N in the topsoil over time, except for the initial drop caused by the change from forest to pasture cover.

Fig. 7 indicates what can happen with properly managed grass/legume pastures. The data is from the humid tropics of Australia and indicate that the initial N contents in the soil can be maintained if a correct grass/legume association is used. Intensive research on this subject by CIAT and others in the humid tropics of Latin America hopefully would produce similar results.

Secondary fallows

Very little is known about the dynamics of N after the land is abandoned and during the development of secondary forest. Fig. 8 indicates a total recovery of N in the arable layer of a 16-year-old secondary forest and a pasture fallow in Carare-Opón, Colombia. Far more data of this type are needed before it will be possible to quantify this process of N regeneration and its influencing factors more precisely.

Discussion

Two major limitations have to be taken into account when interpreting data on N dynamics in shifting cultivation systems and its alternatives in Latin America. First, the data originating from different localities are based on different sampling methods and on studies of varying length. Second, it is difficult to consider N dynamics without taking account of the influence of other soil problems that exert just as much or more influence on plant growth, such as soil acidity and deficiency of P, K, S, Ca, Mg and micronutrients. What is needed is a network of well coordinated studies throughout the humid tropics, tracing in detail the dynamics of N and other elements before and after land clearing and with various systems as alternatives to shifting cultivation in Latin America.

References

1   Benites J R 1981 Nitrogen response and cultural practices for corn-based cropping systems in the Peruvian Amazon. PhD Thesis. Raleigh: North Carolina State University. 148 p.
2   Bruce R C 1965 Effect of *Centrosema pubescens* on soil fertility in the humid tropics. Queensl. J. Agric. Anim. Sci. 22, 221–226.
3   Ewel J, Berish C, Brown B, Price N and Raich J 1981 Slash and burn impacts on a Costa Rican wet forest site. Ecology 62, 816–829.
4   Falesi I C 1976 Ecossistema de Pastagem Cultivada na Amazônia Brasilera (Cultivated pasture ecosystem in the Brazilian Amazon). Belém, Brazil: EMBRAPA, Centro de Pesquisa Agropecuária do Trópico Umido, CPATU, Bol. Tec. 1.
5   Fassbender H W 1977 Ciclos de elementos nutritivos en ecosistemas forestales tropicales y su transformación con la agricultura rotativa (Nutrient cycles in tropical forest ecosystems and their transformation with shifting agriculture). *In* FAO-SIDA Workshop-Meeting on Soil Management and Conservation in Latin America, Lima, Peru. Rome: FAO.
6   Fittkau E J and Klinge H 1973 On biomass and trophic structure of the Central Amazonian forest ecosystems. Biotropica 5, 2–14.
7   Greenland D J and Kowal J M L 1960 Nutrient content of a moist tropical forest in Ghana. Plant and Soil 12, 154–174.
8   North Carolina State University. 1974, 1980, 1982. Research on Tropical Soils Annual Reports. Raleigh: Soil Science Department, North Carolina State University.
9   Salas G de las 1978 El sistema forestal Carare-Opón (The Carare-Opón forest system). CONIF. Ser. Tec. 8, Bogotá, Colombia: Corporación Nacional de Investigación y Fomento Forestal.
10   Salas G de las and Folster H 1976 Bio-element loss on clearing a tropical rainforest. Turrialba 26, 179–186.
11   Sánchez P A 1979 Soil fertility and conservation considerations for agroforestry systems in the humid tropics of Latin America. *In* Soils Research in Agroforestry. Eds., H O Mongi and P A Huxley. pp 79–124. ICRAF Bulletin 001e. Nairobi, Kenya: Int. Council Research Agroforestry.
12   Sánchez P A and Cochrane T T 1980 Soil constraints in relation to major farming systems in tropical America. *In* Priorities for Alleviating Soil-related Constraints to Food Production in the Tropics, pp 107–139. Los Baños, Philippines: IRRI.
13   Sánchez P A, Gichuru M P and Katz L B 1982 Organic matter in major soils of tropical and temperate regions. XII International Soil Science Congress (New Delhi). 1, 99–114.
14   Sánchez P A, Bandy D E, Villachica J H and Nicholaides J J 1982 Amazon soils: management for continuous crop production. Science. 216, 821–827.
15   Serrão E A S, Falesi I C, Viega J B and Texeira J F 1979 Productivity of cultivated pastures in low fertility soils of the Amazon of Brazil. *In* Pasture Production in Acid Soils of the Tropics. Eds. P A Sanchez and L E Tergas. pp 195–226. Cali, Colombia: Centro Internacional de Agricultura Tropical.

16 Seubert C E, Sánchez P A and Valverde C 1977 Effects of land clearing methods on soil properties and crop performance in an Ultisol of the Amazon Jungle of Peru. Trop. Agric. Trinidad 54, 307–321.

17 Silva L F da 1979 Influencia do manejo de una ecossistema nas propiedades edáficas dos Oxissolos de 'Tabuleiro' (Influence of the management of an ecosystem on the edaphic properties of 'Tabuleiro' Oxisols). Itabuna, Bahia, Brazil: Centro de Pesquisas do Cacau, CEPLAC.

18 Watters R F 1971 Shifting Cultivation in Latin America. FAO Forestry Dev. Paper 16. Rome: FAO.

Plant and Soil 67, 105–117 (1982).   0032-079X/82/0671-0105$01.95.   SU-09
© 1982 *Martinus Nijhoff/Dr W. Junk Publishers, The Hague.*

Nitrogen distribution in several traditional agro-ecosystems in the humid tropical lowlands of south-eastern Mexico

Distribución del nitrógeno en varios agroecosistemas tradicionales en los trópicos humedós bajos en el sureste Mexicano

S. R. GLIESSMAN*

Depto. de Ecología, Colegio Superior de Agricultura Tropical, Cárdenas, Tabasco, México

Key words Beans Corn Crop residues Manioc Mexico N distribution Rice Taro Traditional agriculture.

Abstract Nitrogen distribution was examined in five local agro-ecosystems typical of the lowlands of tropical south-eastern Mexico: monoculture corn, corn/bean polyculture, manioc (yuca, cassava), taro (malanga), and upland rice. Total biomass and nitrogen content were determined monthly for standing live, standing dead, and litter biomass of both crop and non-crop components of each system. The crop component was further divided into roots, crown, stem, leaves, fruits, and flowers. Soil nitrogen determinations were also made monthly. Results demonstrated that nitrogen maintenance in the system is highly dependent on the proportion of the net biomass produced which is returned to the system. Leguminous and weed components may reduce net nitrogen losses from these systems.

Resumen Se examinó la distribución de nitrógeno en cinco agroecosistemas locales típicos de la tierra baja del sureste tropical de México: monocultivo de maiz, policultivo maiz/frijol, yuca, malanga y arroz de temporal.

Se determinó mensualmente la biomasa y el contenido de nitrógeno para la biomasa viva, muerta, y hojarasca de los componentes cultivo y no-cultivo en cada sistema. Se dividió el componente cultivo entre raíces, corona, tallo, hojas, frutos y flores para análisis de nitrógeno.

Mensualmente se determinó el nivel de nitrógeno en el suelo también. Los resultados demostraron que el mantenimiento de nitrógeno en cada sistema es altamente dependiente de la proporción de la biomasa neta producida que se regresa al sistema. Componentes leguminosa y maleza podrían reducir pérdidas netas de nitrógeno en estos agroecosistemas.

Introduction

In the lowland tropical region of south-eastern Mexico there is a great diversity of intensive local cropping systems that provide the greater part of the basic food products needed by the rural inhabitants (campesinos) of the area[5,11]. These agro-ecosystems have been developed in and adapted to ecological conditions characteristic of a humid tropical habitat, and have also been designed to satisfy the prevalent cultural needs of the local populations. This development is thought to have been taking place since early Mayan times[10], with the incorporation of considerable technology since the Spanish Conquest in the 16th Century.

* Present address: Environmental Studies Program, University of California, Santa Cruz, California 95064, USA

A very important aspect of the traditional agricultural practices that is still observed today is a strong foundation in ecological theory with a focus on maintaining productivity on a long-term basis. This contrasts with what has become a modern emphasis on maximizing yields on a short-term basis[6]. The structure and functioning of traditional agro-ecosystems is similar to many natural ecosystems, with farmers' capitalizing on local inputs of nutrients and greater nutrient conservation within the system rather than their depending on costly inputs from outside the system[7]. Detailed study of these systems can help provide information for the preservation of the traditional production base of such regions as well as for the improvement of modern agricultural practices[9].

In the present study the distribution of nitrogen is examined in several local agro-ecosystems in order to provide criteria for comparisons with aspects of nitrogen cycling in natural systems. In addition, this study helps to demonstrate the need for an ecological focus for understanding nutrient needs in agricultural systems.

Methods

Study area

This study took place on land belonging to the communal collective of Lázaro Cárdenas, Tacotalpa, Tabasco, Mexico, during 1978. This community of approximately 100 families occupies 1000 ha at the base of the Chiapas Highlands at the southern extreme of Tabasco (17°36' N, 92°49' W), at an elevation of about 80 m. The region is one of the wettest in Mexico, as annual rainfall is above 4000 mm; annual mean temperature is above 25°C. The average rainfall in the dry season months of March–May is above 100 mm, so that the climate classification is Af[4]. Soils are red to brown clay or clay loam, derived from alluvial transport of soil from the uplifted limestone of the nearby mountains. They are moderately leached and acidic (pH 5.0–5.5).

The study site in the community was on 10 ha devoted to a great diversity of agricultural activities; specific land-use is based upon local variations in soil, topography, and land-use history. Activities vary in space and time, and variations include annual and perennial cropping systems, small monocultures and complex polycultures, and upland and lowland crop systems. All experimental agro-ecosystems are managed using local practices and plant varieties, without applications of chemical pesticides or fertilizers. Work groups, rotated by the community periodically, are responsible for crop management. Production is destined for local consumption, although at times excess is sold in surrounding towns. The rest of the land in the community is devoted to market production of cattle and bananas as part of an agricultural reform program in progress in the State.

Agro-ecosystems

Five cropping systems were chosen for study:

1) Corn/bean polyculture On approximately 1.0 ha previously covered by young secondary growth (1.5 yr since last cultivated) on flat terrain with well-drained soil, this site was planted in early December 1977 at a density of 10,000 holes ha^{-1} with 5 seeds of corn (*Zea mays* var. local Sierra) per hole. The corn was interplanted with beans (*Phaseolus vulgaris* var. local black bush) at 60,000 holes ha^{-1} with 3–4 seeds per hole. Site preparation was performed with machetes, and the unburned slash left on the soil surface. The crop was weeded with machetes at 30 and 60 days after planting. Beans were harvested a month before corn. The area was prepared for re-planting in July, but was not planted and instead allowed to become fallow.

2) Corn monoculture 1.5 ha previously covered by 3 year-old secondary growth on hilly terrain with well-drained soil was planted early in December 1977, with the same corn variety as above and at the same density. The site was prepared for planting by clearing vegetation with machetes; unburned slash was left on the soil surface. After harvest the area was left fallow until replanting in October with a legume cover crop (*Canavalia ensiformis*) at a density of 10,000 plants ha^{-1}.

3) Upland rice 2.0 ha previously covered by 8 year-old secondary growth on low-lying, water-logged terrain with heavy clay soil was cleared in April with machetes, burned in early May, and planted shortly thereafter with a mixture of two local varieties of rice (*Oryza sativa*). 60,000 holes ha^{-1} were direct-seeded with 10–12 seeds per hole. The crop was allowed to resprout along with weeds following harvest of the grain in September.

4) Malanga (taro) This site was approximately 0.5 ha of a periodically flooded, continuously damp area along a stream course. It was previously occupied by tall grasses and herbaceous vegetation, and had heavy clay soils. The site was planted in December, 1977, with an introduced variety of taro, locally called malanga (*Colocasia esculenta*) at a density of 10,000 plants ha^{-1}. The planting was weeded periodically with machetes and harvesting began in August. Young shoots were replanted following harvest to maintain a permanent crop cover.

5) Yuca (manioc/cassava) Approximately 1.0 ha of raised ground occupied previously by 1.5 year-old secondary growth on hilly terrain with welldrained soil was planted in mid-December of 1977 with stem cuttings of a local variety of manioc or yuca (*Manihot esculenta* var. local Ceiba). The planting density was 10,000 plants ha^{-1}. The site was prepared with machetes, the slash left unburned, and the crop periodically weeded with machetes for only the first 6 months after planting, *i.e.*, until continuous soil cover was established. Harvest began in June, continued through October, and was followed by a legume cover crop (*Canavalia ensiformis*) at a density of 10,000 plants ha^{-1}.

Sampling

In each of the 5 agro-ecosystem sites soils were sampled monthly from January to December except in March and October. Soil cores were made at 6 randomly-chosen points in each system at depths of 0–15, 15–30, and 30–60 cm. The 6 samples were arbitrarily pooled by pairs to make 3 samples for each depth per site. Soils were then dried, ground, and duplicate determinations made for total-N[1].

Non-crop biomass (weeds, non-crop leaf-litter, and slash residue) was sampled in each system monthly except March. Sampling did not begin in the rice site until early May, after the secondary growth had been cleared. At each date all organic matter on the surface in 5 randomly placed 50 cm × 50 cm squares was retrieved, separated into weed or litter components, dried at 70°C for 48 hours, weighed, ground in a Wiley mill and then two subsamples of each component from each square were analyzed for total-N using micro-Kjeldahl techniques (Chapman and Pratt[3], pp 102–107).

Crop biomass, both above and below ground, was determined by monthly sampling the crops in each ecosystem. For annuals for which more than one seed was planted per hole, all plants in a hole were sampled. Ten holes were collected for beans, rice and the annual cover crop (Canavalia), and 5 for corn, malanga, and yuca. All were chosen at random. All plants were taken to the laboratory, separated into roots, crown (an arbitrary section connecting stems and roots), leaves and stems, standing dead matter, and fruits and flowers. Any crop litter retrieved in the non-crop biomass samples was combined with the crop component for analysis. After washing with filtered tap water, samples were treated as described above.

Results and discussion

Data for total-N in the soils of each agro-ecosystem are presented in Table 1. Except in the case of malanga, total-N showed a tendency to decrease during the first 6–7 months of sampling in the upper 0–15 cm of the soil. For the corn and malanga, total-N in the upper layer increased again during the last 5 months. For the corn/bean and yuca systems the rise did not take place until just towards the end of the year. For the other two depths, the tendency for total-N to gradually increase was observed in all the systems, although less so in the rice system. The rise did not take place in the corn/bean system until very late in the year, but in August in the corn and yuca systems and even earlier in the malanga. Increases approached 80% for the 15–30 cm depth and 90% for the 30–60 cm depth. Rice is the only system which showed a net drop at all depths.

Overall biomass distribution can be seen in Table 2. In most systems total biomass showed a net loss over the year, as might be expected after cutting the secondary growth. Biomass in the malanga system, however, stayed relatively

Table 1. Total Kjeldahl-nitrogen (% N) in soils of 5 traditional agro-ecosystems sampled monthly (exclusive of March and October) during 1978. Numbers are unweighted means of three paired cores per site with 2 analytical replicates per paired core

System	Depth (cm)	J	F	A[e]	M	J	J	A	S	N[e]	D
Corn/Bean	0–15	0.239[c]	0.177	0.210[d]	0.179	0.190	0.192	0.231	0.184	0.221	0.260
	15–30	0.093	0.074	0.081	0.127	0.102	0.088	0.131	0.130	0.130	0.154
	30–60	nd	0.052	0.050	0.049	0.054	0.050	0.060	0.117	0.065	0.091
Corn	0–15	0.238[c]	0.179	0.248[d]	0.230	0.225	0.214	0.275	0.279	0.283[e]	0.310
	15–30	0.136	0.111	0.110	0.164	0.171	0.136	0.229	0.207	0.194	0.220
	30–60	nd	0.080	0.068	0.096	0.098	0.090	0.159	0.143	0.122	0.148
Rice	0–15	nd	nd	0.237[a]	0.209[bc]	0.187	0.201	0.213	0.220[d]	0.235	0.217
	15–30	nd	nd	0.127	0.131	0.115	0.126	0.125	0.144	0.144	0.121
	30–60	nd	nd	0.072	0.079	0.081	0.080	0.079	0.095	0.082	0.093
Malanga	0–15	0.176[c]	0.184	0.206	0.220	0.215	0.200	0.233[d]	0.253[d]	0.273[d]	0.236[d]
	15–30	0.114	0.124	0.113	0.152	0.139	0.125	0.170	0.180	0.139	0.183
	30–60	nd	0.087	0.097	0.113	0.116	0.109	0.137	0.155	0.098	0.153
Yuca	0–15	0.212[c]	0.199	0.210	0.222	0.214[d]	0.214[d]	0.232[d]	0.201[d]	0.268[d]	0.236[d]
	15–30	0.113	0.101	0.110	0.149	0.133	0.120	0.156	0.143	0.187	0.154
	30–60	nd	0.064	0.061	0.060	0.073	0.069	0.087	0.091	0.125	0.119

[a] samples taken in May, after vegetation cut, but before burning.
[b] samples taken in May, following burning of slash.
[c] month in which site was cleared and planted.
[d] month in which crop harvested.
[e] sites not sampled in March or October.
nd = not determined.

constant; the most dramatic drop was observed in the rice system immediately following clearing and burning. For yuca the drop was to almost a third of the initial amount, and for the corn to half. The corn/bean system showed a downward tendency until the last sampling date.

The crop component of the biomass of each system closely reflected the growth, development and harvest patterns for each crop. Net increases for the two corn systems were between 6–8 tons ha^{-1}. Malanga showed the lowest total increase in the crop biomass component (3.69×10^3 kg dry matter ha^{-1}), while yuca showed the greatest (almost 17.0×10^3 kg dry matter ha^{-1}). Rice showed the most dramatic increase (14.0×10^3 kg dry matter ha^{-1} in 4 months).

Final harvest yields were 3.0×10^3 kg air-dried corn grain ha^{-1} and 0.5×10^3 kg air-dried beans ha^{-1} in the corn/bean polyculture; 2.6×10^3 kg air-dried corn grain ha^{-1} in the corn monoculture; 1.5×10^3 kg air-dried rice ha^{-1} in the

Table 2. Biomass distribution (10^3 kg dry matter ha^{-1}) in the crop and non-crop components of 5 traditional agro-ecosystems in Tabasco, Mexico, during monthly samples taken in 1978. Values are means of 5 holes per sample date except for beans and rice (10 holes). P = Biomass present and standing but not harvested until felled

System	J	F	M	A	M	J	J	A	S	O	N	D
Corn/Bean												
Crop	0.43	2.73	7.31	8.01	P	P	0.80	P	0.44	3.72	0.28	0.19
Non-crop	9.90	8.33	nd	6.25	8.99	11.8	3.17	6.72	5.90	4.77	6.73	8.03
Total	10.3	11.1	14.0^b	14.3	8.99	11.8	3.97	6.72	6.34	8.49	7.01	8.22
Corn												
Crop	0.48	2.53	4.94	6.15	P	P	P	P	2.03	2.22^a	0.44^a	4.15
Non-crop	18.5	14.2	nd	12.8	25.7	10.8	8.39	9.59	13.2	8.16	7.28	8.37
Total	19.0	16.7	17.0^b	19.0	25.7	10.8	8.39	9.59	15.2	10.4	7.72	12.5
Rice												
Crop	0	0	0	0	0	0.32	4.53	9.55	14.2	1.66	5.08	2.25
Non-crop	nd	nd	nd	16.3^c	6.61^d	1.62	0.58	2.24	nd	3.42	2.07	1.82
Total	nd	nd	nd	16.3	6.61	1.94	5.11	11.8	14.2	5.08	7.15	4.07
Malanga												
Crop	0.88	0.53	1.04	0.90	2.21	1.07	0.91	3.17	2.60	3.49	3.24	3.69
Non-crop	2.04	1.15	nd	2.36	4.48	4.58	1.75	2.92	2.88	4.72	2.32	1.74
Total	2.93	1.68	2.50^b	3.26	6.69	5.65	2.66	6.09	5.48	8.21	5.56	5.43
Yuca												
Crop	1.02	3.44	4.82	8.53	6.22	13.46	nd	nd	nd	17.0	3.56	4.53
Non-crop	10.5	9.05	nd	8.93	7.36	5.60	2.62	2.81	3.15	1.58	2.26	3.35
Total	11.5	12.5	12.0^b	17.5	13.6	19.1	16.5^b	17.5^b	18.0^b	18.6	5.82	7.88

[a] Canavalia present but not sampled.
[b] approximation.
[c] slash sample in early May before burning.
[d] sample in mid-May following fire.
nd = not determined.

Table 3. Nitrogen distribution (kg N ha^{-1} except as noted) in a corn (*Zea mays*)/bean (*Phaseolus vulgaris*) agro-ecosystem in Tabasco, Mexico, during the 1978 growing season. Values are unweighted means of 5 (corn) or 10 (bean) planting holes per sample date with 2 analytical replicates per hole. % Exported = 100 × (Exported N/Total System N), and % Non-crop = 100 × (Total Non-crop N/Total System N)

System	J	F	M	A	M	J	J	A	S	O	N	D
Crop												
Corn												
Standing dead	0	0.61	5.90	16.2	nd	nd	0	0	0	0	0	0
Root	0	0.79	2.00	2.68	0	0	0	0	0	0	0	0
Crown	0	1.34	1.50	1.98	0	0	0	0	0	0	0	0
Leaves	1.47	26.4	42.0	0	0	0	0	0	0	0	0	0
Fruit	0	0	14.5	48.6	0	0	0	0	0	0	0	0
Total	1.47	29.1	65.8	69.6	nd	nd	0	0	0	0	0	0
Bean												
Standing dead	0.54	0.95	3.37	0	0	0	0	0	0	0	0	0
Roots	0.79	1.87	1.30	0	0	0	0	0	0	0	0	0
Leaves	5.68	6.87	0	0	0	0	0	0	0	0	0	0
Stems	1.38	4.12	0	0	0	0	0	0	0	0	0	0
Fruit	0	10.2	20.9	0	0	0	0	0	0	0	0	0
Total	8.39	24.0	25.5	0	0	0	0	0	0	0	0	0
Crop litter	0	0	0	0	nd	nd	6.0	nd	2.68	37.2	2.49	1.67
Crop total	9.86	53.1	91.3	69.6	nd	nd	6.0	nd	2.68	37.2	2.49	1.67
Non-crop												
Litter	133.	114.	96.7	73.0	121.	135.	46.8	77.7	83.4	51.3	86.6	77.5
Weeds	5.16	1.86	5.60	14.6	5.78	14.1	0.0	5.60	6.51	10.1	8.11	30.2
Total	139.	116.	102.	87.2	127.	149.	45.8	83.3	90.0	61.4	94.7	107.7
System total	148.	169.	194.	157.	127.[a]	149.[a]	52.8	83.3[a]	92.6	98.6	97.2	109.
Exported												
(kg N ha^{-1})	0	0	25.5	48.6	0	0	0	0	0	0	0	0
(Per cent)	0	0	13.2	31.0	0	0	0	0	0	0	0	0
Non-Crop (%)	93.4	68.5	52.8	55.6	100	100	88.4	100	97.1	62.2	97.4	98.5

upland rice monocultures; 3×10^3 kg edible fresh malanga crowns ha^{-1} month^{-1}; and 15×10^3 kg fresh yuca tubers ha^{-1}. Air-dried grain yields contained *ca.* 14% water, and fresh malanga and yuca yields *ca.* 80%.

Nitrogen distributions in the biomass component are presented in Table 3–7. Total-system biomass-N generally increased rapidly as the crops developed, dropped dramatically with the crop harvest, and then again gradually increased.

The nitrogen increase in the crop components of the corn/bean system with the corresponding drop in the non-crop component can be seen in Table 3. Immediately following harvest, accumulated litter kept total system biomass-N at high levels, but once the rains began in June, N-levels fell drastically. Then, as the weed component became established during the fallow period (May–December), and dead material left standing from the previous crop was felled and became part of the litter on the soil surface, biomass-N levels gradually increased again. At the end of the year, the total-system biomass-N was approximately two-thirds of the initial level.

In the corn monoculture (Table 4), the same initial biomass-N increase in the crop component took place. Again, it was not until rains began in June that N-levels in the non-crop component fell, and these reached a low in July. Weeds played an important part in later stages. Although less than a quarter of the nitrogen was in the weed segment at any sampling date, weeds were nevertheless continually growing and adding biomass to the litter. Further, values in the Table represent nitrogen in green-weed biomass only. Once Canavalia entered the system (October), N-levels reached those observed at the outset.

Rice presents an interesting contrast to the two corn systems (Table 5). The large input of nitrogen with the felling of the secondary growth would be expected, as would the great loss following fire in early May. Some nitrogen was recovered by the developing crop, but after harvest, nitrogen in biomass fell off considerably once again. Weeds played an important part in the system, especially grasses tolerant of the water-logged soil. Their rapid growth added biomass continually to the litter component. The final level (35.2 kg ha^{-1}) is the lowest total biomass nitrogen for all of the systems.

The malanga system displayed the most uniform biomass-N distribution of all the systems (Table 6). Nitrogen was constantly contributed in small amounts to the total-system biomass-N pool by weeds cut every 45 days. Once harvest began in August, total biomass-N levels remained constant despite continued crop-N removal. A dry season (April–June) drop in biomass-N was observed, followed by a recovery in August.

The N-levels in the yuca system proved to be interesting (Table 7) for a crop with relatively low levels of nitrogen in the harvestable portion (roots). The immediate N-input with site preparation was expected, and once the crop became established non-crop nitrogen became less important. Most biomass-N was in the crop component by October; N-levels fell drastically after harvest, but increased once again with the introduction of the Canavalia.

Table 4. Nitrogen distribution (kg N ha⁻¹ except as noted in Table 3 legend) in a corn (*Zea mays*) agro-ecosystem in Tabasco, Mexico, during the 1978 growing season. Values are unweighted means of 5 planting holes per sample date with 2 analytical replicates per hole. See Table 3 legend for further explanation

System	J	F	M	A	M	J	J	A	S	O	N	D
Crop												
Corn												
Standing dead	0.58	1.08	3.15	11.8	nd	nd	nd	nd	0	0	0	0.53
Roots	0.88	0.81	1.23	1.98	0	0	0	0	0	0	0	0.33
Crown	0	2.07	1.48	2.40	0	0	0	0	0	0	0	0.58
Stems + Leaves	7.71	34.9	30.6	0	0	0	0	0	0	0	0	6.82
Fruit	0	0	18.5	49.2	0	0	0	0	0	0	0	0
Total	9.17	38.9	54.9	65.4	nd	nd	nd	nd	0	0	0	7.82
Canavalia												
Standing dead	0	0	0	0	0	0	0	0	0	0	0	15.2
Roots	0	0	0	0	0	0	0	0	0	0	0	1.54
Crown	0	0	0	0	0	0	0	0	0	0	0	0.93
Stems + Leaves	0	0	0	0	0	0	0	0	0	0	0	46.4
Fruits + Flowers	0	0	0	0	0	0	0	0	0	0	0	17.0
Total	0	0	0	0	0	0	0	0	0	0	0	81.1
Crop litter	0	0	0	0	0	0	1.00	nd	21.7	q7.6	3.78	8.06
Crop total	9.17	38.9	54.9	65.4	nd	nd	1.00	nd	21.7	17.6	3.78	97.0
Non-crop												
Litter	166.	175.	nd	135.	227.	103.0	74.7	82.2	117.	83.0	90.0	93.0
Weed	13.2	4.81	nd	24.0	10.1	23.4	8.26	21.8	25.2	12.7	14.3	2.28
Total	179.	180.	nd	159.	237.	126.	82.9	104.	142.	95.3	104.	95.3
System total	188.	209.	54.9ᵃ	225.	237.ᵃ	126.ᵃ	83.9ᵃ	104.ᵃ	163.	113.	108.	192.
Exported												
(kg N ha⁻¹)	0	0	0	49.2	0	0	0	0	0	0	0	0
(Per cent)	0	0	0	21.9	0	0	0	0	0	0	0	0
Non-Crop (%)	95.1	86.2	nd	70.9	100	100	98.9	100	86.7	84.4	96.5	49.6

ᵃ approximation.

Table 5. Nitrogen distribution (kg N ha^{-1} except as noted in Table 3 legend) in an upland rice (*Oryza sativa*) agro-ecosystem in Tabasco, Mexico, during the 1978 growing season. Values are unweighted means of 10 planting holes with 2 analytical replicates per hole. See Table 3 legend for further explanation

System	J	F	M	A	M	J	J	A	S	O	N	D
Crop												
Rice												
Standing dead	0	0	0	0	0	0	0.49	0	8.86	0	0	0
Roots	0	0	0	0	0	0	5.14	3.71	4.68	0	0	0
Crown	0	0	0	0	0	0	6.20	5.60	9.12	0	0	0
Leaves	0	0	0	0	0	4.33	38.1	61.2	57.9	0	0	0
Fruits + Flowers	0	0	0	0	0	0	0	10.6	16.8	0	0	0
Total	0	0	0	0	0	4.33	49.9	81.1	97.4	0	0	17.8
Crop litter	0	0	0	0	0	0	0	0	0	14.3	45.8	17.8
Crop total	0	0	0	0	0	4.33	49.9	81.1	97.4	14.3	45.8	17.8
Non-crop												
Litter	nd	nd	nd	180.7[a]	72.3[b]	18.1	1.79	13.1	17.0	20.4	10.3	15.3
Weeds	nd	nd	nd	12.1	0	1.46	6.53	14.4	15.2	14.9	11.7	2.16
Total	nd	nd	nd	193.	72.3	19.6	8.32	27.5	32.2	35.3	22.0	17.4
System total	nd	nd	nd	193.	72.3	23.9	58.2	109.	130.	49.6	67.7	35.2
Exported												
(kg N ha^{-1})	0	0	nd	0	0	0	0	0	16.8	0	0	0
(Per cent)									12.9			
Non-Crop (%)	nd	nd	nd	100	100	81.9	14.3	25.3	24.8	71.2	32.4	49.5

[a] slash sampled in early May before burning.
[b] sampled in mid-May following fire.

Table 6. Nitrogen distribution (kg N ha^{-1} except as noted in Table 3 legend) in a malanga (*Colocasia esculenta*) agro-ecosystem in Tabasco, Mexico, during the 1978 growing season. Values are unweighted means of 5 plants with 2 analytical replicates per plant. See Table 3 legend for further explanation

System	J	F	M	A	M	J	J	A	S	O	N	D
Crop												
Malanga												
Standing dead	0.56	0.25	0.85	0.56	1.13	0.60	0	1.75	1.24	1.05	3.42	1.10
Roots	0.54	0.67	0.46	0.33	0.34	0	0.10	0.53	0.48	1.06	1.84	1.02
Crown	4.01	3.19	4.74	4.12	9.44	6.69	4.74	13.2	10.9	20.1	19.7	31.5
Leaves	8.52	3.06	5.62	3.64	4.59	0.60	1.16	8.96	11.8	9.96	12.9	13.8
Total	13.6	7.17	11.7	8.65	15.5	7.89	6.00	24.4	24.4	32.2	37.9	47.4
Crop litter	0	0	0	0	0	0	0	0	0	5.93	2.18	1.74
Crop total	13.6	7.17	11.7	8.65	15.5	7.89	6.00	24.4	24.4	38.1	40.1	49.2
Non-crop												
Litter	12.5	5.51	nd	17.4	34.4	54.0	10.8	28.2	27.7	46.2	25.9	11.5
Weeds	22.5	19.6	nd	13.4	32.2	10.6	13.8	12.7	16.9	11.8	11.8	11.2
Total	35.0	25.1	nd	30.9	66.6	64.6	24.6	40.9	44.6	58.0	37.5	22.7
System total	48.6	32.2	nd	39.5	82.1	72.5	30.6	65.3	69.0	96.1	77.6	71.9
Exported												
(kg N ha^{-1})	0	0	0	0	0	0	0	13.2	10.9	20.1	19.7	31.5
(Per cent)	0	0	0	0	0	0	0	20.2	16.2	20.9	25.4	43.8
Non-Crop (%)	72.0	77.8	nd	78.1	81.1	89.9	80.0	62.6	64.6	60.3	48.4	31.6

Table 7. Nitrogen distribution (kg N ha^{-1} except as noted in Table 3 legend) in a yuca (*Manihot esculenta*) agro-ecosystem in Tabasco, Mexico, during the 1978 growing season. Values are unweighted means of 5 plants with 2 analytical replicates per plant. See Table 3 legend for further explanation

System	J	F	M	A	M	J	J	A	S	O	N	D
Crop												
Yuca												
Standing dead	0	0.10	0.45	1.96	1.30	2.89	nd	nd	nd	0	0	0
Roots	1.45	15.9	10.1	13.9	14.8	60.7	nd	nd	nd	26.3	0	0
Crown	2.17	5.26	4.34	9.02	9.45	12.4	nd	nd	nd	15.2	0	0
Stems + Leaves	17.0	58.1	67.0	89.4	29.7	95.2	nd	nd	nd	104.	14.5	14.6
Yuca litter	0	0	0	0	0	0	3.28	nd	24.2	17.5	14.5[a]	14.6[a]
Total[b]	20.7[a]	79.4[a]	81.9[a]	114.[a]	55.2[a]	171.[a]	nd	nd	nd	163.[a]	14.5[a]	14.6[a]
Canavalia												
Standing dead	0	0	0	0	0	0	0	0	0	0	2.63	2.22
Roots	0	0	0	0	0	0	0	0	0	0	1.17	0.56
Crown	0	0	0	0	0	0	0	0	0	0	0.51	0.50
Stem/Leaves	0	0	0	0	0	0	0	0	0	0	42.7	31.9
Fruits + Flowers	0	0	0	0	0	0	0	0	0	0	6.90	33.4
Total	0	0	0	0	0	0	0	0	0	0	54.0	68.6
Crop total	20.7	79.4	81.9	114.	55.2	171.	nd	nd	nd	163.	68.4	90.6
Non-crop												
Litter	128.	96.4	nd	94.2	74.4	50.8	18.6	28.1	23.9	10.3	7.01	47.5
Weeds	2.56	8.87	nd	9.80	9.90	27.0	12.3	11.1	11.1	7.68	30.4	3.91
Total	131.	105.	nd	104.	84.3	77.8	30.9	39.2	35.0	18.0	37.4	51.4
System total	151.	185.	nd	218.	140.	249.	34.2[a]	39.2[a]	59.2[a]	181.	106.	142.
Exported												
(kg N ha^{-1})	0	0	0	0	0	60.7	0	0	0	26.3	0	0
(Per cent)	0	0	0	0	0	24.4	0	0	0	14.6	0	0
Non-Crop (%)	86.3	57.0	nd	47.6	60.4	31.2	90.4	100	59.1	10.0	35.5	36.2

[a] approximation.
[b] includes crop (Yuca) litter, unlike other tables.

An important characteristic of all the agro-ecosystems is that a relatively low percentage of the total-system biomass-N was removed with the crop harvest (Tables 3–7). The corn/bean polyculture removed a greater percentage of total biomass-N (31%) than the corn monoculture (22%), but in the polyculture a lower total amount of nitrogen was in the non-crop biomass. The rice system's harvest removed the lowest percentage (12.9% of total biomass-N). The two corn systems had the greatest proportion of nitrogen in the non-crop components throughout the year, while the yuca and rice systems had the lowest.

In total, these results indicate the dynamic nature of nitrogen in these cropping systems. That total-N in the soil actually increased in four of the five systems studied merits detailed study of the mechanisms involved. A great part of the N-increase may come from its mobilization from the original non-crop components, especially secondary-growth slash. The rapid transfer of this nitrogen into the crop component is one way that it stays in the system. Uptake by weeds is another. The probable volatilization of nitrogen upon burning in the rice system further illustrates this: the rice system at the end of the season had less soil-N and biomass-N. The transfer of nitrogen from the crop to the non-crop component, then back again, becomes a potentially important mechanism. The use of weeds that have this capability but that do not interfere with the crop may have been selected for in traditional agro-ecosystems [2].

Cover-crop legumes such as *Canavalia ensiformis* play a very important part in supplying nitrogen to the crop once the transfer from litter has been depleted [8]. These plants are very important for replacing nitrogen removed with the crop harvest. It therefore becomes very important to thoroughly understand what percentage of nitrogen can be removed from tropical agro-ecosystems in relation to that which is retained or recycled. Comparisons of N-cycle properties need to be made with both natural ecosystems and modern mechanized cropping systems. Understanding the mechanisms of N-transfers and inputs in traditional agro-ecosystems offers great potential for the design and management of more efficient and productive systems in the future.

Acknowledgements   The major part of the funding for this study was received from the Programa de Inversiones para el Desarrollo Rural (PIDER-Tabasco), for which support is most gratefully acknowledged. Special thanks for field work are due to Fauso Inzunza and Radamez Bermudez, and for laboratory work to Manuel Ramos, Rubisel Maza, and Zenaida Castro. To Angel Ramos Sanchez I express my appreciation for his support as the former director of the Colegio Superior de Agricultura Tropical. Eric Gliessman helped considerably with data treatment.

References

1   Bremner J M 1965 Total nitrogen. *In* Methods of Soil Analysis. Black C A (Ed.). pp 1149–1172. Amer. Soc. Agron. Madison, Wisc.
2   Chacon J C and Gliessman S R 1982 Use of the 'non-weed' concept in traditional tropical agro-ecosystems in southeastern Mexico. Agro-ecosystems (in press).

3   Chapman H D and Pratt P F 1976 Methods of Analysis for Soils, Plants, and Water. Editorial Trillas, Mexico City. (In Spanish).

4   Garcia E 1973 Modifications of the Climatic Classification System of Koeppen. Mexico National University Press (UNAM), Mexico City. 246 p. (In Spanish).

5   Gliessman S R (Ed.) 1978 Memoirs. Seminar on Agro-ecosystems with an Emphasis on the Study of Traditional Agricultural Technology. Colegio Superior de Agricultura Tropical (CSAT). Cardenas, Tabasco, Mexico. 216 pp. (In Spanish).

6   Gliessman S R 1980 Some ecological aspects of traditional agricultural practices in Tabasco, Mexico: applications for production. Biotica 5, 93–101. (In Spanish, English summary).

7   Gliessman S R and Amador A M 1980 Ecological aspects of production in traditional agro-ecosystems in the humid lowland tropics of Mexico. *In* Tropical Ecology and Development. Furtado J I (Ed.). The International Society of Tropical Ecology, Kuala Lumpur, Malaysia.

8   Gliessmman S R and Garcia E R 1979 The use of some tropical legumes in accelerating the recovery of productivity of soils in the lowland humid tropics of Mexico. *In* Tropical Legumes: Resources for the Future. NAS Publ. No. 27, pp 292–293. U.S. National Academy of Sciences, Washington, D.C.

9   Gliessman S R, Garcia E R and Amador A M 1981 The ecological basis for the application of traditional agricultural technology in the management of tropical agro-ecosystems. Agro-ecosystems 7, 173–185.

10 Harrison P D and Turner II B L (Eds) 1978 Prehispanic Mayan Agriculture. Univ. New Mexico Press. Albuquerque, N.M. 414 p.

11 West R C, Psuty N P and Thom B G 1969 The Tabasco Lowlands of Southeastern Mexico. Louisiana State Univ. Press. Baton Rouge, Louisiana. 193 p.

Plant and Soil 67, 119–127 (1982).   0032-079X/82/0671-0119$01.35.   SU-10
© 1982 *Martinus Nijhoff/Dr W. Junk Publishers, The Hague.*

Nitrogen distribution in hybrid and local corn varieties and its possible relationship to a declining soil nitrogen pool under shifting agriculture at Indian Church, Belize

Distribución de nitrógeno en variedades locales e híbridos de maíz y su posible relacion con la disminucion de las reservas de nitrogeno en el suelo bajo agricultura itinerante en Indian Church, Belize

J. D. H. LAMBERT
Department of Biology, Carleton University, Ottawa, Ontario, Canada.

and J. T. ARNASON
Department of Biology, University of Ottawa, Ottawa, Ontario, Canada.

Key words   Belize   Corn   N-allocation   N-cycling   P-allocation   Traditional agriculture.

Abstract   Nitrogen levels at harvest in hybrid corn and weeds were determined over a three year period. Nitrogen losses due to grain removal accounted for 45% of all assimilated-N in the milpa. Percent N and percent P levels in the hybrid and a local black corn were also determined. Uptake of nitrogen and phosphorus in roots and leaves of both varieties were sigmoid until reproductive growth was initiated, at which time the percentage of both nutrients in the leaves declined rapidly. Levels in the seed at maturity in the hybrid variety accounted for 78% N and 52% P; levels were 58% N and 50% P in the local corn.

Hybrid corn was not considered suitable for local farmers because a high percentage of total-N was removed at harvest and the hybrid had a high susceptibility to weevil attack.

Resumen   Se determinaron los niveles de nitrógeno en maiz local, híbrido y malezas durante tres años. Las pérdidas de nitrógeno por cosecha del grano fueron 45% del nitrógeno total asimilado por la milpa. Los porcentajes de N y P en el híbrido y en el maiz negro local fueron también determinados observándose tasas de absorción sigmoidales en las raices y hojas de ambas variedades hasta alcanzar el período reproductivo. Luego el porcentaje de estos nutrimentos bajó rápidamente en las hojas. Los niveles en las semillas maduras en el híbrido bajaron a 78% y 52% de los valores iniciales de N y P respectivamente. mientras que en la variedad local éstos bajaron a 58% y 50% respectivamente.

El maiz híbrido no se consideró apropiado para los campesinos locales debido al alto porcentaje de nitrógeno total extraido por la cosecha además de la suceptibilidad del grano cosechado a las plagas durante el almacenamiento.

Introduction

A review of available literature led Rodin and Bazilevich[8] to conclude that there was a more intensive accumulation of mineral elements per unit of biomass in tropical forests than in any other vegetation type. The continuous supply of litter to the soil and uptake by the vegetation helps to maintain a nearly closed nutrient cycle[5]. Nitrogen, while mineralized more slowly than other major elements, is tightly held in the nutrient cycle. When a forest is cleared and burned for agriculture, the nutrient balance is disrupted and the chemical composition of the ash leads to changes in the chemical properties of the surface soil.

The generally low yields of native crops in tropical areas, such as those for corn, have resulted in a major effort by agronomists over the past four decades to produce new varieties that give higher yields. One consequence of increased grain production, however, is a greater loss of nutrients from the agro-ecosystem at harvest.

After the initial supply of amino acids in the seed has been exhausted, the corn seedling is primarily dependent on soil nitrogen for protein synthesis. Normally nitrogen is taken up as NO_3^-. Some of this may be reduced to NH_4^+ in the roots and assimilated into amino acids. The assimilation of NO_3^- requires energy and carbon for amino acid skeletons; both are supplied by photosynthate from the leaves transported to the roots *via* the phloem. The amino acids and nitrate can then be transported in the transpiration stream to the leaves, where the remaining nitrate is reduced and assimilated and the amino acids incorporated into leaf protein[7].

Nitrate uptake by the roots increases slowly until reproductive growth is initiated, then it appears to remain constant. The availability of NO_3^- to the roots and the physiological state of the plant at the start of reproductive growth may affect the rate at which photosynthate is transferred to the roots. When reproductive growth is successful, the seed and associated structures are more competitive sinks[11].

The purpose of this paper is to examine changing nitrogen and phosphorus levels in a hybrid (Pioneer 230) and a local variety of black corn during vegetative and reproductive growth stages. Nutrient allocation patterns during growth can have important consequences for the agroecological value of introduced hybrids.

Materials and methods

Field experiments were carried out in a seasonal dry hardwood forest (High Bush) and a Cohune Palm forest (Corosal) at Indian Church in the Orange Walk District of Belize[3]. The study area has not been exposed to rigorous agricultural activity for at least the last 700 years. It is underlain by a creamy pink, moderately hard, amorphous limestone that is high in exchangeable cations. Both the Pioneer 230 hybrid corn and the local black variety used in the experiments were planted according to local practice: four seeds were placed in holes made at 1 m intervals with a planting stick, equivalent to a planting rate of 16 kg fresh seeds ha^{-1}.

Traditionally, corn is planted just prior to the start of the rainy season in late May (the Cosecha planting) and then again in January (the Yashking planting) after all standing corn stalks of the Cosecha have been cut. Beans are planted in November on land cleared of all corn stalks and weeds. We followed the same planting schedule.

Cultivar plots were 25 m × 25 m. At the corn harvests, five 3 m × 3 m plots were sampled to determine the number of individual corn stalks, grain yields, and total corn biomass. Beans were sampled from five 1 m × 1 m plots for seed and total bean biomass. Weeds were sampled in all three harvests from five 1 m × 1 m plots. All above-ground parts and roots were removed and separated by species for nutrient analysis.

In the experiments to examine nutrient allocation patterns, samples of ten individual plants were collected at 2, 8, 10, 14 and 16 weeks. The total plant, above and below ground, was removed. Plants were separated into roots, stalks, leaves, flowers, husk, cob and grain, except for the 2-week samples for which insufficient root material was available for analysis. All samples were oven dried and returned to Carleton University for nutrient analysis. After arrival all samples were again oven dried at 60°C for 24 hours.

Soil samples from the High Bush milpa were collected weekly for the three year study period, and from the Corosal milpa samples were taken monthly between May 1978 and May 1980. Five cores per site (to 10 cm) were pooled, weighed, and then oven dried at 60°C for 24 hours to determine percent moisture. All samples were ground and sieved for nutrient analysis. The organic content of each sample was determined by removing carbonates with HCl and combusting the oven-dried remainder at 450°C for 4 hours. Total-N in soil and plant material was determined using the semi-micro-Kjeldahl technique with K_2SO_4–$CuSO_4$–Se reagents[4]. Available-P was determined colorimetrically in acid extracts[6].

Results and discussion

Yields and N contents of the crops

Nitrogen levels in the hybrid corn and weeds at the time of harvest between 1977–79 are presented in Table 1. Grain yields in the 1977 and 1978 Cosecha crops (August and September harvests) were average for the region whenever the hybrid was grown. However, hybrid corn is not often grown in the region for reasons discussed later.

Weed growth following forest cutting and burning was always low. However, weeds tended to accumulate a much higher proportion of the available nutrients relative to their total biomass than did corn. Percent-N for total corn biomass was below 0.99 on all sampling dates, whereas the N-content of total weed biomass was always greater than 1.35% N. In 1977 weeds contained 10% of the total biomass nitrogen but accounted for only 5% of the total biomass. In 1978 these values increased to 34% and 21%, respectively, and in 1979 weeds controlled 38% of the total nitrogen and 20% of total biomass.

In the third year of milpa use there was a substantial drop in both grain production and total weed biomass. These declines are not readily explainable, though soil nitrogen levels dropped from 0.41 to 0.37% in the third year and mean soil moisture also declined somewhat (Fig. 1). Additionally, the increasing bulk density of the surface soil horizon with continued exposure[9] may make it even more difficult for both cultivars and annual weeds to compete with established weeds and suckers from tree species that had survived both burns. Root systems of these established weeds and suckers commonly inhibit the early growth of the cultivars and are strong competitors for available nutrients and moisture.

The 1977 bean and Yashking (April harvest) corn yields are lower than normal but show where nitrogen accumulated in the crops. Bean yields were poor because of wet weather, but these plants had 60% of the nitrogen in total above ground biomass though only 29% of the dry matter.

Table 1. Biomass and total nitrogen content (kg ha^{-1}) in hybrid corn, beans, and weeds in High Bush milpa harvests. The milpa was established from forest in January 1977. Percent grain-N refers to the proportion of the N in the total biomass (crop + weeds) that was in harvested grain

Component	Aug/77 Corn		Dec/77 Beans		Apr/78 Corn		Sept/78 Corn		Sept/79 Corn	
	kg ha^{-1}	kg N ha^{-1}	kg ha^{-1}	kg N ha^{-1}	kg ha^{-1}	kg N ha^{-1}	kg ha^{-1}	kg N ha^{-1}	kg ha^{-1}	kg N ha^{-1}
Crop										
Grain	2971	34.6	91	3.3	386	6.4	2526	32.6	785	12.0
Cob	555	2.2			349	2.5	606	2.0	431	2.0
Husk	440	1.6	78	1.6	320	1.6	278	0.8	413	1.7
Stalk	4654	25.0	72	11.6	1773	19.1	1778	7.3	1734	3.7
Root	1140	8.1	10	0.1	305	1.3	788	3.8	412	2.1
Total	9760	71.5	251	16.6	3133	30.9	5976	46.5	3775	21.5
Weeds	560	7.6	608	11.1	2153	38.4	1591	23.9	969	13.3
Total	10320	79.1	859	27.7	5286	69.3	7567	70.4	4744	34.8
Percent grain-N	44		12		9		46		34	

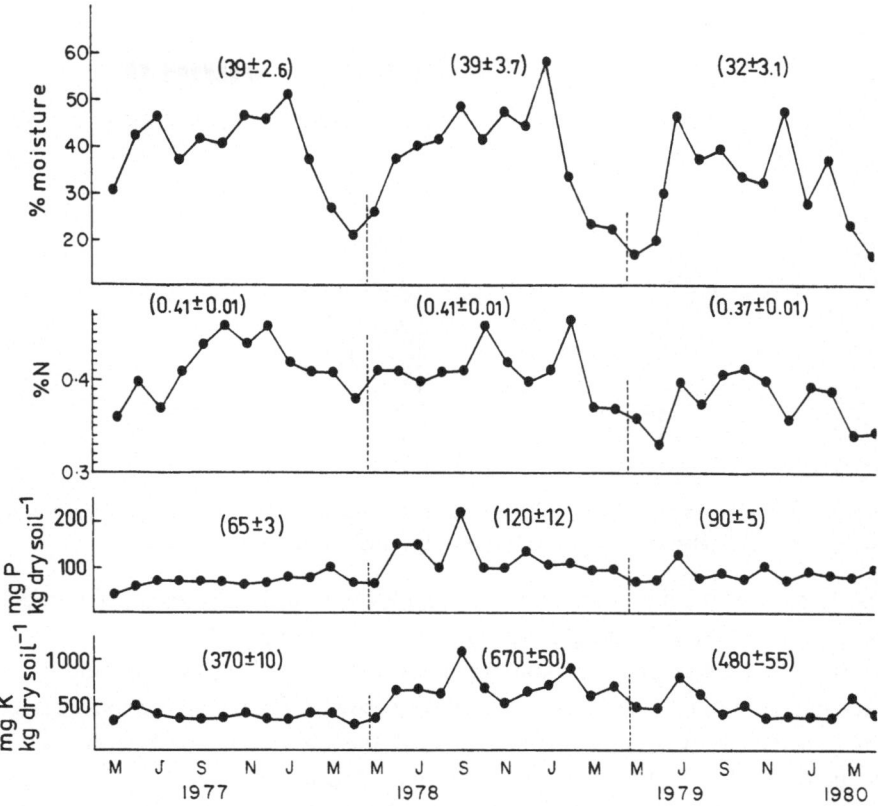

Fig. 1. Monthly levels of moisture and N, P and K in a Belize high bush milpa soil. Values in parentheses are seasonal means ± standard errors.

The low Yashking hybrid corn yield reflects the inability of a four month corn variety to successfully complete grain production under conditions of declining soil moisture and high ambient air temperatures. While no data are presently available for vegetative vs. reproductive growth and for changing plant-N levels, observations of vegetative growth suggest a pattern similar to that for the Cosecha planting. Information is still required for soil nitrate levels during the dry season; the physiological stress placed on the plant at this time may result in a greater than usual proportion of photosynthate directed to the roots rather than to the reproductive structures. Local, two-month maturing varieties planted during this period produce yields only slightly less than hybrid yields during the Cosecha.

During the entire three year sampling period, 28,800 kg ha^{-1} of plant biomass that contained 281 kg N was produced. 70 kg N or 25% of the total N accumulation was removed at harvest (Fig. 2).

Temporal patterns of nitrogen allocation

The importance of examining nutrient uptake rates at different stages of plant growth has been long recognized [2, 10]. Fig. 3 presents dry matter accumulation

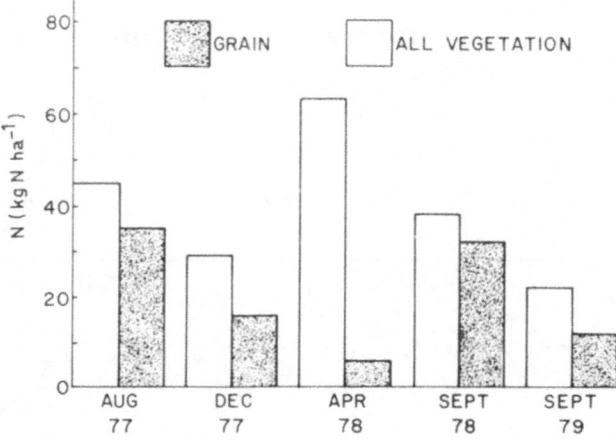

Fig. 2.   The relationship of grain nitrogen to total assimilated nitrogen in a corn-bean agro-ecosystem. 'All vegetation' is total crop biomass plus total weed biomass at harvest.

patterns over the life cycles of hybrid and local corn. The patterns of N and P accumulation at different growth stages are shown in Fig. 4.

Both vegetative and reproductive parts of the plants grow sigmoidally. At approximately eight weeks of age, vegetative growth of both corn varieties ceased and further growth was directed toward the development of reproductive structures. The vegetative growth of the hybrid corn accounted for $< 45\%$ of the total dry matter present at harvest, whereas for the local black variety vegetative growth accounted for $> 60\%$. Reproductive growth began in both hybrid and

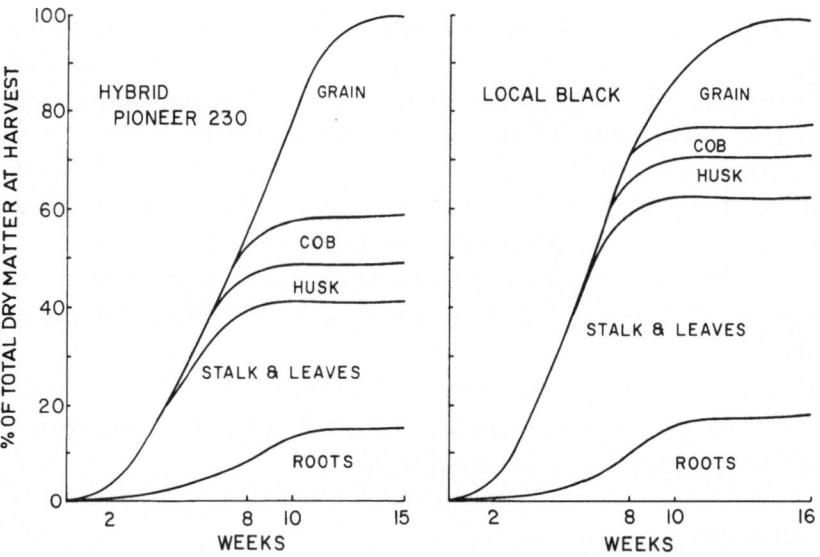

Fig. 3.   Dry matter allocation patterns in an introduced hybrid and a local variety of corn. Data were collected at 2, 8, 10, and 16 weeks.

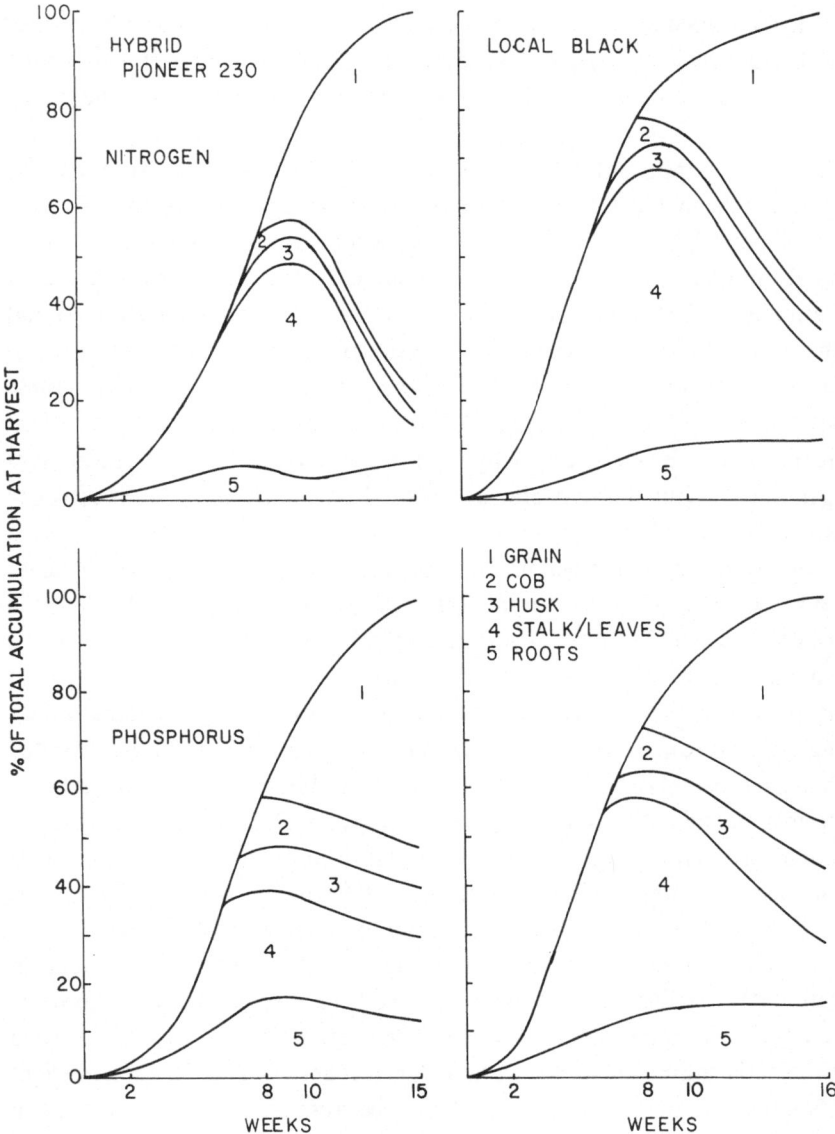

Fig. 4.   Nitrogen and phosphorus allocation pattern in a hybrid and local variety of corn. Data were collected at times noted on x-axis.

local varieties after approximately six weeks of growth. At six weeks the hybrid was 1.5 m tall and the local variety, 2 m. After reproductive growth became obvious, the hybrid husk and cob attained full size rapidly, whereas the local variety took an additional week or two: the growth rate of the hybrid grain at 10 weeks appeared not to be reached by the local variety until the twelfth week, and between thirteen and fifteen weeks the hybrid reached reproductive maturity while the local black variety did not reach maturity until at least sixteen weeks.

When maturity was attained the husks and leaves senesced, although the cob and grain continued to lose moisture. To facilitate grain drying and surface water loss at this time (when rains can still occur) the stalks are broken over. Harvesting occurs when husks are brittle and the grain can be easily removed from the cob by hand.

Differences in nutrient allocations were also striking (Fig. 4). With the emergence of the first leaf, N and P levels increased sharply. During the following eight week period, N and P accumulation in the local variety exceeded that of the hybrid although net uptake in both was sigmoid. At the eight week stage net N and P uptake by the leaves, stalk and roots slowed down, and at ten weeks N and P in the vegetative biomass began to decline sharply. By this time the lower leaves may no longer be functional, and N and P were probably translocated from these to the reproductive tissues. Levels in the roots remained relatively constant during this period. By maturity the hybrid grain contained 78% of the total plant nitrogen and 53% of the phosphorus. In the local black corn, grain nitrogen accounted for 58% of the total nitrogen and 50% of the phosphorus.

Hybrid grain production was greater than local-variety production and faster by two weeks. The hybrid apparently directs a greater proportion of its growth toward grain rather than toward vegetative production. The benefit to the small tropical farmer, however, may be short-lived.

Numerous researchers have documented the expected nutrient losses under shifting agriculture in the tropics [5, 9]. Reasons for such losses have been attributed to enhanced erosion, runoff, annual burning of stubble, or such low initial levels of nutrients that all nutrients quickly become limiting. Our data indicate that soil nutrient levels declined slightly in the third year of cultivation, but it is not likely that the decreases were sufficient to result in the low grain yields. Rather, we believe that yield declines were due primarily to increased weed growth [3].

While soil nutrient levels showed a slight decline in the third year of cultivation, the decline was not likely to have been great enough to cause the reduced grain yields measured. Weeds may be better adapted to the disturbed condition of the milpa. Local corn varieties are less destructive of the soil nutrient pool than the introduced hybrid corn. This is because of the greater amounts of nutrients removed in hybrid corn harvests. For this reason local corn varieties may be better suited to long-term cultivation in milpa ecosystems. Furthermore, the Pioneer 230 hybrid corn is highly susceptibile to Sitophilus attack when in storage (unpublished data). Phenolics in local corn grain that inhibit weevils after harvest have apparently been bred out of the hybrid.

We believe that introduced hybrid corn is only suitable in intensive mechanized agricultural systems. In such systems high nutrient losses at harvest can be replaced by fertilizers and the grain harvested can be quickly removed to storage areas where weevil control is possible. Small farmers with limited means are likely to find hybrid grain uneconomical.

Conclusions

A variety of corn developed locally over several generations of Mayan agriculture in Belize produced less grain than an introduced hybrid variety, but consequently may have placed less strain on soil nutrient reserves. The local variety is also less susceptible to Sitophilus attack. Thus we consider the local variety to be better suited to these agro-ecosystems than the introduced hybrid which is bred mainly for high yield.

Acknowledgements   Financial support was provided by the Canadian International Development Agency. Chemical analyses were completed by J. Gale. Special acknowledgement and thanks are due to A. Urquhart for valuable comments on the paper.

References

1   Hanway J J 1962 Corn growth and composition in relation to soil fertility. II. Uptake of N, K and P and their distribution in different plant parts during the growing season. Agron. J. 52, 217–222.
2   Hanway J J 1963 Growth stages of corn (*Zea mays* L.). Agron. J. 53, 487–492.
3   Lambert J D H and Arnason J T 1980 Nutrient levels in corn and competing weed species in a first year milpa, Indian Church. Belize. Plant and Soil 55, 415–427.
4   McKeague J A 1978 Manual on Soil Sampling and Methods of Analysis. 2nd Edition. Ottawa: Canadian Society of Soil Science. 212 p.
5   Nye P H and Greenland D J 1960 The soil under shifting cultivation. Comm. Bur. Soils Tech. Commun. 51. Harpenden, England; Commonwealth Agricultural Bureaux. 156 p.
6   Olson S R and Dean L A 1965 Phosphorus. *In* Black C A (Ed.). Methods of Soil Analysis, pp 1034–1049. Madison, Wisconsin: American Society of Agronomy.
7   Pate J S 1973 Uptake, assimilation and transport of nitrogen compounds by plants. Soil Biol. Biochem. 5, 109–119.
8   Rodin L E and Bazilevich N I 1967 Production and Mineral Cycling in Terrestrial Vegetation. Edinburgh and London: Oliver and Boyd. 288 p.
9   Sanchez P A 1976 Properties and Management of Soils in the Tropics. New York: J. Wiley and Sons. 618 p.
10   Sayre J D 1948 Mineral accumulation in corn. Plant Physiol. 23, 267–281.
11   Urquhart A A 1980 The Transport, Metabolism and Redistribution of Xylem-borne Amino Acids in Pea Seedlings. Ph. D. Thesis. Ottawa, Canada: Carleton University. 195 p.

Plant and Soil 67, 129–137 (1982).   0032-079X/82/0671-0129$01.35.   SU-11
© 1982 *Martinus Nijhoff/Dr W. Junk Publishers, The Hague.*

¹⁵N-urea transport and transformation in two deforested Amazonian soils under laboratory conditions

Transporte de urea ¹⁵N y sus transformaciones en dos suelos deforestados del Amazonas bajo condiciones de laboratorio

R. L. VICTORIA, P. L. LIBARDI, K. REICHARDT and E. MATSUI
Centro de Energia Nuclear na Agricultura (CENA), Caixa Postal 96, 13. 400 Paracicaba, S.P. Brazil

Key words Amazonas Deforestation Leaching 15-N N-cycling Mineralization Urea.

Abstract Brazilian agriculture is now expanding toward the Amazon region, where large new areas of virgin lands are being brought under cultivation. There is therefore an urgent need to better understand the conditions and characteristics of the soils of that region. In this study a Red Yellow Podzol and a Yellow Latosol were used to examine urea transport and transformation in the laboratory under water-saturated conditions. The soils were collected in an area that was deforested in 1976 and planted to tropical fruits since then.

Soils were subjected to miscible displacement techniques under both continuous feed and pulse applications of urea to mathematically describe urea transport and transformation as functions of depth and time. Transformation mechanisms were considered to be first order kinetics.

Urea was readily leached from both soils. Recovery of urea in the effluent of the 30 cm columns was 91% for the Podzol and 86% for the Latosol. NH_4^+–N from urea hydrolysis was also readily leached and its recovery in the effluent was 4.2% for the Podzol and 11.2% for the Latosol. Very little nitrogen – including exchangeable NH_4^+–N and biomass nitrogen – was left in the columns of either soil at the end of the experiment.

These results emphasize that extremely careful management of these soils is necessary to prevent nitrogen losses, particularly losses of fertilizer-N.

Resumen La agricultura brasileña se expande a la región amazónica donde áreas vírgenes grandes se ponen actualmente bajo cultivo. Es por lo tanto urgente mejorar el conocimiento de las características y condiciones de los suelos de esa región. En este estudio un podzol amarillo rojizo y un latosol amarillo fueren usados para estudiar el transporte de urea y sus transformaciones bajo saturación en el laboratorio. Los suelos fueron colectados en un área deforestada en 1976 y plantada bajo frutales.

Los suelos se sometieron a técnicas de desplazamiento miscible bajo alimentación contínua y aplicaciones pulsadas con el fin de describir matemáticamente el transporte de úrea en función del tiempo y de la profundidad. Se consideró que los mecanismos de transformación obedecían cinéticas de primer orden.

La úrea es rápidamente lixiviada de ambos suelos. La recuperación de úrea en el efluente de las columnas de 30 cm de diámetro fué de 91% para el podzol y 86% para el latosol. El nitrógeno amoniacal, producto de la hidrólisis de úrea se lixivió rápidamente y su recuperación en el efluente fué 4,2% para el podzol y 11,2% para el latosol. Una proporción muy pequeña del nitrógeno, incluyendo el NH_4^+ intercambiable y el nitrógeno en la biomasa, permaneció en las columnas al final del experimento.

Estos resultados indican el extremo cuidado en el manejo de estos suelos necesario para evitar pérdidas de nitrógeno, particularmente del aplicado en fertilizantes.

Introduction

Increasingly rapid development of new lands for agricultural purposes in Brazil, particularly in the Amazonian region where many new farms are being established along the main roads, makes the need to better understand the soils in these regions urgent. Because the Brazilian government has opted to use urea as the main nitrogen fertilizer for Brazilian agriculture, it is especially important to understand the behavior of urea-N in these soils.

Miscible displacement methods provide the opportunity to study transport characteristics and transformation processes of urea in soils. It has been used by several research workers to study transport and transformations of nitrogeneous compounds [4, 7, 13, 17, 19], phosphorous compounds [5, 16] and pesticides [8, 16].

The objective of this study was to examine the transport and transformation of urea in two Amazonian soils from recently-deforested areas. The experiment was conducted in the laboratory under steady-state, water-saturated conditions. This experiment will yield information necessary for improved nitrogen management for future commercial agriculture in this region.

Methods

Soil samples were collected from the 0–30 cm layer of an experimental area of the Instituto Nacional de Pesquisas da Amazônia (INPA), located at Km 60 of the Manaus-Caracarai highway (BR-174) in the Amazonian Rainforest region of Brazil (2°30′ S, 60°00′ W, 2200 mm mean annual precipitation). The area was deforested in 1976 and planted to tropical fruits. The soils collected were a Red Yellow Podzol and a Yellow Latosol. Selected chemical and physical characteristics for both soils are shown in Table 1.

Air-dried soils were passed through a 2 mm sieve and packed into acrylic plastic cylinders, 30 cm long and 5.65 cm internal diameter (Fig. 1). The soil was initially saturated from the bottom plate with 0.01 M $CaSO_4$, after which steady state flow of 200 ppm urea-N in 0.01 M $CaSO_4$ was established through a constant-head burette. During each of the experiments a pulse of ^{15}N-labelled urea (200 ppm N) was admitted to the column. Time for admittance was arbitrary and not the same for both soils. Column data for both experiments are presented in Table 2.

Effluent was collected in an automatic fraction collector at the base of the column and analyzed for urea-N, NH_4^+–N and ^{15}N in both urea and NH_4^+. Urea was determined by the Dyacetil Monoxime method[10] and NH_4^+ by steam distillation in the presence of MgO[2]. ^{15}N isotopic analyses were performed in an Atlas-Varian CH-4 mass spectrometer with samples prepared by oxidation with lithium hypobromite[12].

Near the end of each experiment soil solution samples were withdrawn from the columns using syringes inserted into holes made previously at 5 cm depth increments. The samples were subsequently analyzed for urea. At the end of the experiment each soil column was cut into 10 layers of 3 cm each. After mixing, a portion of each layer was extracted with 2 N KCl (5 : 1 wet weight : volume ratio). The extract was analysed for NH_4^+–N and ^{15}N as described earlier, and the remaining soil for organic-N[3] and ^{15}N.

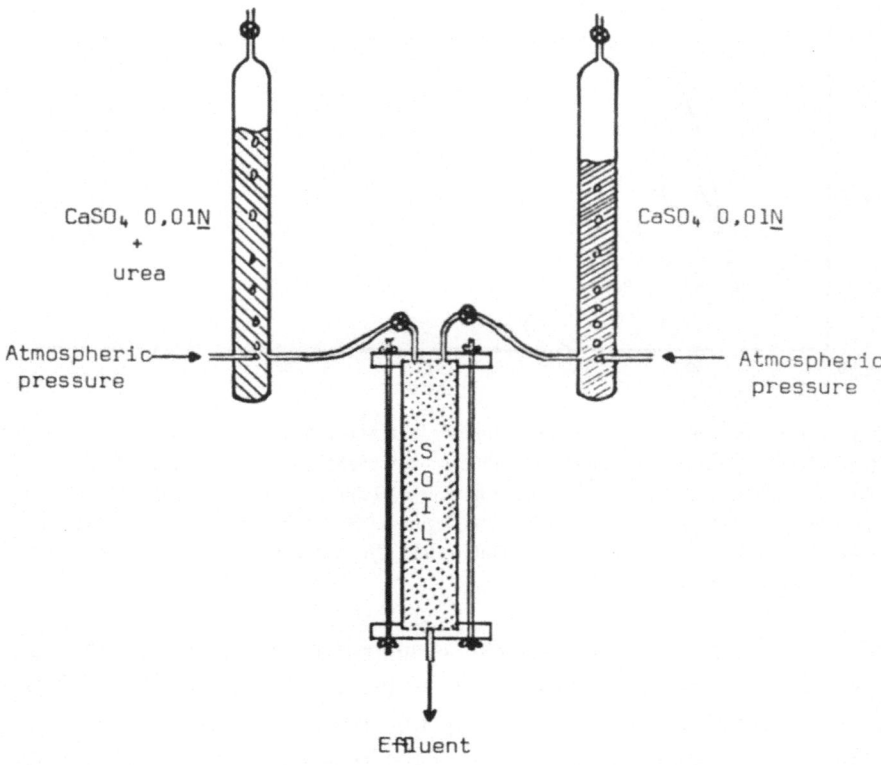

CaSO₄ 0,01N
+
urea

CaSO₄ 0,01N

Atmospheric →
pressure

← Atmospheric
pressure

S
O
I
L

Effluent

Fig. 1.   Schematic diagram of the experimental set-up (after Corey *et al.*[7]).

Results and discussion

Fig. 2 shows the experimental data and theoretical breakthrough curves for both continuous feed and pulse applications of urea in the Red Yellow Podzol soil column. The theoretical breakthrough curve (the ratio of effluent concentration to initial concentration plotted against time[6]) for continuous feed in this soil, plotted with $K = 0$, almost perfectly fit experimental data for the first 160 min; after this, experimental data lay below the theoretical curve, probably

Table 1. Physical and chemical characteristics of the soils used in the miscible displacement analysis. pH was determined in a 1 : 1 fresh soil : water slurry

Soil	Organic matter (%)	pH	Al^{+++}	Ca^{++}	Mg^{++} meq/100 g	K	P µg/g	Clay %	Sand %	Textural class
Red yellow podzol	2.2	5.1	0.60	0.44	0.14	26.0	5.8	24	76	Sandy-clay loam
Yellow latosol	1.6	4.7	0.98	0.41	0.16	9.8	1.6	46	53	Sandy-clay

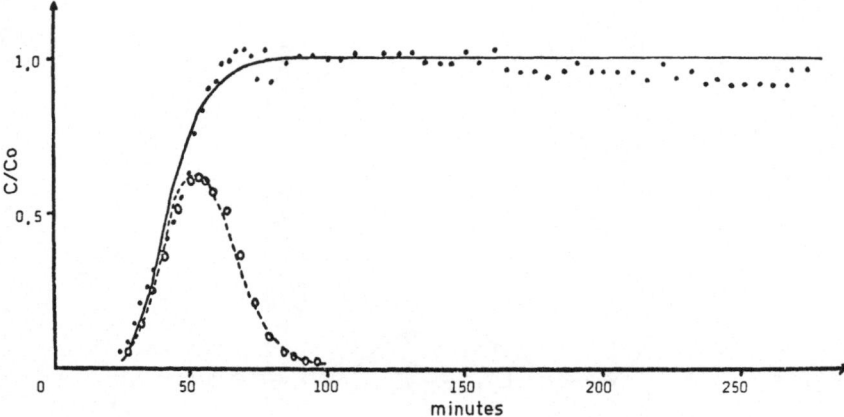

Fig. 2. Breakthrough curves for both continuous feed (●) and pulse (○) applications of urea in the Red Yellow Podzol Soil. Solid lines are theoretical curves based on conditions listed in Table 2 and K (the rate coefficient for irreversible transformation) = 0 min^{-1} (continuous feed) or 0.002 min^{-1} (pulse feed). The 23.5 minute pulse of ^{15}N was begun at 200 minutes but its theoretical curve was plotted assuming $t_0 = 0$ at the beginning of infiltration. C/C_0 is the ratio of effluent concentration to initial (pre-addition) concentration.

indicating urea transformation. This is supported by results from the pulse application, for which it was necessary to use a rate coefficient (K) of 0.0002 min^{-1} to obtain a good fit between experimental and theoretical curves.

Fig. 3 shows breakthrough curves for the Yellow Latosol soil column. The continuous feed curve is plotted with $K = 2 \times 10^{-4}$ min^{-1}, and it can again be observed that the theoretical curve does not perfectly fit the experimental after the pulse of ^{15}N-urea. This indicates that the rate coefficient for the pulse feed is

Table 2. Column data for the miscible displacement experiments. Porosity in both soils was calculated rather than measured; this may lead to small errors in water content and consequently in mean flow velocities

	Red yellow podzol	Yellow latosol
Soil weight (g column^{-1})	1030	920
Particle density (g cm^{-3})	2.42	2.32
Bulk density (g cm^{-3})	1.37	1.22
Porosity (P)	0.434	0.474
Mean flow (ml min^{-1})	8.4	2.4
Mean flux (ml cm^{-2} min^{-1})	0.334	0.094
Mean velocity (cm min^{-1})	0.769	0.198
Urea concentration (mg N l^{-1})	200	200
^{15}N enrichment in the urea pulse (Atom % ^{15}N)	10.438	10.438
Pulse time (min)	23.5	84
Apparent diffusion coefficient	0.81	0.291
Retention coefficient (R)	1.13	1.12

actually higher than the value used for plotting the continuous feed curve. Indeed, the pulse curve was plotted with $K = 9 \times 10^{-4} \, min^{-1}$. These differences between continuous and pulse feed rate coefficients (K) in both soil columns suggest, to some extent, that urease activity may be varying with time, being very low at the beginning of urea infiltration and then building up slowly. Such temporal variation has been noted in other soils by Ardakani et al.[1] and Victória[18]. To confirm changes in urease activity with time for the soils studied in this paper the length of the experiments should be extended.

Recovery of applied ¹⁵N-urea in the effluent was 91% for the Red Yellow Podzol and 86% for the Yellow Latosol. Calculated urea retention coefficient (R) was found to be small and about the same for both soils (Table 2). These values are about twice that found by Wagenet et al.[19] for a Tyndal silt clay loam, possibly because of a lower cation exchange capacity and organic matter content of the Tyndal soil.

Figures 4 and 5 show urea distributions in the soil solution within the column profiles at a time near the end of the experiments. There is good agreement between experimental and theoretical values, and values of the rate coefficients confirm those used for the breakthrough curves, particularly those used for the ¹⁵N-urea pulse (Figs. 2 and 3). Although urease activity may vary with time[1,18], both Fig. 4 and 5 indicate that first order kinetics describe urea hydrolysis well in both soils.

Table 3 presents data for the fate of the urea applied to the soils based on the ¹⁵N applications. Urea was readily leached from both soils without undergoing hydrolysis. Additionally, the rate coefficient (K) for urea transformation is much higher for the Podzol than the Latosol (0.002 vs 0.0009, respectively), although the quantity of urea transformed is smaller in the Podzol. This can be explained by differences in transport velocity, however: the smaller velocity of the Latosol (Table 2) allows more time for the action of urease. Wagenet et al.[19] used the same rate coefficient to describe urea transport and transformation in various columns of a Tyndal silt clay loam. Rough calculations for two of their columns show that for the one with lower velocity the quantity of urea hydrolysed is higher; this may explain some of their imperfect curve fits.

For both soils described in this paper, it can be seen that almost no nitrogen

Table 3. Fate of ¹⁵N–urea–N pulses applied to soil columns as described in Fig. 2 and 3 legends. Values are percent of applied urea–¹⁵N

Soil	Effluent urea–N	Effluent NH_4–N	Soil NH_4–N	Soil organic–N	Total recovery (%)
Red yellow podzol	91	4.2	2.3	0.6	98.1
Yellow latosol	86	11.2	0.1	0.6	97.9

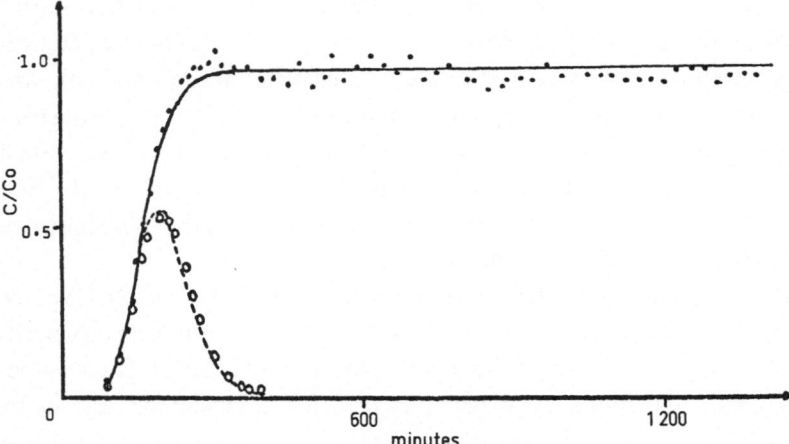

Fig. 3. Breakthrough curves for both continuous feed (●) and pulse (○) applications of urea in Yellow Latosol soil. $K = 2 \times 10^{-4}$ min^{-1} (continuous feed) or 9×10^{-4} min^{-1} (pulse feed), and the 84 minute ^{15}N pulse was begun at 600 minutes, but its theoretical curve was plotted assuming $t = 0$ at the beginning of its infiltration. See Fig. 2 legend and Table 2 for further explanations.

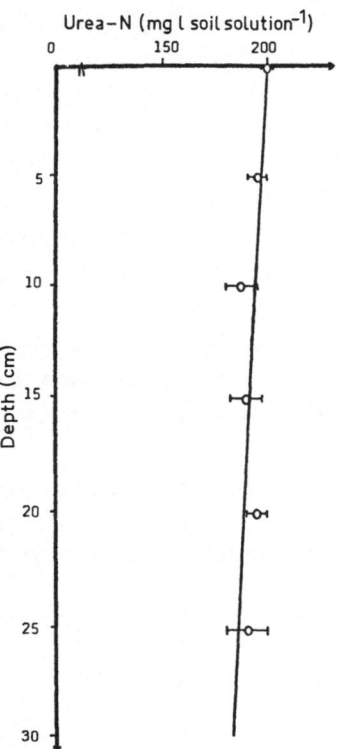

Fig. 4. Urea distribution with depth in the Red Yellow Podzol soil column at $t = 300$ min. The solid line is the theoretical curve based on the conditions listed in Table 2 and $k = 0.002$ min^{-1}. Horizontal bars denote standard deviations ($n = 4$).

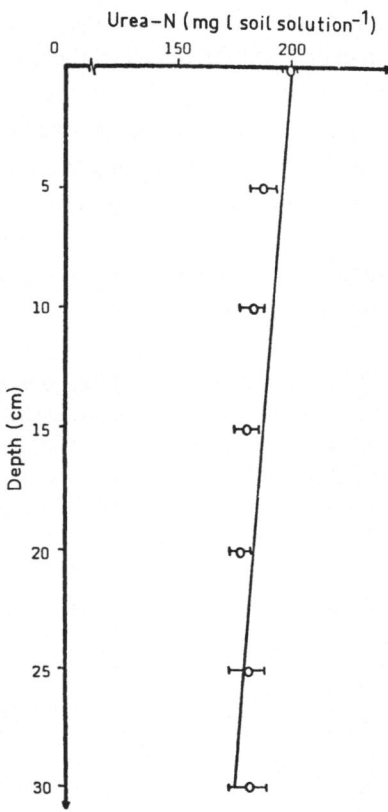

Fig. 5. Urea distribution with depth in the Yellow Latosol soil column at t = 1442 minutes. k = 9 × 10⁻⁴ min⁻¹. See Fig. 4 legend and Table 2 for further explanation.

was left in the soil after the experiments, either as soil biomass or exchangeable NH_4^+ (Table 3).

Higher exchangeable NH_4^+–¹⁵N was found in the Podzol, despite its lower cation exchange capacity. This is probably due to the fact that the ¹⁵N pulse in this soil was admitted at a time near the end of the experiment (see arrows in Fig. 2). Consequently, there was not sufficient time for the NH_4^+–¹⁵N formed to leave the columns as it did in the Latosol (Table 3 and Fig. 3). NH_4^+–N in the effluent was higher in the Latosol, which is in accordance with the higher quantity of urea transformed in this same soil. Total recovery of nitrogen was found to be about 98% for both soils. The small error observed is probably due to errors in the analytical process.

Conclusions

As pointed out by Wagenet et al.[19], a great deal of caution should be exercised when extrapolating laboratory studies to field situations. Nevertheless, results of this study lead us to believe that effective management of soil and of N-fertilizer will be very important for the success of future agriculture in the Amazon region. Urea is apparently readily leached below 30 cm in both soils. Even when the urea hydrolysed to NH_4^+ quickly, the NH_4^+ was not held by the soils. Thus, although the water-saturated conditions used in this study will overestimate dry-season losses of urea-N, during the rainy season there is clearly a potential for great loss of fertilizer nitrogen in these soils, and mechanisms of nitrogen protection should be used.

Further studies of these soils taken from under forest vegetation and studies performed under nonsaturated conditions would be valuable.

Acknowledgements   This paper is a contribution of CENA with funding provided by Financiadora de Estudos e Projetos (FINEP), Conselho Nacional de Desenvolvimento Cientifico e Tecnologico (CNPq), and Comissao Nacional de Energia Nuclear (CNEN). P. L. Libardi and K. Reichardt are CNPq fellows. The first author wishes to acknowledge SCOPE-UNEP for funding the trip to Cali.

References

1   Ardakani M S, Volz M G and McLaren A D 1976 Consecutive steady state reactions of urea, ammonium and nitrite-nitrogen in soil. Can. J. Soil Sci. 55, 83–91.
2   Bremner J M 1965 Inorganic forms of nitrogen. *In* Black C A (Ed.). Methods of Soil Analysis, pp 1179–1237. Madison, Wisconsin: American Society of Agronomy.
3   Bremner J M 1965 Total nitrogen. *In* Black C A (Ed.). Methods of Soil Analysis, pp 1149–1178. Madison, Wisconsin: American Society of Agronomy.
4   Cassel D K, Krueger T H, Schroes F W and Norum E B 1974 Solute movement through disturbed and undisturbed soil cores. Soil Sci. Soc. Am. Proc. 38, 36–40.
5   Castro C L and Rolston D E 1977 Organic phosphate transport and hydrolysis in soil: theoretical and experimental evaluation. Soil Sci. Soc. Am. J. 41, 1085–1092.
6   Cho C M 1971 Convective transport of ammonium with nitrification in soils. Can. J. Soil Sci. 51, 339–350.
7   Corey J C, Nielsen D R and Kirkham D 1967 Miscible displacement of nitrate through soil columns. Soil Sci. Soc. Am. Proc. 31, 497–503.
8   Davidson J M, Mansell J M and Baker D R 1972 Herbicide distributions within a soil profile and their dependence upon adsorption. Soil Crop Sci. Soc. Florida 32, 36–41.
9   Davidson J M, Ou L T and Rao P S C 1976 Behaviour of high pesticide concentrations in soil-water systems. Tuscon, Proc. of Hazardous Waste Research Sum. Washington, D.C.: EPA. 7 p.
10   Douglas L A and Bremner J M 1970 Colorimetric determination of microgram quantities of urea. Anal. Lett. 3, 79–87.
11   Hoffman D L and Rolston D E 1980 Transport of organic phosphate in soil as affected by soil type. Soil Sci. Soc. Am. J. 44, 46–52.
12   IAEA 1976 Tracer Manual on crops and soils. Vienna. International Atomic Energy Agency. 227 p.
13   Misra C, Nielsen D R and Biggar J W 1974 Nitrogen transformations in soil during leaching. II. Steady-state nitrification and nitrate reduction. Soil Sci. Soc. Am. Proc. 38, 294–299.
14   Misra C, Nielsen D R and Biggar J W 1974 Nitrogen transformations during continuous leaching. III. Nitrate reduction in soil columns. Soil Sci. Soc. Am. Proc. 38, 300–304.
15   Selim H M, Kanchanasut P, Mansell R S, Zelasny L W and Davidson J M 1974 Phosphorus and chloride movement in a spodosol. – Soil Crop Sci. Soc. Florida 34, 18–23.
16   Selim H M, Mansell R S and Elzeftawy 1976 Distributions of 2.4–D and Water in soil during infiltration and redistribution. Soil Sci. 121, 176–183.

17 Starr J L, Broadbent F E and Nielsen D R 1974 Nitrogen transformations during continuous leaching. Soil Sci. Soc. Am. Proc. 38, 283–289.
18 Victória R L 1980 Miscible displacement of urea in some Amazonian soils. Ph. D. Thesis. Escola Superior de Agricultura Luiz de Queiroz (ESALQ), University of São Paulo, Piracicaba-SP, Brazil. 90 p. (In Portuguese.)
19 Wagenet R J, Biggar J W and Nielsen D R 1977 Tracing the transformations of urea fertilizer during leaching. Soil Sci. Soc. Am. J. 41, 896–902.

11. Lane, P. W. and Nelder, J. A. (1982) Analysis of covariance and standardization as instances of prediction. *Biometrics*, **38**, 613-21.

12. Goldstein, H. and Blatchford, P. (1986) Inferences about the association between the characteristics of schools and their pupils: the problem of aggregation. *J. Educ. Statist.*, **11**, 67-84.

13. Aitkin, M. and Longford, N. (1986) Statistical modelling issues in school effectiveness studies. *J. R. Statist. Soc. A*, **149**, 1-43.

Plant and Soil 67, 139–146 (1982).   0032-079X/82/0671-0139$01.20.   SU-12
© 1982 *Martinus Nijhoff/Dr W. Junk Publishers, The Hague.*

Nitrogen cycling in sugarcane

Ciclo de nitrógeno en caña de azucar

A. PUPPIN RUSCHEL and P. B. VOSE

Centro de Energia Nuclear na Agricultura (CENA) 13.400 Piracicaba, São Paulo, Brasil

Key words Beans Crop rotation Intercropping Leaching Legume N-cycling N_2-fixation N-fertilizer

Summary Sugarcane has been grown extensively in Brasil for more than 50 years, and in the northeast from the time Brasil was discovered. Use of N-fertilizer started in the 1940's with applications of sodium nitrate, and little yield improvement was obtained in most cases. Average yield is around 70 t ha^{-1} yr^{-1} (4-harvest mean), with the first harvest 1.5 year from planting and a ratoon harvest each year thereafter. Nitrogen responses are obtained only with ratoon crops. Nitrogen inputs to the plant come from native soil-N, fertilizer-N, and biological fixation. Sources of loss include N-leaching from leaves and decomposing roots and loss of stems and leaves at harvest. There are technical and economic problems with returning factory waste (*vinhoto*) to the fields as fertilizer. A reasonably conservative estimate of biological nitrogen fixation holds that 17% of total plant nitrogen is fixed by the plant, or 16.6 kg N ha^{-1} for a harvest of 70×10^3 kg ha^{-1}. Rotation and intercropping of legumes with sugarcane could increase N_2-fixation by 35 kg N ha^{-1} yr^{-1} (soybean rotation) and 25 kg N ha^{-1} yr^{-1} (Phaseolus beans intercropping).

Resumen La caña de azucar se cultiva extensamente en Brasil desde hace mas de cincuenta años y en el noreste del pais desde la época del descubrimiento. El uso del fertilizante nitrogenado empezó en 1940 con aplicaciones de nitrato de sodio que resultaban en poco aumento de los rendimientos en general. El rendimiento promedio actual es de 70×10^3 kg ha^{-1} año^{-1} (promedio de cuatro cosechas); se realiza la primera cosecha al año y medio de la siembra y en los años sucesivos se cosecha la soca. Se obtienen respuestas positivas a la fertilización nitrogenada solamente en estas últimas cosechas de socas. Las entradas de nitrógeno a la planta provienen del nitrógeno nativo del suelo, el añadido por fertilizante y por fijación biológica. Las pérdidas incluyen la lixiviación de nitrógeno de las hojas y por descomposición de raices y la exportación de hojas y tallos por cosecha. El retorno de los residuos de la fábrica al campo presenta aun ciertos problemas técnicos y económicos. La fijación biológica de nitrógeno puede ser razonablemente estimada asumiendo que el 17% del nitrógeno total de la planta proviene de esta fuente; o sea unos 16,6 kg N ha^{-1} para una cosecha de 70×10^3 kg ha^{-1}. Los cultivos de leguminosas intercalados o en rotación con la caña de azucar podrían elevar la fijación biológica de nitrógeno hasta valores de 35 kg N ha^{-1} año^{-1} (rotación con soya) o 25 kg N ha^{-1} año^{-1} (cultivo intercalado de frijol, Phaseolus).

Introduction

Sugarcane (*Saccharum* sp.) in Brasil does not respond to N-fertilizer in the year it is planted, although subsequent ratoon crops do respond. That microorganisms fix nitrogen in the rhizosphere of sugarcane[15] is one indication that fixation may play an important role in the nitrogen self-sufficiency of this crop.

Sugarcane in some places has been grown without fertilizer for more than 50 years, without a decrease in yield. Although 130 kg N are removed with a 100×10^3 kg ha^{-1} yr^{-1} yield[7], only in the last 20 years have crops received fertilizer-N, usually at rates of 50–60 kg N ha^{-1} yr^{-1}.

Evaluation of the N-balance in this crop is very difficult for a number of reasons. First, some of the nitrogen in the system may circulate from the roots of the first year's crop back to the subsequent ratoons, and at present there is no accurate way to evaluate the origin of the nitrogen in the plant. The nitrogen could be derived from soil, from fertilizer, or from the roots of the original plants. Second, accumulation of organic-N through decomposition of leaves is usually disturbed by fire loss during harvest. Third, leaching of nitrogen from leaves by rain has been observed but it is not yet possible to determine accurately the rate and relate it to growth and N-release. Finally; the rate of N-immobilization after the first harvest is difficult to estimate.

This paper presents an overview of N-cycling in the sugarcane crop in Brasil. In particular, we discuss the possible substitution of N-fertilizer with N_2-fixation, and intercropping or rotation of sugarcane with legumes to improve the acquisition of nitrogen by the system. Understanding the N-cycle in sugarcane is becoming increasingly important because of the increasing cost of N-fertilizer and petroleum-derived energy.

Nitrogen in tropical soils

Nitrogen in soil under sugarcane, as in natural ecosystems, is constantly being lost and replenished by natural processes (Fig. 1). In most agricultural systems the amount of nitrogen readily available is usually insufficient to support high crop productivity; certain legume systems (*e.g.* soybeans), which receive nitrogen from biological N_2-fixation, are exceptions.

Tropical soils generally have low levels of organic matter (0.5 to 0.2%), and a low turnover of N through mineralization. Fertilizer-N additions increase yields of many small grains as well as increase soil-N cycling. Libardi et al.[6] used ^{15}N-urea to follow mineralized-N and organic N in three subsequent crops of Phaseolus beans grown on an alfisol (Terra Roxa misturada) commonly used for sugarcane. They observed that after the first crop, tagged mineral-N decreased with soil depth at four levels down to 120 cm, as did the tagged organic-N fraction (Table 1). With the second and third croppings tagged mineral-N increased with depth below 60 cm, while tagged organic-N was higher at 0–30 cm than at 30–120 cm. Losses in their system at first cropping (taking into account the high degree of statistical error) were not detectable. However, in subsequent cropping, 30 and 20% of the nitrogen added could not be accounted for, indicating losses that could be due to denitrification or leaching. The ^{15}N in soil organic matter decreased to only half of that observed at the first cropping, indicating that mineralization of the native organic-N was not comparable to that of fertilizer-N.

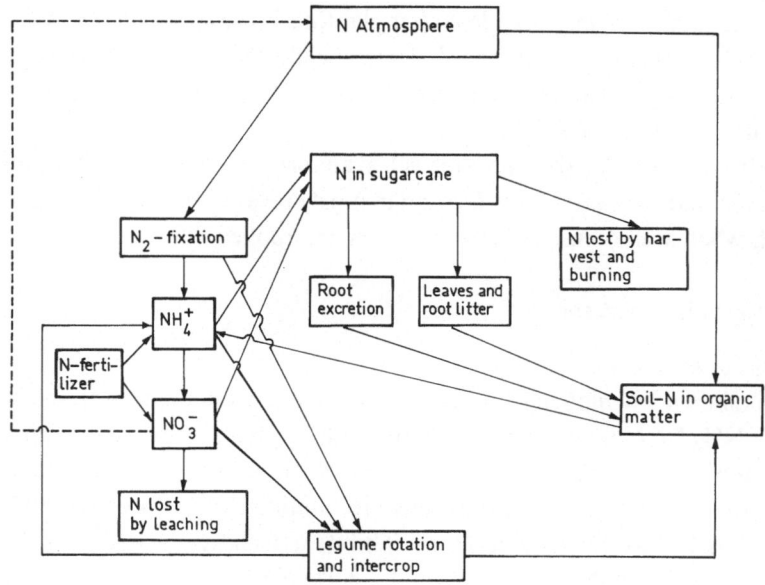

Fig. 1. Principal pathways for nitrogen gained and lost by sugarcane ecosystems. Except for N-fixation, inputs to and outputs from the system *via* legumes (rotation and intercropping) are not included.

Table 1. Fertilizer-derived soil and plant nitrogen following three successive Phaseolus bean crops grown on a typical Brazilian sugarcane soil. Values are the percent of the soil mineral-N (N_{min}) or of the soil organic-N (N_{org}) or of the plant total-N (N_{tot}) that is derived from ^{15}N-fertilizer applied at the first cropping (Cervellini *et al.*[2])

Component	Crop 1		Crop 2		Crop 3	
	N_{min}	N_{org}	N_{min}	N_{org}	N_{min}	N_{org}
Soil						
0– 30 cm	18.7	23.0	1.5	13.7	1.1	15.8
30– 60 cm	8.2	8.9	1.4	4.4	1.2	4.5
60– 90 cm	2.1	9.1	4.0	1.4	2.1	3.6
90–120 cm	0.9	1.5	4.7	1.6	3.2	3.8
Plant	38.5[a]		4.3[a]		1.0[a]	

Apparent nitrification in these soils was also studied to 40 cm depth[2]. At 20 cm from the bean stems 20 days after planting, both soil ammonium and nitrate decreases with depth, although soil ammonium more quickly than nitrate. NH_4^+ values at 40 cm depth were only 3% of those in the upper layer whereas NO_3^- values were 28%, suggesting that nitrate moved more quickly through the profile than did ammonium. Forty days after planting, however, soil NH_4^+ values 20 cm from the plants were near 0 at all depths, while NO_3^- values in the upper layer were near day 20 values and at 40 cm were *ca.* 20% of these. This indicates active nitrification in these soils.

Although similar data are not yet available for sugarcane crops, one might assume that the rates of nitrification in most areas cultivated with sugarcane are high when ammonium or urea fertilizers are applied but otherwise low.

Nitrogen in sugarcane

Gains of nitrogen

As shown schematically in Fig. 1, sugarcane can assimilate both soil NO_3^--N and NH_4^+-N. In nutrient solution studies, plants grow well with levels of ammonium below 28 ppm NH_4^+-N, at which level toxicity symptoms appear and the growth of roots and shoots are inhibited[1,20]. Moreover, NH_4^+-N at levels as low as 3 ppm completely inhibits associative N_2-fixation. Sources of mineral-N include fertilizers and, particularly for ratoon crops, both decomposing roots and litterfall plus harvest residue. However, leaves are largely burned at harvest; N-additions from the decomposition of leaves dropped before harvest are not known. Nitrogen can also come from biological nitrogen fixation, discussed below, as well as from bulk precipitation, which adds about 2–5 kg N ha^{-1} yr^{-1} to systems in the São Paulo area (Ruschel and Vose, unpublished).

Losses of nitrogen

Vinhoto, the factory by-product of sugarcane fermentation, is at present often returned to the soil as fertilizer. Only a 10-km radius from a mill can be covered economically by this practice. Although pipes and pumps can be used to transport *vinhoto* to more distant fields they are commonly destroyed within 3 months by heavy precipitations of salts.

Increase of N by rotation and intercropping

The soil-N pool can be increased substantially by the rotation and intercropping of legumes with sugarcane. By planting soybeans before planting sugarcane and by planting Phaseolus beans in between previously-planted sugarcane rows, soil nitrogen can be increased by associative biological fixation without harming sugarcane yields. This has occurred in Colombia[4], and these practices are common in southern China and parts of India[11,14]. In Brasil, 2.3 million ha are planted to sugarcane; 1/4 of this, or 575×10^3 ha, is renewed annually and could be available for a 6-month soybean rotation. After sugarcane is planted, 60% of the area planted to sugarcane could be used for intercropping with Phaseolus beans (380×10^3 ha). With average yields of 70×10^3, 1.0×10^3, and 0.6×10^3 kg ha^{-1} for sugarcane, soybean and Phaseolus beans respectively, with nitrogen contents of 1.4% for sugarcane, 5.0% for soybeans and 3.0% N for

Phaseolus beans, and 50% of the N in legumes and at least 17% of the N in sugarcane derived from symbiotic fixation[13], additions of nitrogen into the system via stems and leaf residue alone could be of the order of 16.6 kg N ha^{-1} from sugarcane, 17.5 kg N ha^{-1} from soybeans, and 6.3 kg N ha^{-1} from Phaseolus beans (Table 2). Including N_2 inputs via the legume grains of 42.5 and 21.3 kg N ha^{-1} for the rotation and intercropped phases, respectively (Table 2) brings the total fixed to 40–80 kg N ha^{-1}. Thus, legume rotations plus intercropping could add at least 20–63 kg N ha^{-1} during the first sugarcane crop and 6–21 kg N ha^{-1} during subsequent ratoon crops.

Nitrogen from N_2-fixation

Dinitrogen fixation in sugarcane mainly occurs in the rhizosphere[15,17], although there are also N_2-fixing micro-organisms in the roots, stalks, and phyllosphere[3,8]. The rate of fixation is still not known with certainty, but varies with plant variety[12] and soil moisture and season. Nitrogenase activity in the rhizosphere is higher during summer[16].

Tentative calculations of fixation rates, based on extrapolation from rates of ^{15}N uptake by non-amended intact root systems under simulated normal-atmosphere conditions, suggest that 3.4 kg N ha^{-1} yr^{-1} is fixed in or on the plant and 50 kg N ha^{-1} yr^{-1} in the rhizosphere[17]. N-balance experiments in large containers plus δ ^{15}N ‰ techniques (described by Vose[18], p. 170) suggest that 17% of the plant's nitrogen is derived from N-fixation directly[13]. This estimate is for a sugarcane variety that had one of the poorest nitrogenase activities of five Brazilian varieties tested. Working with a ratoon crop in South Africa, Purchase[9] has suggested that about 25 kg N ha^{-1} yr^{-1} might be derived from nitrogen fixation.

A field study using δ ^{15}N ‰ values[19] showed clear evidence of ^{14}N isotope dilution in 8 month old sugarcane, and differences among five cultivars. Recently we determined δ ^{15}N ‰ values for a variety of sugarcane grown either with fertilizer for 7 years or without fertilizer for 7 and 15 years (Table 3). Substantial isotopic dilution of atmospheric ^{14}N occurred in plants from all the treatments, implying assimilation of recently fixed N_2, although we need more knowledge of discrimination effects to fully evaluate the results. The difference between soil ^{15}N and plant ^{15}N is greater the longer the field is without fertilizer. The reason for this is not at present clear, because denitrification tends to lead to higher $^{15}N/^{14}N$ ratios[5,10].

Conclusion

The N-cycle in sugarcane is complex. We need much more detailed experimentation before we can assign reliable values to nitrogen losses and gains and before we can adequately evaluate the roles of nitrogen fixation and various cultural practices in the nitrogen economy of this crop.

Table 2. Potential inputs of nitrogen from N_2-fixation in renewed (first year) sugarcane ecosystems that incorporate rotation with soybeans and intercropping with Phaseolus beans. One-fourth of the total area planted to sugarcane in Brazil $(2.30 \times 10^6$ ha) is renewed each year

Component	Yield[a] (10^3 kg DM* ha^{-1})	Renewed area (10^3 ha)	(10^6 kg DM)	N in regional yield		N_2 fixed			
				(% N)	(10^6 kg N)	(%)	(10^6 kg N)	(kg N ha^{-1}) plant**	seed
Sugarcane	70.0	575	4025[b]	1.4	56.4	17	9.58	16.6	–
Soybeans (rotation)	1.0	575	402[c]	5.0	20.1	50	10.1	17.5	42.5
Phaseolus (intercropped)	0.6	380	159[c]	3.0	4.79	50	9.56	6.3	21.3

[a] Low average yield obtained in Brasil; does not include roots, litterfall, or harvest residue.
[b] Assumes 10% of yield remains in soil after harvest.
[c] Assumes 70% of yield remains in soil after harvest.
[d] Assumes 66% of renewed area $(575 \times 10^3$ ha) is planted to Phaseolus.
* DM, dry matter.
** Above ground residue.

Table 3. Natural δ ^{15}N ‰ of soil and sugarcane variety CB 45-3 grown with N-fertilizer for 7 yrs and without N-fertilizer for 7 or 15 years (Vose et al.[19])

Component	Nitrogen fertilizer		
	With 7 yrs	Without 7 yrs	15 yrs
Young bud	2.55	1.0	0.60
Leaves	6.88	5.57	2.75
Soil	12.65	14.29	9.11

Acknowledgements We thank the UNDP/IAEA project BRA/78/006 and the Centre de Energia Nuclear na Agriculture, Universidade de S. Paulo for funding and support.

References

1 Alejo N O 1980 Efeito do Nitrogênio Nítrico, Ammoniacal e de Uréia Sobre o Crescimento, Carbohidrato e Compostos Nitrogenados em Cana-de-Açúcar (*Saccharum* spp. v. NA 56–79) Cultivada em Solução Nutritiva – M.S. Dissertação Piracicaba, S.P. Brasil: ESALQ.

2 Cervellini A, Ruschel A P, Victória R L and Reichardt K 1980 Fate of ^{15}N applied as ammonium sulphate to bean crop. *In* Soil Nitrogen as Fertilizer or Pollutant, IAEA Panel-Proceedings Series STI/PUB/535, pp 23–36. Vienna: IAEA.

3 Graciolli L A and Ruschel A P 1981 Microorganisms in the phyllosphere and rhizosphere of sugarcane. *In* Vose P B and Ruschel A P (Eds.). Associative N$_2$-fixation. Vol. II. pp 91–101. Boca Raton, Florida: CRC Press.

4 Gutiérrez D J B 1981 Inoculation con *Rhizobium japonicum* en soya (*Glycine max* L.) Merril, intercalada a caña-de-azucar (*Saccharum officinarum* L.) en el Valle del Cauca-Colombia. *In* Vose P B and Ruschel A P (Eds.). Associative N$_2$-fixation. Vol. II. pp 169–175. Boca Raton, Florida: CRC Press.

5 Kohl D H, Shere C B and Commover B 1971 Fertilizer nitrogen: contribution to nitrate in surface water in corn belt watershed. Science 174, 1331.

6 Libardi P L, Victória R I, Reichardt K and Cervellini A 1982 Nitrogen cycling in a^{15}N-fertilized tropical bean (*Phaseolus vulgaris* L.) crop. *In* Robertson G P, Herrera R and Rosswall T (Eds.). Nitrogen Cycling in Ecosystems of Latin American and Caribbean. Plant and Soil 67, 193–208.

7 Malavolta E, Haag H P, Mell F A F and Brasil Sobro M D C (Eds.) 1974 Nutrição mineral de plantas cultivadas. Nutrição mineral e adubação de cana-de-açúcar, pp 259–292. São Paulo, Brazil: Bivraria Pioneira Editora.

8 Patriquin D G, Graciolli L A and Ruschel A P 1980 Nitrogenase activity of sugarcane propagated from stem cutting in sterile vermiculite. Soil Biol. Biochem. 12, 413–417.

9 Purchase B S 1980 Nitrogen fixation associated with sugarcane. *In* Collingwood, D (Ed.). Proceedings of the South African Sugar Technologists Association, pp 173–176. Natal, South Africa: SASTA.

10 Rennie D A and Paul E A 1974 Nitrogen isotope ratios in surface and sub-surface soil horizons. *In* Proceedings of Symposium on Isotope Ratios as Indicators of Pollutant Sources and Behaviour Indicators. IAEA Publ. No. SM-191/31, pp 441–453. Vienna: IAEA.

11 Ruschel A P 1980 As alternativas energéticas na agricultura tropical brasileira. Energia 2, 31–34.

12   Ruschel A P and Ruschel R 1978 Varietal differences affecting nitrogenase activity in the rhizosphere of sugarcane. *In* Reis, F S and Dick J (Eds.). Proc. XVI Congress International Society of Sugarcane Technologists 2, 1941–1948.

13   Ruschel A P and Vose P B 1977 Present situation concerning studies in associative N-fixation sugarcane, *Saccharum officinarum* L., Boletim Científico 045. Piracicaba, São Paulo, Brazil: CENA. 28 p.

14   Ruschel A P and Vose P B 1980 Nitrogen fixation as a source of energy in tropical agriculture. *In* Areus P L (Ed.). FAO/SIDA Workshop on Organic Recycling in Agriculture, San Jose, Costa Rica. Rome: FAO (*In press*).

15   Ruschel A P, Henis Y and Salatai E 1975 Nitrogen-15 tracing of N-fixation with soil-grown sugarcane seedlings. Soil Biol. Biochem. 7, 181–182.

16   Ruschel A P, Orlando F V and Zambello Jr E 1978 The effect of nitrogen, phosphorus and potassium fertilization and irrigation on nitrogenase activity and yield of sugarcane. *In* Reis F S and Dick J (Eds.). Proc. XVI Congress International Society of Sugarcane Technologists 2, 1903–1912.

17   Ruschel A P, Victória R L, Salati E and Henis Y 1978 Nitrogen fixation in sugarcane (*Saccharum officinarum* L.). *In* Granhall U (Ed.). Environmental Role of Nitrogen-Fixing Blue-Green Algae and Asymbiotic Bacteria. Ecol. Bull. Stockholm 26, 297–303.

18   Vose P B 1980 Introduction to Nuclear Techniques in Agronomy and Plant Biology. London: Pergamon International Library.

19   Vose P B, Ruschel A P and Salati E 1978 Determination of N_2-fixation especially in relation to the employment of nitrogen-15 and of natural variation. *In* Abstracts. XXVI Congresso Nacional de Botânica, pp 89–90. Brasília, D.F. Univ. Brasilia.

20   Vose P B, Ruschel A P, Victória R L and Matsui E 1981 Potential N_2-fixation by sugarcane (*Saccharum* sp.) in solution culture. I. Effect of ammonium *vs* NO_3^- variety and nitrogen level. *In* Vose P B and Ruschel A P (Eds.). Associative N_2-fixation. Vol. II. pp 119–125. Boca Raton, Florida: CRC Press.

Plant and Soil 67, 147–156 (1982).   0032-079X/82/0671-0147$01.50.   SU-13
© 1982 *Martinus Nijhoff/Dr W. Junk Publishers, The Hague.*

Nitrogen gains and losses in sugarcane (*Saccharum* sp.) agro-ecosystems on the coast of Peru

Ganacias y perdidas de nitrógeno en un agroecosistema de caña de azucar en la costa del Perú

S. VALDIVIA V.

Instituto Central de Investigaciones Azucareras (ICIA), Trujillo, Perú

Key words   Burning Deserts Irrigation N-fertilization N-cycling Peru Sugar cane

Abstract   Irrigated sugarcane crops on the Peruvian coastal desert require 300 kg N-fertilizer ha^{-1} for economically-optimum yields. About 24 kg N ha^{-1} is added to the system *via* irrigation water from rivers and deep wells. 1.2% of the total N in the soil is in available form during the growing season. N extracted by the aerial part of the plants ranges from 210–246 kg N ha^{-1} crop^{-1}; *ca.* 70% of this is removed in the harvest, and most of the remainder is lost *via* pre-harvest burning.

Resumen   La caña de azucar bajo riego en la zona desértica de la Costa Peruana requiere fertilización con 300 kg N ha^{-1} para que se obtengan rendimientos económicos máximos. Cerca de 24 kg N ha^{-1} se añaden al sistema provenientes del agua de riego tomada de rios y pozos. El 1,2% del nitrógeno total del suelo se encuentra en forma disponible durante la época de crecimiento. La parta Aérea de la planta extrae entre 210 y 246 kg N ha^{-1} por cosecha; el 70% de esta cantidad es exportado con la caña que va al molino y el resto se pierde por quema antes del corte.

Introduction

In Peru the major sugarcane plantations are situated on the northern coast, in a climatic region which is classified as hyper-arid[16] or subtropical desert (Holdridge, cited by Tosi[15]). Within this coastal desert Zamora[21] distinguishes three geomorphic or physiographic units: a) irrigated alluvial valleys, b) coastal plains or terraces, and c) ridges, hills and low slopes. Cane plantations are situated in the irrigated alluvial valleys, which have very similar soils and climatic characteristics and are suitable for the growth of many different kinds of crops.

In this study we evaluate N gains and losses in sugarcane systems within the Chicama valley, which has a cultivated land area of 72×10^3 ha, 37% of which are affected by salinity or poor drainage[7]. This valley contains the Casa Grande (22000 ha) and Cartavio (12500 ha) sugar cooperatives which, together with the individual growers, account for 50% of the total area planted with sugarcane. From the point of view of N fluxes, three distinct cultivation zones can be clearly differentiated: non-saline zones irrigated with river and well water (normal soils), zones irrigated with 'filter press mud water"[17], and saline zones. We limit the present examination to systems within normal zones.

Table 1. Soil characteristics of Chicama valley fields planted to sugarcane

Field	Depth (cm)	Texture (USDA)	pH	Specific conductivity (mmho cm⁻¹)	Na[a] (%)	Cations (meq 100 g⁻¹)	CaCO₃ (%)	Organic matter (%)	Total N (%)	C/N	Available N[b] (kg ha⁻¹)	Available P[c] (mg kg⁻¹)	Available K₂O[d] (mg 100 g⁻¹)
1 Mocollope	0–30	Clay loam	7.4	0.78	1.3	22.5	1.8	1.92	0.10	13.0	51	10	12.5
(n = 40)	30–60	Clay loam	7.5	0.60	1.3	22.5	2.0	1.40	0.09	11.0	45	4	10.6
	60–90	Clay loam	7.5	0.59	1.3	21.9	2.2	0.96	0.07	10.0	36	3	9.8
2 Chacarilla	0–30	Loam	7.6	1.26	2.0	22.0	2.1	1.73	0.09	13.7	45	9	10.5
(n = 40)	30–60	Loam	7.6	0.85	1.7	20.9	1.8	1.13	0.06	12.6	34	3	8.1
	60–90	Loam	7.6	0.72	1.9	17.9	1.3	0.82	0.05	11.9	27	3	7.7
3 Campo Nuevo	0–30	Silt loam	7.4	2.65	1.4	22.5	4.4	2.80	0.13	15.1	59	5	24.0
(n = 48)	30–60	Loam	7.4	1.13	1.0	20.8	4.8	1.55	0.08	13.4	40	2	10.9
	60–90	Loam	7.5	0.83	1.0	20.0	5.4	0.99	0.05	12.5	30	1	9.7
4 Chiclin 3	0–30	Silt loam	7.3	0.63	1.0	20.3	1.2	2.20	0.11	14.3	51	3	8.5
(n = 48)	30–60	Silt loam	7.4	0.49	1.0	20.1	0.3	1.91	0.11	12.8	51	2	8.2
	60–90	Silt loam	7.4	0.52	1.1	18.5	0.7	1.51	0.09	11.7	49	1	8.2
5 Chiquitoy	0–30	Loam	7.9	2.03	2.2	20.9	5.0	3.07	0.14	16.1	69	9	18.0
(n = 30)	30–60	Loam	7.9	1.26	1.9	18.5	6.3	1.76	0.09	14.1	47	3	10.8
	60–90	Loam	8.0	1.28	2.1	17.5	6.5	1.28	0.07	13.3	38	2	9.4
6 Chiclin 2B	0–30	Silt clay loam	7.9	1.20	1.6	19.9	1.1	1.99	0.12	11.1	59	7	11.2
(n = 45)	30–60	Silt loam	7.9	0.68	1.5	19.5	0.7	1.52	0.11	9.6	55	6	9.2
	60–90	Loam	8.0	0.58	1.7	17.3	0.6	0.90	0.08	7.4	48	5	8.2

[a] Per cent of total cations.
[b] Estimated as 1.2% of total-N.
[c] Olsen's method.
[d] Ammonium acetate method.

Study sites

The results presented below have been taken from experiments carried out in the Chicama valley with the cultivar H32-8560. Annual rainfall in the valley is generally less than 20 mm, the mean mid-day temperature is 20.1°C (range = 15–25°), mean relative humidity is 82.5% (range = 74–90%), and mean daily evaporation is 4.5 mm.

Sugarcane is planted from September to March, at the bottom of furrows spaced 1.5 m apart. Surface irrigation is the rule, and urea is the only commercial fertilizer applied, usually rates of 300 kg N ha^{-1} for newly planted cane, and somewhat less for ratoon cane. In some cases the fertilizer applications are divided into two-part doses, with the first dose usually incorporated during mechanized soil preparation, and the second dose applied from a knapsack to the surface of dry soil followed by irrigation. The age of the cane at harvest varies from 18 to 22 months and harvest occurs after the cane is left without water for 60 to 120 days. The crop is harvested with push rakes after the fields have been burned.

Under the FAO soil classification system the soils are classified as Fluvisols. Under the U.S. classification system these soils belong to the order Entisol, sub-order Fluvent and major group Ultifluvent. They are formed from recent alluvial deposits with a profile of incipient edaphogenetic development, and have a smooth relief, medium or moderately fine texture, a high total exchange capacity, an alkaline reaction, much calcium, low levels of organic matter and N, medium to low levels of available P, and moderate to good supplies of available K. In the low parts of the valley, saline soils or Solonchaks predominate; the areas with saline-sodic soils are small and those with sodic soils are insignificant. Table 1 gives some of the principal physical and chemical soil characteristics of fields where nitrogen fertilization experiments have been carried out.

Methods

Samples of irrigation water were taken monthly over a period of a year in the river outlets and in each of the wells. In both the water and the extracts of soil samples, nitrates were determined colorimetrically by the brucine method and ammonia was determined by the Nessler method. Total-N was determined by Kjeldahl techniques. To calculate the N added to the soil via irrigation water we used the volume recommended for obtaining maximum yields, which for 18 months is equivalent to 32 600 m^3 water per hectare[3].

Data from 22 fields (n = 436) were used for regression and correlation analyses of total N and available N. N lost *via* harvested cane was calculated with data from the Mocollope and Chacarilla fields (Table 1) alone. To ascertain the percentage of plant-N that is lost through burning, 8 plots (each 15 m^2) were sampled in 9 selected fields for greencane and burned-cane weights and N-contents[8].

Calculations of N immobilization in underground parts of the sugarcane was based on preliminary data (2 sample pits) from a root experiment in which a soil pit 1.5 m long, 0.5 m wide, and 0.9 m deep was sampled for roots in 10 cm deep sections. The dried-out roots were separated from moist rootlets and both were then weighed and analysed for N.

Table 2. Experimental conditions for sugarcane N-balance studies described in text

	Mocollope	Chacarilla	C. Nuevo	Chiclin 3	Chiquitoy 1	Chiclin 2B
Date of planting	Feb 1969	Jan 1969	May 1965	March 1971	Oct 1971	Oct 1975
Crop status	3rd cut	2nd cut	1st cut	1st cut	1st cut	4th cut
Parching period (months)*	4.5	4.3	3.5	3.5	4.7	5.5
Age at harvest (months)	19.6	21.5	22.0	20.7	22.7	18.2
Water applied (m^3 ha^{-1})	36 740	37 021	36 000	41 000	21 454	23 265
Area of plots (m^2)	2 500	2 500	180	2 500	2 500	2 500
Fertilizer range (kg N ha^{-1})	100–400	119–454	0–400	200–400	175–375	0–600
Replications	4	4	5	4	6	5
Fertilization rate for maximum yield (kg N ha^{-1})	391	390	373	337	375	567
Optimum fertilization rate (kg N ha^{-1})	348	342	318	332	na	446

na, not available.

* Parching period, period prior to harvest without irrigation.

Results and discussion

Principal nitrogen gains by the soil

In the Chicama valley, the only source of intentionally-applied N is urea. Results from numerous field experiments have demonstrated that maximum production of recoverable sugar is obtained with applications of more than 300 kg N ha^{-1}. Fig. 1 shows yield results obtained in 6 field fertilization experiments[6,11,13,14,18,19]. It can be seen that the yield of recoverable sugar rises with increasing doses of N applied to the soil, with optimum yields attained with doses of N between 318 and 348 kg N ha^{-1}, except in the case of field Chiclín 2B where the optimum was found to be 446 kg N ha^{-1}. As will be noted from Table 2, these experiments include spring, summer and autumn plantings, range

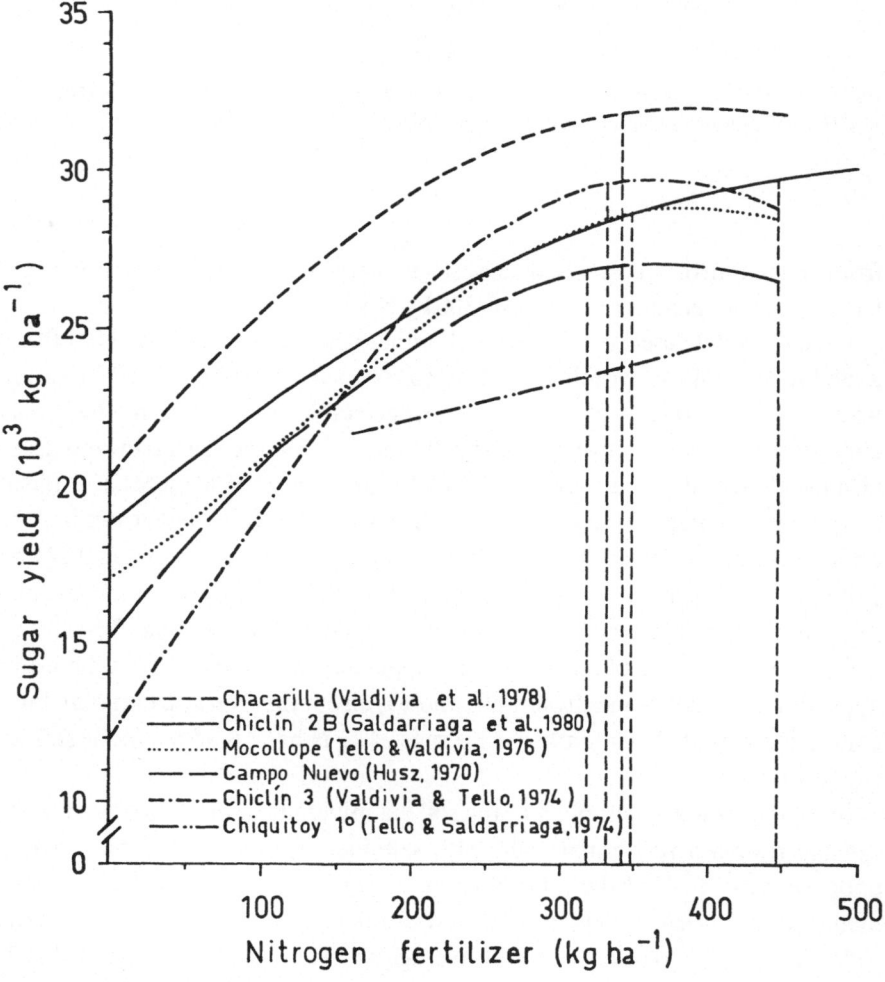

Fig. 1. Effect of nitrogen fertilization on yields of recoverable sugar and economically-optimum doses for 6 fields in the Chicama valley.

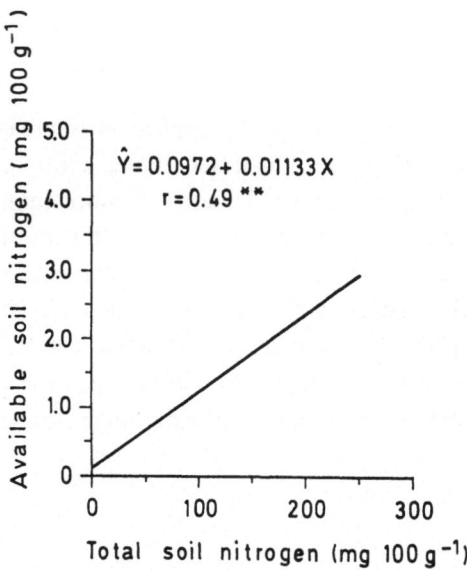

Fig. 2. Relationship between the percentage of total soil N and available soil-N ($NO_3^- $–N + NH_4^+–N) in normal Chicama valley soils (n = 436).

from first to fourth cut cane, and harvest ages vary from 18 to 22 months. Current fertilizer recommendations are now 300 kg N ha^{-1}.

Under coastal desert conditions with rainfall almost zero, almost all of the N added to the soil via hydrologic pathways comes from river and well irrigation water[10]. In the river water there is no detectable NH_4^+–N, and NO_3^-–N concentrations vary between 0.4 and 0.8 mg l^{-1} for the Cartavio and Casa Grande Cooperatives respectively. This gives an average of 20 kg N ha^{-1} added to systems irrigated with river water on the basis of the volume of water used to obtain maximum yields[3]. N concentrations in well water vary between 0.92 and 1.05 mg l^{-1} (90% as NO_3^- and 10% as NH_4^+) for Casa Grande and Cartavio respectively, which extrapolates to an average input of 32 kg N ha^{-1}.

Given that in years of normal water supply approximately 35% of the water applied to the fields comes from wells and 65% from the river, we find that, for a cultivation period of 18 months, on average 24 kg N ha^{-1} is added to the soil *via* irrigation water.

Inorganic N is also supplied to the soil via mineralization of roots and other organic residues. In normal soils, with salinities of less than 4 mmho cm^{-1}, approximately 1.2% of the total N is present in available form (Fig. 2). Gros[4] mentions that in calcareous soils mineralization of organic matter is slow, in the order of 0.8–1.2% annually. As an average for the 6 fields listed in Table 1, available (mineral) N was found to be 55.7 kg N ha^{-1} for the 0–30 cm layer, 45.3 kg N ha^{-1} for the 30–60 cm layer, and 38.0 kg N ha^{-1} for the 60–90 cm layer.

Principal pathways of nitrogen loss from sugarcane soils

There is a close relationship between the concentration of N in sugarcane and the amounts of N-fertilizer applied to the soil[19,20]. If the data from the experiments in the Mocollope and Chacarilla fields are expressed as N in harvested cane and correlated with the N applied to the field, we find a strong statistical significance. It can be seen from Fig. 3A that for fields fertilized with 300 kg N ha^{-1}, 147–172 kg N ha^{-1} is transported in burned cane to the factory for processing. There is also a strong correlation ($r = 0.75$) between the % of N lost through burning and the N applied to the soil (Fig. 3B). An amount equivalent to 30% of the 300 kg N ha^{-1} applied to the fields is lost through burning. These values indicate that the aerial part of the plant takes up 210– 246 kg N ha^{-1}, of which 63–74 kg N ha^{-1} are lost by burning the cane before harvest.

Preliminary data from a study still in progress show that the stubble left behind in the field after harvest contains approximately 90 kg N ha^{-1}. Rhizomes retain 45 kg N ha^{-1} after harvest, and roots retain 100 kg N ha^{-1}. Thus, the quantity retained by the unharvested part of the cane after harvest is 235 kg N ha^{-1}, a good part of which is in the form of dead organic matter.

The only information we have with respect to leaching losses of nitrogen is a field study with systems of drains at a depth of 1.8 m on a soil with a moderately fine texture. Under these conditions, losses of NO_3–N in drainage water show the N loss to be *ca.* 22 kg N ha^{-1} per year[11], which represents a loss of 33 kg N ha^{-1} for a period of 18 months.

Losses of N through leaching can be diminished by controlling irrigation, but in the period of active growth and in soils of moderate permeability, N losses through leaching need give no cause for concern[2].

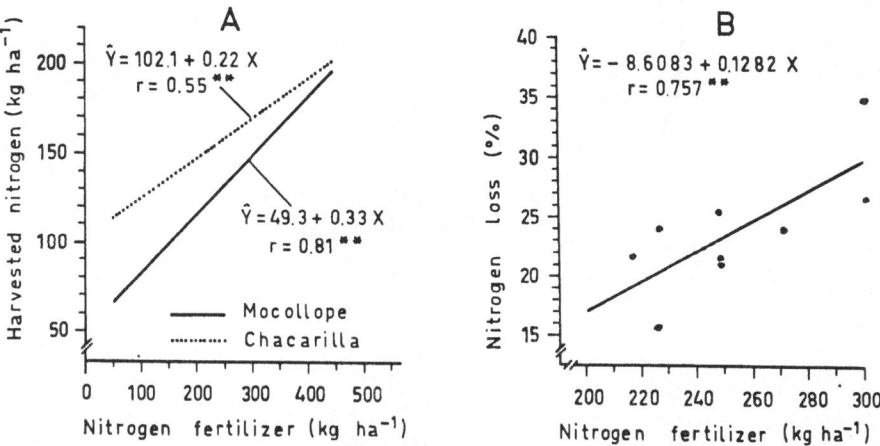

Fig. 3.   Effects of increasing levels of nitrogen fertilizer on the quantity of N harvested (A) (n = 40) and on the percentage of aerial-biomass N lost through burning (B) (n = 9).

The nitrogen balance

The principal gains and losses of nitrogen in normal Chicama valley sugarcane soils have been represented in Fig. 4; total input into the soil is estimated to be 463 kg N ha^{-1} and extraction at 478–514 kg N/ha. The balance is in favour of N-losses, with an apparent average loss of 33 kg N ha^{-1}. However, because a large proportion of the fine roots left in the soil after harvest will add a large quantity of mineralizable organic N to the soil, the system probably does not exhibit a net N-impoverishment.

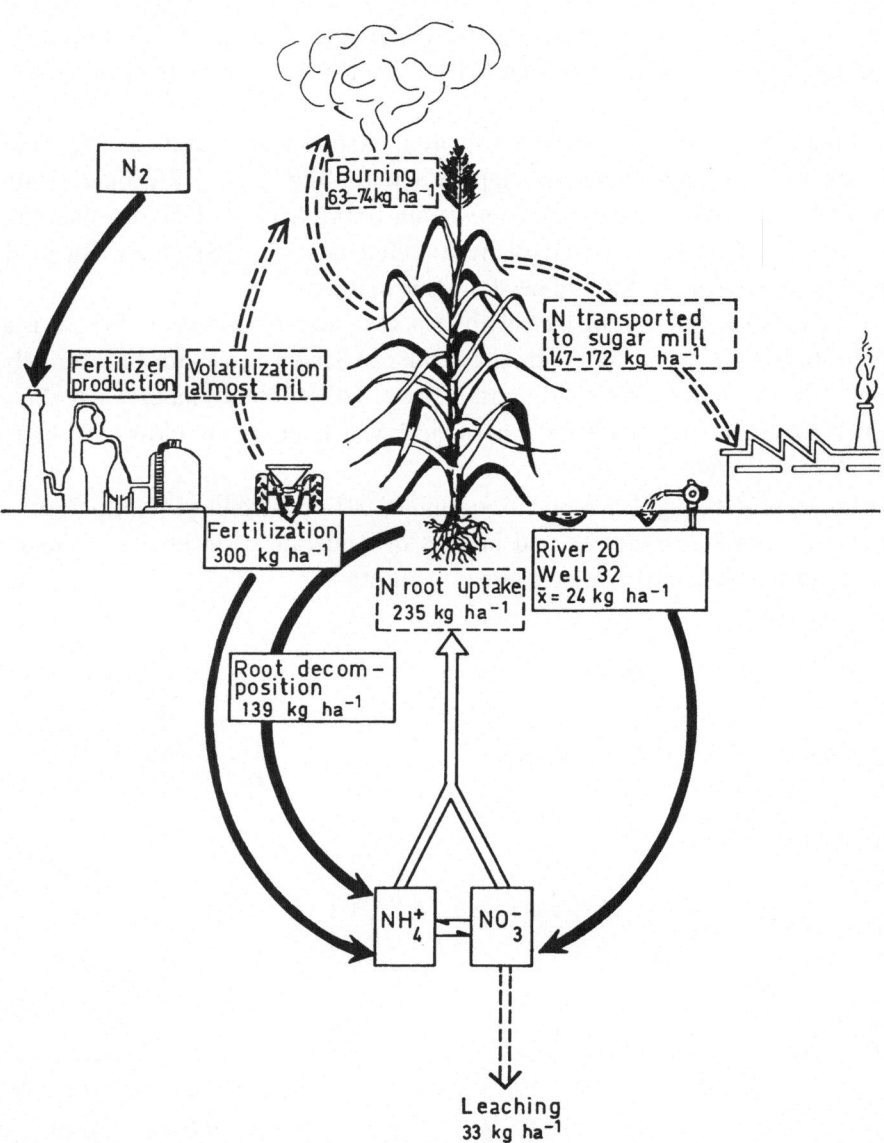

Fig. 4.   Nitrogen cycle of H32-8560 sugarcane in normal soils of the Chicama valley, for a growth period of 18 months. N$_2$-fixation, denitrification, and NH$_3$-volatilization fluxes are unknown.

It should be stressed that losses through NH_3-volatilization have been assumed to be insignificant, owing to the increasingly widespread tendency to apply urea below the soil surface. Volatilization losses are usually almost zero when fertilizer is applied in this manner[1,5,9,12].

It should also be stressed that, for lack of information, no account has been taken of the gain in N through non-symbiotic fixation (*e.g.* by Azotobacter), nor of losses through denitrification, net biological immobilization, and NH_4^+-fixation in the crystalline structure of clays.

Conclusions

1. Irrespective of sugarcane age at harvest, of whether sugarcane is newly planted or a ratoon crop, or whether the growth period has spanned 1 or 2 summers, for maximum yield more than $300 \, kg \, N \, ha^{-1}$ must be applied to sugarcane fields in the coastal desert of Northern Peru.

2. The average input of N via irrigation water is $24 \, kg \, N \, ha^{-1}$. 1.2% of the total N in the soil is in available form at any given sample date; this is equivalent to $139 \, kg \, N \, ha^{-1}$.

3. N extracted by the aerial part of the plant ranges from $210–246 \, kg \, N \, ha^{-1}$. Of this total but excluding harvest residue, 30% is lost through burning and 70% removed from the system in the form of cane transported to the factory. N in the harvest residue (sugarcane stumps, rhizomes and roots) was $235 \, kg \, N/ha$.

References

1　Allison F E 1966 The fate of nitrogen applied to soils. Adv. Agron. 18, 219–258.
2　Demolón A 1966 Principios de Agronomía. 1. Dinámica del suelo. Omega, Barcelona. 527 p.
3　Eppink D L 1974 Valor económico del agua. *In* Aspectos agrohidrológicos del cultivo de la caña de azúcar en el Perú. Boletín Técnico. Instituto Central de Investigaciones Azucareras (ICIA), 3, 10–23.
4　Gros A 1967 Abonos. 4° edición. Mundi Prensa, Madrid. 445 pp.
5　Hauck R D 1968 Soils and fertilizer nitrogen – A review of recent work and commentary. Trans. 9th. Int. Cong. Soil. Sci. Australia, Vol. 2, 475–486.
6　Husz G 1970 Maximale Stickstoffgaben zu Zuckerrohr. Die Bodenkultur 21, 219–230.
7　Masson M L 1973 Evaluación de la salinidad en el Perú. *In* Evaluación y control de degradación de tierras en zonas áridas de América Latina. Proyecto Regional FAO/PNUD RLA 70/457. Boletín Latinoamericano sobre fomento de tierras y aguas. 6, 363–384.
8　Morales C M, Valdivia V S and Pinna C J 1982 Pérdidas de nutrimentos por efecto de la quema de la caña de azúcar.
9　Olson R A, Frank K D and Dreier A P 1964 Controlling losses of fertilizer nitrogen from soils. Trans. 8th. Int. Cong. Soil Sci., Romania, Vol. 4, 1023–1032.
10　Pinna C J and Valdivia V S 1977 Contribution of nutrients by irrigation water in sugarcane areas of Peru. Proc. International Society of Sugar Cane Technologists 16, 1441–1453.
11　Saldarriaga A S, Paz-Vergara P E and Angulo A E 1982 Determinación de pérdidas de nitrato y amonio por lixiviación de un campo cultivado con caña de azúcar.
12　Takahashi D 1967 Volatilization losses of applied N found at Ewa. Annu. Report Exp. Sta. HSPA, pp 8–9.

13   Tello A H and Saldarriaga A S 1974 Respuesta del cultivar de caña H32-8560 a la aplicación de dosis ascendentes de nitrógeno. Saccharum 2, 30–54.

14   Tello A H and Valdivia V S 1976 Efecto de la aplicación tardía del nitrógeno en el cultivar de caña H32-8560. I. Acción de la misma, con relación a la aplicación temprana en el rendimiento. Saccharum 4, 1–18.

15   Tosi J 1960 Zonas de Vida Natural en el Perú. Ed. I.I.C.A.-OEA. Boletín Técnico No 5, 271 p.

16   UNESCO 1977 Un nuevo mapa de la distribución mundial de las regiones áridas. La naturaleza y sus recursos. 13, 2–3.

17   Valdivia V S and Pinna C J 1980 Aporte de nutrimentos al suelo por el riego con 'agua de cachaza' en las Cooperativas Azucareras del Perú. Saccharum 8 (1). *In press.*

18   Valdivia V S and Tello A H 1974 Efecto del abonamiento NP en el rendimiento y calidad de la caña de azúcar. Saccharum 2, 35–69.

19   Valdivia V S, Tello A H and Pinna C J 1978 Efecto de la aplicación tardía del nitrógeno en el cultivar de caña H32-8560. II. Influencia de la dosis crecientes, en el rendimiento, calidad y nutrientes, así como en su variación con la edad. Saccharum 6, 146–177.

20   Valdivia V S, Pinna C J and Tello A H 1980 Efecto de la aplicación tardía del nitrógeno en el cultivar de caña H32-8560. III. Acción de la misma con relación a la aplicación temprana, en la calidad y nutrientes en la planta. Turrialba 30, 3–8.

21   Zamora J C 1974 Los suelos de las tierras bajas del Perú. Oficina Nacional de Evaluación de Recursos Naturales. 21 p.

Plant and Soil 67, 157–165 (1982).   0032-079X/82/0672-0157$01.35.   SU-14
© 1982 Martinus Nijhoff/Dr W. Junk Publishers, The Hague.

Potential growth of alfalfa (*Medicago sativa* L.) in the desert of Southern Peru and its response to high NPK fertilization

Potencial de crecimiento de alfalfa (Medicago sativa L.) en el desierto del sur del Peru y su Respuesta a la alta fertilizacion NPK

M. N. VERSTEEG, I. ZIPORI, J. MEDINA and H. VALDIVIA
*FAPROCAF Project, Apartado 1319, Arequipa, Perú**

Key words Alfalfa   Deserts   K-fertilization   N-cycling N_2-fixation   Peru   P-fertilization   Rhizobium   Urea

Abstract   Irrigated and highly fertilized alfalfa growing in the deserts of Southern Peru reached maximum growth rates of about 200 kg dry forage $ha^{-1} d^{-1}$ during the summer period and of 150 kg $ha^{-1} d^{-1}$ during winter. These high rates were maintained for 10 to 20 days, after which growth rates declined. 'Ceiling' yields of about 5000 kg dry forage ha^{-1} in summer and 3500 kg ha^{-1} in winter were obtained in a growth period of 53 days. Simulations with an adapted model indicate that a decreased photosynthetic rate for aging leaves is a probable cause for the decrease. Growth curve simulations were also very sensitive to the level of carbohydrate reserves in the root system at harvest.

High NPK fertilization (420, 280 and 420 kg $ha^{-1} yr^{-1}$ of urea–N, P and K respectively) increased NO_3–N in soil 2.5 fold, available–K 1.6 fold, and available–P 4.3 fold. NH_4–N content did not increase. The higher amounts of available nutrients resulted in only about 10 percent increases in maximum growth rate and maximum yields. With respect to plant composition (%N, %P and %K), a significant response only to the higher P level was observed and a very slight, non-significant response to the higher K level was also noted. High N-fertilization did not increase the N-content of the plant, indicating that the rhizobia present are able to fix up to 900–1000 kg N $ha^{-1} yr^{-1}$ in the above-ground herbage. Commercial inoculants did not improve this N-fixation capacity; even in virgin desert soils after only a few harvests, yields as well as N-contents of non-inoculated alfalfa were of the same order of magnitude as inoculated alfalfa.

Resumen   La alfalfa irrigada y altamente fertilizada cultivada en los desiertos del Sur del Perú, ha alcanzado tasas máximas de crecimiento de cerca de 200 kg ha^{-1} día^{-1} de forraje seco, durante el período de verano, y de 150 ka ha^{-1} día^{-1} durante el invierno. Estas altas tasas fueron mantenidas por 10–20 días, después de los cuales las tasas de crecimiento bajaron. Rendimientos tope de cerca de 5000 kg ha^{-1} de forraje seco en verano y 3500 kg ha^{-1} en invierno, fueron obtenidos en un período de crecimiento de 53 días. Simulaciones con un modelo adaptado, indican que una tasa decreciente de fotosíntesis de hojas viejas, es una probable causa para esta baja. Simulaciones de curvas de crecimiento fueron también sensibles al nivel de reservas de carbohidratos en el sistema de raíces en la cosecha.

La alta fertilización NPK (420, 280 y 420 kg ha^{-1} año^{-1} Urea–N P y K respectivamente) aumentó el nivel de NO_3–N en el suelo 2,5 veces, K disponible 1,6 veces y P disponible 4,3 veces. El contenido de NH_4–N no aumentó. Las altas cantidades de nutrientes disponibles resultaron en solo cerca de 10% de aumento en tasas máximas de crecimiento y en máximos rendimientos. Con respecto a la composición de plantas (%N, %P y %K) se observó una respuesta significativa solo al alto nivel de P, y una muy ligera, no significativa, respuesta al más alto nivel de K fué también notada. Alta

* FAPROCAF is a joint Peruvian-Dutch and Israeli agricultural research project in Arequipa, Peru.

fertilización con N no aumentó el contenido de N de la planta; indicando que las cepas de Rhizobium presentes son capaces de fijar hasta 900–1000 kg N ha^{-1} anño^{-1} en el forraje. Inoculantes comerciales no mejoraron esta capacidad de fijación de N; aún en suelos vírgenes desérticos después de solo algunas cosechas, rendimientos tanto como contenidos de N de alfalfa no inoculada fueron del mismo orden de magtitud que alfalfa inoculada.

Introduction

Present insight into the basic processes that govern crop growth is such that a reasonable assessment of the production capacities of crops and pastures under varying circumstances may be made[1]. Some authors[10,11] state that under optimal conditions only small differences in productivity and water-use efficiency exist between species having similar photosynthetic pathways. On the other hand, for some crops strongly conflicting data are available, especially in water use efficiency. One of these crops is alfalfa, for which much data exist showing it to have a high water consumption relative to that of other crops[2]. There are also indications, however, that a higher water use efficiency can be obtained[4].

In 1978, a Peruvian-Dutch-Israeli agricultural research project started in the desert of Southern Peru (Pampa de La Joya), with the aim of improving forage production under the prevailing conditions. Since precipitation at the site is practically zero, all plant production depends on the application of irrigation water. Thus, any improvement in the efficiency of water use may lead to higher production under limited water supply. Because of the absence of rain, water supply can be closely controlled, and this provides favorable conditions to investigate soil-plant-water relationships. An important additional objective of the project is the validation and improvement of plant-growth simulation models, such as those developed by de Wit *et al.*[19] and van Keulen[10], so that the insights and experience acquired in the present project can be more easily extrapolated to and used for predicting crop growth in other areas.

Since alfalfa is a major component of crop rotations in the southern deserts of Peru, much attention is paid to this crop. In order to establish the production capacity of the crop and its potential growth rate under the prevailing environmental conditions, one of the first experiments established was one in which alfalfa was grown under optimal conditions of water and nutrient supply.

In this paper we report on an experiment designed to define growth curves (dry matter accumulation through time) for highly-fertilized and non-fertilized alfalfa. Since there are various reports of positive responses of this legume to N-fertilization[5,12], and because alfalfa has a very high nitrogen requirement of around 800 kg N ha^{-1} for a good crop that produces around 25000 kg dry matter ha^{-1}, we included nitrogen in the fertilizer application. In addition, some results of modelling activities and experiences with non-inoculated alfalfa in virgin desert soil (with no previous vegetation) are reported.

Methods

In a well-established alfalfa field (cv. Tambo) at the San Camilo experimental station (16°42′ S, 71°11′ W), half of a 24 × 196 m area was fertilized with 420 kg urea–N, 280 kg triple super-phosphate-P and 420 kg potassium sulphate-K ha^{-1} yr^{-1} administered in seven equal applications at the start of each of several experimental growth curve determination. The other half was not fertilized during the experiment. The field had been fertilized with 40 kg N, 45 kg P (both as ammonium phosphate) and 65 kg potassium sulphate-K when the alfalfa was planted in virgin desert soil 1.5 years before the start of the experiment. Each half was divided into 4 equal blocks. These blocks were further subdivided into 3 sub-blocks; each sub-block was used to determine one growth curve based on cutting 8 strips about 10 m^2 each in area. These cuttings were made at one-week intervals following the initial general harvest of the total field, except for the first two cuttings that were made at days 7 and 11, respectively.

Within a given sub-block, the location of a plot for a certain cutting was chosen randomly. After the final sampling at 8 weeks, the whole field was harvested (general harvest) and the next growth curve was determined in the second sub-block. After the third growth curve, the first sub-block was used again for the fourth curve, the second sub-block for the fifth, and so on. This repetition scheme assumes that the effects of cutting a sub-block in 8-weeks portions (on subsequent growth-rate determinations) are negligible after two general harvests. For each growth curve the locations of weekly-harvest plots within the sub-block were newly randomized. Harvesting was done with an Agria power-mower which left a stubble of about 5 cm. All harvested alfalfa was removed from the field.

The experimental field was on sandy soil (90% sand, 5% silt, 5% clay) with a pH of 7.2–7.5. Soil analyses before the start of the experiment showed a total N content of 0.02–0.03 %N (Kjeldahl), an available-P content of 5 mg P kg dry soil^{-1} (Olsen), and a K content of 14.5 mg K kg dry soil^{-1} based on extraction in a 1:1 soil:water mixture. Later determinations showed that K content was equivalent to 170 kg of ammonium-acetate-extractable-K. Irrigation water contained 3.2–3.8 mg N l^{-1}.

In several series plants and soils were sampled and analyzed for nitrogen, phosphorus and potassium. Determinations of nitrogen were done using the Kjeldahl method; phosphorus and potassium in plants were determined following digestion in sulphuric and nitric acids (Schouwenburg and Walinga[15]). Potassium in soils was extracted in a 1:1 soil:water solution and phosphorus extracted by Olsen's method[9]. Extracts were then analyzed by colorimetric techniques (P) or by flame spectrophotometry (K). The leaf-area-index (LAI) was also measured in some series from samples of 0.5 m^2, using an electronic surface-area meter.

The growth curve obtained in January–February 1980 (summer growth curve) and the one obtained in mid-June–August 1980 (winter curve) were used as a basis for simulations, which were based upon the 'ARID CROP' simulation model described by van Keulen[10].

In addition, a non-inoculated field of alfalfa, sown on virgin soil in August 1979 together with barley, was sampled in December 1980 for nitrogen determinations. The results were compared to those obtained in an inoculated field on virgin soil. Both fields received the same initial fertilization as the experimental field noted above.

Results and discussion

Growth curves

Growth-rate curves for alfalfa grown during different periods of the year are shown in Fig. 1. Maximum growth rates for above-ground biomass varied between 150 kg (winter) and 200 kg (summer) dry matter ha^{-1} d^{-1}, but high rates could generally be maintained only during a relatively short period of 10–20 days; thereafter forage growth rates declined in 4 of the 7 growth periods. This was to such an extent that above-ground dry matter accumulation remained constant or even declined by the time of final harvest, thus exhibiting 'ceiling yields'.

In a descriptive model adapted for these conditions (van Keulen and Zipori, unpublished), this phenomenon of strongly-decreasing rates of above-ground dry matter accumulation could be achieved by assuming that the photosynthetic capacity of the leaves decreases during this period and concurrently a substantial part of the carbohydrates formed are translocated from aboveground organs to the roots (refilling). Incorporation of these assumptions into the simulation model of van Keulen[10] resulted in a reasonable agreement between simulated results and the measured curves obtained both in summer and winter (Fig. 2). Also, the simulated values of LAI were not far from those measured for the summer crop (Fig. 3).

The accumulated maximum yields obtained in the 7 growth periods gives a total annual yield of 27400 kg ha^{-1} of dry herbage. This is approximately the

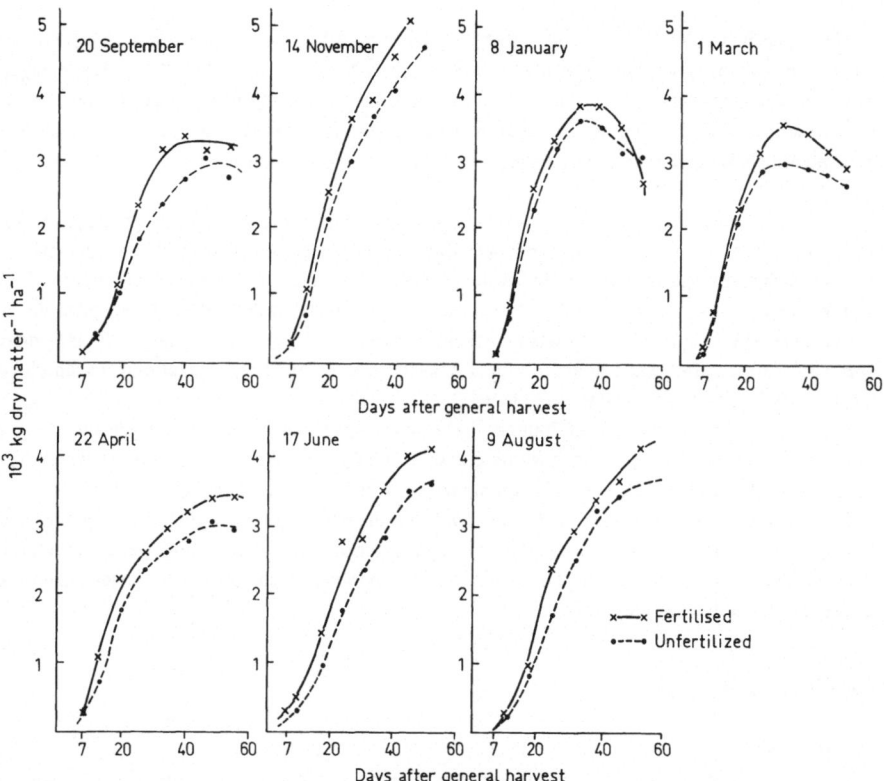

Fig. 1.   Growth curves for previously-established, irrigated alfalfa grown with and without NPK-fertilizer in a Southern Peru desert during different parts of the year. Dates in upper-left corners denote beginnings of regrowth after general, whole field mowings. Values are for above-ground biomass.

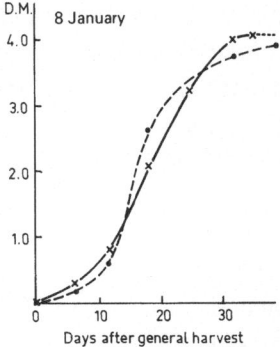 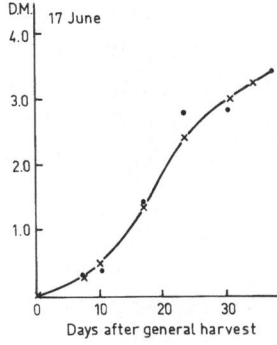

Fig. 2.   Predicted (× —— ×) vs actual (●----●) growth curves for summer (8 January) and winter (17 June) grown alfalfa. Predicted curves are from van Keulen and Zipori (unpublished). See Fig. 1 legend for further explanation.

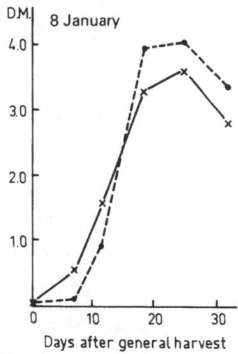

Fig. 3.   Predicted (× —— ×) vs actual (●-----●) changes in leaf area with time for summer-grown alfalfa. See Fig. 2 legend for further explanation.

same as the total annual dry herbage production obtained in two experiments with the same variety at a cutting frequency of 10 harvests per year (26.5–27.7 · 10^3 kg) dry matter ha^{-1} yr^{-1}. However, if the average yields from the growth curve obtained at 30 days in summer (3.5 · 10^3 kg ha^{-1}) and in winter (2.95 · 10^3 kg ha^{-1}) could be maintained at a cutting frequency of every 30 days during the whole year, an annual yield of 39.3 · 10^3 kg ha^{-1} could theoretically be possible. However, generally too-frequent cutting leads very quickly to decreased yields[6, 17]. In fact, we have found that yields on day 35 in winter-grown fields harvested every 35 days are much lower (2000 kg ha^{-1}) than yields obtained on day 35 in fields harvested every 56 days (a practice which was employed in the growth curve experiment; Fig. 1).

Available experimental evidence in the literature shows a low soluble carbohydrate level in the roots at the same time as yields decrease in crops frequently cut[17]. This is in close agreement with results obtained in a sensitivity analysis on reserve levels in the roots at the moment of harvest (using the adapted 'ARID CROP' simulation model). Fig. 4 shows that in the simulation of the 17 June crop (cf. Fig. 2), a decrease in initial reserve level, from 40% soluble carbohydrates to 30%, led to a decrease of $1.1 \cdot 10^3$ kg ha^{-1} dry matter at 35 days, or about 34%.

Plant and soil nutrient levels

A striking phenomenon is the observation of the relatively small differences in nitrogen, phosphorus and potassium biomass concentrations between the highly fertilized plants and the non-fertilized ones. Average plant composition during three growth periods (1 March, 17 June, and 9 August) showed the largest differences between fertilized and non-fertilized alfalfa to be in P-contents (Fig. 5). Yet the average P-content of non-fertilized alfalfa at 10% flowering (0.27% P at 37 days) is above the level of 0.23% P considered by Nelson and Barber[13] to be upper limit of phosphorus deficiency in alfalfa, but slightly below the 0.30% P considered indicative of healthy plants. Other authors consider 0.25% as sufficient[16].

Differences in K-content were much less pronounced, and the level of 3.4% K at 10% flowering is still above the highest critical level of 3.3% K reported by Rhykerd and Overdahl[16].

Even smaller differences were found in N-content between fertilized and non-fertilized alfalfa, the latter containing 3.8% N at 10% flowering and the former, 4.0% N. These values are far above the critical level of 3.0% at 10% flowering reported by Nelson and Barber[13]. The values of 3.8–4.0% N in the dry matter indicate that a good crop in this environment ($30 \cdot 10^3$ kg ha^{-1} yr^{-1} dry matter yield, obtained several times at the site) removes around 1100–1200 kg N in harvests annually. If it is assumed that of this amount about 60–75 kg N comes

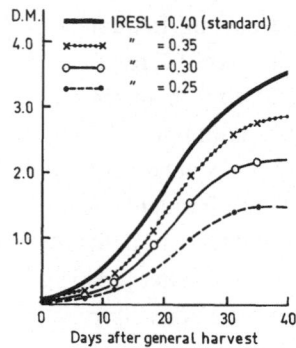

Fig. 4.   Predicted dry matter accumulation over time for crops with various initial levels of soluble carbohydrates in roots (IRESL) at day 0. The predicted growth curve of 17 June (Fig. 2), for which a 40% IRESL was assumed, is used as a model standard.

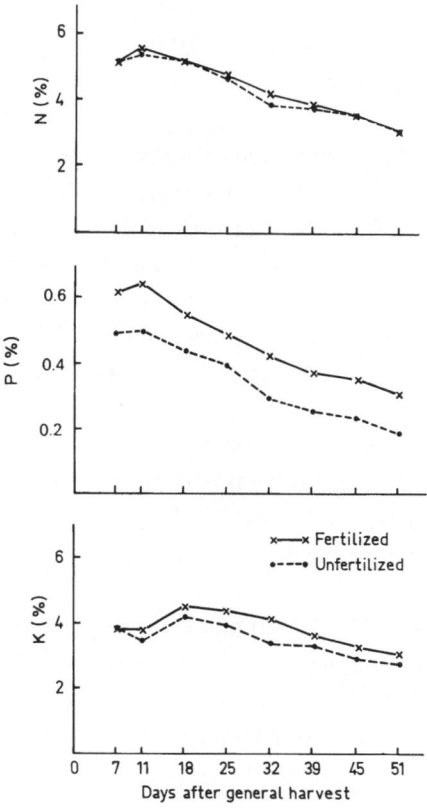

Fig. 5.   Nitrogen, phosphorus, and potassium in above ground biomass of alfalfa grown with and without NPK-fertilizer. Values plotted are means of three cropping periods. (1 March, 17 June and 9 August.) See Fig. 1 legend for further explanation.

from irrigation water and about 100–200 kg N comes from mineralization of organic-N (estimated from the uptake of N by unfertilized maize after alfalfa), then alfalfa apparently fixes around 900–1000 kg N ha^{-1} yr^{-1} in this environment, a value that is appreciably higher than the range of 50–465 kg N ha^{-1} yr^{-1} reported in the literature[3,14].

Table 1. Soil mineral-N, available-P (Olsen), and potassium in 1:1 soil-water extract in the upper 60 cm of the fields used for growth curve determinations. Values are means of 20 samples taken between 30 April and 1 August, and all are reported as mg kg dry soil^{-1}

Layer	NPK-fertilized				Not fertilized			
	NO_3^--N	NH_4^+-N	P_2O_5-P	K^+	NO_3^--N	NH_4^+-N	P_2O_5-P	K^+
0–30	19.4	1.4	27.9	24.1	10.3	1.7	4.8	15.3
30–60	24.1	1.7	15.2	19.6	4.8	1.4	5.2	12.3
Average	21.8	1.6	21.6	21.9	7.6	1.6	5.0	13.8

This high N-fixation capacity may partly be explained by special growing conditions, any of which can enhance N-fixation:

i)   a year-round growing season;
ii) high solar radiation [7,8];
iii) a light sandy soil that provides good aeration [15]; and
iv) an adequate and completely controlled water supply [15].

Analyses of soil samples from the root zone (Table 1) show that here also the most important effect of fertilization was in the phosphorus pool sizes. There was a more than four-fold increase in P_2O_5-P level in the fertilized part of the experimental field relative to the unfertilized part. NO_3^--N and potassium contents increased very significantly as well, whereas NH_4^+-N remained at the same level in both parts of the field.

Plant-N concentrations of samples from a non-inoculated alfalfa crop sown in virgin desert soil were not different from N-concentrations in samples from an inoculated field also sown with alfalfa cv. Tambo (3.4 [$\pm$ 0.1] %N and 3.4 [$\pm$ 0.2] %N, respectively). All samples were taken in September, 1980, 32 days after previous cutting.

Further, yields of both fields were at about the same level, $25-30 \cdot 10^3$ kg dry forage ha^{-1} yr^{-1}. Root samples from both fields showed good nodulation, indicating a very efficient natural inoculation at this site, probably by bacteria supplied from irrigation water and from dust from neighboring fields.

Acknowledgements   The authors thank the FAPROCAF laboratory staff, M. de Wijs, E. Soto, G. Sanz, V. Frisancho and E. González for the skilfull analysis of the samples.

References

1   Anon 1975 Actual and potential herbage production in arid regions. Wageningen, The Netherlands: Dept. of Theoretical Production Ecology Agric. Univ. and Centre for Agrobiological Research (CABO) 14 p.
2   Briggs L J and Shantz H L 1913 The water requirements of plants. I. Investigation in the Great Plains in 1910 and 1911. U.S. Dept. Agr. Bur. of Plant Ind. Bull. 284. Washington, D.C.: USDA 49 p.
3   Burton J C 1972 Nodulation and symbiotic nitrogen fixation. *In* Hanson C H (Ed.). Alfalfa Science and Technology, pp 229–246. Madison, Wisconsin: Am. Soc. of Agron. 229–246.
4   Evenari M, Tadmor N H and Shanon L 1970 Consumptive water use of range plants under desert conditions. Bi-annual Res. Rep. 4, Hebrew Univ. Dept. of Botany, project No. A 10-SWC-20. Jerusalem, Israel: Hebrew University. 16 p.
5   Feigenbaum S, Barzily S, Levin I and Kafkafi U 1976 Studies on fertilization of field-grown irrigated alfalfa. II. Effect of N and K supply on growth rates, dry matter and protein yield. Final report to Israel Chemicals Ltd., Proj. No. 301/010 ARO. Bet Dagan, Israel: Agric. Research Organization (AZO) 16 p.
6   Feltner K C and Massengale M A 1966 Influence of temperature and harvest management on growth, level of carbohydrates in the roots, and survival of alfalfa (*Medicago sativa* L.). Crop Sci. 6, 585–588.

7   Ham G E, Lawn R J and Brun W A 1976 Influence of inoculation, nitrogen fertilizers and photosynthetic source-sink manipulations on field-grown soybeans. *In* Nutman P S (Ed.). Symbiotic Nitrogen Fixation in Plants, International Biological Programme 7, pp 239–253. Cambridge: Cambridge Univ. Press.

8   Hardy R W F and Havelka U D 1976 Photosynthate as a major factor limiting nitrogen fixation by field-grown legumes with emphasis on soybeans. *In* Nutman P S (Ed.). Symbiotic Nitrogen Fixation in Plants, International Biological Programme 7, pp 221–239. Cambridge: Cambridge Univ. Press.

9   Hesse P R 1971 A Textbook of Soil Chemical Analysis. London: John Murray, Ltd. 520 p.

10   Keulen H van 1975 Simulation of Water Use and Herbage Growth in Arid Regions. Wageningen, The Netherlands: Centre for Agric. Publishing and Documentation. 176 p.

11   Lof H 1975 Water use efficiency and competition of arid – zone annuals with special reference to the grasses *Phalaris minor* Retz and *Hordeum murinum*. Agric. Res. Report 853. Wageningen, The Netherlands: Centre for Agric. Publishing and Documentation. 109 p.

12   Malm N R 1972 Fertilization of established alfalfa with nitrogen and potassium. Agric. Exp. Stn. Res. Rep. 240, New Mexico USA: New Mexico State Univ. 7 p.

13   Nelson W L and Barber S A 1964 Nutrient deficiencies in legumes for grain forage. *In* Sprague H B (Ed.). Hunger Signs in Crops, pp 143–179. N.Y.: David McKay Company.

14   Nutman P S 1976 IBP field experiments on nitrogen fixation by nodulated legumes. *In* Nutman P S (Ed.). Symbiotic Nitrogen Fixation in Plants, International Biological Programme 7, pp 211–237, Cambridge: Cambridge Univ. Press.

15   Pate J S 1976 Physiology of the reaction of nodulated legumes to the environment. *In* Nutman P S (Ed.). Symbiotic Nitrogen Fixation in Plants, International Biological Programme, pp 335–360. Cambridge: Cambridge University Press.

16   Rhykerd C L and Overdahl C J 1972 Nutrition and fertilize use. In Hanson C H (Ed.). Alfalfa Science and Technology, pp. 437–468. Madison, Wisconsin: Am. Soc. of Agron.

17   Robinson G D and Massengale M A 1968 Effect of harvest management and temperature on forage yield, root carbohydrates, plant density, and leaf area relationships in alfalfa (*Medicago sativa* L. cv. Moapa). Crop. Sci. 8, 147–151.

18   Schouwenburg J Ch and Walinga J 1978 Methods of Analysis for Plant Material. MSc Course on Soil Science and Water Management. Wageningen, The Netherlands: Agricultural University. 92 p.

19   Wit C T de, Goudriaan J, Laar H H van, Penning de Vries F W T, Rabbinge R, Keulen H van, Louwerse W, Sibma L and Jonge C de 1978 Simulation of assimilation, respiration and transpiration of crops. Simulation Monographs, ISBN No. 90-220-0601-8. Wageningen, The Netherlands: Centre for Agric. Publishing and Documentation. 141 p.

Plant and Soil 67, 167–186 (1982).   0032-079X/82/0672-0167$01.50.   SU-15
© 1982 *Martinus Nijhoff/Dr W. Junk Publishers, The Hague.*

Crop utilization and fixation of added ammonium in soils of the West Indies

Utilización y fijación de nitrógeno amoniacal en cultivos de las Indias Occidentales

N. AHMAD, E. D. REID, M. NKRUMAH, S. M. GRIFFITH and L. GABRIEL
Department of Soil Science, University of the West Indies, St. Augustine, Trinidad, W.I.

Key words Ammonia-fixation   Fulvic acids   Humic acids   Maize   N-fertilizer   Sorghum
Tropical soils   West Indies

Abstract NH_4^+-fixation by inorganic and organic soil components and crop utilization of fertilizer nitrogen was studied in a number of Caribbean soils using ^{15}N fertilizers.

At moderate rates of nitrogen application, NH_4^+-fixation by clays during several-week laboratory incubations was rapid and highly variable, ranging from less than 10% to over 70% of the NH_4^+ added. The 2:1 lattice types were the most reactive, and the process was almost complete by one week after fertilization. Fixation increased with rate of NH_4^+–N application and was higher at elevated temperatures in soils that were allowed to air-dry during incubation. NH_4^+–N fixation was more active in the fulvic fractions of the soil organic matter than in the humic fractions (25–69% vs 0–3% of the added NH_4^+ was fixed in each, respectively). There was little incorporation of fertilizer-N by the N-containing fractions of soil organic matter.

Plant uptake of added NH_4^+–N in greenhouse pot experiments showed that a greater percentage of fertilizer-N was taken up by Sudan grass (*Sorghum sudanense*) at a fertilizer rate of 40 kg NH_4^+–N ha^{-1} than at a rate of 200 kg NH_4^+–N ha^{-1}. However, the recovery was low, ranging from 10 to 25 percent of that applied.

In field experiments with maize (*Zea mays*), urea-N was rapidly lost when applied to soils in a wet tropical environment. At normal rates of application (100 kg urea-N ha^{-1}), only about half of the fertilizer was utilized by the crop. Mulches did not significantly affect the fate of added nitrogen; however, mulching did result in increased yields for dry-season cropping, due probably to water conservation effects.

There is good indication that for conditions in Trinidad, NH_4^+–N is better utilized and less subject to unidentified losses than is urea. Addition of fertilizer-N resulted in crop uptake of important quantities of native soil nitrogen.

The Caribbean Andepts were outstanding in that they showed very little NH_4^+-fixation under all experimental conditions and very little tendency for apparent nitrification of added NH_4^+–N.

Resumen   Cuando se aplicaron tasas moderadas de nitrógeno se observó una rápida y muy variable proporción de fijacion de NH_4^+ en las arcillas durante incubaciones de varias semanas en el laboratorio. Los valores de fijación iban de 10 a 70% del NH_4^+ añadido. Las arcillas del tipo 2:1 fueron las que mostraron mayor actividad y el proceso se completó en una semana después de la fertilización. La fijación aumentó con la tasa de aplicación de N amoniacal y fué mayor a altas temperaturas en aquellos suelos que se sometieron a secado durante la incubación. La fijación de amonio fué asi mismo mayor en las fracciones fulvicas de la materia orgánica que en las húmicas (25–29% vs 0–3% del NH_4^+ añadido, respectivamente). Hubo poca incorporación de nitrógeno proveniente del fertilizante en las fracciones nitrogenadas de la materia orgánica del suelo.

La absorción del NH_4 añadido en experimentos en potes bajo condiciones de invernadero, fué mayor cuando se aplicaron 40 kg NH_4^+–N ha^{-1} que cuando se aplicaron 200 kg NH_4^+–N ha^{-1} a *Sorghum sudanense*. Sin embargo la recuperación fue baja, entre 10 y 25% del total aplicado.

En experimentos de campo con maiz el nitrógeno proveniente de la aplicación de úrea se perdió rapidamente en suelos tropicales húmedos recuperándose solo cerca de la mitad del total aplicado (100 kg úrea-N ha^{-1}). La adición de material vegetal al suelo no alteró significativamente el comportamiento del nitrógeno aplicado aunque si aumentó los rendimientos en cultivos de secano, probablemente debido a efectos de conservación de agua.

Existe una buena evidencia de que en Trinidad la utilización de nitrógeno amoniacal es mejor y menos sujeta a pérdidas no identificadas que las aplicaciones de úrea. La fertilización nitrogenada produjo importantes incrementos en la absorción del nitrógeno nativo del suelo.

Los Andepts del Caribe mostraron una notable baja capacidad de fijación de $NH_4{}^+$ bajo todas las condiciones experimentales y poca tendencia a la nitrificación del nitrógeno añadido.

Introduction

Due to a long history of mainly exploitive agriculture in the Caribbean, soil deterioration characterized by considerable erosion of the topsoil and depletion of plant nutrients has occurred. One of the consequences of this is that total-N is low on the whole (*ca.* 0.17–0.37% N), except for the Andepts, which have high organic-C (10–25% C) and total-N (0.68–1.50% N) contents. Amounts of soil-N are generally positively related to rainfall. Transformations of nitrogen in Caribbean soils are such that even with the addition of ameliorants, in particular inorganic fertilizers, there is a very short lived effect on the concentration of nitrogen in the soil system, and initial conditions are rapidly regained.

In the rainforests of Trinidad, about 60 kg N ha^{-1} is added to the forest floor annually in the 7,000 kg leaf-fall ha^{-1}. Where natural vegetation is removed, there is a rapid loss of soil-N. For example, as much as 70% of the nitrogen in the forest floor of a Mora forest (*Mora excelsa*) can be lost when this forest is cleared and Caribbean pine (*Pinus caribbea*) planted. The rate of decomposition of plant litter is rapid and rates of 0.44% and 0.45% per day have been measured for cacao leaves[18] and Mora leaves[9], respectively.

De Souza[12] noted the widespread occurrence of nodulation in local rainforest legumes, and more recent research[8,23] has confirmed the importance of biological N-fixation in the N-cycle of these systems. Studies on the mineralization of organic-N indicate that the rate is dependent more on the nature of the nitrogen than on the total amount present. Available mineral-N ($NO_3{}^-$-N and $NH_4{}^+$-N) in West Indian soils under natural vegetation or cropping is usually low (10–40 mg N kg dry soil^{-1}). Accumulation of $NO_3{}^-$-N occurs throughout the dry season and could reach a concentration of over 150 mg $NO_3{}^-$-N kg dry soil^{-1} at the commencement of the rainy season[16,20]. $NH_4{}^+$-N as well as urea-N are rapidly nitrified in many Caribbean soils[7,26]. However, in poor soils, $NH_4{}^+$ added as $(NH_4)_2SO_4$ could be partially recovered in up to 6 weeks[26]. Hardy and Rodrigues[17] showed that in some sugar cane soils in Guyana, large quantities of $NH_4{}^+$-N exist in easily exchangeable form; these could have accumulated from the long use of $(NH_4)_2SO_4$ or from the decomposition of swamp residues.

Rodrigues[25] showed that some West Indian soils contain an appreciable amount of fixed NH_4^+-N which could range from 14 to 78% of the total N-content. Levels of this fixed ammonium increased progressively with depth and could lead to abnormally low C/N ratios for subsoils. NH_4^+-N thus fixed has been shown to be particularly associated with micaceous or illitic clay minerals. Ahmad et al.[4] found that K^+ ions could be preferentially fixed relative to NH_4^+. Ammonium may also be immobilized by microbial activity, and there is evidence to indicate that the active fractions of organic matter can absorb and retain the NH_4^+ cation[5,6]. NH_4^+-N may also be volatilized from Caribbean soils if it is produced too rapidly or added in large amounts as fertilizer[7,11,21].

Nitrification inhibitors (such as pyrimidine compounds and thiourea) and slow-release fertilizers have been used in West Indian soils in an attempt to reduce the amount of NO_3^--N leached from the soil. The recovery of applied-N by grass[30,31] and the response of rice to nitrogen applications[3] are increased with the use of nitrification inhibitors, but the amount of NH_3 volatilized when sulphate of ammonia is applied to grass is also increased. Ahmad and Shand[2] showed that nitrification is more rapid in soils of high pH and that liming of acid soils increased the rate. However, NO_2^--N accumulated in high pH soils or in acid soils which were limed.

Results of empirical experiments on the utilization of N-fertilizers in the Caribbean region have indicated that recovery by the crop to which the fertilizer is added range from 20 to 100%, depending upon the crop, the nature of the soil, the rate of N-application, and rainfall conditions. Ahmad et al.[4] obtained high recoveries for pasture crops growing on soils with good drainage when N-application rates were less than 100 kg N ha^{-1} per application. Conversely, Fletcher[14] found that such crops grown on poorly drained soils recovered just over 20% of the nitrogen added. Fletcher[14] also observed that fertilizer use did not result in increased total soil-N, and therefore concluded that losses either in gaseous form or through leaching may have been very important.

To learn more about the exact fate of added-N, we carried out fertilizer-use studies using labelled-N sources to follow directly the various pathways of the added fertilizer. Accordingly, we performed a series of laboratory, greenhouse and field experiments to assess ammonium fixation by clay and organic fractions of representative soils and N-uptake and utilization by crops. The results are reported and discussed in this paper.

Methods and materials

Laboratory experiments

For studies on the fixation of NH_4^+-N by soil clay, the upper 15 cm of 15 Caribbean soils which represent a range of texture, mineralogy, land-use histories and other properties (Table 1) were collected, air-dried, and after 4 weeks of storage treated as follows:

i) Fixation of added NH_4^+-N over time 2,000 mg NH_4^+-N kg dry soil^{-1} was added as $(^{15}NH_4)_2SO_4$ with 11.2% excess ^{15}N to 300 g of 7 of the soils, and each was incubated at room temperature (*ca.* 30°C) for 8 weeks. The soils (replicated 3 times) were brought to 50% waterholding-capacity (WHC) with distilled-H_2O and rewetted every four days. After one, four, six and eight weeks, 10 g sub-samples were analyzed for exchangeable-N and fixed-N[27]. Humic and fulvic acids, extracted from the soils with 0.5 N NaOH solution in an N_2 atmosphere[15] were analyzed for total-N.

ii) Effects of different levels of added-NH_4^+ on NH_4^+-fixation by clay fractions Clay fractions from seven of the selected soils were separated by dispersion with sodium hexametaphosphate and centrifugation[19] and to 100 mg of each, 200, 400, 800 and 1,600 mg N kg dry soil^{-1} as $(^{15}N-NH_4)_2SO_4$ (11.2% ^{15}N excess) were added. The samples (replicated 3 times) were wetted to saturation and re-wetted daily with 1 ml distilled-H_2O. Fixed-NH_4^+ was determined as before after two and six weeks.

iii) Effects of temperature and moisture on NH_4^+-fixation by clay fractions Clay fractions obtained as above were incubated with 800 mg NH_4^+-N kg dry soil^{-1} as $(^{15}NH_4)_2SO_4$ (11.2% ^{15}N excess) at 25°C and 40°C. After wetting the soils to saturation, 3 replicates of each were either a) stoppered to prevent moisture loss (closed system), b) re-wetted when dry (repeated wetting and drying), or c) not re-wetted after the initial wetting (open system). All samples were analyzed for fixed-NH_4^+ as above after a 4-week incubation period.

iv) Effects of removing organic matter on NH_4^+-fixation by clay fractions One g samples of clay obtained as described above were treated with NaOBr solution to remove organic carbon and washed successively with a 0.1 N of NaCl and 600 g C_2H_5OH l^{-1} as described by Griffith[15]. NH_4^+ $-N$ (800 mg N kg dry soil^{-1} as $(^{15}NH_4)_2SO_4$ (11.2%^{15}N excess) were added and the samples incubated in a saturated condition at 25°C. All samples were re-wetted daily and at eight weeks were analyzed for fixed-NH_4 as above.

v) NH_4^+-fixation in the humic and fulvic acid fractions Organic-C in the soil was fractionated into humic and fulvic acids[15]. One g samples of each fraction were placed in test tubes and treated with 2000 mg $^{15}NH_4^+-N$ kg dry soil^{-1} and enough water to saturate the soils, and then incubated for two weeks. Water loss was monitored gravimetrically and lost water replaced every 3 days. After incubation, 0.1 g subsamples were removed for ^{15}N analyses.

Fate of nitrogen added to Sudan grass: greenhouse experiments

Three two-kg replicate subsamples of each Caribbean soil (Table 1) were put into clay pots and treated with either 20 or 100 mg N kg dry soil^{-1} as $(^{15}NH_4)_2SO_4$ (11.2% ^{15}N excess). Each pot was further treated with 100 mg P kg dry soil^{-1} as $CaHPO_4$ and 100 mg KCl-K kg dry soil^{-1}, as well as trace elements[26]. The pots were sown with Sudan grass (*Sorghum sudanense* var. Piper) and the soils were kept continuously moist thereafter by placing the pots in saucers to which water was added daily. The crops were harvested after 35 days and the stools allowed to tiller to give a subsequent crop which was harvested at day 70. After the second crop, the roots were removed from the pots. Following a 70-day fallow, during which no water was added, the pots were again sown to Sudan grass, and P, K, and trace nutrients added as outlined above. The crop was harvested 35 days later. All crop harvests were dried, weighed, and ground for total-N and ^{15}N analyses.

Soils in the pots were sampled 14 days after each fertilization and at the conclusion of the experiment (175 days). Samples were air-dried and analyzed for exchangeable-N, fixed NH_4^+-N, total-N, hexosamine-N, serine and threonine-N, and amino-N, and the ^{15}N components of each. Fulvic and humic acids were also extracted from the soils and analyzed for their total-N and ^{15}N contents.

Table 1. Properties of selected soils used in NH$_4^+$-N fixation studies

Soil	Texture % silt	Texture % clay	pH	C$_{organic}$	N$_{organic}$	CEC (meq 100 g^{-1})	Mineralogy[a]	Soil taxonomy (USDA)	Land-use[b]
Barataria peat	nd	nd	4.8	22.90	1.44	nd	peat	fluvaquentic tropohemists	c
Montserrat clay	18	47	6.4	3.14	0.24	26.0	verm, mont, kao.	tropic hapludolls	c
Princes Town clay	8	90	7.1	1.41	0.17	35.0	mont.	aquentic chromuderts	c
Sevilla clay	20	55	6.5	0.76	0.09	23.0	mont, kao.	aquentic chromuderts	c
Talparo clay	15	65	4.2	0.78	0.13	35.0	mont, kao., ill.	aquentic chromuderts	c
Arena fine sand	7	11	5.0	0.84	0.06	4.7	kao.	orthoxic quartzipsamments	f
Valencia fine sand	17	20	4.8	1.90	0.13	3.0	kao.	typic troporthods	f
Montreal loam A	13	10	5.5	7.82	0.74	15.0	all. B	typic dystrandepts	s
Montreal loam C	21	22	5.5	2.43	0.25	6.6	all. A	typic dystrandepts	s
Soufriere loam (top)	11	15	5.8	1.95	0.18	10.5	all. B	typic vitrandepts	c
Soufriere loam (buried)	24	14	6.2	1.88	0.22	11.5	all. B	typic troporthents	c
Akers loam	12	20	5.8	1.67	0.15	11.6	hal.	udic haplustalfs	c
Bellevue loam	14	11	6.0	1.57	0.10	7.7	hal, kao.	typic tropudalfs	c
St. Anns clay loam	14	62	7.0	3.29	0.38	4.4	gib, boeh.	typic eutrorthox	s
Cuffy Gully stony loam	16	28	6.6	1.67	0.14	24.9	mont, ill.	typic dystropepts	s

[a] verm = vermiculite, mont = montmorillonite, kao = kaolinite, ill = illite, all = allophane, hal = hallosite, gib = gibsite, boeh = boehmite.

[b] c = cropped, f = forest, s = scrub.

nd = not determined.

CEC = cation exchange capacity.

Fate of nitrogen added to maize: field experiments

Experiments with maize (*Zea mays* L. var. Pioneer × 304) were conducted on St. Augustine loam (pH 5.2, CEC 4.2, total-N 0.11%, total-C 1.2%, WHC 25%, wilting point 8%, and classified as an Orthoxic Tropudult by Smith[28]) to study the fate of added nitrogen, the forms of the added nitrogen that might be incorporated into the organic/inorganic fraction of the soil, and in particular the subsequent availability of the added-N to a succeeding crop. The effects of mulching on the fate of added-N were studied as well.

The experimental area has a mean annual rainfall of about 1600 mm limited almost entirely to the period June to December. In the first season, urea ^{15}N at 1, 3 and 10% N atom excess were respectively added to 2-m^2 plots. Just before planting, eight plots for each level of enrichment were selected randomly. Four of the plots at each level of enrichment were treated with 10×10^3 kg bagasse mulch (dry weight) ha^{-1} (to an average depth of two cm). Labelled urea fertilizer was applied to the soil at 100 kg N ha^{-1} and superphosphate-P and KCl-K were added at rates of 100 and 150 kg ha^{-1}, respectively. Maize was then hand planted in three rows 60 cm apart at 30 cm spacing. After the first crop only unlabelled N-fertilizer was applied to the plots, and four control plots (two mulched and two unmulched. all with added P and K but none with added-N) were established. In addition, the eight plots that were treated with 1% excess ^{15}N were fallowed since the start of the third season (dry season, 1978 to 1979) and plots on which 3% excess ^{15}N was applied, were fallowed during the fourth 1979–1980 dry season.

Soil samples were taken at harvest (108–120 days after planting), air-dried, and crushed to 2 mm for analyses for NO_3^--N, NH_4^+-N and organic-N. The samples were also analyzed for ^{15}N. Plant material taken at harvest was prepared for analysis by sorting into stems, leaves, and seeds, and these fractions were then oven-dried, ground and analyzed for total-N and ^{15}N. From this data, the utilization of nitrogen by the 1977–78 wet season crop was calculated as well as the incorporation of the fertilizer N into the various plant components.

Laboratory analyses for ^{15}N

After performing total-N determinations for soil and plant material[27], the distillate was further acidified, concentrated by evaporation and then made up to volume (50 ml). An aliquot was placed in a Rittenberg flask and under high vacuum and after immediate freezing and thawing during which atmospheric-N was excluded, the sample was reacted with alkaline NaOBr. The released-N was collected in a discharge tube under high vacuum and the tube was sealed. A Statron NO-15 Emission Spectrometer was used to estimate the isotope ratios ^{14}N/^{15}N from which the ^{15}N content was calculated[13].

^{15}N content was obtained by subtracting the natural abundance (0.366%) from the measured values, and the recovery of ^{15}N in any sample was calculated from the total-N contents and enrichment values[22].

Results and discussion

Laboratory experiments

i) Fixation of added-NH_4^+ over time Fixation of added-$(^{15}NH_4)SO_4$ was greatest in the clay soils rich in smectoid minerals, for example Montserrat clay and Princes Town clay, but the soils which were texturally coarser (Arena fine sand, Valencia fine sand, Cuffy Gully stony loam) maintained a more constant level of fixed-NH_4^+ over longer periods (Table 2). The amounts of $(^{15}NH_4)_2SO_4$ fixed by the organic soil (Barataria peat) approximated those fixed by the sandier soils but were more consistent. Over the eight-week period there was a rapid

Table 2. Fixed-NH_4^+ in Caribbean soils (0–15 cm) following additions of 2000 mg $^{15}NH_4$–N kg dry soil^{-1}

Soil	Initial NH_4–N*	Week 1 NH_4–N	Week 1 % df*	Week 2 NH_4–N	Week 2 % df	Week 4 NH_4–N	Week 4 % df	Week 8 NH_4–N	Week 8 % df
Barataria peat	323	470	14	420	16	350	14	340	12
Montserrat clay	279	740	60	730	57	720	57	680	52
Princes Town clay	349	1010	59	950	57	690	50	660	45
Sevilla clay	280	450	45	480	43	470	41	450	41
Talparo clay	490	1040	45	990	48	970	46	910	45
Arena fine sand	18	20	9	21	14	30	21	25	13
Valencia fine sand	104	200	36	200	36	190	35	190	31
Cuffy Gully stony loam	104	560	24	560	23	560	18	550	22

* Values are mg NH_4^+–N kg dry soil^{-1} and % df is per cent of fixed NH_4^+–N derived from fertilizer as determined by ^{15}N analysis.

return to initial levels of fixed-NH_4^+ in the organic soil. Isotope discrimination in the migration and entrapment of NH_4^+ in the lattices of the alumino-silicates could be a contributory factor to low levels of fixed fertilizer NH_4^+ in the inorganic soils, but this is not likely to explain the low levels in the organic soil. Isotopic replacement of ^{15}N in the active fraction may also have occurred in both inorganic and organic soils, but the importance of this process was not determined.

ii) Effects of different levels of added-NH_4^+ on NH_4^+-fixation by clay fractions   An eight fold increase in the level of added-NH_4^+ resulted in variable changes in the amounts of NH_4^+ fixed two and six weeks later by the clay-fractions of the 7 soils studied (Table 3). The effect was greater in the heavy clays containing relatively higher amounts of $2:1$ minerals (montmorillonite, illite, vermiculite), particularly at the higher concentrations of added-NH_4^+. However, there was more fixed-NH_4^+ at six weeks than at two in all the samples studied. Incubation conditions ($40°C$, re-wetted daily) could have induced a high capacity for fixing NH_4^+-N. In all treatments, but particularly at the 200 and 400 mg added-NH_4^+-N kg dry soil^{-1} levels, substantial N-mineralization occurred during incubation and this could be the source of the unlabelled NH_4^+-N that was fixed as a result of the treatments. The Akers soil, an Andept, behaved unusually in that it fixed much less NH_4^+-N than the other soils at all concentrations, although like the other soils, the amount fixed increased with time.

iii) Effects of temperature and moisture on NH_4^+-fixation by clay fractions   In the closed system where the samples were prevented from losing moisture, NH_4^+-fixation by the clay fractions incubated at $40°C$ was usually lower than in those at $25°C$ (Table 4). Levels of NH_4^+-fixation at $25°C$ were least in the closed system. In the open system in which moisture loss was not prevented, much higher levels of NH_4^+-N were fixed at $40°C$ than at $25°C$; in the fractions that were repeatedly wet and dried, this difference was especially marked. At the higher temperature and in the two non-closed water regimes, clay fractions containing larger proportions of $2:1$ layer silicates fixed the most NH_4^+-N. The percent of fixed-NH_4^+ derived from the fertilizer was much higher at $40°C$ than at $25°C$ regardless of the water regime. Where the clays were dried and re-wetted, a consistently higher level of fixation occurred in all the samples. Initial adsorption of $(^{15}NH_4)_2SO_4$ to the clay fractions, then, was particularly temperature dependent, *i.e.* generally higher at the higher temperature.

Isotopic replacement by $^{15}NH_4^+$-N is dependent on mass action, and a period of drying increases the concentration of $^{15}NH_4^+$-N in the solution phase. Thus, the percentage of fixed-NH_4^+ from fertilizer-N was consistently higher in the two systems which included a dry period than in the constantly-moist closed system, regardless of the temperature. Generally, therefore, NH_4^+-fixation in the field

Table 3. Effects of different levels of added $^{15}NH_4^+$ (mg NH_4^+-N kg^{-1} dry soil) on NH_4^+-fixation by the clay fractions of seven Caribbean soils (0–30 cm)

Soil	Initial NH_4^+-N	200				400				800				1600			
		2 wks		6 wks		2 wks		6 wks		2 wks		6 wks		2 wks		6 wks	
		NH_4^+-N	%df*	NH_4^+-N	%df	NH_4^+-N	%df	NH_4^+-N	%df	NH_4^+-N	%df	NH_4^+-N	%df	NH_4^+-N	%df	NH_4^+-N	%df
Montserrat clay	558	900	36	1300	50	950	40	1440	59	1040	40	1440	52	1100	37	1490	53
Princes Town clay	350	1150	40	1580	53	1300	40	1690	50	1310	49	1850	71	1290	55	1820	77
Sevilla clay	560	1020	31	1290	41	1060	33	1300	39	1010	29	1410	40	1020	30	1460	41
Talparo clay	700	1300	39	1700	51	nd	nd	1650	47	1220	36	1770	54	1100	34	1770	52
Valencia fine sand	520	720	22	880	30	730	20	880	27	730	21	990	27	690	22	1030	36
Akers loam	nd	140	13	180	16	100	15	150	25	100	9	150	10	110	10	150	16
Cuffy Gully stony loam	320	980	51	1270	63	970	47	1290	61	880	48	1340	76	880	51	1340	75

nd = not determined.
* See Table 2.

Table 4. Effects of organic matter removal, temperature, and water content on NH_4^+-fixation in clay fractions of seven Caribbean soils (0–30 cm) incubated with 800 mg $^{15}NH_4^+$-N kg dry soil^{-1}

	Organic matter				Moisture regime								
	Present		Removed		Closed system[a]				Open system[b]				Rewetted[c]			
					40°C		25°C		40°C		25°C		40°C		25°C	
	NH_4^+-N*	% df*	NH_4^+-N	% df	NH_4^+-N	% df	NH_4^+-N	% df	NH_4^+-N	% df	NH_4^+-N	% df	NH_4^+-N	% df	NH_4^+-N	% df
Montserrat clay	460	25	700	42	450	24	530	nd	780	nd	500	nd	1270	29	460	31
Princes Town clay	460	36	570	39	530	34	630	19	980	57	590	22	1616	51	570	34
Sevilla clay	490	16	740	27	600	26	660	26	660	nd	570	30	1270	59	620	16
Talparo clay	530	14	700	26	620	19	660	11	940	38	620	13	1680	48	670	51
Valencia fine sand	440	11	530	27	470	14	480	14	510	22	440	15	840	54	480	11
Montreal loam	50	10	40	11	50	nd	100	31	40	64	nd	nd	130	20	nd	nd
Cuffy Gully stony loam	350	13	500	nd	390	35	350	9	900	63	340	9	1300	34	350	31

[a] stopped to prevent evaporation; [b] not rewetted; [c] rewetted when dry; nd = not determined.
* See Table 2.

should occur to a greater extent where the soil temperature is high and where the soil undergoes rapid cyclic wetting and drying. These are conditions to which all of the studied soils are normally exposed in the field.

iv) Effects of removing organic matter on NH_4^+-fixation   The removal of organic matter by NaOBr treatment resulted in increased NH_4^+-fixation in the clay fractions of all but one of the soils (Table 4). Increased NH_4^+-fixation ranged from 20 to 50% except in the Montreal loam, an Andept, in which NH_4^+-fixation decreased 20%. There were similar changes with respect to the isotopic exchange for the predominantly 2:1 layer silicates in these clay fractions. The Montreal loam soil contained much organic matter but in association with allophane.

These results show that removal of the organic matter associated with clay greatly increases the ability of the clays to fix NH_4^+, particularly if the clay is an expanding-lattice type. Under field conditions such soils may fix more NH_4^+ as soil erosion progresses and as they lose organic matter, which occurs during continual cropping.

v) Ammonium fixation in the humic and fulvic acid fractions   Chemical and isotopic analysis of the nitrogen contents of the humic and fulvic acid fractions of 14 fertilized soils indicated that most of the added-NH_4^+-N was incorporated into the more water-soluble fulvic acid fraction; very little isotopic exchange occurred with the humic acid fraction (Table 5). The fulvic acid fractions of soils developed on recent volcanic ash containing larger proportions of allophane and halloysite than kaolinite and montmorillonite have consistently more nitrogen in the organic matter analysed. All the mineral soils containing high proportions of 2:1 layer silicates and kaolinite have appreciably lower levels of nitrogen in the fulvic acid fractions as compared to peat and allephanic soils.

The fulvic acid-N in the Princes Town clay may in part be the result of NH_3-volatilization when the fertilizer was added to and incubated with this soil. This soil had the highest pH (pH $\geq$ 7) of all soils studied (Table 1). However, even when other extraction techniques (0.1 M $Na_4P_2O_7$ and ultrasonic dispersion) were used, the percentage of fulvic acid-N derived from the fertilizer ranged from 42 to 53%, a small increase compared to the value in Table 5 (40%), and absolute fulvic acid-N ranged from 15 to 51 mg N kg dry soil^{-1} in the organic fraction. Thus, the organic fraction in the Princes Town clay soil is apparently destabilized with difficulty; this is a characteristic of many Vertisols[1].

Fate of nitrogen added to Sudan grass: greenhouse experiment
i) Utilization of native and applied nitrogen by Sudan grass   In all 5 of the soils studied there was greater N-uptake with higher rates of fertilization (Table 6). For Montserrat clay, Princes Town clay and Montreal loam, this N-uptake difference disappeared by the third crop (Table 7). However, in all soils and for

Table 5. Total-N in the humic and fulvic acid fractions of 14 Caribbean soils fertilized with 2000 mg $^{15}NH_4$–N kg dry soil^{-1} before a two-weeks incubation

Soil	Fulvic acid-N		Humic acid-N	
	N*	% df*	N	% df
Barataria peat	1840	25	7144	3
Montserrat clay	1290	55	220	nd
Princes Town clay	240	40	nd	nd
Sevilla clay	1280	61	nd	nd
Talparo clay	1270	63	nd	nd
Arena fine sand	1220	66	110	3
Montreal loam A	2260	40	700	nd
Montreal loam C	1570	47	90	3
Soufriere loam (top)	1680	52	320	nd
Soufriere loam (buried)	2510	51	160	3
Akers loam	1830	66	160	nd
Bellevue loam	1120	69	80	4
St. Anns clay loam	1170	48	220	1
Cuffy Gully stony loam	1140	64	nd	nd

nd = not determined.
* Values are final levels of nitrogen (mg N kg dry soil^{-1}) and % df is the percent of the nitrogen derived from the fertilizer as determined by ^{15}N analysis.

Table 6. N-uptake and fertilizer-N recovered by Sudan grass (*Sorghum sudanense*) grown in pots at two levels of fertilization with $(^{15}NH_4)_2SO_4$*

Soil	Fertilizer rate	Total plant-N	Fertilizer-N in plant	Fertilizer-N recovered (%)
Montserrat	20	50.5	4	10
clay	100	101.5	21	10.5
Princes Town	20	39.5	5	22.5
clay	100	46.5	18	9
Sevilla clay	20	17.5	2.5	6.5
	100	38.5	17	9
Talparo clay	20	27.5	3.5	9
	100	54.5	22.5	21.5
Cuffy Gully	20	39.5	2	5
stony loam	100	66.5	16	8

* Values are mg N kg dry soil^{-1} except fertilizer-N recovered (per cent).

Table 7. N-uptake and fertilizer-N recovered by Sudan grass (*Sorghum sudanense*) grown in pots at 2 levels of $(^{15}NH_4)_2SO_4$ fertilization over 3 cropping periods. N-fertilizer was applied only before crop 1

Soil	N-fertilizer rate	Crop 1		Crop 2		Crop 3	
		N*	%df*	N	%df	N	%df
Montserrat clay	20	6.6	4	11.3	5	22.4	13
	100	46.7	19	20.7	20	22.4	32
Princes Town clay	20	8.8	9	12.5	6	18.1	23
	100	9.7	33	13.9	30	18.0	51
Sevilla clay	20	6.0	10	12.1	5	4.3	26
	100	10.9	46	19.5	40	7.9	54
Talparo clay	20	6.4	12	15.2	4	9.1	27
	100	12.9	54	18.2	32	23.3	49
Montreal loam A	20	nd	nd	4.2	3	3.6	6
	100	nd	nd	11.5	14	3.7	25
Cuffy Gully stony loam	20	7.7	9	12.4	3	19.2	6
	100	14.0	33	27.3	25	25.0	20

* N values are mg N kg dry soil^{-1} and % df is percent of total-N obtained from fertilizer-N based on ^{15}N analysis.
nd = not determined.

each crop data from the isotopic analyses show that at the higher level of fertilization a greater percentage of the nitrogen taken up was derived from the fertilizer.

Recovery of the fertilizer-N by Sudan grass was unexpectedly low (Tables 6 and 7), and this could be because of denitrification and volatilization losses of fertilizer-N or because biological N-fixation in the rhizosphere of the grass may have been considerable, thus making the plant less dependent on fertilizer-N. Biological N_2-fixation to 30 kg N ha^{-1} yr^{-1} has been measured in Sudan grass systems (Ruschel, A. P., personal communication).

The Montreal soil, an Andept, maintained an exceptionally high level of exchangeable-N up to 175 days with apparently no increase in fixation during this time (Table 8). This indicates that this soil has little capacity to fix NH_4^+–N and perhaps low rates of nitrification at low fertilizer levels. In other soils exchangeable-N was very much lower after 175 days. There were significant decreases in the fixed-NH_4^+ fraction and most probably fixed-NH_4^+. There were substantial decreases in fixed-NH_4^+ over the three cropping periods (175 days) in Talparo clay and Cuffy Gully stony loam soils; only Princes Town clay

Table 8. Soluble + exchangeable mineral-N (Sol. + Exch.) and fixed-NH_4^+-N in soils described in Table 1

Soil	N-fertilizer rate	14 days			175 days		
		Sol. + Exch.	Fixed-NH_4^+		Sol. + Exch.	Fixed-NH_4^+	
		N*	N	% df*	N	N	% df
Montserrat clay	20	120	278	1	30	286	1
	100	330	298	5	43	296	2
Princes Town clay	20	130	358	2	10	444	4
	100	322	383	9	40	449	12
Sevilla clay	20	100	345	7	30	358	8
	100	180	360	11	115	372	11
Talparo clay	20	210	522	3	20	461	1
	100	478	535	6	25	462	2
Montreal loam C	20	270	27	1	240	30	1
	100	592	29	2	255	32	1
Cuffy Gully stony loam	20	79	102	1	5	98	1
	100	484	138	21	15	103	1

* See Table 7.
nd = not determined.

showed a substantial increase in fixed-NH_4^+ over this period. These trends indicate that in some soils a part of the fixed-NH_4^+ may be available to plants over time. However, it was not possible to completely separate the fertilizer-N contribution to crop N-uptake from that derived from the native organic-N, since ^{15}N-containing organic-N may also have been taken up by the crop either as simple amines or through organic matter breakdown.

ii) Fertilizer-N in the various soil-N fractions of cropped soils The amount of fertilizer-N detected in the organic fractions of seven soils after three crops of Sudan grass had been grown was very low (Table 9), except for the Valencia fine sand soil. In this soil *ca.* 8% of the fulvic acid-N was fertilizer-derived. The Montreal loam contained the most organic-N (total amino-N and NH_4^+-N) in its humic and fulvic acid fractions following the three crops of Sudan grass. Results from uncropped soils (Table 5), where the level of added-N was 2,000 mg NH_4^+-N kg dry soil^{-1}, indicated much higher levels of fertilizer-N in the fulvic acid fractions of the soils studied. The enrichment level of the fertilizer used with cropped soils may have been too low for reasonably accurate estimates of fertilizer-N in the soil N-fractions at the end of the experiment.

175 days following fertilizer-N application, 2 to 18% was in the fixed and exchangeable NH_4^+–N fraction, less (1–3%) was in the hexosamine fraction, and little was detected in the serine, threonine and other amino fractions, even though these fractions generally contain a high percentage of the total-N in soils.

Fate of nitrogen added to maize: field experiment

i) Crop yield   Mean yields in response to the nitrogen and mulching treatments in the field plots are shown in Fig. 1. Nitrogen additions apparently resulted in a 50% yield increase, and in both the fertilized and unfertilized plots mulching appeared to be beneficial in the two dry seasons but depressed yields in the two wet seasons. Soil moisture measurements showed that mulching increased water storage in the soil and so presumably benefited the crop in the dry season; however, mulching probably contributed to periodic waterlogging of the soil following heavy rainfall in the wet seasons, with consequent ill-effects on yield.

ii) Mineral nitrogen   Concentrations of mineral-N in the fertilized mulched and unmulched plots decreased rapidly with depth for all seasons and the

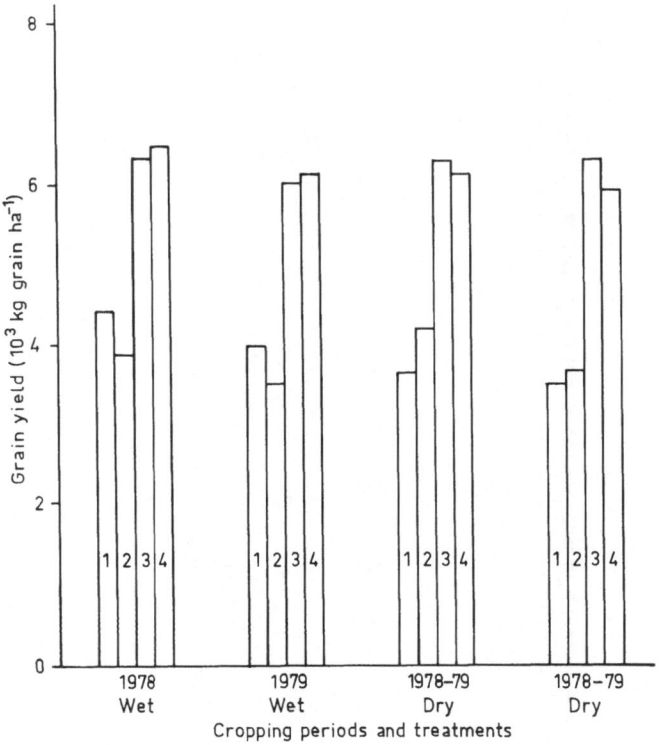

Fig. 1.   Field-grown maize yields (10^3 kg grain ha^{-1}) as affected by N-fertilizer (100 kg N ha^{-1}) and mulch (10^4 kg dry bagasse ha^{-1}). Within each seasonal column, bar 1 = unmulched control, 2 = mulched without N, 3 = mulched plus N, and 4 = unmulched plus N.

Table 9. NH$_4^+$–N and organic-N in seven soils (0–30 cm) after three crops of pot-grown Sudan grass initially fertilized with 100 mg N kg dry soil^{-1} as ^{15}(NH$_4$)$_2$SO$_4$

Soil	Humic acid-N		Fulvic acid-N		NH$_4$–N (fixed + exch.)		Hexosamine-N		Serine and threonine-N		Amino-N	
	N*	%df*	N	%df	N	%df	N	%df	N	%df	N	%df
Montserrat clay	158	1	404	2	563	3	350	3	168	1	394	1
Princes Town clay	12	1	205	1	444	3	144	1	106	1	352	1
Sevilla clay	9	3	235	2	nd	nd	nd	nd	nd	nd	nd	nd
Talparo clay	66	2	285	1	492	2	134	1	118	1	292	1
Valencia fine sand	175	3	193	8	258	18	88	2	110	1	302	1
Montreal loam C	2180	1	1616	1	1662	3	1034	1	666	1	2318	1
Cuffy Gully stony loam	34	2	207	1	nd	nd	nd	nd	nd	nd	nd	nd

* See Table 7.
nd = not determined.

Table 10. Mineral soil-N (mg N kg^{-1} dry soil^{-1}) with depth at harvest of mulched and unmulched dry season maize

Soil depth (cm)	Unmulched			Mulched		
	NO_3^--N	NH_4^+-N	Total	NO_3^--N	NH_4^+-N	Total
			1977-78 dry season			
0– 30	10.8	11.1	21.9	7.5	10.9	18.4
30– 60	4.7	3.8	8.5	5.5	5.0	10.5
60– 90	4.7	0.0	4.7	4.6	0.0	4.6
90–120	3.6	0.0	3.6	4.3	0.0	4.3
120–150	4.8	0.0	4.8	3.6	0.0	3.6
			1978–79 dry season			
0– 30	12.8	12.5	25.3	11.9	13.8	25.7
30– 60	6.7	3.4	10.1	4.2	3.9	8.1
60– 90	4.8	1.8	6.6	4.8	3.1	7.9
90–120	5.2	1.4	6.4	4.6	1.1	5.7
120–150	3.6	0.0	3.6	3.9	0.0	3.9

Table 11. Mineral soil-N (mg exchangeable-N kg^{-1} dry soil) with soil depth at harvest of mulched and unmulched 1979 wet season maize

Soil depth (cm)	Unmulched			Mulched		
	NO_3^--N	NH_4^+-N	N_{min}	NO_3-N	NH_4-N	N_{min}
0– 30	11.0	11.3	22.3	9.8	12.7	22.6
30– 60	6.3	4.0	10.3	4.6	3.6	8.2
60– 90	4.6	2.4	7.0	4.6	1.9	6.5
90–120	3.2	1.4	4.6	4.3	1.8	6.1
120–150	3.0	0.0	3.0	2.9	0.0	2.9

amounts were low (Tables 10 and 11). These small amounts have probably resulted from a long history of cultivation; the range (0–26 mg N kg^{-1} dry soil) is similar to that reported by Cornforth[10] and Stanford and Legg[29] for long-cultivated soils. Most of the mineral-N was found in the surface layer; there was a tendency for NH_4^+-N to be more prevalent than NO_3^--N here and for NO_3^--N to be more prevalent than NH_4^+-N at depth.

In the dry season (1977–78), no exchangeable NH_4^+-N was recovered in either the mulched or unmulched plots below 60 cm depth, but measurable amounts were recovered after wet-season crops. This probably indicates that nitrification and NH_4^+-fixation at these lower soil depths were greater in the dry season than in the wet season because of reduced aeration in the wet season. Mulching had little effect on the total amount of NO_3^--N and NH_4^+-N and their distribution with soil depth regardless of season.

iii) Total organic nitrogen   The pattern of organic-N distribution for the dry season (1977–78) and the wet season (1978) (Table 12) shows that amounts in the surface layer did not exceed 1560 mg organic-N kg dry soil^{-1} in the dry season, but there was a noticeable decrease (10–12% in the surface soils) in concentrations for both the mulched and unmulched plots during the following wet season. The lower concentrations in the wet season most likely resulted from losses due to crop removal, soil erosion, and possibly leaching and volatilization of NH_4^+. Evidence of leaching was obtained in another experiment not reported here, although larger dressings of urea (400 kg urea-N ha^{-1}) were used.

iv) Nitrogen uptake   Uptake of fertilizer-N was apparently greater from the mulched plots (Table 13). Amounts of ^{15}N-fertilizer accumulated by some plant parts (seeds and leaves; Table 14) also appeared to be increased by the mulch treatment, although accumulation in the stems was similar for both treatments. These results seem to indicate that the mulch treatment increased the availability and uptake of both fertilizer and soil nitrogen.

v) Effect of added-N on N-uptake   Both urea and NH_4^+-N when added to the soil at practical farming levels resulted in the mineralization and crop uptake of native soil nitrogen probably through enhanced microbiological activity. Some data to show this for urea is presented in Table 13 for a field experiment, and for $(NH_4)_2SO_4$ in Table 6 for a greenhouse experiment. Recoveries for NH_4^+-N were even higher at the levels used – although the uptake of the added form might be quite low (Table 14). It can therefore be concluded that addition of N-fertilizers at medium levels stimulated the release of native soil nitrogen.

Table 12. Organic N (mg N kg dry soil^{-1}) with soil depth at harvest of unmulched and mulched maize

Soil depth (cm)	1977–78 dry season		1978 wet season	
	Unmulched	Mulched	Unmulched	Mulched
0– 30	1560	1500	1300	1300
30– 60	883	894	943	844
60– 90	756	793	635	629
90–120	630	621	604	623
120–150	585	581	570	596

Table 13. Uptake of soil and fertilizer nitrogen by maize (grain + above-ground residue) in unmulched and mulched field plots (1977–78 dry season). Fertilizer rate = 100 kg urea − ^{15}N ha^{-1} (10% excess ^{15}N)

Treatment	Soil-N uptake (kg N ha^{-1})	Fertilizer-N uptake (kg N ha^{-1})
Unmulched	175	45
Mulched	189	56

Table 14. Effect of fertilizer (100 kg urea − ^{15}N ha^{-1} (10% excess ^{15}N)) and mulch (10 × 10^3 kg dry bagasse ha^{-1}) treatment on fertilizer-N accumulation in plant parts of maize (1977–78 dry season)

Treatment	Fertilizer-N (kg N ha^{-1})		
	Seed	Stem	Leaves
Unmulched	3.36	0.49	0.63
Mulched	4.19	0.48	0.89

Acknowledgements The authors are pleased to acknowledge financial assistance from the Organization of American States (Department of Scientific and Technological Affairs) and the International Atomic Energy Agency.

References

1 Ahmad N 1982 Vertisols. *In* Wilding L P and Smeck N E (Eds). Pedogenesis and Soil Taxonomy. Vol. 2, Ch. 12. New York: Elsevier. *In press.*
2 Ahmad N and Shand C 1974 The effect of nitrate inhibition and slow release nitrogen fertilizers of nitrification rates in some Trinidad soils. Trop. Agric. Trinidad 51, 167–178.
3 Ahmad N and Whiteman P T S 1972 Comparisons of sulphate of ammonia and slow release nitrogen fertilizers for rice in Trinidad. Agron. J. 61, 730–734.
4 Ahmad N, Tullock Reid L T and Davis C E 1969 Fertilizer studies on Pangola grass (*Digitaria decumbens* Stent) 1. Description of the experiments and effect of nitrogen. Trop. Agric. Trinidad 46, 173–178.
5 Banerjee S K and Dasak R K 1978 Studies on ammonia and urea treated humic substances and their interaction with some heavy metal ions. J. Ind. Soc. Soil Sci. 26, 347–351.
6 Broadbent F I and Tyler K B 1962 Laboratory and greenhouse investigations of nitrogen immobilization. Soil Sci. Soc. Am. Proc. 26, 459–462.
7 Chesney H A D 1967 Denitrification in some Trinidad soils. M. Sc. Thesis. Univ. West Indies, St. Augustine, Trinidad.
8 Collins P 1979 Non-symbiotic nitrogen fixation in some soils of Trinidad and Tobago. Ph.D. Thesis, UWI, St. Augustine, Trinidad.
9 Cornforth I S 1970 Reafforestation and nutrient reserves in the humid tropics. J. Appl. Ecol. 7, 609–615.

10   Cornforth I S 1974 A review of work on nitrogen in West Indian soils. Trop. Agric. Trinidad 51, 145–153.

11   Cornforth I S and Davis J B 1968 Nitrogen transformation in tropical soils. 1. Mineralization of nitrogen-rich organic materials added to soil. Trop. Agric. Trinidad 45, 211–221.

12   De Souza D N A 1966 Nodulation of indigenous Trinidad legumes. Trop. Agric. Trinidad 43, 265–267.

13   Fiedler R and Proksch G 1975 The determination of nitrogen-15 by emission and mass spectrometry in biochemical analysis: a review. Anal. Chim. Acta 78, 1–62.

14   Fletcher R E 1971 Fertilizer studies with two forage grasses on Piarco fine sand. M.Sc. Thesis. Univ. West Indies, St. Augustine, Trinidad.

15   Griffith S M 1977 Chemical characteristics of humic materials from West Indian Andepts. Ph. D. Diss., University of the West Indies, St. Augustine, Trinidad.

16   Hardy F 1946 Seasonal fluctuations in soil moisture and nitrate in a humid tropical country. Trop. Agric. Trinidad 23, 40–49.

17   Hardy F and Rodrigues G 1951 The nitrogen enigma of the sugar-cane soils of British Guiana. Proc. of the Meeting of British West Indies. Sugar Technologists, Georgetown, Brit. Guiana, pp 97–100. Port of Spain, Trinidad: Sugar Association of the Caribbean, Inc.

18   Humphries E C and Rodrigues G 1945 Decomposition of cacao leaves under natural conditions. J. Agric. Sci. Camb. 35, 247–253.

19   Jackson M L 1958 Soil Chemical Analysis. Englewood Cliffs, New Jersey: Prentice-Hall, Inc.

20   Macklin M C 1964 Seasonal fluctuations of nitrate and ammonia nitrogen under leguminous cover in some soils of Central Trinidad. DTA (Diploma in Tropical Agriculture) Report, St. Augustine, Trinidad: University of West Indies.

21   Medford D L 1963 Nitrogen fertilizer trials on sugar cane. Bull. Min. of Agric. Barbados 36, 60–63.

22   Myers R J K and Paul E A 1971 Plant uptake and immobilization of ^{15}N-labelled ammonium nitrate in a field experiment with wheat. In Nitrogen-15 in Soil-Plant Studies, pp 55–64. Vienna: IAEA.

23   Quilt P and Donawa F 1979 The response of pigeon pea to Rhizobium inoculation. Proc. of Grain Legume Workshop, Univ. West Indies, St. Augustine, Trinidad.

24   Rao M S 1977 Urea absorption and urease activity in some Trinidad soils. Ph.D. Thesis, Univ. West Indies, St. Augustine, Trinidad.

25   Rodrigues G 1954 Fixed ammonia in tropical soils. J. Soil Sci. 5, 264–274.

26   Shand C R 1973 Transformation and availability of fertilizer nitrogen in some Trinidad soils. Ph.D. Thesis. Univ. West Indies, St. Augustine, Trinidad.

27   Silva J A and Bremner J M 1966 Determination and isotope-ratio analysis of different forms of nitrogen in soils. 5. Fixed ammonium. Soil Sci. Soc. Am. Proc. 30, 587–594.

28   Smith G D 1974 Study on the correlation of soils of the former British Territories in the West Indies. Mimeo Report Dept. of Soil Sci., Univ. West Indies, St. Augustine, Trinidad 74 p.

29   Stanford G and Legg J O 1969 Correlation of soil N-availability indices with N-uptake by plants. Soil Sci. 105, 320–326.

30   Weir C C 1965 Effect of mixing thiourea and ammonium sulphate on nitrification of ammonium in a tropical soil. Adv. Frontiers Plant Sci. 13, 195–201.

31   Weir C C and Davidson J G 1968 The effect of retarding nitrification of added fertilizer nitrogen on yield and nitrogen uptake of Pangola grass (Digitaria decumbens). Trop. Agric. Trinidad 45, 301–306.

Plant and Soil 67, 187–191 (1982).   0032-079X/82/0672-0187$00.75.   SU-16
© 1982 *Martinus Nijhoff/Dr W. Junk Publishers, The Hague.*

Distribution of ^{15}N fertilizer in field-lysimeters sown with garlic (*Allium sativum*) and foxtail millet (*Setaria italica*)

Estudios de fertilización y producción de cultivos realizados con [15]*N*

M. A. LAZZARI

Laboratorio de Humus y Biodinámica del Suelo
Departamento de Ciencias Agrarias, Universidad Nacional del Sur, 8000 Bahia Blanca, Argentina

Key words Ammonium sulphate Argentina Foxtail millet Garlic N-assimilation N-cycling
N-sources 15-N Urea

Abstract We examined the distribution of residual ^{15}N and its uptake by a foxtail millet crop grown in field lysimeters following a previous garlic crop fertilized with either ^{15}N-urea or ^{15}N-ammonium sulphate. Garlic apparently removed more N from the lysimeters treated with urea-N than from those treated with $(NH_4)_2SO_4$. Fertilizer-N in the lysimeters was similar (*ca.* 32% of original) following millet harvest. About 16 per cent of both fertilizers in the lysimeters was removed by the millet.

Resumen Se examinó la distribución y absorción de nitrógeno en moha de Hungría (*Setaria italica*) cultivada en lisímetros de campo a continuación de un cultivo de ajo colorado (*Allium sativum*) el cual había sido fertilizado con urea o con sulfato de amonio marcados con [15]N. Aparentemente el ajo extrajo mas nitrógeno de los lisímetros que habían sido fertilizados con urea que de los tratados con sulfato de amonio. La cantidad de nitrógeno remanente en los lisímetros luego de la cosecha de la moha fue similar para las dos fuentes de nitrógeno (*ca.* 32% del original). La moha recuperó alrededor del 16% del nitrógeno aplicado al cultivo de ajo.

Introduction

During recent years the application of N-15 techniques in agricultural research has been increasing as a tool to evaluate the use of fertilizers with better precision. Most research is carried out with compounds enriched with N-15. The number of laboratory or greenhouse studies greatly exceeds that of field studies, the advantages of the former being that they are less costly and permit better control of variables.

Field and lysimeter experiments enable a more realistic estimation to be made of natural transformations of nitrogen compounds[8]. A direct measurement of N recovery is thus possible. Long-term field studies indicate that crops can recover around 50% of the applied nitrogen. Losses by denitrification can be estimated as the deficit shown in the N balance. Losses of up to 40% by denitrification have been reported[8].

In N-balance sheets usually the amounts of N in soil and crop are measured directly[11]. Sometimes N is also determined in leachate water[4,12]. Generally the losses by runoff and leaching are negligible[8], although depending on amount of fertilizer, rainfall, soil porosity, and other factors, it may reach high amounts. Table 1 shows the results obtained using different cultivation practices but under the same experimental conditions. Differences within a given experiment can be ascribed to soil type, crop, humidity and type of fertilizer.

A complete N-balance can be obtained by using gas lysimeters[10]. Carter *et al.* designed an experimental system that permits a nitrogen balance to be obtained under field conditions. Microplots were examined consisting of soil in cylinders 30–60 cm in diameter, which were then planted with Sudan grass or left bare. In such cylinders precautions are usually taken to prevent losses by leaching. This experimental system was further explored by Myers and Paul[11].

This paper describes results from a programme initiated in 1977 to study the economy, balance and dynamics of nitrogen in order to obtain better yields. The experiments, which are still in progress, are carried out in automatic differential-weighing lysimeters, which also permit the assessment of evapotranspiration.

Materials and methods

Lysimeters

Three automatic lysimeters previously described by Donnari *et al.*[5] were used. They are installed in the fields of Universidad Nacional del Sur in Bahía Blanca, Argentina, the soil being that of the region. Details of the site as well as the soil profile are given in a previous paper[9], which also presented the results on the effects of nitrogen fertilization and water regime on a variety of red garlic (*Allium sativum*). The present paper reports on results obtained from May 1978 to April 1979.

Cultural practices

The soil in the lysimeters was hoed and raked. Fertilizer was applied at a rate equivalent to 100 kg P_2O_5 ha^{-1} and 60 kg K_2O ha^{-1}. On May 22, 1978, garlic cloves were planted in the same manner as earlier described[9]. On August 1 (end of the dry winter season) the first dose of nitrogen fertilizer was applied consisting of 100 kg N ha^{-1} in two of the lysimeters. In one of them, ($^{15}NH_4)_2SO_4$ was applied while in the other ($^{15}NH_2)_2CO$ was used. A second fertilizer treatment with the same rate was applied on September 26. The third lysimeter was left as a control. No replications were possible due to the complexity of the system and the cost of running the experiments.

The garlic crop was severely damaged by a pathogen, and consequently we decided to study the residual effect of the applied nitrogen on the production of a subsequent crop of foxtail millet (*Setaria italica*). For this purpose the remaining parts of the garlic plants were removed from the lysimeters

Table 1. Balance of nitrogen in different experimental systems

Type of study	Crop	% of N-fertilizer recovered by crop-soil system	Reference
Greenhouse	Rye grass	87– 93	Gasser et al.[7]
Greenhouse	Sudangrass, tomato, corn	71– 93	Broadbent and Nakashima[2]
Lysimeter	Corn	67	Owens[12]
Lysimeter	Corn	65– 92	Chichester and Smith[4]
Cylinders in soil	Sudangrass	96–102	Carter et al.[3]
Cylinders in soil	Wheat	64– 83	Myers and Paul[11]
Gas lysimeter	Rhodesgrass	99–101	Martin and Ross[10]
Field	Sudangrass	79– 83	Westerman et al.[14]

and the soil was disinfected with gaseous methyl bromide. After soil preparation, Setaria was planted on 26 December 1978 at a density equivalent to 14 kg seed ha^{-1} in 6 furrows per meter. The crop was harvested on April 10. During the vegetation period a total of 332 mm rainfall was recorded; irrigation was therefore not necessary.

Fertilizers

The fertilizers (ammonium sulfate fertilizer with 8.12 atom % ^{15}N and urea fertilizer with 10.2 atom % ^{15}N) were each dissolved in 10 litres of water and uniformly applied. To minimize losses by NH$_3$ volatilization the solution was applied when the soil was dry, so that infiltration was rapid.

Soil and plant sampling

Soil samples were taken immediately before sowing and after harvest. Sampling was carried out with a screw-type auger at three places in each lysimeter. The samples were collected and mixed for each 20 cm layer down to 100 cm. The holes were refilled with the same soil to prevent abnormal water movement in the profile. Samples were air-dried, ground and stored in closed plastic containers. All plants were harvested, dried at 60°C and separated into grain and leaves plus stems. The samples were ground and kept in closed containers. After harvest the roots in the fertilized lysimeters were extracted to 60 cm, dried and analysed. All samples were analysed for total-N following the micro-Kjeldahl procedure[1]. The ^{15}N/^{14}N ratio was determined after oxidation with sodium hypobromite[6] with a Q24-Zeiss (Jena) prism spectrograph[13]. Total nitrogen was averaged from two determinations and those for isotopic ratios from four determinations. The amount of N in the fertilizer was calculated using equation 1.

$$X = A(B - C)/D \tag{1}$$

where X = amount of N-fertilizer in the soil,
   A = total-N in sample,
   B = atom % ^{15}N in sample,
   C = atom % ^{15}N in control,
   D = excess atom % ^{15}N in applied fertilizer.

When the crop was sown and harvested, the spigots at the bottom of the lysimeters were open (130 cm depth). No percolation water was collected.

Results and discussion

Table 2 presents the results obtained for total- and residual-nitrogen for each soil layer. Nitrogen contents were obtained using the values for soil bulk density. When millet was sown the lysimeter which was fertilized with (^{15}NH$_4$)$_2$SO$_4$, showed a higher content of residual nitrogen (70%) from the fertilizer N added than that in which urea was used (51%). The garlic crop could have absorbed the remaining nitrogen or it is also possible that some urea-N could have been lost by ammonia volatilization due to the soil pH (8.2).

After harvesting the millet, both soils showed a similar amount of residual nitrogen (*ca.* 32%) but in the treatment with ammonium sulfate it showed larger concentrations towards the surface.

Table 3 shows the observed yields for different parts of millet in the fertilized lysimeters and the control. The crop clearly responded to the fertilization applied to the garlic crop 3 and 4 months before sowing the millet. The total dry weight in the treatment with urea was more than double that of the control. While not as

Table 2. Total nitrogen and fertilizer N-15 recovery (nitrogen derived from fertilizer, Ndff) in soils of lysimeters planted to millet following an N-15 fertilized garlic crop. Fertilizer was either $(^{15}NH_4)_2SO_4$ or urea

Depth (cm)	Before millet				After millet			
	$(^{15}NH_4)_2SO_4$		$(^{15}NH_2)_2CO$		$(^{15}NH_4)_2SO_4$		$(^{15}NH_2)_2CO$	
	% Total-N	% Ndff	% Total-N	% Ndff	% Total-N	% Ndff	% Total-N	% Ndff
0– 20	0.087	21.6	0.097	14.2	0.093	17.3	0.088	12.9
20– 40	0.084	23.8	0.073	13.1	0.092	3.9	0.076	10.2
40– 60	0.073	15.6	0.054	12.0	0.056	3.2	0.045	5.0
60– 80	0.034	3.7	0.057	7.7	0.056	4.4	0.042	2.6
80–100	0.071	5.5	0.075	3.7	0.073	3.4	0.054	2.0

Table 3. Dry matter yield, total-N, and fertilizer N-15 recovery (% N derived from fertilizer, Ndff) in lysimeter-grown millet following a garlic crop fertilized with N-15 ammonium sulphate or urea

Crop component	Control	$(^{15}NH_4)_2SO_4$			$(^{15}NH_2)_2CO$		
	Yield (g m^{-2})	Yield (g m^{-2})	Total-N		Yield (g m^{-2})	Total-N	
			% N	% Ndff		% N	% Ndff
Leaves + stalks	749	1384	0.28	5.85	1616	0.34	7.18
Grain	244	396	1.62	10.44	498	1.91	8.30
Roots	na	1.02	0.42	0.05	1.72	0.73	0.01

Control, non fertilized lysimeter; na, not available.

Table 4. Added-nitrogen balance of cropped lysimeters sown with garlic followed by millet and fertilized with either ^{15}N-ammonium sulphate or urea for the period August 1978 to April 1979

	$(^{15}NH_4)_2SO_4$ (g N·m^{-2})	$(^{15}NH_2)_2CO$ (g N·m^{-2})
Initial nitrogen fertilizer	20.00	20.00
Fertilizer extracted by the garlic crop and/or lost	5.98	9.89
Remaining fertilizer	14.02	10.11
Fertilizer extracted by the millet	3.27	3.10
Remaining fertilizer	6.43	6.51
Deficit	4.32	0.50

high, the treatment with ammonium sulfate also produced yields that were clearly higher than in the control. Both treatments showed similar values for the uptake of fertilizer-N (Ndff), *i.e.*, 16.3% for the treatment with ammonium sulfate and 15.5% for that with urea. The value for Ndff for the leaves plus stems was higher in the urea treatment while in grain it was slightly lower than for the ammonium sulfate treatment. Taking into account the higher yields for all three plant parts in the case of urea fertilization, and observing also that total nitrogen was higher than in the treatment with ammonium sulfate, it seems that the application of urea produces a greater absorption of native soil nitrogen (priming effect).

Table 4 presents an estimate of the nitrogen balance for the garlic-millet crop rotation. No losses through leaching were recorded.

This study is being continued using the same nitrogen sources on a new garlic cycle which will be then followed by Setaria.

References

1 Bremner J M 1965 Total nitrogen. *In* Black C A (Ed.). Methods of Soil Analysis, Part 2. Agronomy 9, 1149–1178. Madison, Wisc.: Am. Soc. Agron.

2 Broadbent F E and Nakashima T 1968 Plant uptake and residual value of six tagged nitrogen fertilizers. Soil Sci. Soc. Am. Proc. 32, 388–392.

3 Carter J N, Bennet O L and Pearson R W 1967 Recovery of fertilizer nitrogen under field conditions using nitrogen-15. Soil Sci. Soc. Am. Proc. 31, 50–56.

4 Chichester F W and Smith S J 1978 Disposition of ^{15}N labelled fertilizer nitrate applied during corn culture in field lysimeters. J. Environ. Qual. 7, 227–233.

5 Donnari M A, Rosell R A and Torre L 1978 Productividad de ajo. II. Evapotranspiración real y necesidad de agua. Turrialba 28, 331–337. (In Spanish, English summary.)

6 Fiedler R and Proksch G 1975 The determination of nitrogen-15 by emission and mass spectrometry in biochemical analysis: a review. Anal. Chim. Acta 78, 1–62.

7 Gasser J K R, Greenland D J and Rawson R A G 1967 Measurement of losses and during subsequent growth of ryegrass, using ^{15}N-labelled fertilizers. J. Soil Sci. 18, 289–295.

8 Hauck R D 1971 Quantitative estimates of nitrogen-cycle processes. Concepts and review. *In* Nitrogen-15 in Soil-Plant Studies, pp 65–80. Vienna, Austria: IAEA.

9 Lázzari M A, Rosell R A and Landriscini M R 1978 Productividad del ajo. I. Fertilización nitrogenada y riegos. Turrialba 28, 245–251. (In Spanish, English summary.)

10 Martin A E and Ross P J 1968 A nitrogen balance study using labelled fertilizers in a gas lysimeter. Plant and Soil 28, 182–186.

11 Myers R J K and Paul E A 1971 Plant uptake and immobilization of ^{15}N labelled ammonium nitrate in a field experiment with wheat. *In* Nitrogen-15 in Soil Plant-Studies. pp 55–64. Vienna, Austria: IAEA.

12 Owens L D 1960 Nitrogen movement and transformations in soils as evaluated by lysimeter study utilizing isotopic nitrogen. Soil Sci. Soc. Am. Proc. 24, 372–376.

13 Sommer K and Kick H 1965 Auf eine Eichkurve bezogene emissionsspektrographische ^{15}N-Bestimmung mit dem Spektrographen Q24-Zeiss. Z. analyt. Chem. 220, 21–26.

14 Westerman R L, Kurtz L T and Hauck R D 1972 Recovery of ^{15}N-labelled fertilizers in field experiments. Soil Sci. Soc. Am. Proc. 36, 82–86.

Plant and Soil 67, 193–208 (1982). 0032-079X/82/0672-0193$02.40.   SU-17
© 1982 *Martinus Nijhoff/Dr W. Junk Publishers, The Hague.*

Nitrogen cycling in a ^{15}N-fertilized bean (*Phaseolus vulgaris* L.) crop

Ciclo del nitrógeno en cultivo de frijol (Phaseolus vulgaris *L.*) *fertilizado con* ^{15}N

P. L. LIBARDI, R. L. VICTORIA, K. REICHARDT and A. CERVELLINI

Centro de Energia Nuclear na Agricultura (CENA), Caixa Postal 96, 13400 Piracicaba, Brazil

Key words Beans Brasil Crop N-recovery K-fertilizer N-cycling N-fertilizer 15-N P-fertilizer Urea.

Abstract To increase our understanding of the fate of applied nitrogen in *Phaseolus vulgaris* crops grown under tropical conditions, ^{15}N-labelled urea was applied to bean crops and followed for three consecutive cropping periods. Each crop received 100 kg urea-N ha^{-1} and 41 kg KCl-K ha^{-1}. At the end of each period we estimated each crop's recovery of the added nitrogen, the residual effects of nitrogen from the previous cropping period, the distribution of nitrogen in the soil profile, and leaching losses of nitrogen.

In addition, to evaluate potential effects of added phosphorus on nitrogen cycling in this crop, beans were treated at planting with either 35 kg rock-phosphate-P, 35 kg superphosphate-P, or 0 kg P ha^{-1}.

Results showed that 31.2% of the nitrogen in the first crop was derived from the applied urea, which represents a nitrogen utilization efficiency of 38.5%. 6.2% of the nitrogen in the second crop was derived from fertilizer applied to the first crop, and 1.4% of the nitrogen in the third crop. Nitrogen utilization efficiencies for these two crops, with respect to the nitrogen applied to the first crop, were 4.6 and 1.2%, respectively. In total, the three crops recovered 44.3% of the nitrogen applied to the first crop. The remainder of the nitrogen was either still in the soil profile or had been lost by leaching, volatilization or denitrification. ^{15}N enrichment of mineral-N (NO$_3$ + NH$_4$) suggests that at the end of the second crop, the pulse of fertilizer applied to the first crop had probably passed the 120 cm depth. ^{15}N enrichment of organic-N suggests that root activity of beans and weeds transported nitrogen to 90–120 cm (or deeper).

We could account for 109 kg fertilizer-N ha^{-1} in harvested biomass, crop residue, and soil at the end of the first cropping period. This indicates an experimental error of about 10% if no nitrogen was lost by volatilization, denitrification, or leaching below 120 cm. At the end of the second and third crops, 76 and 80 kg N ha^{-1}, respectively, could be accounted for, suggesting that 20 to 25% of the applied-N was lost from the system over a 2-crop period. The two types of added phosphorus did not significantly differ in their effects on bean yields.

Resumen Con el fin de aumentar la comprensión del destino del nitrógeno aplicado a *Phaseolus vulgaris* bajo condiciones tropicales, se aplicó úrea marcada con ^{15}N y se siguió la marcha del experimento por tres períodos de cultivo sucesivos. Cada cultivo recibió 100 kg N ha^{-1} en forma de úrea y 41 kg K ha^{-1} en forma de KCl. Al final de cada período se estimó la recuperción del nitrógeno añadido, los efectos residuales del nitrógeno aplicado en el período anterior, la distribución del nitrógeno en el perfil del suelo y las pérdidas por lixiviación. Adicionalmente, para evaluar los efectos del fósforo añadido sobre el ciclo del nitrógeno, se fertilizó el cultivo con 35 kg P en forma de roca fosfatada ha^{-1} o con 35 kg P como superfosfato ha^{-1} y un tercer experimento sin P como control.

Los resultados mostraron que el 31,2% del nitrógeno en el primer cultivo provenía de la úrea aplicada, lo cual representa una eficiencia de utilización de 38,5%. En los dos períodos subsiguientes el 6,2% y el 1,4% del nitrógeno provenía del fertilizante anteriormente aplicado, respectivamente. Las eficiencias de utilización fueron en estos casos de 4,6 y 1,2 porciento respectivamente. En total los tres cultivos recuperaron 44,3% del nitrógeno aplicado al primero. El nitrógeno restante estaba en el suelo a había sido perdido por lixiviación, volatilización o desnitrificación.

El aumento en ^{15}N en el nitrógeno mineral ($NH_4 + NO_3$) indicó que al fin del segundo período de cultivo, el frente de nitrógeno aplicado había ya pasado los 120 cm de profundidad. El aumento en ^{15}N en materia orgánica indicó que la actividad de las raices del cultivo y las malezas transportó e incorporó el nitrógeno a 90–120 cm y mas.

Podemos calcular que del total aplicado como fertilizante, 109 kg N ha^{-1} se hallaban en la biomasa cosechada, en los residuos de cosecha y en el suelo al final del primer período de cultivo. Si se considera que no hubo pérdidas por lixiviación, volatilización o desnitrificación, nuestro error experimental sería de un 10%. El mismo cálculo para el segundo y el tercer períodos de cultivo dió 76 kg N ha^{-1} y 80 kg N ha^{-1} respectivamente, indicando asi pérdidas de 20 a 25% en los dos primeros años de cultivo. No se observaron diferencias entre las dos fuentes diferentes de fósforo.

Introduction

The important role of nitrogen in plant nutrition makes it essential to fully understand its cycling in agricultural systems. Many recent studies have considered this question in temperate agriculture (e.g.[3,4,5,7]), but very little information is available for tropical areas where the mobility and mineralization rates of nitrogen are very high. The increasing cost of nitrogen fertilizers demands that management practices result in a high utilization efficiency of nitrogen by crops. Further, the tendency to overfertilize has in some places resulted in groundwater contamination by nitrate's leaching below the root zone.

The objective of the present study was to investigate the fate of nitrogen applied to bean crops grown under tropical conditions. Urea labelled with ^{15}N was used to estimate crop recoveries of fertilizer nitrogen and nitrogen utilization efficiencies, the residual effects of nitrogen applied to one crop on succeeding crops, N-distribution in the soil profile after cropping, and leaching losses of nitrogen.

Materials and methods

Three consecutive crops of bean (*Phaseolus vulgaris*, L. var. Carioca) were grown on Terra Roxa Estruturada (Paleudalf) soil at experimental fields of the Escola Superior de Agricultura 'Luiz de Queiroz', University of São Paulo, Piracicaba, SP. The *ca.* 10 ha field site is 580 m above sea level, 22°42'30'' S and 47°38'00'' W. Meteorological data collected at a permanent meteorological station located at the site since 1917 show that mean annual rainfall is 1247 mm, mean annual air temperature is 20.8°C, mean annual relative humidity is 69%, mean annual wind speed is 2.5 m s^{-1}, and mean annual daily sunshine is 6 h day^{-1}.

Crops were grown from 15 March 1978 to 29 June 1978, from 21 November 1978 to 19 February 1979, and from 25 September to 8 January 1980. Between crops, soil was left unweeded. During this period without weed control, mainly deep-rooted grasses grew on the area, and these were incorporated into the soil before planting the next crop. Bean plants were spaced at 40 × 10 cm intervals, for a stand density of 250,000 plants ha^{-1}.

Each crop received 100 kg urea-N ha^{-1}, one third at planting and two thirds thirty days later, 35 kg P ha^{-1}, and 41 kg K–KCl ha^{-1}. The experimental design included 24 experimental plots: 3 levels of

^{15}N enrichment (1.116, 2.505, and 10.457 atom % ^{15}N) × 2 types of phosphorus (superphosphate and rock phosphate 'Araxa') × 4 replications. Urea was ^{15}N-enriched for only the first cropping period, and plots were sampled for soil ^{15}N in order of increasing ^{15}N enrichment: plots with 1.116% ^{15}N were sampled after the first crop, plots with 2.505% ^{15}N after the second crop, and plots with 10.457% ^{15}N after the third. Superphosphate contained 20% citric-acid-soluble P_2O_5, and rock phosphate, 4%. Plots were separated from each other by 2 m buffer strips that were planted with beans but unfertilized, and that served as control plots.

Prior to the start of the experiment, the soil was sampled in 30 cm layers to 120 cm and analysed for mineral-N (NO_3^- + NH_4^+ in 2 M KCl extract [2], organic-N (by Kjeldahl wet digestion of KCl-extracted soils [1]), pH (in a 1 : 1 wet-soil : water slurry), soil texture (by the pipette technique), and bulk density. Special care was taken with bulk density determinations because these values are very critical for calculations of total soil volume per hectare. Data for soil water retention and hydraulic conductivity were taken from Reichardt et al.[6], who worked at the same site.

During the experimental period, soil water potential was measured with tensiometers at depths of 105, 120 and 135 cm for estimates of soil water fluxes at the lower boundary of the control volume (120 cm). Soil water flow at this depth was estimated by Darcy's equation:

$$q_w = -K(h_{120})\frac{\psi_{105} - \psi_{135}}{30}$$

where q_w = soil water flux (cm day^{-1}), $K(h_{120})$ = soil hydraulic conductivity (cm day^{-1}) at 120 cm depth for soil matric potential h, and ψ = hydraulic potential (cm H_2O), which is the sum of matric potential h (cm H_2O) and gravitational potential z (cm H_2O). Subscripts (135 and 105) indicate the depth (cm) at which ψ was measured.

After each harvest, crop weight, total N content, and the ^{15}N/^{14}N ratio were measured for straw and grain separately but not for roots. In addition, soil samples were collected from each plot at depths of 0–30, 30–60, 60–90 and 90–120 cm with a 2.5 cm diameter auger. In each plot 12 samples were obtained at each depth interval and composited for mineral-N and organic-N determinations as described above.

The percent of crop biomass-N derived from fertilizer-N (Ndff%) and the nitrogen utilization efficiency (NUE%) were calculated as:

$$Ndff\% = \frac{Atom\ \%\ ^{15}N\ excess\ in\ sample}{Atom\ \%\ ^{15}N\ excess\ in\ fertilizer} \times 100, \quad and$$

$$NUE\% = \frac{(Ndff\%)\,(kg\ biomass\text{-}N\ ha^{-1})}{(kg\ fertilizer\text{-}N\ ha^{-1})}$$

Results and discussion

Table 1 presents soil data obtained prior to the start of the experiment. In Fig. 1–5 are data for rainfall, irrigation and soil water potentials below the root zone for the entire experimental period. The first crop was grown in the dry season and the second and third were grown in the wet season.

Table 2 shows that yields were increased by phosphorus for at least the first two crops, but that the phosphorus source (superphosphate vs rock phosphate) made little difference. Since both sources of phosphorus affected crop yield similarly, the two P-groups were treated as one in subsequent analyses.

Table 3 presents the results of plant biomass and plant-N analyses for each crop and ^{15}N level. The percent of biomass-N derived from fertilizer-N and the nitrogen use efficiency (NUE) calculations (Table 4) were robust for ^{15}N-

Table 1. Physical and chemical soil characteristics before experimental cropping and fertilization

Depth (cm)	Texture (%)			Bulk density $(g\,cm^{-3})$	pH	Mineral-N** $(mg\,N\,kg\,dry\,soil^{-1})$	Organic-N (%)
	Sand	Silt	Clay				
0– 30	43.2	13.0	43.8	1.44 (0.17)*	5.7	10.1	0.048
30– 60	33.1	12.0	54.8	1.45 (0.09)	5.8	6.3	0.043
60– 90	30.1	11.6	58.3	1.32 (0.09)	5.7	4.1	0.041
90–120	31.5	12.6	55.9	1.26 (0.06)	6.0	2.3	0.028

* Values in parentheses are standard deviations (n = 9).
** Mineral-N is NH_4^+-N plus NO_3^--N.

Table 2. Grain yield* (10^3 kg fresh weight ha^{-1}) for three consecutive Phaseolus crops treated with phosphorus as noted, 100 kg urea-N ha^{-1} and 41 kg KCl–K ha^{-1} before each crop

Crop	Phosphorus treatment		
	0 (control)	35 kg rock-phosphate-P ha^{-1}	35 kg superphosphate-P ha^{-1}
First	1.66 (0.26)	2.04 (0.12)	2.02 (0.14)
Second	0.55 (0.09)	1.16 (0.12)	1.35 (0.16)
Third	0.94 (0.28)	1.15 (0.11)	1.38 (0.33)

* Fresh grain contains about 15% H_2O. The control treatment received no fertilizer. Values are means ($\pm$ standard deviations) of four replicated plots.

enrichment levels: the three levels of enrichment yielded similar estimates, although the lower enrichment (1.116 atom % ^{15}N excess) may be too low for studies extended over long periods of time, e.g. two years. The first crop took up 31.2% of its N from the fertilizer, for a utilization efficiency of 38.5%.

Of the biomass-N in the second crop, 6.2% was derived from the fertilizer applied to the first crop, and in the third crop, 1.4%. These values represent nitrogen utilization efficiencies of 4.6 and 1.2%, respectively, for the nitrogen applied to the first crop. Thus, the three crops recovered 44.3% of the nitrogen applied to the first crop. The greatest source of error for this estimate comes from the dry matter determinations (Table 3); total-N and atom % ^{15}N estimates vary much less. The large variation in nitrogen utilization efficiency (Table 4) can be traced to variation in biomass among replicate plots.

The 56% of the original fertilizer-N not taken up by the crops was either still in the soil profile or had been lost by leaching, volatilization or denitrification. Soil data from analyses after crops 1, 2 and 3 are presented in Tables 5, 6 and 7, respectively. Enrichment data show that after the first crop, mineral-N reached the 90–120 cm layer, and that at the end of the second crop the pulse of fertilizer

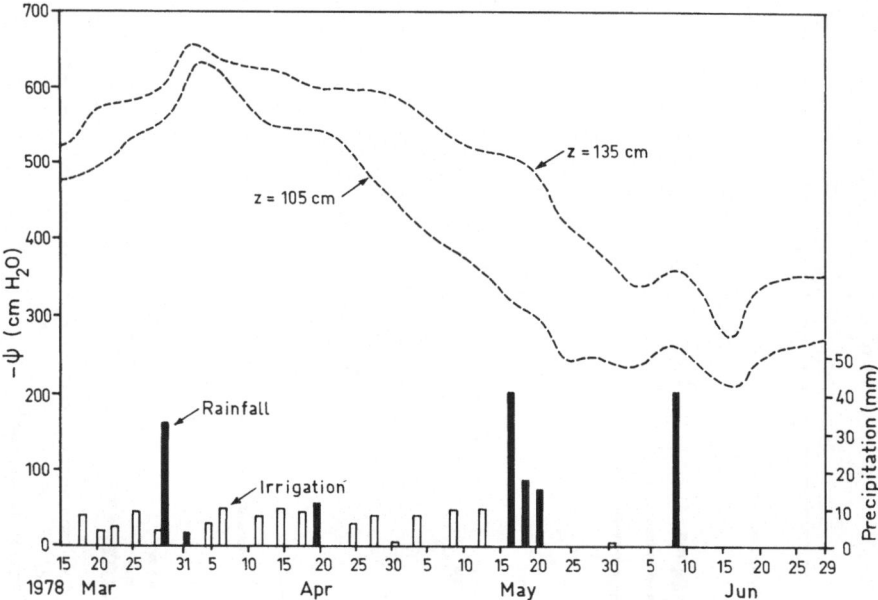

Fig. 1.   Soil water potential at 105 and 135 cm depths (dashed lines), rainfall, and irrigation during the first bean crop (15 March–29 June 1978).

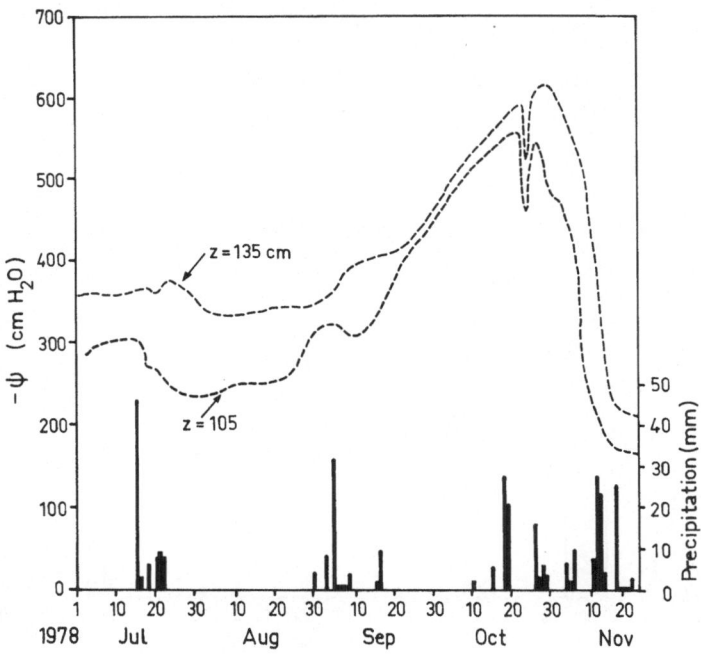

Fig. 2.   Soil water potential at 105 and 135 cm depths (dashed lines), and rainfall during the first fallow period (no irrigation occurred).

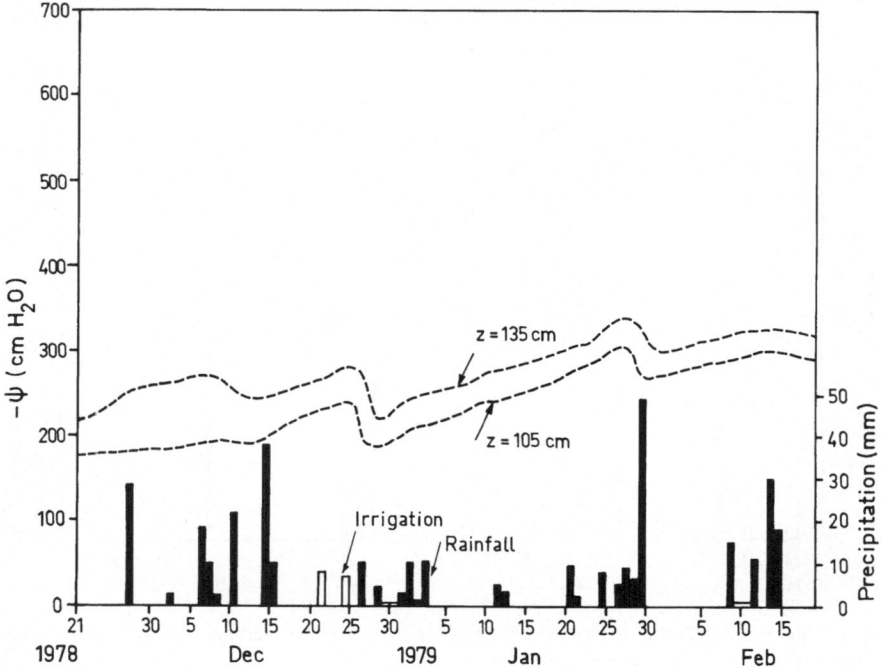

Fig. 3.   Soil water potential at 105 and 135 cm depths (dashed lines), rainfall, and irrigation during the second bean crop (21 November 1978–19 February 1979).

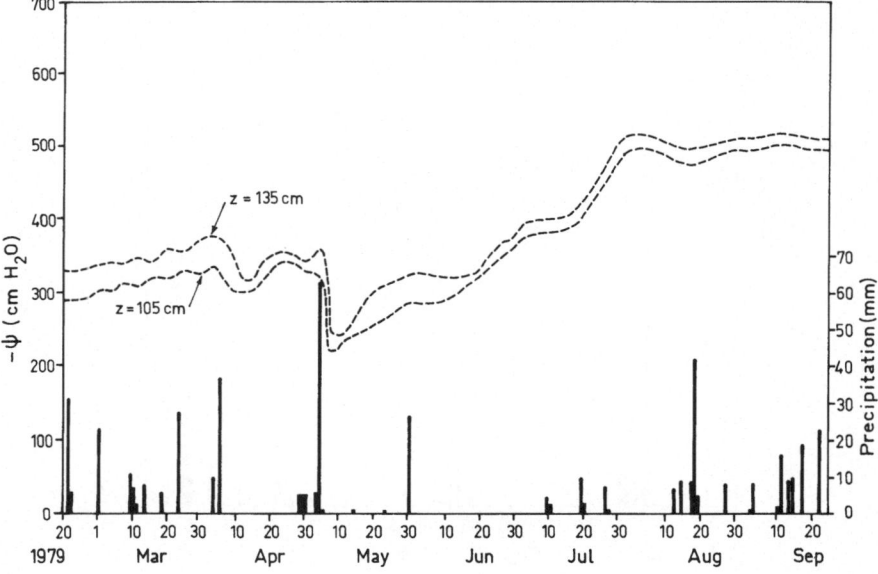

Fig. 4.   Soil water potential at 105 and 135 cm depths (dashed lines), and rainfall during the second fallow period (no irrigation occurred).

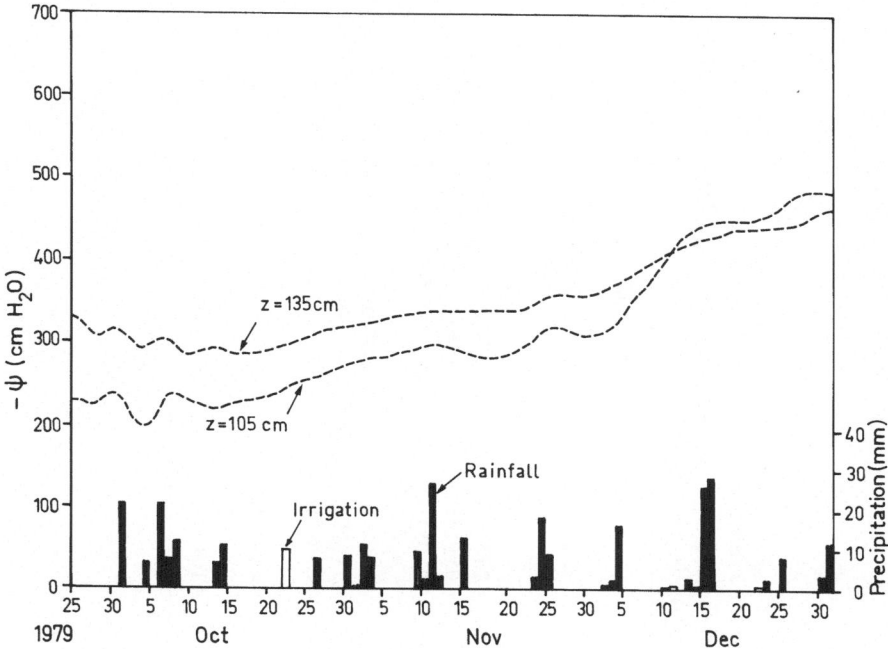

Fig. 5. Soil water potential at 105 and 135 cm depths (dashed lines), rainfall, and irrigation during the third bean crop (25 September–31 December 1979).

applied to the first crop had probably passed the 120 cm depth. The downward movement of mineral-N continued at least until the third crop was harvested. ^{15}N enrichment data for organic-N show that root activity (of both beans and weeds) probably transported ^{15}N to the 90–120 cm layer or deeper: organic-^{15}N was distributed evenly throughout the profiles.

Table 8 presents data for the proportion of soil-N that was derived from fertilizer-N for the first cropping period. Table 9 shows this distribution after the third crop. The high standard deviations in Table 8 indicate that enrichment of the order of 1% ^{15}N is too low to accurately study soil nitrogen, even for as short a period as 120 days (one bean crop). Data for high enrichment (ca. 10% ^{15}N) are not available, since for the first cropping period only low-enrichment plots were sampled, but presumably much of this experimental error would disappear with higher enrichments. Standard deviations for analyses following the third crop (Table 9) are not substantially lower despite the higher initial enrichment, probably because of dilution over time.

Calculations for organic-N are especially sensitive to inadequate enrichment. In the upper 30 cm of soil there is almost 3000 kg organic-N ha^{-1}, and results show very small enrichments in ^{15}N after the application of 100 kg N ha^{-1} even at ^{15}N enrichments of 10%. Atom % ^{15}N data for organic-N (Tables 5, 6 and 7) show enrichments only slightly above the natural (background) atom % of ^{15}N. Thus, the percent of soil organic-N derived from fertilizer-N was very small (Tables 8 and 9), and low enrichment levels resulted in extremely high standard deviations.

Table 3. Above-ground dry matter, total-N content and ^{15}N-content of three consecutive Phaseolus crops grown with different enrichments of ^{15}N applied in 100 kg urea-N ha^{-1} before the first crop. Subsequent crops were fertilized with unlabelled urea. All crops were also fertilized with 41 kg KCl-K ha^{-1} and 35 kg P ha^{-1} (as either superphosphate or rock phosphate). Straw includes pods without seeds but not roots. Natural atom% ^{15}N = 0.368

Crop	^{15}N-treatment (atom%)	Above-ground biomass (10^3 kg dry matter ha^{-1})		Biomass-N (%)		^{15}N (atom%)	
		Straw	Seed	Straw	Seed	Straw	Seed
First	1.12	2.34* (0.64)	2.10 (0.48)	1.80 (0.35)	4.20 (0.07)	0.579 (0.035)	0.624 (0.069)
	2.51	1.93 (0.37)	2.01 (0.20)	1.63 (0.18)	4.05 (0.35)	1.016 (0.119)	0.048 (0.131)
	10.46	1.90 (0.38)	1.81 (0.64)	1.80 (0.21)	4.65 (0.57)	3.26 (0.63)	3.77 (0.46)
Second	1.12	0.82 (0.27)	0.99 (0.40)	1.73 (0.11)	4.03 (0.11)	0.424 (0.018)	0.430 (0.019)
	2.51	1.25 (0.30)	1.42 (0.40)	1.68 (0.32)	4.13 (0.04)	0.502 (0.045)	0.491 (0.030)
	10.46	1.20 (0.29)	1.36 (0.49)	1.80 (0.07)	4.63 (0.74)	0.876 (0.145)	0.932 (0.167)
Third	1.12	1.31 (0.96)	1.18 (0.83)	1.56 (0.22)	4.44 (0.04)	0.380 (0.005)	0.387 (0.005)
	2.51	1.19 (0.23)	1.26 (0.51)	1.60 (0.06)	4.91 (0.24)	0.402 (0.007)	0.405 (0.008)
	10.46	1.20 (0.36)	1.36 (0.63)	1.71 (0.33)	4.10 (0.18)	0.510 (0.045)	0.499 (0.062)

* Values are means (± standard deviations) of 8 replicate field plots.

Table 4. Biomass-N derived from fertilizer (Ndff), total nitrogen in biomass, and nitrogen utilization efficiency [(proportion of N from fertilizer]) × (biomass-N) for three Phaseolus crops. See Table 3 for treatments.

Crop	^{15}N-treatment (atom % ^{15}N)	Ndff (%)		Biomass-N (kg N ha^{-1})		N-utilization efficiency (%)			
		Straw	Seed	Straw	Seed	Straw	Seed	Total	Mean (n = 24)
First	1.12	28.2* (4.6)	34.3 (9.2)	42.1 (14.2)	88.3 (20.3)	11.9 (4.5)	30.3 (10.7	42.1 (11.6)	38.5 (6.0)
	2.51	30.3 (5.6)	31.8 (6.1)	31.3 (6.8)	81.4 (13.6)	9.5 (2.7)	25.9 (6.6)	35.4 (7.1)	
	10.46	28.6 (6.2)	33.7 (4.6)	34.2 (7.9)	84.2 (31.6)	9.8 (3.1)	28.4 (11.3)	38.1 (11.7)	
Second	1.12	7.5 (2.4)	8.3 (2.5)	14.1 (4.8)	39.8 (16.1)	1.1 (0.5)	3.3 (1.7)	4.4 (1.8)	4.6 (1.0)
	2.51	6.3 (2.1)	5.8 (1.4)	20.9 (6.4)	58.6 (16.4)	1.3 (0.6)	3.4 (1.3)	4.7 (1.4)	
	10.46	5.0 (1.4)	5.6 (1.7)	21.6 (5.3)	73.0 (24.8)	1.1 (0.4)	3.5 (2.0)	4.6 (2.0)	
Third	1.12	1.6 (0.7)	1.7 (0.7)	20.3 (15.2)	52.2 (36.7)	0.3 (0.3)	0.9 (0.7)	1.2 (0.8)	1.2 (0.3)
	2.51	1.6 (0.3)	1.7 (0.4)	19.1 (3.8)	62.1 (25.0)	0.3 (0.1)	1.1 (0.5)	1.4 (0.5)	
	10.46	1.4 (0.4)	1.3 (0.6)	20.4 (7.8)	51.4 (24.1)	0.3 (0.1)	0.7 (0.4)	1.0 (0.4)	

* See Table 3.

Table 5. Soil nitrogen after first Phaseolus harvest. All plots received 33 kg urea-N ha^{-1} enriched with 1.12 atom % ^{15}N immediately before planting and 67 kg ^{15}N-urea-N 30 days later. Samples were collected on 29 June 1978

Depth (cm)	Mineral-N (NO$_3^-$-N plus NH$_4^+$-N)			Organic-N		
	+ Urea-N		No urea-N	+ Urea-N		No urea-N
	(mg N kg dry soil^{-1})	(atom % ^{15}N)	(mg N kg dry soil^{-1})	(% N)	(atom % ^{15}N)	(% N)
0– 20	21.73* (7.98)	0.517 (0.089)	11.61 (4.25)	0.067 (0.033)	0.374 (0.018)	0.091 (0.020)
30– 60	12.23 (4.81)	0.496 (0.081)	6.30 (2.48)	0.040 (0.021)	0.372 (0.015)	0.080 (0.030)
60– 90	8.94 (5.02)	0.412 (0.018)	3.51 (1.96)	0.037 (0.018)	0.371 (0.009)	0.062 (0.030)
90–120	8.69 (5.70)	0.389 (0.017)	2.03 (1.33)	0.030 (0.011)	0.369 (0.012)	0.040 (0.010)

* See Table 3.

Table 6. Soil nitrogen after second Phaseolus harvest. Initial fertilization (before first crop harvest) included 2.51 atom % ^{15}N; fertilizer applied to this crop (100 kg urea-N ha^{-1}) was unlabelled. Samples were collected on 19 February 1979

| Depth (cm) | Mineral-N (NO$_3$$^-$-N plus NH$_4$$^+$-N) | | | Organic-N | | |
| | + Urea-N | | No urea-N | + Urea-N | | No urea-N |
	(mg N kg dry soil^{-1})	(atom % ^{15}N)	(mg N kg dry soil^{-1})	(% N)	(atom % ^{15}N)	(% N)
0– 30	14.80* (13.24)	0.417 (0.021)	9.6 (3.51)	0.068 (0.025)	0.378 (0.015)	0.085 (0.02)
30– 60	8.97 (5.91)	0.448 (0.051)	5.5 (2.17)	0.036 (0.013)	0.374 (0.013)	0.050 (0.02)
60– 90	14.90 (10.90)	0.513 (0.094)	3.8 (2.12)	0.038 (0.019)	0.370 (0.014)	0.010 (0.005)
90–120	13.6 (5.11)	0.563 (0.043)	4.0 (2.62)	0.023 (0.008)	0.372 (0.008)	0.010 (0.002)

* See Table 3.

Table 7. Soil nitrogen after third Phaseolus harvest. Initial (first crop) fertilizer included 10.46 atom $\%$ ^{15}N; fertilizer applied to second and third crops (100 kg urea-N ha^{-1} each) was unlabelled. Samples were collected on 8 January 1980

Depth (cm)	Mineral-N ($NO_3^- $-N plus NH_4^+-N)			Organic-N		
	+ Urea-N		No urea-N	+ Urea-N		No urea-N
	(mg N kg dry soil^{-1})	(atom $\%$ ^{15}N)	(mg N kg dry soil^{-1})	($\%$ N)	(atom $\%$ ^{15}N)	($\%$ N)
0– 30	13.94* (10.40)	0.542 (0.100)	7.49 (4.02)	0.074 (0.028)	0.425 (0.041)	0.053 (0.015)
30– 60	10.94 (3.71)	0.639 (0.201)	4.60 (2.47)	0.044 (0.011)	0.392 (0.005)	0.066 (0.021)
60– 90	12.45 (7.26)	0.766 (0.241)	4.84 (2.66)	0.038 (0.010)	0.392 (0.007)	0.038 (0.010)
90–120	13.12 (6.37)	1.030 (0.401)	6.21 (3.34)	0.032 (0.009)	0.400 (0.011)	0.019 (0.025)

* See Table 3.

Table 8. Percent and amount of soil-N derived from fertilizer-N (Ndff) for the first cropping period

Soil layer (cm)	Mineral-N			Organic-N		
	Total	Ndff		Total	Ndff	
	(kg N ha^{-1})	(%)	(kg N ha^{-1})	(kg N ha^{-1})	(%)	(kg N ha^{-1})
0– 30	93.8* (36.1)	20.0 (11.9)	18.8 (13.3)	2890 (1460)	0.802 (2.4)	23.1 (83.8)
30– 60	53.0 (21.1)	17.1 (10.8)	9.1 (6.8)	1734 (917)	0.534 (2.0)	9.2 (34.7)
60– 90	35.5 (20.1)	5.9 (2.4)	2.1 (1.5)	1467 (720)	0.401 (1.2)	5.9 (17.9)
90–120	32.7 (21.5)	2.8 (2.3)	0.9 (0.9)	1130 (417)	0.134 (1.6)	1.5 (17.9)

* See Table 3.

Despite these limitations, important observations can be made from this study. Fig. 6 shows the partition of the 100 kg N ha^{-1} applied to the first crop in three components: i) crop-N removal, ii) soil mineral-N (NO_3^- + NH_4^+) and iii) soil organic-N. The mass balance of nitrogen at the end of the first crop totals 109.1 kg N ha^{-1}, which is an error of about 10% if no nitrogen was lost by volatilization, denitrification or leaching. At the end of the second and third crops the balances total 75.8 and 80.1 kg N ha^{-1}, respectively, indicating that 20 to 25% of the applied-N was lost from the system. Volatilization and denitrification losses are probably small under the conditions of the experiment, so it is reasonable to conclude that about 20–25 kg N ha^{-1} of fertilizer-N were leached below 120 cm. This is supported by data in Table 10, which show rainfall, irrigation, and soil water flux data for the entire experimental period. Of 1784 mm of water entering the system, 155 mm drained below the 120 cm depth, mainly after the first crop and during the second crop. During this period, as suggested by data in Tables 5, 6 and 7, fertilizer-N must have passed the 120 cm depth.

Although the data in Fig. 6 have large standard deviations, they show the general dynamics of nitrogen in the system. Mineral-N is clearly moving to deeper layers. Probably the pulse of fertilizer applied to the first crop passed the 120 cm depth at the end of the second crop. Due to the low N-utilization of the crop, much nitrogen remains in the soil profile, where it is available for leaching and consequently for fertilizer-pollution of ground water.

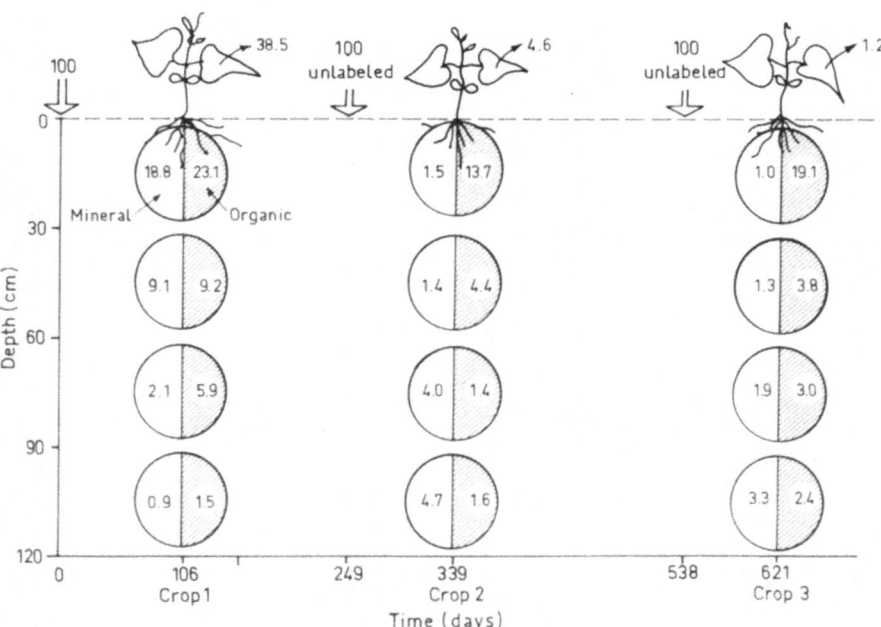

Fig. 6.   Fate of 100 kg N ha^{-1} ^{15}N-labelled urea applied during the first of the three successive Phaseolus crops. All values are in kg original-fertilizer-N ha^{-1}.

Table 9. Percent and amount of soil-N derived from initial fertilizer-N after the third crop. Fertilizer-N was applied to the first (initial fertilizer-N), the second and the third crop

Soil layer (cm)	Mineral-N			Organic-N		
	Total	Ndff		Total	Ndff	
	(kg N ha^{-1})	(%)	(kg N ha^{-1})	(kg N ha^{-1})	(%)	(kg N ha^{-1})
0– 30	60.2* (45.4)	1.7 (1.0)	1.0 (0.9)	3194 (1261)	0.6 (0.4)	19.1 (14.8)
30– 60	47.4 (16.3)	2.7 (2.0)	1.3 (1.0)	1907 (491)	0.2 (0.05)	3.8 (1.4)
60– 90	48.2 (29.0)	3.9 (2.4)	1.9 (1.6)	1507 (408)	0.2 (0.07)	3.0 (1.3)
90–120	49.4 (24.1)	6.6 (4.0)	3.3 (2.6)	1206 (344)	0.3 (0.1)	2.4 (1.1)

* See Table 3.

Table 10. Water balance components for the three Phaseolus cropping periods during 1978–79. Q_{120} = total soil water draining past 120 cm depth. All values are in mm H_2O

Period	Precipitation	Irrigation	Q_{120}
First crop 15 March–29 June	164	114	10
Uncropped 30 June–20 November	337	0	43
Second crop 21 November–19 February	371	15	82
Uncropped 20 February–24 September	490	0	6
Third crop 25 September–30 December	283	10	14

Acknowledgements This paper is a contribution from the Center for Nuclear Energy in Agriculture (CENA), Piracicaba, São Paulo, Brazil; research was conducted under the joint Food and Agriculture Organization (FAO)/International Atomic Energy Agency (IAEA)/Gesellschaft für Strahlung und Umweltforschung (GSF) Coordinated Research Programme (Contract RC-1597), with funding from the National Atomic Energy Commission (CNEN), Financiadora de Estudos e Projetos (FINEP) and the National Council for Scientific and Technological Development (CNPq).

References

1 Bremner J M 1965 Total nitrogen. *In* Black C A (Ed.). Methods of Soil Analysis, Part 2, pp 771–1149. Madison, Wisconsin: Am. Soc. Agron.
2 Bremner J M and Keeney D R 1966 Determination of isotope ratio analysis of different forms of nitrogen in soils. 3. Exchangeable ammonium, nitrate and nitrite by extraction distillation methods. Soil Sci. Soc. Am. Proc. 30, 577–582.
3 Burford J R, Dowdell R J and Webster C P 1978 The fate of fertilizer nitrogen applied to permanent grassland: lysimeter studies. *In* Letcombe Laboratory, Annual Report. pp 46–48. Wantage, U.K.: Letcombe Laboratory.
4 Burford J R, Dowdell R J and Crees R 1978 Crop uptake and soil immobilization of fertilizer nitrogen after direct drilling and ploughing. *In* Letcombe Laboratory, Annual Report, pp 48–50. Wantage, U.K.: Letcombe Laboratory.
5 IAEA 1980 Proceedings of a Seminar on Isotope Techniques in Studies of the Useful Conservation and the Pollutant Potential of Agricultural Nitrogen Residues. IAEA Special Report 48. Vienna: International Atomic Energy Agency. 29 p.
6 Reichardt K, Grohmann F, Libardi P L and Queiroz S V 1976 Spatial Variability of Physical Properties of a Tropical Soil. II. Soil Water Retention Curves and Hydraulic Conductivity. CENA Report BT-004. Piracicaba, Brazil: Centro de Energia Nuclear na Agricultura. 24 p.
7 Strebel O, Grimme H, Reniger M and Flaige H 1980 Field study with nitrogen-15 of soil and fertilizer nitrate uptake and water withdrawal by wheat. Soil Sci. 130, 205–211.

Plant and Soil 67, 209–220 (1982).   0032-079X/82/0672-0209$01.80.

© 1982 *Martinus Nijhoff/Dr W. Junk Publishers, The Hague.*

Nitrogen cycling in a flooded-soil ecosystem planted to rice (*Oryza sativa* L.)*

Ciclo de nitrógeno en arroz (Oryza sativa) *cultivado bajo innundación*

K. R. REDDY

Agricultural Research and Education Center, Institute of Food and Agricultural Sciences, University of Florida, Sanford, FL 32771, USA

Key words Ammonification Crop residues Denitrification Flooded soil 15-N N-fertilizers N_2-fixation Nitrification Rice Volatilization

Abstract ^{15}N studies of various aspects of the nitrogen cycle in a flooded rice ecosystem on Crowley silt loam soil in Louisiana were reviewed to construct a mass balance model of the nitrogen cycle for this system. Nitrogen transformations modeled included 1) net ammonification (0.22 mg $NH_4{}^+$–N kg dry soil^{-1} day^{-1}), 2) net nitrification (2.07 mg $NO_3{}^-$–N kg^{-1} dry soil^{-1} day^{-1}), 3) denitrification (0.37 mg N kg dry soil^{-1} day^{-1}), and 4) biological N_2 fixation (0.16 mg N kg dry soil^{-1} day^{-1}). Nitrogen inputs included 1) application of fertilizers, 2) incorporation of crop residues, 3) biological N_2 fixation, and 4) deposition. Nitrogen outputs included 1) crop removal, 2) gaseous losses from NH_3 volatilization and simultaneous occurrence of nitrification-denitrification, and 3) leaching and runoff. Mass balance calculations indicated that 33% of the available inorganic nitrogen was recovered by rice, and the remaining nitrogen was lost from the system. Losses of N due to ammonia volatilization were minimal because fertilizer-N was incorporated into the soil. A significant portion of inorganic-N was lost by ammonium diffusion from the anaerobic layer to the aerobic layer in response to a concentration gradient and subsequent nitrification in the aerobic layer followed by nitrate diffusion into the anaerobic layer and denitrification into gaseous end products. Leaching and surface runoff losses were minimal.

Resumen Se revisaron varios aspectos del ciclo de nitrógeno estudiados con ^{15}N en un ecosistema de arroz de innundación en suelos franco limosos Crowley en Louisiana, USA, con el fin de construir un balance de masas para el nitrógeno.

Las tranformaciones que se incluyeron en el modelo fueron: 1) amonificación neta (0,22 mg NH_4–N kg^{-1} suelo seco dia^{-1}), 2) nitrificación neta (2,07 mg NO_3–N kg^{-1} suelo seco dia^{-1}), 3) desnitrificación (0,37 mg N kg^{-1} suelo seco dia^{-1}) y 4) fijación biológica de nitrógeno (0,16 mg N kg^{-1} suelo seco dia^{-1}). Las entradas de nitrógeno al sistema serían aquellas por aplicación de fertilizantes, incorporación de residuos de cosecha, fijación biológica de nitrógeno, deposición. Las salidas serían por cosecha, perdidas gaseosas por volatilización de NH_3 y la ocurrencia simultanea de nitrificación y desnitrificación, lixiviación y escorrentía. El balance de masas indicó que el 33% del nitrógeno inorgánico disponible fué recuperado por el arroz y el resto se perdió del sistema. Las pérdidas por volatilización de NH_3 fueron mínimas porque el fertilizante fué incorporado al suelo. Una proporción significativa del nitrógeno inorgánico se perdió por difusión de NH_4 de la capa anaeróbica a la aeróbica en respuesta al gradiente de concentraciones; luego ocurre nitrificación en la capa aeróbica, difusión y finalmente desnitrificación y pérdida en forma gaseosa. Las perdidas por lixiviación y escorrentía fueron mínimas.

* Florida Agricultural Experiment Station, Journal series No. 3855.

Introduction

Lowland rice is a major food for approximately 50% of the world's population. Throughout the rice-growing regions of the world, nitrogen is most commonly the nutrient limiting rice growth and yield.

Nitrogen in flooded-soil ecosystems occurs in inorganic and organic forms, with the latter form predominant. Organic-N includes compounds from amino acids, amines, proteins, and humic compounds with low N-content; inorganic forms include ammonium, nitrate, and nitrite. Ammonium dominates the inorganic-N pool at any given time; it comes from the mineralization of organic-N and the application of fertilizers. Gaseous forms of nitrogen that occur in this agro-ecosystem include NH_3, N_2 and N_2O.

The gains and losses of nitrogen in the flooded-soil ecosystem are regulated by a series of biochemical and physicochemical processes that transform one form of nitrogen to another (Fig. 1). Nitrogen inputs (gains) to the systems come principally from 1) the application of fertilizers, 2) the incorporation of crop residues, 3) biological N_2 fixation, 4) interflow, runoff, and irrigation water, 5) atmospheric NH_3, N_2O, and NO absorption by soil and plants, and 6) dry deposition and rainfall. Nitrogen losses from the flooded soil system mainly result from 1) crop removal, 2) simultaneous occurrence of nitrification-denitrification reactions, 3) NH_3 volatilization, 4) non-reversible fixation of ammonium-N by clay minerals, and 5) leaching and surface runoff.

In this report I examine the agronomic and ecologic significance of the processes controlling nitrogen utilization by rice and the role of these processes in determining the N lost or gained by the ecosystem. In particular, I synthesize the considerable ^{15}N data that has recently accumulated for a flooded Crowley silt loam soil ecosystem in Louisiana. This soil type is typical of those used for rice cultivation in this region. Some characteristics of this ecosystem appear in Table 1.

Characteristics of flooded soil

Flooded soils are generally characterized by the absence of oxygen when compared to upland soils. In most rice fields, the dissolved oxygen content of the overlying water column remains relatively high due to a low density of oxygen-consuming organisms and photosynthetic oxygen production by algae. In contrast, oxygen is slowly renewed and the demand is usually high in the underlying soil, especially in those soils with high organic matter content. The greater potential consumption of oxygen at the soil-water interface compared to the renewal rate through the floodwater results in the development of two distinct soil layers: an oxidized or aerobic top layer and an underlying reduced or anaerobic layer. The thickness of the aerobic zone can vary considerably (Fig. 2).

Rice plants growing in anaerobic soil systems have a unique mechanism for

Table 1. Characteristics of flooded Crowley silt loam soil ecosystems

	Units
Climate	
Mean annual temperature	20°C
Mean temperature during growing season	27°C
Annual precipitation	1400 mm yr^{-1}
Precipitation during growing season	520 mm 4 months^{-1}
Growing season for rice	120 days yr^{-1}
Soil	
Carbon (dry weight)	0.70%
Nitrogen (dry weight)	0.08%
Soil organic N (30 cm depth)	2090 kg ha^{-1}
Soil inorganic N (30 cm depth)	68 kg ha^{-1}
Bulk density (upper 30 cm)	1.15 g cm^{-3}
Soil type	Typic albaqualfs
Texture	Silt loam
pH	5.6
Cation exchange capacity (meq 100 g dry soil^{-1})	9.4
C:N ratio	8.75

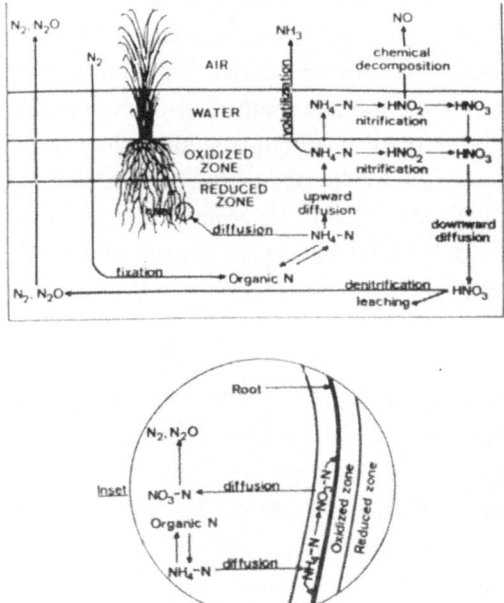

Fig. 1.   Nitrogen cycle in a flooded soil ecosystem.

transporting atmospheric oxygen through the stems to the roots, and some of this oxygen diffuses into the adjacent root rhizosphere. This results in oxygen zonation in the soil around the root surface (Fig. 1). The aerobic soil layer supports the aerobic microflora and is the site for nitrogen processes that require oxygen, while the anaerobic soil layer supports anaerobic microflora and is the site for nitrogen processes that can occur in the absence of oxygen.

Lowland rice is normally grown under flooded soil conditions, but in certain situations these soils are subjected to alternate flooding and draining cycles. In the United States, most rice is grown only during the summer months (May through September) under flooded conditions, and for the remainder of the year these soils are drained. Under this type of management, anaerobic soil conditions predominate during flooding and aerobic conditions during the rest of the year.

Nitrogen-cycle processes

Nitrogen-cycle processes that control gains and losses of nitrogen in the flooded soil ecosystem include ammonification, NH_3 volatilization, nitrification, denitrification, assimilatory nitrate reduction, and N_2 fixation.

Ammonification

Ammonification (or N-mineralization) is the conversion of organic-N to ammonium-N. This process proceeds at a much slower rate in flooded-soil systems than in drained-soil systems. Ammonification supplies about 60% of the nitrogen requirement of rice crops grown on Crowley silt loam soils in Louisiana[11]. Dolmat et al.[2] observed that ammonium-N mineralized under anaerobic conditions was a good estimate of nitrogen available to rice during the growing season. Laboratory incubation studies showed that ammonium-N accumulation in flooded Crowley silt loam soil occurred at a rate of 0.22 mg NH_4-N kg dry soil^{-1} day^{-1} (0.575 kg N ha^{-1} day^{-1}), which is equivalent to approximately 3.4% of the total soil organic-N and crop residue organic-N over one cropping season.

Ammonia volatilization

Ammonia volatilization is a pH-dependent reaction. Alkaline pH favors the presence of aqueous forms of NH_3 in solution, while at acidic or neutral pH the NH_3 is predominately in the ionic NH_4^+ form. Losses of NH_3 through volatilization are insignificant if pH is below 7.5, and often losses are not serious if the pH is below 8.0. However, considerable NH_3 loss can occur if the pH of the system is in the range of 8 to 10, or above. In flooded soils planted to rice, NH_3 volatilization is not considered an important mechanism of nitrogen loss except in specialized cases where high ammonium-N concentrations exist in conjunction with high pH at the floodwater-soil interface. High pH conditions in floodwater can develop during sunlight hours as a result of an imbalance between photosynthesis and respiration of algae and submerged aquatic macrophytes. Under these conditions, the pH of the water column can increase by 2 to 3 units during mid-day when the photosynthetic process is actively with-drawing CO_2 from the system, and can fall at night when respiratory activities liberate free-CO_2 into water. When urea or ammoniacal fertilizers are applied to the surfaces of these systems, significant volatilization losses can be observed. However, these losses can be reduced by incorporating the fertilizers into soil[4, 19].

Nitrification-denitrification

Nitrification and denitrification are known to occur simultaneously in flooded soil systems. Nitrification occurs in the aerobic zone while denitrification occurs predominantly in the underlying anaerobic zone. The magnitude of nitrification is controlled by oxygen diffusion rates, the thickness of the aerobic zone, the ammonium-N concentration, and levels of inorganic-C. Net nitrification rates in the surface aerobic soil layer of a flooded Crowley silt loam soil were found to be 2.07 mg NO_3–N kg dry soil^{-1} day^{-1} (ref.[18]). Nitrification is generally active during most of the day except under special conditions of high pH (>9.0) and free NH_3 at the floodwater-soil interface. In flooded Crowley silt loam soil, pH of the aerobic soil zone was in the range of 5 to 6.

Rates of denitrification are influenced by readily oxidizable-C, temperature, pH, denitrifying populations, and nitrate-N concentrations. In a flooded Crowley silt loam, denitrification rates of 0.37 mg N kg dry soil^{-1} day^{-1} in the anaerobic soil layer have been measured[17]. These reactions involve both oxidation (nitrification) and reduction (denitrification), with corresponding N valence changes of from -3 (for NH_4–N) to $+5$ (for NO_3–N), followed by reductions to $+1$ (for N_2O) or to zero (for N_2). Four moles of oxygen are required to produce one mole of N_2 gas. A combined nitrification-denitrification reaction can be written as:

$$24\,NH_4^+ + 48\,O_2 \rightarrow 24\,NO_3^- + 24\,H_2O + 48\,H^+$$
$$24\,NO_3 + 5\,C_6H_{12}O_6 + 24\,H^+ \rightarrow 12\,N_2 + 30\,CO_2 + 42\,H_2O$$

$$24\,NH_4^+ + 5\,C_6H_{12}O_6 + 48\,O_2 \rightarrow 12\,N_2 + 30\,CO_2 + 66\,H_2O + 24\,H^+$$

In this reaction nitrate-N is an intermediate product and does not appear in the final reaction. Laboratory and field investigations using ^{15}N have indicated very little or no nitrate-N accumulation in these systems. The major source of ammonium-N to the aerobic zone comes from fertilizers, mineralization of organic-N in the aerobic zone, and diffusion of ammonium-N from the underlying anaerobic zone. The nitrate-N formed in the aerobic layer is the only nitrogen that is constantly supplied to the anaerobic zone. Nitrate-N moves to the anaerobic zone by diffusion, and is subsequently removed by denitrification (Fig. 3)[6]. In lowland rice fields nitrate-N is not used as a fertilizer N-source. In a

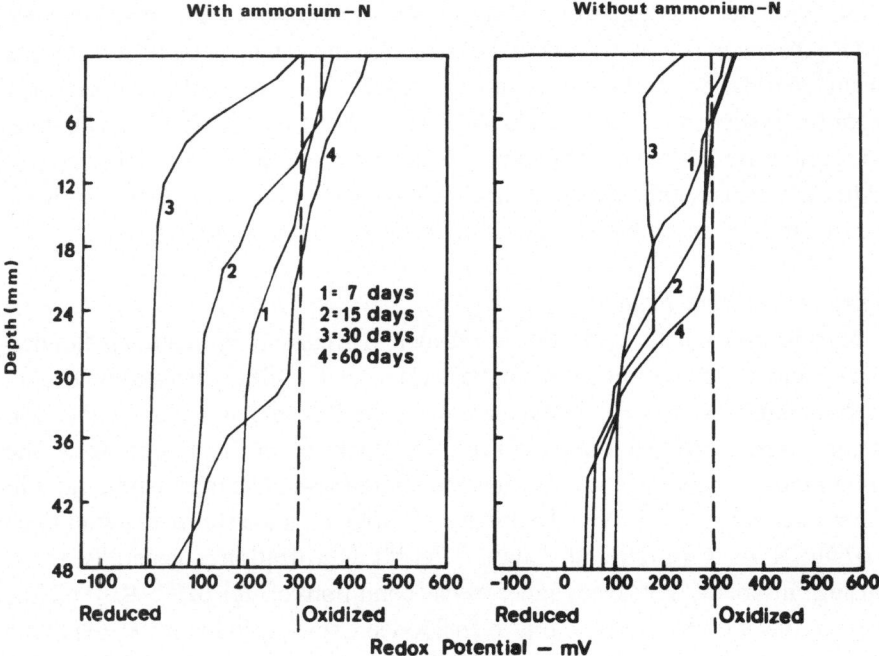

Fig. 2.   Depth of aerobic layer in a flooded soil after different periods of incubations with and without added ammonium nitrogen.

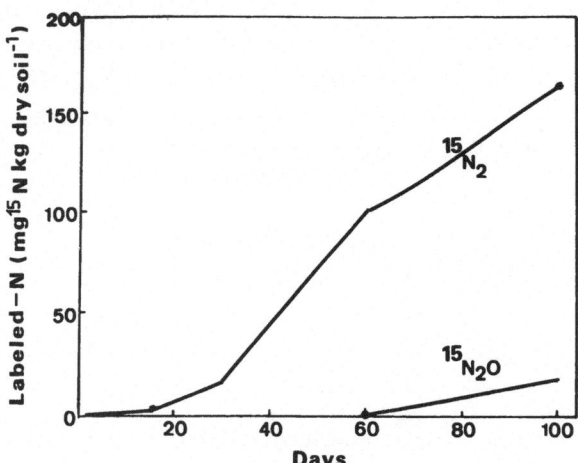

Fig. 3.   Gaseous losses of added ammonium-N in an incubated and flooded Crowley silt loam treated with $^{15}(NH_4)_2SO_4$.

flooded field subjected to alternate flooding and draining, nitrification can occur when soil is drained and then the accumulated nitrate be denitrified upon reflooding. In Louisiana and other rice growing areas of the United States, this practice is not followed during the growing season. However, nitrate-N accumulates in these systems when flooded fields are drained after the rice harvest and this nitrate-N is readily denitrified upon reflooding for the next rice crop.

Nitrogen fixation

Atmospheric N_2 fixation in flooded rice fields occurs in the floodwater and aerobic soil layer, in the anaerobic soil layer, and in the root rhizosphere. Laboratory studies[14] with ^{15}N showed that in a flooded Crowley silt loam soil containing no plants, N_2 fixation occurred at 0.16 mg N kg dry soil^{-1} day^{-1} (0.78 µg N cm^{-2} day^{-1}), and that most of this fixation was by algae in the surface 0.5 cm of the aerobic soil layer. The amount of N_2 fixed by intact soil cores obtained from a rice field was in the range of 0.12–0.15 mg N kg dry soil^{-1} day^{-1} (0.58–0.75 µg N cm^{-2} day^{-1}). In this study, N_2 fixation occurred primarily in the surface 0.5 cm soil layer. However, a field experiment using the acetylene reduction method showed a higher fixation rate of 0.41–0.74 mg N kg dry soil^{-1} day^{-1} (2–3.6 µg cm^{-2} day^{-1}). The amount of N_2 fixed in this system is low and not enough to support high yields of rice, but may be an important contribution to the nitrogen economy of the soil.

Nitrogen movement

Nitrogen transport in flooded soil systems can occur by 1) ion diffusion, 2) leaching and interflow, and 3) surface runoff.

Ammonium-N diffusion

Ammonium-N diffusion from the anaerobic soil layer to the aerobic soil layer provides ammonium-N to the nitrifying organisms. The rate of this movement is governed by 1) the concentration gradient established as a result of ammonium-N consumption in the aerobic zone, 2) the ammonium-N regeneration rate in the anaerobic zone, 3) the ammonium-N concentration in the pore water, 4) other cations on the soil exchange complex, 5) the cation exchange capacity of the soil, and 6) the relative volume of pore space which is a function of the soil's bulk density. The quantity of ammonium-N transferred by diffusion per unit area per unit time is proportional to the diffusion coefficient and the concentration gradient. Studies on flooded Crowley silt loam soil[16] showed that ammonium-N flux was about 10.4 µg NH_4–N cm^{-2} day^{-1}, and diffusion of ammonium-N from the anaerobic soil layer accounted for more than 50% of this (5.97 µg cm^2 day^{-1}); the ammonium-N diffusion coefficient was 0.216 cm^2 day[18].

Nitrate-N diffusion

In a flooded rice field, nitrate-N in the floodwater and aerobic soil layer readily diffuses into the anaerobic soil layer in response to the downward concentration gradient. The flux of nitrate-N from the floodwater and the aerobic soil layer to the anaerobic soil layer is controlled by 1) the oxidizable-C supply in the anaerobic zone, 2) the thickness of the aerobic zone, 3) floodwater depth, 4) the nitrate-N concentration in the floodwater and in the aerobic zone, 5) temperature, and 6) mixing and aeration in the floodwater. For flooded Crowley silt loam soil, the measured diffusion coefficient for nitrate-N's moving from the floodwater and aerobic zone to the anaerobic zone[18] was 1.33 cm^2 day^{-1}. This value is about six times higher than the diffusion coefficient value for ammonium-N diffusion.

Leaching, interflow, and runoff

In flooded rice fields of Louisiana, losses of nitrogen through leaching, interflow and runoff are insignificant compared to other sources of loss. In ^{15}N field studies[7] very little or no fertilizer-N moved deeper than 20 cm in a flooded Crowley silt loam. In drained fields, nitrate-N accumulates and subsequently a significant portion can be lost through leaching. Very little quantitative information is available on the importance of leaching, interflow, and runoff to the nitrogen balance of rice paddies.

Plant uptake

Rice is noted for its poor utilization of fertilizer-N. Several ^{15}N field studies of Crowley silt loam rice systems[9,11,13] have shown that *ca.* 63% of total nitrogen uptake by rice is derived from native soil-N and fixed N$_2$, with the remainder derived from fertilizer-N.

Another study using ^{15}N[7,13] revealed that over one cropping cycle approximately one-third of the fertilizer-N applied to the rice was recovered in the grain, approximately one-fifth to one-fourth was recovered in the straw, approximately one-fourth remained in the soil and roots, and the remaining one-fifth to one-fourth was lost from the system. Incorporation of crop residues resulted in less than 10% of the added-N released during the second year. These studies also showed that during early growth, the rice plant largely utilized fertilizer-N rather than native soil-N, while during later growth native soil-N was the major source of nitrogen for the plant. ^{15}N studies conducted in several other countries have also showed poor recoveries (12 to 41%) of fertilizer-N by rice[3]. DeDatta *et al.*[1], using ^{15}N-labelled ammonium sulfate, observed recoveries of 15 to 71% of the added-N by rice. In another field study, Reddy and Patrick[15] found that about 26 and 31% of added-N was removed by rice from plots receiving surface applications of ^{15}N-labelled urea and ammonium sulfate, respectively, while only 10% of the added-N was recovered from plots treated with ^{15}N-labelled rice straw.

Nitrogen losses

Studies conducted with paddy soils in different parts of the world have indicated that the low recovery of ammoniacal fertilizers by rice is largely due to losses of nitrogen through nitrification and subsequent denitrification. Laboratory studies using ^{15}N on a flooded Crowley silt loam indicated that substantial nitrogen losses occurred following ammonium-N application[6, 8, 12, 15]. The studies demonstrated that nitrification-denitrification reactions were controlling nitrogen loss, and also showed that more ammonium-N was lost from a flooded soil than was actually present in the aerobic soil layer at any one time. Apparently, ammonium-N diffused from the anaerobic zone to the aerobic zone where it underwent nitrification, and nitrate-N thus formed diffused into the anaerobic zone where it underwent denitrification[6, 16]. Losses in the laboratory studies amounted to as much as 80% of the added-N. In a field study about 25% of the applied ammonium-N was unaccounted for[11], even when the fertilizer was applied at 7.5 cm depth. This indicates that a significant portion of the nitrogen lost was probably *via* this nitrification-diffusion-denitrification pathway.

Severe N losses were also shown to occur in a Crowley silt loam soil subjected to alternate draining and flooding. Organic-N is converted to ammonium-N in both aerobic and anaerobic soils, with the ammonium-N thus formed oxidized to nitrate-N under aerobic conditions and the nitrate-N then denitrified under anaerobic conditions. Such losses can be high in soils planted to lowland rice where management practices require draining and flooding. Patrick and Wyatt[5] observed a large nitrogen loss (up to 20% of total N) following repeated cycles of flooding and drying to field moisture. Reddy and Patrick[10] observed a loss of 24% of total-N in a Crowley silt loam soil which underwent incubation for 2 days under aerobic conditions followed by 2 days under anaerobic conditions for 4 months.

The agronomic significance of the high pH that develops in the floodwater of rice fields has largely been neglected as a factor favouring direct volatilization loss of NH_3. Losses due to volatilization can be serious when fertilizers are broadcast into floodwater rather than incorporated into soil. Volatilization losses of surface-applied urea have ranged from 8 to 50%, while losses from surface applied ammonium sulfate have ranged from 4 to 15 percent[4, 19, 20]. Losses of ammonium-N through volatilization were less than 1% when fertilizers were incorporated into the soil. In flooded soil ecosystems of Louisiana (Crowley silt loam), losses of added-N through volatilization were observed to be less than 3% (unpublished results, W. H. Patrick, Louisiana State University, Baton Rouge, LA).

Table 2. Nitrogen budget for a flooded Crowley silt loam soil planted to rice. The growing season is 4 months (May–August) and the effective soil profile depth is 30 cm. Asterisks (*) indicate estimated values. Potential mineralizable-N = the maximum amount of soil organic N available for plant uptake. Fertilizer N was incorporated into the soil.

	kg N ha^{-1} season^{-1}
Inputs	
Potential mineralizable-N*	69
Inorganic N in the soil profile	68
Crop residue N	35
Fertilizer N added	100
Rainfall	2
Biological N$_2$ fixation	10
Total IN	284
Outputs	
Crop removal	93
Biological immobilization	30
Ammonia volatilization	5*
Nitrification-denitrification	102*
Inorganic N in the soil profile	26
Leaching	< 1*
Interflow	< 1*
Surface runoff	< 1*
Total OUT	259

Nitrogen budget

A list of input and output values obtained from ^{15}N studies on various components of the nitrogen cycle are shown in Fig. 4 and Table 2 for a flooded Crowley silt loam soil planted to rice. The major nitrogen inputs are ammonification of soil organic-N, fertilizer-N, and biological N$_2$ fixation. Ca. 49% of the added fertilizer-N is recovered by rice during the cropping season. The remainder of the plant-N was derived from native soil organic-N. A large portion of the fertilizer-N did not appear in any of the plant or soil components. Possible loss mechanisms include NH$_3$ volatilization, nitrification followed immediately by denitrification, and leaching and surface runoff. Fertilizer-N was incorporated into the soil, thus decreasing the possibility of NH$_3$ volatilization losses. Some of the nitrogen was probably lost in leaching or surface runoff, although ^{15}N was not recovered in the soil profile beyond the 25 cm depth. Denitrification probably accounted for large losses of nitrogen from this ecosystem. Soil organic-N is converted to gaseous end products by ammonification of soil organic-N to ammonium-N, followed by the upward

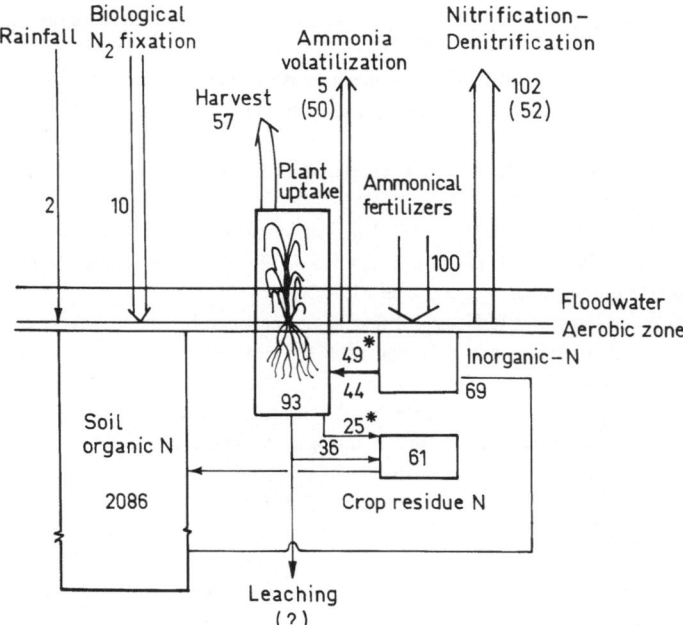

Fig. 4.   Schematic representation of the main aspects of the nitrogen cycle during the rice growing season. Arrows are fluxes (kg N ha^{-1} growing season^{-1}). Boxes represent pool sizes (kg N ha^{-1}) at harvest. Soil N is to 30 cm depth. Asterisks indicate fertilizer derived-N.

diffusion of ammonium-N to the aerobic layer, its oxidation to nitrate in this layer, the downward diffusion of nitrate to the anaerobic soil layer, and finally the reduction of nitrate-N to gaseous end products such as N_2 and N_2O (Fig. 1). Nitrogen losses in flooded soils can also be enhanced by the aerobic layer around the root zone. Although no experimental evidence is available, it is possible that nitrogen in the aerobic root rhizosphere and the adjacent anaerobic soil acts in much the same way as in the interface described above. Unless all types of nitrogen losses are identified and measured quantitatively under field conditions, it is difficult to obtain a reliable mass balance for nitrogen in a flooded soil ecosystem.

References

1   DeDatta S K, Magnaye C P and Magbanua J T 1969 Response of rice varieties to time of nitrogen application in the tropics. *In* Proc. Symp. Trop. Agric. Res., pp 73–87. Manila, Philippines: Inter. Rice Research Institute.

2   Dolmat M T, Patrick Jr W H and Peterson F J 1980 Relation of available soil nitrogen to rice yield. Soil Sci. 129, 229–238.

3   IAEA 1970 Rice Fertilization. – IAEA. Tech. Rep. Ser. No. 108, Vienna: International Atomic Energy Agency. 177 p.

4   Mikkelsen D S, DeDatta S K and Obcemea W N 1978 Ammonia volatilization losses from flooded rice soils. Soil Sci. Soc. Am. J. 42, 725–730.
5   Patrick Jr W H and Wyatt R 1964 Soil nitrogen loss as a result of alternate submergence and drying. Soil Sci. Soc. Am. Proc. 28, 647–653.
6   Patrick Jr W H and Reddy K R 1976 Nitrification-denitrification reactions in flooded soils and sediments: Dependence on oxygen supply and ammonium diffusion. J. Environ. Qual. 5, 469–472.
7   Patrick Jr W H and Reddy K R 1976 Fate of fertilizer nitrogen in flooded soil. Soil Sci. Soc. Am. J. 40, 678–681.
8   Patrick Jr W H and Tusneem M E 1972 Nitrogen loss from flooded soil. Ecology 53, 735–737.
9   Patrick Jr W H, De Laune R D and Peterson F J 1974 Nitrogen utilization by rice using ^{15}N depleted ammonium sulfate. Agron. J. 66, 819–820.
10   Reddy K R and Patrick Jr W H 1975 Effect of alternate aerobic and anaerobic conditions on redox potential, organic matter decomposition and nitrogen loss in a flooded soil. Soil Biol. Biochem 7, 87–94.
11   Reddy K R and Patrick Jr W H 1976 Yield and nitrogen utilization by rice as affected by method and time of application of labelled nitrogen. Agron. J. 68, 965–969.
12   Reddy K R and Patrick Jr W H 1977 Effect of placement and concentration of applied $^{15}NH_4$–N on nitrogen loss from flooded soil. Soil Sci. 123, 142–147.
13   Reddy K R and Patrick Jr W H 1978 Residual fertilizer nitrogen in a flooded rice soil. Soil Sci. Soc. Am. J. 42, 316–318.
14   Reddy K R and Patrick Jr W H 1979 Nitrogen fixation in a flooded rice soil. Soil Sci. 128, 80–86.
15   Reddy K R and Patrick Jr W H 1980 Losses of applied $^{15}NH_4$–N, urea-^{15}N and organic ^{15}N in flooded soil. Soil Sci. 130, 326–330.
16   Reddy K R, Patrick Jr W H and Phillips R E 1976 Ammonium diffusion as a factor in nitrogen loss from flooded soil. Soil Sci. Soc. Am. J. 40, 528–533.
17   Reddy K R, Patrick Jr W H and Phillips R E 1978 The role of nitrate diffusion in determining the order and rate of denitrification in flooded soil: I. Experimental results. Soil Sci. Soc. Am. J. 42, 268–272.
18   Reddy K R, Patrick Jr W H and Phillips R E 1980 Evaluation of selected processes controlling nitrogen loss in flooded soil. Soil Sci. Soc. Am. J. 44, 1241–1246.
19   Ventura W H and Yoshida T 1977 Ammonia volatilization from a flooded tropical soil. Plant and Soil 46, 521–531.
20   Vlek P L G and Craswell E T 1979 Effect of nitrogen source and management on ammonia volatilization losses from flooded rice soil systems. Soil Sci. Soc. Am. J. 43, 352–358.

Plant and Soil 67, 221–226 (1982).   0032-079X/82/0672-0221$00.90.   SU-19
© 1982 *Martinus Nijhoff/Dr W. Junk Publishers, The Hague.*

The nitrogen balance of vegetable crops irrigated with untreated effluent

Efecto del riego con efluentes municipales no tratados en el ciclo del nitrógeno

E. SCHALSCHA B. and I. VERGARA F.
Departamento de Química Inorgánica y Analítica, Facultad de Ciencias Químicas y Farmacológicas, Universidad de Chile, Santiago, Chile

Key words   Chile Drinking water Irrigation N-cycling Nitrate Sewage Vegetables

Abstract   In the agricultural areas near Santiago, Chile, *ca.* 780 kg N ha^{-1} yr^{-1} are added to vegetable crops *via* irrigation with untreated sewage effluent draining from the metropolitan area. Nitrate levels in surface wells in the area, from which drinking water is derived, often exceed established limits for human consumption. Of the 779 kg N ha^{-1} added to crops in one year, 161–287 kg N ha^{-1} yr^{-1} were removed by crop harvest and much of the remainder apparently eventually leached to the 1–15 m deep water table.

Resumen   En areas dedicadas al cultivo de hortalizas cercanas a Santiago de Chile se añaden a los cultivos *ca.* 780 kg N ha^{-1} año^{-1} con el agua de riego proveniente de efluentes municipales no tratados. De esta cantidad entre 161 y 287 kg N ha^{-1} año^{-1} es extraido por cosecha y gran parte del remanente aparentemente alcanza el agua freática que está a profundidades entre 1 y 15 m. El nivel de nitratos en los pozos utilizados para el consumo humano excede los valores máximos permisibles.

Introduction

The use of effluent for irrigating agricultural land is widespread throughout the world. This practice supplies a range of essential nutrients to the soil and consequently to crops also. Together with the positive effects of fertilization, however, there is a potential for numerous negative effects associated with effluent application. Usually an attempt is made to minimize deleterious effects by treating such effluent in special plants, but this is not always the case.

Untreated effluent, for example, is used in a highly productive agricultural sector west of Santiago, Chile. There, the crops are irrigated exclusively with untreated sewage and industrial waste water taken from the main sewer known as the Zanjón de la Aguada (ZA), which collects around two-thirds of the wastes discharged by a population of *ca.* four million inhabitants.

The effluent from the ZA contains microbial pathogens, whose presence gives rise to epidemiological problems when this water is used for irrigating horticultural crops that are eaten raw[3]. The presence of heavy metals in the effluent also implies a potential danger for subsequent contamination of soils, crops, and the subsurface water[8,10].

The study reported here is focused specifically on the effects of nitrogenous substances that the effluent from the ZA introduces into the soil/plant system.

Although part of this nitrogen will serve as fertilizer, there is an excess which, in nitrate form, will migrate toward deeper layers and pollute subsurface and surface water. The water table varies from 1.5–15 m deep, and is the principal source of drinking water for most of the population of the area. In earlier studies[10] the water in area wells was found to have levels of NO_3^-–N that exceed limits set by the U.S. Health Service[6].

Description of the area

Hydrogeologically, the study area is part of the Mapocho River watershed. The Mapocho River thus receives the untreated effluent from the ZA that is not used for irrigation.

The ZA is approximately 40 km long, and receives along its length the domestic sewage of around two-thirds of the municipal and industrial wastewater discharged by households, processing plants, tanneries, electroplating industries, and others. Secondary channels that lead off from this main sewer are used for irrigating the area.

The water table varies between 1 and 15 m along the east-west run. Drainage to groundwater is blocked by an impermeable layer of volcanic material some 80 m thick, which encloses an aquifer that is apparently isolated from pollution. Wells sunk for drinking water are fed from surface water.

For the study reported here, we selected a sector, with soils that are classified as Aquic Chromuderts, representative of the area. The water table on the sampling site varies between 1 and 2.2 m. Principal crops are lettuce (*Lactuca sativa*), spinach beet (*Beta cicla* L.) and celery (*Apium graveolens dulce*). The root zone of these crops is *ca.* 0.3 m. Yields are very high, with three lettuce harvests per year, two celery harvests, and four cuts of spinach beet. No commercial fertilizer is used.

The rainy season is four months long; annual rainfall is not more than 350 mm. The remainder of the year is dry, and irrigation water comes exclusively from the ZA at an average annual rate of 2.2×10^4 m^3 H$_2$O ha^{-1} yr^{-1}.

Methods

In the area under study, soil samples were taken in accordance with Rible *et al.*[7] at depths to 190 cm with a manual auger. Parallel samples were also taken from a nearby soil that was not irrigated with effluent.

Soil samples were air-dried, ground and screened through a 2-mm mesh. Field moisture was determined on the basis of air-dried samples; residual moisture was determined from samples dried at 105°C.

The air-dried samples were analysed using methods described by Jackson[5] and Black[2]. pH was measured in a saturation paste, total N was determined by the modified Kjeldahl method, NO_3–N was extracted with a $CuSO_4$–Ag_2SO_4 solution and determined by the colorimetric method with phenol disulphonic acid, exchangeable NH_4–N was determined by Nessler's method, cation exchange capacity was determined by saturation with Na, and chlorides were determined in the saturation extract by Mohr's method.

Plant samples collected were washed, dried at 65°C, and ground in a Wiley mill. In this material total-N was determined by the modified Kjeldahl method, nitrate content was measured with a nitrate-specific electrode in an Al sulphate extract obtained in accordance with the Beker and Smith technique[1], and chloride content was determined potentiometrically with $AgNO_3$ in the extract with HNO_3[4].

Water samples were collected from 4 sources: 1) effluent used for irrigation, 2) water from wells situated in the area irrigated with this effluent, 3) water from surface wells away from effluent-irrigated areas, and 4) water from control water-channels. The samples were transported in polyethylene containers to the laboratory, where they were analysed in accordance with APHA–AWWA–WPCF methods[11]. pH was measured with a Beckman pH meter, and total N by the modified Kjeldahl method; chemical oxygen demand (COD) was determined by oxidation with $K_2Cr_2O_7$, NO_3-N by the colorimetric method with phenol disulphonic acid, chlorides by potentiometric argentometry, and total and suspended solids gravimetrically.

Results and discussion

Table 1 shows some of the properties of the untreated ZA effluent used for irrigation alongside those of the water from a channel not fed by effluent; values are also presented for analyses of water from wells inside and outside the zone of effluent irrigation.

Marked differences were observed between the effluent from the ZA and water from the control channel, especially with regard to the contents of suspended solids, Cl^-, COD, total-N and NH_4-N, which was ca. 90% of total-N. The differences between the water samples from the wells within and out of the area are less marked, except that the nitrogen content in effluent-area wells was much higher, particularly NO_3-N contents.

Table 2 groups some properties of the soil irrigated with effluent from the ZA vs those of the nearby untreated control soil; marked differences were seen in total-N and organic-C contents, especially in the root zone (0–30 cm depth).

Table 3 presents data for contents of total N, NO_3-N and Cl^- in plants cultivated on areas irrigated with and without effluent. The total-N contents of the plants grown with effluent were high, exceeding those of the control plants in all cases.

We have attempted in Table 4 to establish a nitrogen balance for treated fields. A balance of this type has obvious limitations since it is not possible to evaluate all the parameters involved. However, the reliability of the data relating to those parameters covered by the study should be reasonable since values are based on replications of up to 6 samples for each horizon and the variability is relatively low (< 20–25% in the case of NO_3^--N determinations). The factors taken into account for this balance included the nitrate nitrogen in the soil below the root zone, the input of N originating from the effluent used for irrigation, and the N taken up by the respective crops.

N input was high (around 800 kg N ha^{-1} yr^{-1}) and, as expected, represents a large excess with respect to the amount that the crops are able to take up, despite notably high yields. The total amounts extracted by crops do not exceed 290 kg N ha^{-1} yr^{-1} (Table 4). A further large proportion of the nitrogen introduced by the

Table 1. Some properties of the effluent water used for irrigation near Santiago, Chile, and water from surface wells. Control values are from a stream channel and well not subject to effluent. All values are expressed in mg l^{-1} and are the means of 3 replicate samples

	Surface water		Subsurface water		
	Effluent	Control	Well 1	Well 2	Control
Suspended solids					
Total	465	29	33	15	3
Volatile	225	8	29	10	3
Dissolved solids					
Total	1425	885	1397	970	1203
Volatile	325	102	221	195	239
Cl$^-$	248	111	213	202	194
COD	268	191	68	92	207
Total-N	35.4	3.7	17.0	7.8	2.8
NH$_4$$^+$–N	33.8	1.4	0.8	1.3	2.4
NO$_3$$^-$–N	0.2	0.3	16.5	11.5	0.9
pH	7.3	7.6	7.8	8.0	7.6

irrigation water was nitrified in the soil, and the nitrates formed apparently percolated to the soil below the root zone, from where they drain away from the system. Some of the nitrates may have been denitrified before percolation, but NO$_3$–N in the unsaturated soil below the root zone reached concentrations of around 25 mg l^{-1} in the soil solution. It is useful to note that the yields of the

Table 2. Some properties of soils near Santiago irrigated with untreated effluent. Control soils were not treated with effluent

Depth (cm)	pH	Field H$_2$O (%)	Total N (%)	Organic C (%)	CEC (meq 100 g^{-1})
Treated soil					
0– 5	6.4	10.5	0.28	2.22	23.0
5– 30	7.1	18.5	0.28	2.18	27.6
30– 70	7.3	22.5	0.18	0.58	35.5
70–150	7.3	19.0	0.02	0.14	n.d.
150–190	7.4	20.1	0.02	0.15	31.9
Control soil					
0– 30	7.4	11.6	0.05	0.46	28.5
30– 70	7.5	20.4	0.06	0.56	33.4
70–110	7.6	19.5	0.07	0.57	45.4

Table 3. Total-N, NO_3^-–N and Cl^- in plants cultivated on agricultural soils near Santiago irrigated with untreated effluent. Control soils were not treated with effluent

Crop	Component	Total N (%)	NO_3–N (%)	Cl^- (%)
Treated soils				
Lettuce	Leaves	4.09	0.28	2.33
	Roots	1.05	0.03	nd
Spinach beet	Leaves	4.80	0.39	2.03
Celery	Leaves	3.80	0.15	3.28
	Stalks	1.30	0.22	4.90
	Bulbs	2.10	0.09	1.50
Control soils				
Lettuce	Leaves	2.92	0.18	3.41
	Roots	1.84	0.25	2.20
Spinach beet	Leaves	3.88	0.08	4.07

agricultural vegetable crops were high: *ca.* 8000 kg dry matter ha^{-1} yr^{-1} equivalent to approximately 120×10^3 kg of green aerial material ha^{-1} yr^{-1}.

Because agricultural yields obtained on effluent-irrigated soils are high, the use of an alternative form of irrigation water would entail a high economic cost since it would be very expensive to replace with commercial fertilizers the N that this effluent supplies. The effluent may also supply a large number of other nutrients.

Table 4. Nitrogen balance and dry matter yield of crops near Santiago irrigated with untreated effluent. Mean values assume an equivalent area of cultivation for each species. Unless otherwise noted, values = kg ha^{-1} yr^{-1}

	Lettuce	Spinach beet	Celery	Mean values
Nitrogen				
Irrigation-N	779	779	779	779
Nitrogen in plants	161	223	287	224
Difference	618	556	492	555
Difference (loss) per cent	79	71	63	71
Yield				
Dry matter	3930	4900	15577	8136

References

1   Beker A S and Smith R 1969 Extracting solutions for potentiometric determination of nitrate in plant tissue. Agric. Food Chem. 17, 1284.

2   Black C A (Ed.) 1968 Methods of Soil Analysis. Agronomy No. 9. American Society of Agronomy, Inc. Publisher, Madison Wisconsin, USA.

3   Castillo G and Cordano A M 1975 Enterobacterias en una corriente fluvial. Rev. Lat. Am. Microbiol. 17, 213–219.

4   Chapman H D and Pratt P F 1961 Methods of Analysis for Soils, Plants and Waters. Univ. Calif., Div. Agric. Sci. 309.

5   Jackson M L 1965 Análisis Químico de Suelos. Ed. Omega S.A., Barcelona.

6   Lee D H K 1970 Nitrates, nitrites, and methemoglobinemia. Environ. Res. 3. 484–511.

7   Rible J M, Nash P A, Pratt P F and Lund L J 1976 Sampling the unsaturated zone of irrigated land for reliable estimates of nitrate concentration. Soil Sci. Soc. Am. J. 40, 566–570.

8   Schalscha B E, Vergara F I and Schirado G T 1978 Irrigation with untreated sewage waters and its effects on the content of heavy metals in soils and crops. In State of Knowledge of Land Treatment of Wastewater. Vol. 1, 275–281. CRREL, Hanover, USA.

9   Schalscha B E, Vergara F I, Schirado G T and Morale P M 1979 Nitrate movement in a chilean agricultural area irrigated with untreated domestic sewage waters. J. Environ. Qual. 8, 27–30.

10 Schalscha B E and Vergara F I 1980 Nitrate-nitrogen status and movement in soils irrigated with untreated sewage waters. In Soil Nitrogen as Fertilizer or Pollutant, pp 267–276. International Atomic Energy Agency, Vienna.

11 Standard Methods for the Examination of Water and Waste-Water 1975 14th Ed. APHA-AWWA-WPCF.

Plant and Soil 67, 227–239 (1982).   0032-079X/82/0672-0227$01.95.   SU-20

Research into the Rhizobium/Leguminosae symbiosis in Latin America

Investigación sobre la simbiosis Rhizobium-Leguminosa en América Latina

J. R. JARDIM FREIRE
Department of Soils, Universidade Federal do Rio Grande do Sul, Porto Alegre, Brazil

Key words   Extension Inoculant production Latin America N_2-fixation Rhizobium Training

Abstract   More than 60 institutions and 100 researchers were involved in Rhizobium research in 1978 in Latin America. Half of these researchers were located in Argentina and Brazil. Research activity and the application of research findings vary widely among countries.

Problems that plague research include 1) inadequate training of research personnel and insufficient attention paid to the Rhizobium/Legume symbiosis at agriculture schools; 2) poorly-established research priorities that do not sufficiently weigh the immediate needs for the farmers such as the identification of limiting environmental factors (*e.g.* nutritional deficiencies), techniques for small-scale inoculant production, and quality control of available inoculants; 3) isolation of the researchers and a lack of adequate library support; 4) poorly integrated research teams (*e.g.* in many institutes researchers are either microbiologists with no agricultural background or agronomists lacking microbiological training); and 5) insufficient dissemination of research findings.

Problems with inoculant production and control include 1) a local dependence on national or imported inoculants rather than on locally-selected strains, 2) poor inoculant quality control which results in low inoculation success rates and subsequent discredit to the inoculation practice, and 3) high prices for inoculants.

Extension problems include 1) lacking or deficient legume-promotion programs by government agencies, 2) poor contact between research and extension workers, and 3) administrators, leaders, extension workers and agronomists working in the field that lack adequate knowledge of the Rhizobium/Legume symbiosis.

Immediate measures to foster extension and legume promotion programs and informal and/or official quality control are needed in Argentina, Uruguay, Brazil, Mexico, and probably Colombia. Countries where combined efforts should primarily be directed toward stimulating research and extension include Peru, Venezuela, Costa Rica, and Chile. In Ecuador, Paraguay, Bolivia, Nicaragua, Honduras, Guatemala, the Dominican Republic and Panama, priority should be given to research. Colombia should also be included in this group as national research institutions need to be strengthened. Table 2 lists these priorities more fully.

Resumen   Más de 60 instituciones y 100 investigadores están trabajando de la investigación con Rhizobium en América Latina. La mitad de los investigadores están localizados en Argentina y Brasil. La actividad de investigación y la aplicación de los conocimientos científicos varían ampliamente de acuerdo con los países.

Los problemas de la investigación incluyen: 1) Entrenamiento inadecuado de los investigadores y poca atención para la simbiosis Rhizobium/Leguminosas en las escuelas de Agronomía; 2) investigaciones de baja prioridad sin consideración para las necesidades más inmediatas para los agricultores tales como identificación de los factores limitantes ambientales (por ejemplo: deficiencias nutricionales), técnicas para la producción de inoculantes en pequeña escala y poco control de calidad de los inoculantes disponibles; 3) aislamiento de los investigadores e insuficiente apoyo de literatura; 4) baja interdisciplinaridad en las investigaciones (por ejemplo: en muchas instituciones los investigadores son microbiólogos sin conocimientos de agronomía, o agrónomos sin entrenamiento en microbiología y 5) insuficiente diseminación de los conocimientos científicos.

Los problemas de la producción y control de inoculantes incluyen: 1) las cepas empleadas en los inoculantes (nacionales o importadas) no son seleccionadas localmente; 2) poco control de calidad de los inoculantes y como resultado, inoculantes malos traen descrédito para la práctica de la inoculación, y 3) precios muy altos de los inoculantes.

Los problemas de la extensión incluyen: 1) falta o deficiencia de los programas de promoción de leguminosas por las organizaciones gubernamentales, 2) poco contacto entre los investigadores y los extensionistas y 3) administradores líderes, extensionistas y agrónomos que trabajan en el campo no poseen adecuados conocimientos sobre la simbiosis Rhizobium/Leguminosas.

Algunas medidas inmediatas para promover la extensión y programas de promoción de las leguminosas y/o control oficial de la calidad de los inoculantes son necesarias en Argentina, Uruguay, Brasil, México y posiblemente Colombia. Perú, Venezuela, Costa Rica y Chile necesitan esfuerzos combinados dirigidos prioritariamente para promover la investigación y extensión. En Ecuador, Paraguay, Bolivia, Nicaragua, Honduras, Guatemala, República Dominicana y Panamá, la prioridad debe ser dada para la investigación. Colombia debe ser incluída en este grupo por la razón de que las instituciones nacionales deben ser fortalecidas. La tabla 2 relaciona estas prioridades con mas detalles.

Introduction

Many cereals such as corn (*Zea mays*) can attain much higher levels of high-protein productivity than can legumes. However, high cereal productivity is dependent upon the use of nitrogen fertilizers, while this is not the case for leguminous crops in symbiosis with Rhizobium. This advantage of legumes is of great importance for developing countries with limited energy resources.

Legume cultivation has been little exploited in relation to potential production in these countries, despite the fact that production could meet the rising demand for protein over much of the world at a lower cost than could cereal production. Legumes, however, present special cultivation problems that are not encountered with cereals. Legumes must be matched with the right Rhizobium symbiont, and for high productivity, nutritional and environmental requirements of host and symbiont must be met.

Rhizobia-legume research in Latin America

In 1978, the Porto Alegre Microbiological Resources Centre (MIRCEN) Project began to assemble information concerning the Rhizobium/Legume symbiosis in Latin America. A questionnaire to more than 60 institutions and other sources[1, 2, 3, 6] was used for this compilation, summarized in Table 1. Major research efforts in the region are concentrated in the following countries.

Argentina
Legumes play a large role in the economy of Argentina and in the national diet. Alfalfa (*Medicago sativa*) is the largest legume crop, with over four million ha devoted to its growth, and is the main basis of beef production together with

Table 1. The status of Rhizobium/Legume research and application in Latin America in 1979

Feature	Argentina	Chile	Uruguay	Brasil	Paraguay	Bolivia	Ecuador	Peru	Colombia	Venezuela	Costa Rica	El Salvador	Nicaragua	Dominican Rep.	Mexico	Guatemala	Panama
Institutions	11	4	2	14	nd	4	nd	3	3	5	1	2	0	nd	5	nd	1
Researchers	35	7	7	35	nd	4	nd	5	20	5	1	3	0	8	12	nd	2
Inoculants production																	
public	+***	+	+	+	—	—	—	+	+	—	—	—	nr	—	—	—	—
private	+	+	+	+	—	—	—	—	—	—	—	—	nr	—	+	—	—
importation	+	+	—	—	+	+	+	+	+	nr	nr	nr	nr	—	+	nr	nr
quality control																	
informal	+	+	+	+	—	+	—	+	—	—	—	—	—	—	+	—	nr
official	—	—	+	+	—	—	—	—	—	—	—	—	—	—	—	—	nr
acceptance*	—	—	+	+	—	—	—	—	—	—	—	—	—	—	—	—	—
intensity of use	L	L	H	H	nr	nr	L	L	L	L	nr	nr	nr	nd	L	nr	nr
Legume importance																	
economic	H	L	H	H	H	L	L	L	L	nd	nd	L	L	L	L	L	nd
dietary	H	M	H	H	H	L	L	H	M	nd	nd	nd	nd	nd	nd	M	nd
Important legumes**	1,2,3,4	1,3,5	3,1	2,8,6	2,7,6	1,8	8,4	8,4,1,3	8	7,8	8	8	8	8	8,1	8,4	nd

* inoculation accepted by farmers and widespread

** 1, alfalfa; 2, soybeans; 3, clovers; 4, peas; 5, lens; 6, peanuts; 7, cowpeas; 8, beans.

*** +, present; —, absent; nr, no record; nd, not determined; H, high; M, moderate; L, low.

clover. Soybeans, planted to 1.5 million ha, are the second largest legume crop, followed in importance by peanuts (*Arachis hypogaea*), beans (*Phaseolus vulgaris*), peas (*Pisum sativum*), and lens (*Lens esculenta*) among the grain legumes. In spite of the importance of legumes and the fact that there are nine inoculant factories in Argentina[4] as well as many institutions and researchers that have long worked on Rhizobium, the use of inoculants by farmers is relatively low.

According to available information, Argentina was the first Latin American country to begin Rhizobium research. Around 1940 the 'Unitad de Symbiosis' was established at the Instituto Nacional de Technologia Agropecuaria (INTA) near Buenos Aires, under the leadership of Enrique Schiel, who retired 10 years ago. The main activities of the Unidad de Symbiosis, still the main institution for this work, include the selection of efficient Rhizobium strains, research on the effects of pesticides, and the production of inoculants (on agar) for grain and forage legumes.

Another group is working in northern Argentina at the Universidad del Nordeste in Resistencia; their main activity is the production of peat inoculants. At the Faculdad the Ciencias Exactas, Universidad de La Plata, a dynamic group works on technological aspects of inoculant production. Activity has recently been started at the Universidad de Rio Cuarto and at the Universidad de Cordoba, as well. Another long-established group has been working sporadically on Rhizobium at the Universidad de Buenos Aires, Catedra de Microbiologia. At the same university, a new group working on forage research is informally evaluating the quality of inoculants for forage legumes. This is the only place that evaluates inoculant quality.

Chile

Chile has a low-density population, beef production is low, and seafood is an important source of protein. However, temperate grain legumes such as peas, lens and chickpeas (*Cicer arietinum*) are also important in the Chilean diet and alfalfa and clovers (*Trifolium* spp.) are important pasture crops.

Chile also has a long-standing tradition of Rhizobium research. Research started at the Universidad de Concepcion in southern Chile, by Luis Longeri, who is still actively leading a small group. Other researchers are working on Rhizobium at the Universidad Tecnica del Estado, the Universidad Catolica de Chile, and the Instituto de Investigaciones Tecnologicas in Santiago. Of these, the Universidad Tecnica del Estado program is the most comprehensive.

Small-scale production of inoculants was started at the Universidad de Concepcion in the 1960's and production continues there and at a private company. Quality control is performed on an informal basis by two university laboratories.

Uruguay

Pastures and forage production are very important parts of the Uruguay economy, which is based on beef and wool production. Plan Agropecuario was successfully established around 1962 to improve agricultural production. The program included an FAO-supported laboratory for Rhizobium research and for checking the quality of inoculants. The national leader was Carlos Batthyany who, together with Richard Date from Australia, succeeded in establishing active programs of research, extension and cooperation with inoculant factories. The leader of this group is now Carlos Labandera. Another group, led by Gloria M. de Drets, is researching the biochemistry and physiology of Rhizobium at the Instituto de Investigaciones en Ciencias Biologicas.

Inoculants have been used both extensively and intensively for establishing pastures of clovers, *Lotus* and other temperate legumes, mostly on soils with no native Rhizobium. Inoculants from the two private factories have been controlled since 1963 by analyzing the broth and the final product according to government regulations.

In 1964, Uruguay researchers promoted the first Latin American meeting of rhizobiologists (RELAR). Since then the meeting has been held without interruption in different countries of the region, initially every year but now every two years.

Brazil

General   In economic terms, by far the most important legume in Brazil is soybeans (*Glycine max*), planted mostly in the south of the country to over twelve million ha. Common beans (*Phaseolus vulgaris*) are planted to smaller areas than soybeans but are very important since they are a major part of the every day diet for many people with low incomes. Cowpeas are also an important part of the diet in the northeast. Peanuts are grown in the State of Sao Paulo, mostly for oil production. Temperate forage legumes such as clovers, alfalfa and *Lotus* are important in southern Brazilian pastures, and tropical legumes such as perennial soybeans (*Glycine max*), siratro (*Macroptilium atropurpureum*), centrosema (*Centrosema pubescens*), and others are important in the southeast.

Around 95% of the inoculant produced in Brazil is for soybeans. In 1980, Brazil's six inoculant factories produced over six million 200 g packages of soybean inoculant, which means that around 30% of the total soybean seed was inoculated. More than 50% of soybean seeds were inoculated in areas where the crop was expanding rapidly, and it is reasonable to assume that inoculation helped with the expansion and subsequent crop productivity. If it is assumed that nodules make a modest contribution of 50 kg N ha^{-1} planted to soybeans, at the present price of mineral nitrogen (US$0.70 kg^{-1} N), Rhizobium-soybean nitrogen fixation is equivalent to 420 million dollars of nitrogen per year.

Inoculation is also used for temperate forage legumes in the State of Rio Grande do Sul and for some tropical forage crops, mostly in the states of Sao

Paulo and Parana. Peanuts and beans are rarely inoculated. Both are usually grown by poorly-educated farmers on small farms, though peanuts are said to fail to respond to inoculation and beans to respond erratically in field experiments. The price of Rhizobium inoculants in Brazil is very low compared to prices in other countries; it has been the government's policy to promote the large-scale use of inoculants since the first inoculant factory started production.

Effective work on Rhizobium started in Brazil at the Biological Institute of the State of Sao Paulo in 1947 by Julio Amaral. Interest at other institutions soon arose and the groups grew stronger with the growing importance of soybeans and the greater interest in pasture legumes. One such group is at Rio de Janeiro, EMBRAPA (km 47), under the leadership of Johanna Dobereiner, and the emphasis of its research program is on associative N_2-fixation in grasses. Another group is based at Piracicaba, in the State of Sao Paulo, under the leadership of Alaides Ruschel, and this group works mostly on common beans and soybeans. Yet another strong group is based at Campinas under Eli Lopes, and works mostly with peanuts and tropical forage legumes. A group also works at the Cerrado Research Center of EMBRAPA in Brasilia, and a new group has formed at the University of Vicosa, Minas gerais State. Other researchers are scattered in different institutions; 35 researchers in Brazil are presently engaged in Rhizobium work.

The IPAGRO/UFRGS group   In 1949, pressed by the demand for inoculants by the growing number of soybean farmers, I started a program of strain collection, strain selection, and the small-scale production of inoculants first on agar and then in peat powder at the Agronomic Research Institute of the State of Rio Grande do Sul (IPAGRO). The group grew and merged with the staff of Soil Microbiology of the Department of Soil Science of the University of Rio Grande do Sul, and now nine scientists direct the group's research, culture collection, inoculant production, inoculant quality control, and training and extension programs.

The major objective of the research program is to select effective and competitive strains of Rhizobium for legumes of economic or experimental importance. Eighty-four of the 650 strains in the Rhizobium collection are from threatened collections (O. N. Allen's of the University of Wisconsin and L. B. Erdman's of the U.S. Dept. of Agriculture). More than 100 strains have been recommended for inoculant manufacture, and since 1956 the majority of the inoculants produced by private industry in Brazil have used strains supplied by the IPAGRO Laboratory. With the establishment of MIRCEN in 1978, the culture collection has undergone reorganization in order to better characterize and preserve inoculants and to better dissiminate the information gathered.

The research program also includes the search for a strain of *R. phaseoli* that is tolerant to stress conditions, the identification and evaluation of soil factors that limit successful inoculation, the evaluation of the need for inoculation in some

Table 2. Promotional needs for Rhizobium/Legume use in Latin America. Visiting consultants are important for all categories

Need	Argentina	Chile	Uruguay	Brazil	Paraguay	Bolivia	Ecuador	Peru	Colombia	Venezuela	Costa Rica	El Salvador	Nicaragua	Dominican Rep.	Mexico	Guatemala	Panama
National institutes*	x				x	x	x	x	x***		x		x	x		x	x
Training programs	x				x	x	x	x	x		x		x	x		x	x
Inoculants production						x**		x	x	x		x			x		
Quality control informal	x	x		x		x			x	x					x		
Quality control official	x	x													x		
National workshops and seminars	x	x	x	x				x	x	x					x		x
Culture collection improvement	x	x	x	x		x		x	x	x					x		

* strengthening of research capability.

** initiated production of laboratory inoculant in late 1978.

*** CIAT responsible only for *P. vulgaris* and tropical forage legumes; national institute necessary for other legumes.

tropical forage legumes, and examination of inoculant-application technology (see Mircen[5]). For many years, Rhizobium-soybean research has been emphasized. Emphasis is now changing to include common beans and alfalfa and other forage legumes.

The IPAGRO Laboratory also supplies inoculants to researchers and experiment/demonstration stations directly in order to encourage research and extension work. This service has contributed significantly to the dissemination of inoculant information. The IPAGRO team has helped the private inoculant industry in Brazil since the industry's start in 1956 by supplying technological advice, informal quality control, and personnel training in addition to selected strains of Rhizobium. Microbiologists at four of the most important companies were trained at IPAGRO laboratories.

Since 1977, on behalf of the federal Ministry of Agriculture, the IPAGRO Laboratory has been in charge of official quality control for inoculants produced in Brazil or imported. This is in accordance with a 1975 law that regulates the production and trade of legume inoculants. The law established a minimum standard and requires that all inoculants contain only Rhizobium strains recommended by the official laboratories. This assures that the strains that are produced by the inoculant factories and subsequently used by the farmers are strains selected at research institutes. Before 1975 inoculant quality was maintained only by informal quality control carried out by laboratories cooperating with factories and examining samples collected at markets and from farmers.

Inter-regional cooperation in the Rhizobium field became a working objective of the group with the establishment of MIRCEN. The training program has been formalized, the culture collection improved, and the inoculant/technology program expanded in order to cover all other countries in Latin America.

Developing a regional research capability has been one of MIRCEN's high priorities. In 1979 and 1980, 45 microbiologists from 13 countries were trained in applied Rhizobium technology *via* 2 short formal courses plus internal practical training. Others have been trained before 1979 and during the regular course for the M.Sc. degree at the Department of Soil Science. These researchers will undoubtedly benefit their institutions of origin, especially those in countries where previously little or no Rhizobium research was performed (*e.g.* Bolivia, Ecuador, El Salvador, and the Dominican Republic). An additional objective of the training program is to train administrators and extension workers. There are few agronomists in the extension services and experiment stations that have adequate knowledge of the benefits of the Rhizobium/legume symbiosis.

Disseminating information to Rhizobium workers in the region is another objective of MIRCEN; this helps to break scientific isolation and stimulate cooperative research. General newsletters, special research bulletins containing suggestions for research, and a culture catalogue have been issued and distributed to workers in the MIRCEN network.

Funding for the activities of the group comes from national sources for the research, inoculant production, and inoculant quality-control programs. The culture collection, training activities, and publication programs are supported by UNEP.

Paraguay

Soybeans are the most important grain legume in Paraguay (400,000 ha planted to soybeans in 1979), and the grain produced is almost totally exported. Cowpeas and peanuts are the main sources of protein for the poor people in rural areas. Forage legumes are rarely grown. There is no Rhizobium research activity nor are there inoculation trials, and so far there has been no success in attracting personnel for training at MIRCEN. Inoculants for soybeans are imported from Brazil.

Bolivia

This country has a low population density and a relatively high availability of animal protein. Consumption of legumes as a source of proteins is thus low. However, soybean cultivation is expanding at a high rate (15,000 ha planted to soybeans in 1976 *vs* over 100,000 ha in 1980). Alfalfa is traditionally used for hay and in pastures but no data on the actual extent of its use are available.

Rhizobium research in Bolivia has started only recently. The Universidad de Santa Cruz de la Sierra in connection with the FAO/UNDP/BOL. Abapo Izozog Project for the Development of the Bolivian Chaco has a soil specialist trained by MIRCEN who expects to undertake the small-scale production of inoculants. At both the Universidad de Chochabamba and the Research Station there is interest in starting work on Rhizobium.

Ecuador

Grain legumes play an important role in the supply of protein to people in Ecuador, although the roles of maize and wheat are more important. The potential for cropping forage legumes has yet to be exploited. There is enthusiasm for soybean research and a good research program exists, though on an Rhizobium/legume research there is little activity apart from an occasional inoculation trial. Inoculant is imported. The government has proposed a factory for inoculant as part of a larger special program for the development of grain legumes for human consumption, but trained personnel are lacking.

Agricultural research institutions are well equipped and have qualified personnel. Two researchers from the INIAP S. Catalina Experiment Station have been trained at MIRCEN, and research is being initiated there.

Peru

Of all Latin American countries, Peruvians have the highest per capita consumption of legumes in the human diet. Grains of many temperate and tropical legumes are consumed. Pasture legumes are important both in the mountains and in the tropical lowland regions of the Amazon. Soybeans have been recently introduced and their cropped area is expanding rapidly. Seed inoculation, however, is not a common practice.

Rhizobium research is presently under way at two centers. One is at the Universidad Nacional Mayor de San Marcos in Lima, where a staff of one M.Sc. trained Agronomist and two technicians are performing strain selection experiments and starting a project to produce a peat inoculant. The other center is at the Universidad de San Cristobal de Huamango, Ayacucho, where a small team of professionals is producing a peat inoculant. Laboratory facilities in Lima are adequate, but no greenhouse is available. In Ayacucho, facilities are poor to medium. There is no private inoculant production and no official or informal quality control of the imported inoculants. There is little if any contact among researchers and extension workers. There are also restrictions on the work undertaken due to scarcities of funds and of trained personnel, although three researchers from the University of San Marcos and one from Ayacucho have attended MIRCEN training courses. Dissemination of available information among extension workers and farmers is poor.

Due to the importance and potential of legumes in Peru, efforts are being made to develop Rhizobium work. In 1980 the FAO/Biological Nitrogen Fixation (BNF) Project sent Carlos Batthyany on an advisory visit to Lima, and in 1981 the USAID-BNF Program offered a training course for extension workers and administrators.

Colombia

Cowpeas and common beans are probably the most important grain legumes in Colombia, but the cropped areas are not large. Soybean is being promoted and the area planted to this crop is expanding.

The availability of Rhizobium strains and inoculants for experimental use in the country is good due to research activities at CIAT. Nevertheless, this international institution works mainly on tropical forage legumes and common beans. Work by nationals is being conducted by a small group at the National University of Colombia under the leadership of Nery Gonzales, and includes two researchers trained at MIRCEN. At Palmira Institute, inoculation trials are run by Varella, also trained at MIRCEN. Some inoculant is imported but there is no quality control.

Venezuela

Protein production within Venezuela is insufficient for the needs of the population. Beef, soybeans, chickpeas, dry beans and lentils are imported. Nitrogen fertilizer is subsidized and applied at high rates on pastures, which also receive large doses of leaf herbicides which kill the abundant native tropical forage legumes.

Some research on Rhizobium is under way at Faculdad de Agronomica, UVC, in Maracay, at CENIAP in Maracay, at Facultad de Agronomia in Maracaibo, and at IVIC in Caracas. There is hope the the Tenth Latin American Meeting on Rhizobium (RELAR) held in September 1980 at Maracay will result in increased attention for Rhizobium/Legume work in Venezuela.

Mexico

The human diet in Mexico is based on maize. However, dry beans are also important and large areas are planted to alfalfa for milk production. Soybean production is expanding rapidly.

Before the Ninth RELAR held in Mexico in 1978, there were only two centers working with Rhizobium. One is at the Departmento de Microbiologia, IPN, where a group has long been working under the leadership of Maria Valdes. The second group works at Fertilizantes Mexicanos S.A., a government enterprise. The Ninth RELAR awoke great interest in biological nitrogen fixation and now eight other centers are working with Rhizobium. A private company produces inoculants but the quality control is thus far informal and many problems related to the quality of the inoculants have occurred.

Other countries

More interest in Rhizobium has developed in Central America during recent years; work is currently underway in Costa Rica, El Salvador, Guatemala and Panama. Nine researchers from Central America and Mexico have been trained at MIRCEN. Common beans are important everywhere in the region and are the staple food in many areas. In the Dominican Republic, interest in Rhizobium is recent but is developing quickly; two researchers have been trained at MIRCEN.

Legume inoculation is apparently low in all Central American and Caribbean countries except Mexico. This region can greatly benefit from research at the University of Puerto Rico where a staff of three under E. Schroeder have developed a strong research program in connection with USAID-BNF and INTSOY.

Regional meetings

The Latin American Meeting on Rhizobium (RELAR) has had and continues to have an important influence on Rhizobium research in the region. Ten meetings have been held since 1964: in Uruguay (1964, 1971), Argentina (1965, 1974), Chile (1967), Brazil (1968, 1970), Colombia (1976), Mexico (1978) and Venezuela (1980). There has been an upsurge of interest in biological N-fixation research and its application in the host countries. Attendance has been growing since the first meeting, which attracted 15 participants; at the Mexico RELAR there were 150 participants. Though many of these participants were students, this nonetheless indicates growing interest in this area at the universities. An unwritten policy of the RELAR organization committee is to select host countries that could benefit from the resulting attention paid to Rhizobium work.

The Latin American Association of Rhizobiologists (ALAR) was created at the 1964 RELAR in Porto Alegre, and since has been coordinated by Carlos Batthyany. The Association had no formal structure until the Tenth RELAR.

International cooperation

A number of international programs promote cooperation among researchers and institutions in the region:

1. The NIFTAL/Univ. of Hawaii/USAID Project offers training opportunities via short courses and internal practical training, selected Rhizobium strains, bibliographic services, and laboratory manuals. Additionally, a network of cooperative experiments has been established for research on tropical legumes.

2. CIAT in Colombia offers opportunities for internal practical training, selected Rhizobium strains, bibliographic services, inoculants for experimental use, and the 'Noticias' newsletter with a bibliography of great value to the researchers. An international network of experiments with common beans has been established as well.

3. The U.S.-AID/Biological N-Fixation (BNF) Consortium includes the Universities of Cornell, North Carolina, Hawaii and Puerto Rico in a BNF program for the tropics that offers training courses and operates mainly through the NifTAL and Puerto Rico Projects.

4. The FAO/BNF Project, based in Rome, is directed at forage legumes and has extension-oriented objectives. Demonstration experiments have been established in Mercedes, Argentina, and in Pucalpa, Peru.

5. The UNEP/UNESCO/ICRO-MIRCEN Project based at Porto Alegre, Brazil, offers training opportunities via short courses and internal practical training, selected Rhizobium strains, inoculants for experimental use, inoculant quality control, bibliographic services, and grants for research and libraries to some key laboratories of the region. A regional network for evaluation of soybean Rhizobium strains is being established, and a network for alfalfa strains is being planned.

Coordination, exchange of information, and cooperation among the different institutions and projects that are involved in Rhizobium work will greatly facilitate the general objective of promoting Rhizobium nitrogen fixation in the region.

References

1 Batthyany C and Freire J R J 1976 Report Prepared to the Food and Agricultural Organization of the United Nations for the FAO/UNEP Programme in Biological Nitrogen Fixation involving Legume Inoculation by Rhizobium in Latin America. Rome: FAO, 63 p. (Not published.)
2 Habit M A 1977 The Need to Increase Food-Legume Production. Rome: FAO.
3 Halliday J 1979 Current work on N-Fixation by Tropical Legumes in Latin America Niftal Network Planning Workshop. Hawaii: University of Hawaii/NifTAL.
4 Microbiological Resources Center (MIRCEN) 1980 Rhizobium MIRCEN Informativo no 2. Porto Alegre, Brazil.
5 Microbiological Resources Center (MIRCEN) 1980 Progress Report. Porto Alegre, Brazil. (Not published.)
6 Reynaert F E 1977 Development of a Coordinated International Programme on Biological Nitrogen Fixation by Microorganisms. FAO. (Not published.)

Plant and Soil 67, 241–246 (1982). 0032-079X/82/0672-0241$00.90. SU-21
© 1982 *Martinus Nijhoff/Dr W. Junk Publishers, The Hague.*

Nitrogen cycling in coffee plantations

Ciclo de nitrógeno en plantaciones de café

E. BORNEMISZA
Department of Soil Science, Universidad de Costa Rica, Ciudad Universitaria, Costa Rica

Key words Coffee Legumes Mineralization N-cycling N-fertilization N-losses Shade trees Weeds

Abstract Nitrogen inputs to the coffee ecosystem are dominated by additions of fertilizer-N (100–300 kg N ha^{-1} yr^{-1}). Small nitrogen inputs from rains and variable from inputs fixation by the leguminous shade trees can amount to 1–40 kg N ha^{-1} yr^{-1}. Organic matter mineralization can be an important nitrogen source also.

Nitrogen losses from the system include removal of N in the harvest (15–90 kg N ha^{-1} yr^{-1}), the removal of coffee and shade tree prunings for firewood, losses from erosion, leaching losses and gaseous losses. Unfortunately, very little information exists for leaching and gaseous losses and for the factors that regulate these processes.

The overall nitrogen cycle in shaded coffee plantings includes three interrelated subsystems. These are the coffee, shade and weeds subcycles.

Resumen La forma principal en que el nitrógeno es añadido al ecosistema del cafetal es por medio de abonos (100–300 kg N ha^{-1} año^{-1}). Pequeñas cantidades del elemento ingresan de la precipitación atmosférica y por fijación de los árboles de sombra que son leguminosos (1–40 kg N ha^{-1} año^{-1}). La mineralización de la materia orgánica puede ser una fuente importante de nitrógeno.

Las pérdidas de este elemento del sistema incluyen lo que se extrae de la cosecha (15–90 kg N ha^{-1} año^{-1}), lo que se pierde por la leña sacada que resulte de la poda del cafeto y de la sombra, pérdidas por erosión, lixiviación y en forma de compuestos gaseosos.

El ciclo global de un cafetal con sombra incluye tres subsistemas correlacionados, los del cafeto, de la sombra y de las malas hierbas.

Introduction

Coffee represents a considerable share of the agriculture of five Latin American countries and is important to the economies of at least six others. The crop is especially important because it can be successfully grown in small single-family units as well as in large plantations.

Among the soil nutrients taken up by the coffee plant, nitrogen is the most important based on the amounts of nutrients required by the crop. Up to 90 kg N ha^{-1} yr^{-1} can be exported in a good crop of 1800 kg of clean beans; even moderate yields of 300 kg coffee yr^{-1} can export about 15 kg of nitrogen[4,5]. Of this amount, slightly more than two-thirds is in the coffee used for commerce, with the remainder in the pulp which usually remains at the site of processing.

When pulp is returned to the field, it contributes to the soil's organic matter, but more often it is discarded due to the considerable labor and other costs involved in its handling and spreading. In addition to the N removed by the harvest, the plants need at least 30 kg N ha^{-1} yr^{-1} for vegetative growth[7].

Nitrogen inputs and uptake

If the recommendations of the different coffee advisory agencies are followed, the largest amount of nitrogen entering the system is fertilizer nitrogen in the form of nitrate, ammonium or urea. Field experiments indicate that coffee uses these different forms of fertilizer with similar efficiencies[3]. Due to appreciable harvest and leaching losses, field experiments usually indicate a need for heavy applications of nitrogen. Malavolta et al.[8], for example, observed that more than 80% of the unfertilized coffee plantations in the state of Sao Paulo showed symptoms of nitrogen deficiency. In Costa Rica, about 80–90% of the coffee-growing area is fertilized with nitrogen. Fertilizer application rates depend on the usual agricultural factors, principally soil properties, climate, planting density, water availability and intensity of management. Application rates commonly fluctuate between 100 and 300 kg N per hectare per year[4,5,8,9,17].

Organic matter mineralization is another important nitrogen source in soils which do not have allophane as their main clay mineral. Allophane reduces the mineralization of organic soil matter in general, and of soil nitrogen in particular[1]. In shaded coffee, organic matter as litter comes from the shade trees as well as from the coffee plants. The shade trees can contribute from 4.6 to 13×10^3 kg dry matter ha^{-1} yr^{-1} under Colombian conditions[15]. In traditional intensively shaded coffee-growing systems, the amount of organic matter entering the system can be considerably higher. Under these conditions mineralization may be the main source of nitrogen used by the coffee. Parra[9] found no response to added nitrogen so long as low plant-densities and abundant shade were present.

Nitrogen in rain in coffee-growing areas is usually quite low, except for sites near industrial centers where air pollution can contribute appreciable amounts of nitrogen to precipitation[13].

Shade is a particularly important factor that affects nitrogen uptake in large areas where highland coffee is grown. It is used throughout the area except where water is limiting, e.g. in Brazil. Shade generally results in reduced nitrogen uptake by the coffee crop. However, Mexican studies[11] have recently quantified the longstanding belief that leguminous shade trees contribute to the nitrogen economy of coffee plantations by fixing nitrogen. They showed that *Inga jinicuil* can fix about 40 kg N ha^{-1} yr^{-1}, while other species examined can contribute around 1 kg N ha^{-1} yr^{-1}. These results are very important since recently it has been possible to reduce the yield gap between shaded and non-shaded trees through careful shade management, and so it is possible to obtain in the presence of shade the high yields associated with non-shaded coffee.

A hectare of coffee plantation may thus receive about 100 kg N ha^{-1} yr^{-1} from non-artificial sources, mainly through symbiotic fixation by leguminous shade trees and through the mineralization of organic matter. This amount covers the nitrogen requirements of low and medium-density plantings, but cannot supply the high nitrogen needs of densely planted (> 5000 plants ha^{-1}), high-yield plantations, which require at least twice this amount.

Nitrogen uptake can be limited by its reduced mobility in soils too dry for adequate movement[8]. However, research in East Africa[10] has shown that during dry spells an upward movement of nitrate can occur, thus improving nitrogen availability since coffee absorbs the bulk of its nutrients from the top soil layers[12].

Nitrogen losses

Losses of nitrogen from the coffee ecosystem have received considerable attention[4,5,7,17]. Nitrogen removed in the harvest of commercial coffee typically varies between 25 and 120 kg N ha^{-1} crop^{-1}. With increasing frequency, the fleshy part of the berry is composted after processing and returned to the soil. This may represent about one-third of the nitrogen in the total bean, or the equivalent of 8 to 40 kg N ha^{-1} crop^{-1}.

In addition to the beans, a significant amount of wood is removed from shaded coffee plantations when the shade trees are pruned. About 1000 to 5000 kg of dry wood ha^{-1} yr^{-1} may be removed, but this is only an estimate since quantitative information is unavailable. As the nitrogen content of wood is about 0.5%, this could result in an additional N loss to the plantation of 1 to 5 kg N ha^{-1} yr^{-1}. As a large fraction of cultivated coffee grows in areas where intensive rain showers are common, leaching and run-off are also important ways nitrogen is lost. This is particularly true now that mechanization is being introduced; mechanization requires more widely-spaced rows, which means that more soil is exposed directly to the impact of rain drops, increasing erosion.

Nitrate-leaching constitutes one of the main channels of nitrogen loss from coffee plantations[6]. Nitrate movement is closely related to soil-water movement, and much coffee is cultivated in areas where heavy showers are common. Other variables also favor nitrate loss from coffee systems: i) most of the fertilizer nitrogen that is added to coffee is very soluble and so is easily leached; ii) much coffee is grown in well-drained soils with high infiltration capacities[9]; iii) rates of nitrogen uptake vary seasonally[7]; and iv) periods of intensive rain are common when nitrogen uptake is low[7]. Unfortunately, little data concerning nitrate losses from coffee plantations have been published.

Soil erosion can also be a major factor regulating nitrogen loss, especially if the soil is not covered with vegetation or a natural mulch. In coffee plantations that are cultivated under shade or have high planting density, a natural mulch formed by litterfall is common. Erosion losses are especially important for highland-grown coffee, often planted on slopes as steep as 50°. New coffee plantations are also vulnerable to erosion loss. Research from Colombia documents yearly losses of nitrogen from unprotected areas that exceed the amount extracted by a good crop of coffee[15]. These losses are much higher than losses reported for temperate zones. On a well-developed coffee plantation that is adequately shaded or that has a high planting density, erosion can be reduced to less than 2% of the losses that occur on unprotected plots[15]. Part of this loss is in the form of soil organic matter, easily carried away by water run-off.

Considerable amounts of nitrogen may be lost by gaseous pathways[2]. In particular, urea is widely applied as a nitrogen source for coffee on the soil surface and is severely affected by volatilization losses in both acid and alkaline soils[16]. If the fertilizer is covered with soil these losses can be minimized, but this is time-consuming and is often not done. No field data that quantify these losses are available. Information for other ecosystems[16] indicates wide inter-system variation that precludes simple extrapolation of these data to possible coffee systems.

Nitrogen subcycles in coffee plantations

Nitrogen cycling in coffee plantations can be broken into three subcycles: a coffee cycle, a shade cycle if shade is present, and a weed cycle (Fig. 1). Nitrogen makes its principal entrance to the overall system in the form of fertilizer, and its largest loss is via the crop harvest. Litter produced by the coffee plants deposits nitrogen in the organic matter of the soil. Up to one-third of the nitrogen in the coffee berries may be returned in the form of processing waste. These residues can be composted and may also be used for biogas production while composting without serious nitrogen loss, and thus supply a source of energy for the coffee processing installations.

The shade trees represent the second important subcycle. If shade trees are present, they may introduce up to 40 kg N ha^{-1} yr^{-1} by symbiotic N-fixation[11]. Colombian studies have shown that shaded plantations can produce 4 to 13 × 10^3 kg litter (dry matter) per hectare per year[15]. Unfortunately, no data is available for how much of this dry matter is contributed by the shade trees. Since little shade tree material is removed from the area most of the nutrients used by these trees move in a closed cycle within the plantation. The only important outlet is wood removed during pruning or elimination of the trees. Deep-rooted varieties of shade trees may remove nutrients from the deeper soil layers that are little used by coffee.

Weeds represent the third subcycle. During their growth period they can take up an appreciable amount of nitrogen. As with shade trees, however, this nitrogen is not removed from the system but only immobilized. But this results in inconvenient competition between coffee and weeds, both of which depend on nitrogen in the top layer of the soil, particularly at the beginning of the rainy season when the coffee is increasing its vegetative biomass. Again, no specific

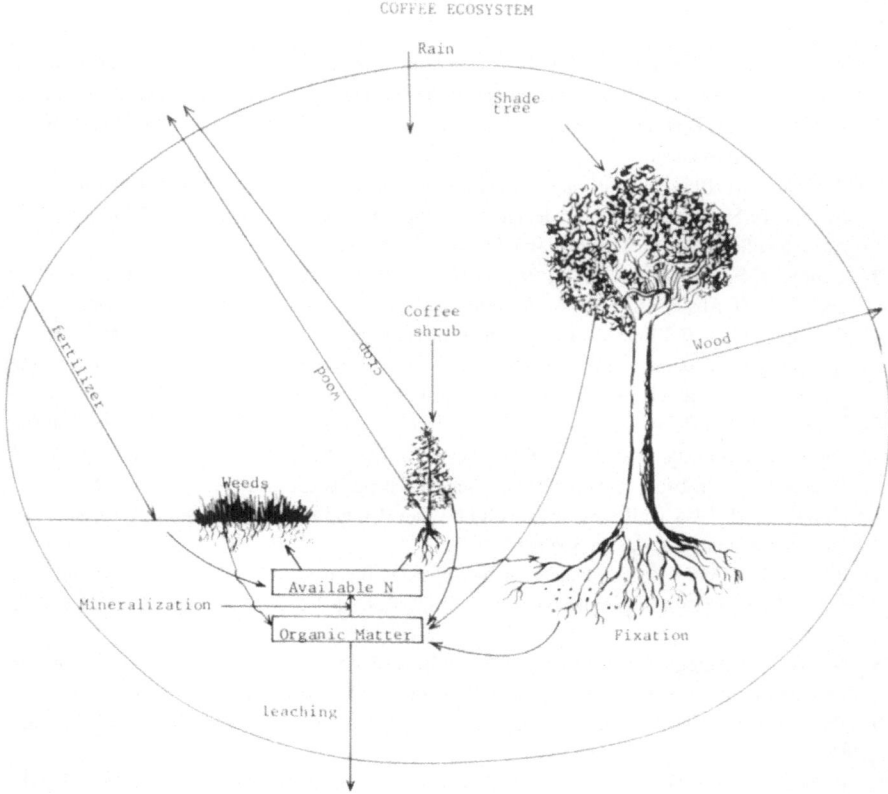

Fig. 1.   A generalized nitrogen cycle for a shaded coffee plantation.

data for this part of the system is available in the literature. For this reason the extent of this competition can only be estimated by noting the yield improvement in many coffee fields when weeds are eliminated.

As can be seen from the foregoing, the coffee plantation ecosystem is complex and much information is still needed for its understanding. The methodology for gathering this information exists. It is only necessary to consider agricultural operations more in an ecological context. The continued increase in the price of agricultural chemicals may lead practicing agronomists and agricultural research workers to take such an ecosystem approach.

Acknowledgements   This study has been partially supported by the Consejo Nacional de Investigaciones Científicas y Tecnológicas de Costa Rica.

References

1 Bornemisza E and Pineda R 1969 Amorphous minerals and nitrogen mineralization in volcanic ash derived soils. *In* Proceedings of the Workshop on Volcanic Ash-Derived Soils in Latin America. Turrialba, Costa Rica: Institute Interamericano de Ciencias Agricolas (IICA)- FAO. 7 p. (In Spanish).

2 Bremner J M 1968 The nitrogenous constituents of soil organic matter and their role in soil fertility. *In* Semain d'Etude Sur le Theme Matière Organique et Fertilité du Sol. Pontificia Academia Scientiarum. Scripta Varia. (Rome) 32, 143–193.

3 Campos C F, Gutiérrez G and Pérez J 1976 Responses of coffee to different nitrogen sources and levels. *In* Abstracts, 2nd Natl. Agronomy Congress, Costa Rica, p. 7. (In Spanish).

4 Carvajal J F 1959 Mineral nutrition and crop requirements of the coffee crop. Ministerio de Agricultura y Ganadería (MAG) – Servico Técnico Interamericanode Cooperacin Agricola (Stica), Techn. Bull. No. 9. San José, Costa Rica 16 p. (In Spanish).

5 Cooil B J and Fukunaga E T 1959 Mineral nutrition of coffee. II. Intensive fertilizer application and its effects. *In* Progress in Coffee Production Techniques, pp 91–95. Interamerican Institute of Agricultural Sciences, Turrialba, Costa Rica. (In Spanish).

6 Cowgill W H 1955 Coffee planting without shade. Coop. Agric. Service Agric. Bull. No. 3. La Aurora-Guatemala, 18 p. (In Spanish).

7 Kupper A 1976 Monthly nitrogen use by coffee and the form and timing for fertilizer nitrogen use. *In* Summary, 4th Brasilian Congress on Coffee Res., pp 215–217. Caxambu, M.G., Brasil. (In Portuguese).

8 Malavolta E, Haag H P, De Mello F A F and Brasil Sobr M O C 1974 Mineral nutrition and fertilization of cultivated plants. Pioneira Ed., Sao Paolo, Brasil. 840 p. (In Portuguese).

9 Parra J 1960 The physical, chemical and fertility properties of Quindio Soils. Cenicafé, Colombia 1960, 323–355. (In Spanish).

10 Robinson J B D and Gaocoka P 1962 Evidence of the upward movement of nitrate during the dry season in Kikuyo red loam coffee soil. J. Soil Sci. 13, 133–139.

11 Roskoski J 1980 The importance of fixation in the nitrogen economy of coffee plantations. *In* First Symposium on Ecological Studies in the Coffee Agroecosystem. Natl. Inst. of Res. Biological Resources, Xalapa, México. 4 p. (In Spanish).

12 Saiz del Rio J F, Fernández C E and Bellavita O 1961 Distribution of absorbing capacity of coffee roots determined by radioactive tracers. Am. Soc. Hortic. Sci. Proc. 77, 240–244.

13 Sanhueza E 1982 The role of the atmosphere in nitrogen cycling. Plant and Soil 67, 61–71.

14 Suárez de Castro F and Rodríguez A 1955 Organic matter equilibria in coffee plantations. Natl. Coffee Growers Assoc. Tech. Bull. Colombia 2(5), 5–28. (In Spanish).

15 Suárez de Castro F and Rodríguez A 1955 Nutrient losses by erosion as affected by different plant covers and soil conservation practices. Natl. Coffee Growers Assoc. Tech. Bull. Colombia 2(14), 1–24. (In Spanish).

16 Terman G L 1980 Volatilization losses of nitrogen as ammonia from surface-applied fertilizers, organic amendments and crop residues. Adv. Agron. 31, 189–223.

17 Vicente-Chandler J, Abruna F and Silva S 1959 A guide to intensive coffee culture. USDA Production Res. Rept. No. 31. Washington, D.C.: U.S. Government Printing Office, 52 p.

Plant and Soil 67, 247–258 (1982). 0032-079X/82/0672-0247$01.80.   SU-22

Nitrogen cycle of tropical perennial crops under shade trees

I. Coffee

Ciclo de nitrógeno en cultivos tropicales permanentes bajo árboles de sombra 1. Café

J. ARANGUREN*, G. ESCALANTE and R. HERRERA
Centro de Ecología, Instituto Venezolano de Investigaciones Científicas (IVIC), Apartado 1827, Caracas 1010A, Venezuela

Key words Coffee Decomposition Litterfall N-cycling N_2-fixation Shade trees Venezuela

Abstract The distribution and fluxes of nitrogen in some parts of a coffee plantation under shade were studied at a typical mountain (1380 m a s l) location in Venezuela. The amounts of nitrogen in the soil to 60 cm depth are by far the largest nitrogen store, reaching a total of 49 000 kg ha^{-1}. The nitrogen flow associated with litterfall was dominated by the shade-tree fraction accounting for a transfer of 86 kg ha^{-1} yr^{-1} of the total 189 kg ha^{-1} yr^{-1}. The rapid decomposition of this litter, although showing a phase of nitrogen accumulation, is an important source of nitrogen to the roots of coffee which occupy preferentially the upper 30 cm of soil and even the litter layer itself. Some evidence of synchrony was found between the peaks of nitrogen transfer to the soil by litter and the periods of high nitrogen demand by the crop plants. It is proposed that the system can amply compensate the nitrogen outputs by harvest (17 kg ha^{-1} yr^{-1}) with a subsidy from the shade trees. *Erythrina* sp. and *Inga* sp. are potential nitrogen fixers although we found no active sites during the dry period sampled. The average litter decomposition constant, k, expressed in terms of nitrogen, was estimated as 4.5, equivalent to a half-life of approximately two months.

Resumen La distribución y flujos de nitrógeno en algunos componentes de una plantación de café bajo sombra fueron estudiados en un cafetal de montaña (1380 m altitud) en Venezuela. Las reservas mayores de nitrógeno estaban en el suelo que hasta los 60 cm tenía 49 × 10^3 kg N ha^{-1}. El flujo de nitrógeno asociado con la caída de hojarasca estaba dominado por la fracción de hojas de los árboles de sombra que contribuyeron con 86 kg N ha^{-1} año^{-1} del total de 189 kg N ha^{-1} año^{-1}. La rápida descomposición de esa hojarasca, aun cuando mostró una fase de acumulación de nitrógeno, es fuente importante de nitrógeno para las raices del cafeto que ocupan preferentemente los primeros 30 cm del suelo y aun en la hojarasca misma. Se encontró evidencia de sincronización entre los picos de transferencia de nitrógeno por la hojarasca y los periodos de mayor demanda por el cultivo. Se encontró que el sistema puede compensar ampliamente le salida de nitrógeno por cosecha con el subsidio proveniente de los árboles de sombra. *Erythrina* sp e *Inga* sp son fijadores potenciales de nitrógeno aun cuando en determinaciones hechas durante el período de sequía no se detectó actividad de fijación biológica. La constante de descomposición (k) promedio para la hojarasca fue de 4,5, equivalente a una vida media de unos dos meses.

* Present address: Departamento de Biología y Química, Instituto Universitario Pedagógico, Caracas, Venezuela.

Introduction

Coffee and cacao are very important crops in tropical America, where they are traditionally grown under shade trees. Both crops originated in the tropics as understory plants in forested areas. Coffee in extensive plantations is grown commonly without the protection of shade trees and more recently there is the tendency to decrease shading in cacao plantations. The nutrient balance of these crops grown in the open probably resembles more the throughflow pattern of annual crops, in which large amounts of fertilizer nitrogen are applied, a varying proportion of nitrogen is extracted by harvest, and the remaining nitrogen is either lost or accumulates in the soil.

Although in some countries large amounts of nitrogen fertilizers are applied to coffee plantations under shade[4], in Venezuela most coffee plantations are managed with minimal fertilization and with shade trees. These agroecosystems resemble, from the point of view of nitrogen cycling, forests in which litterfall and decomposition play an important role in the maintenance of fertility[10,12]. The role of shade trees has long been considered important in this regard for coffee and cacao[6,8]. Suarez de Castro and Rodriguez[18] provide one of the few N-cycle studies of coffee in Latin America. In tropical perennial crops which can be grown under shade trees, the actual amounts of nitrogen and other nutrients exported from the system with the harvest is small when compared to amounts removed by high-yield annual crops. This relatively small output could be balanced by nitrogen inputs from N_2 fixation by shade trees or from nitrogen in shade tree litterfall. The litterfall-N is that which is taken up by deep-rooted shade trees and that would otherwise be inaccessible to the shallow-rooted crop plants.

In this paper we attempt to assess the roles of both shade-tree and crop litterfall in the nitrogen cycle of coffee grown under shade trees. The spatial as well as temporal distribution and transfers of organic matter and nitrogen in mountain-grown coffee were studied for one year in a site where no chemical fertilizers had been used for at least the past ten years.

Study site

The experimental area was located within a north-facing $922 \, m^2$ coffee plantation (1380 m elevation) in 'Hacienda La Cumaca', 40 km west of Caracas, Venezuela. The area was occasionally weeded previous to the study but no fertilizer had been applied in the last ten years. The shade trees form a canopy at 15–20 m. *Erythrina* sp. and *Inga* sp. are by far the most abundant trees but Musa, Clethra, Ficus, Cedrela and Heliocarpus are also among the 178 shade trees present in the plot. A total of 596 adult coffee plants (*Coffea arabica*, predominantly var 'Caturra' with some 'Mundo Nuevo') were counted in the plot. The area receives an annual rainfall of 1200 mm; the first three months of the year are very dry while the last quarter receives slightly less than 100 mm per month. Annual mean temperature is 20°C with a monthly average maximum of 21.2°C in May and a minimum of 18.7°C in December.

The soils of the area are shallow Ultisols developed on deeply weathered mica-schists. The terrain is quite steep; the overall slope in the experimental plot is 25%.

Methods

Coffee phytomass

Four typical fully grown coffee plants were cut at soil level and separated into stem, branches, leaf, flowers and fruit components in the field. Each fraction was weighed and subsamples taken for dry matter determinations and chemical analyses. Root biomass was sampled using a method similar to that of Franco and Inforzato[7] and Suarez de Castro[17]. Four soil blocks 30 × 60 cm in area were cut at distances from 0 to 30 cm measured from each of the cut coffee plants. Initially the soil and roots from the first 10 cm depth were separated, then another layer 10 cm deep and a third layer of the same thickness were sorted for roots. The fourth depth increment was 20 cm thick. The same process was repeated for four more blocks separated from the cut stumps to include distances from 30 to 60 cm. The same depth intervals were sampled. Roots from each of these blocks and layers were hand-sorted, washed and classified into two diameter classes: > 1 mm and < 1 mm. Subsamples were then taken for dry weight determinations and chemical analyses.

Shade trees were not sampled for total phytomass, but leaves, twigs and flowers were collected for chemical analyses.

Litter sampling

Litterfall was intercepted by means of ten 0.5 m² screen trays 50 cm from the soil surface. The collected litter was harvested monthly. Soil-surface litter was collected at the end of the dry season in March, at the height of the rainy season in June and in December, before the start of the dry period, by pressing a 50 × 50 cm metal frame into the litter layer at 50 locations around the plot. All litter collected, both that intercepted by the trays and that from the soil surface was sorted into the following fractions: shade-tree leaves, coffee leaves, twigs, and flowers and fruits. Each sample was dried, weighed, ground in a steel mill and subsampled for analyses.

Soil sampling

Soil samples were taken by depth (0–20, 20–30, 30–60 cm) at 10 locations in the plot. Samples were air-dried, sieved through a 2-mm screen and kept in plastic bags prior to analyses. At four locations and for each soil layer bulk density was determined. This was done by measuring the excavated volume with a thin plastic bag filled with water and then relating this volume to the corresponding dry weight of soil.

Water samples

During October and November bulk precipitation as well as throughfall were sampled with uncovered polypropylene collectors. A spring and the stream draining the plot were also sampled. All samples were treated with methyl mercuric acetate and refrigerated prior to analyses.

Litter decomposition

Nylon bags with a 5 mm mesh were filled with 112 g oven-dried litter composed of 8 g of each of the following fractions: coffee leaves, twigs, flowers and fruits and shade-tree leaves from the various types. Forty eight such bags were placed on the plantation floor. Four replicate bags were collected monthly, their contents sorted, weighed and prepared for analyses as described for other litter samples.

Table 1. Monthly litterfall rates and litter nitrogen concentrations (% N) in a traditional (non-fertilized) Venezuelan coffee plantation

Month	Litterfall (kg ha^{-1} mo^{-1})	Per cent N			
		Leaves		Twigs	Flowers and fruits
		Shade-tree	Coffee		
Jan	467	1.99	1.78	1.47	1.15
Feb	739	2.03	1.69	1.33	1.65
Mar	1672	1.44	1.51	1.50	1.38
Apr	692	1.59	1.97	1.50	1.15
May	816	1.64	1.47	1.62	1.66
Jun	1026	1.81	0.88	1.51	1.13
Jul	1412	2.10	1.50	1.35	1.89
Aug	1160	1.65	1.64	2.05	1.55
Sep	nd	nd	nd	nd	nd
Oct	637	2.05	0.42	1.89	1.45
Nov.	958	1.83	1.66	1.49	0.73
Dec	650	2.02	1.53	1.23	1.41

nd, not determined.

Chemical analyses

All plant subsamples were digested in a perchloric-sulfuric acid 2:8 mixture to which 0.16 g vanadium pentoxide l^{-1} had been added. Nitrogen in the digestate was then determined by automated colorimetry. Soil pH was measured in a 1:1 water:soil mixture. Soil organic carbon was determined by wet oxidation and soil nitrogen by Kjelldahl techniques. Water samples were analyzed for NH_4-N and NO_3-N by automated colorimetry.

Nitrogen fixation and nitrifying bacteria

Sampling of these parameters was initiated during the dry season following the main experimental period. For nitrogen fixation potentials, shade trees, coffee plants and soil samples were analyzed by the acetylene reduction technique[9]. Leaves, bark, roots and soil under the trees were incubated separately. Eight soil samples were also taken along a 30 m transect to quantify the nitrifiers Nitrosomonas and Nitrobacter spp. using the most probable number (MPN) technique of Rowe *et al.*[16]

Results and discussion

Net inputs of nitrogen to any coffee plantation in the absence of fertilization are by rainfall and biological N_2 fixation. Outputs occur by harvest, leaching and gaseous pathways. A potential source or sink of nitrogen is the slow turnover process of NH_4 fixation in the interlayer of some clay minerals (*e.g.*[1, 13]). Internal transfers include uptake and release of nitrogen by the vegetation and soil micro-organisms leading to mineralization, nitrification and immobilization. In traditional coffee plantations under dense shade, the amount of organic matter reaching the soil can range[6, 18] from 5×10^3 to 20×10^3 kg dry matter ha^{-1}

Table 2. Nitrogen transfer rates (kg ha^{-1} month^{-1}) in litterfall fractions in a Venezuelan coffee plantation. Total does not include N in September litterfall (see text)

Month	Leaves		Twigs	Flowers and fruits	Total
	Shade-trees	Coffee			
Jan	5.8	1.3	1.4	0.2	8.7
Feb	10.9	0.3	1.7	1.1	14.0
Mar	11.2	3.1	3.3	4.0	21.6
Apr	3.8	2.3	2.3	2.5	10.9
May	6.2	2.3	3.1	2.5	14.1
Jun	5.4	0.8	6.7	1.6	14.5
Jul	7.9	4.9	6.8	5.2	24.8
Aug	7.7	4.2	8.2	1.1	21.2
Sep	nd	nd	nd	nd	nd
Oct	6.7	0.7	4.4	0.7	12.5
Nov	7.1	4.6	4.5	1.7	17.9
Dec	5.8	3.4	1.3	1.3	11.8
Total (kg ha^{-1} yr^{-1})	78.5	27.9	43.7	21.9	172.0

nd, not determined.

yr^{-1}. This source of nitrogen and other nutrients can vary both seasonally and spatially. In our study plot litterfall totaled 11.2×10^3 kg dry litter ha^{-1} yr^{-1}.

This total was calculated from the data presented in Table 1 corrected for the missing data for September when collected material was tampered with in the field. We found that shade-tree leaves made up 43.4% of total litterfall, coffee leaves 14.8%, twigs 28.6% and flowers and fruits 12.8%. The remaining 0.4% was not identifiable. Litterfall rates peaked both immediately before and in the middle of the rainy season; a third minor peak occurred a month after the rainy season ended. Not only was the amount of litter highly seasonal, but its nitrogen concentration varied as shown in Table 1. The peaks in nitrogen concentrations in shade-tree leaves occurred slightly earlier than the peaks in litterfall rates. This indicates some degree of nitrogen retranslocation before leaf-shedding, and was particularly evident for the peak in the dry season. Presumably an important fraction of litterfall during the rainy period is associated with high winds that take down young leaves still firmly attached to twigs. This scenario is supported by the twig:leaf ratio in litterfall dry weight, which reached a maximum value of 1.66 in June-July while in the dry months was only 0.47.

The amounts of nitrogen in litterfall by months is given in Table 2. The total transfer of nitrogen by this pathway to the forest floor was 172 kg N ha^{-1} for the eleven months sampled. If one corrects for the missing September data by interpolation of the preceding and following month, total litterfall was ca. 189 kg ha^{-1} yr^{-1}. Of this total, shade-tree leaves made up 46%, coffee leaves 16%, twigs, 25% and flowers and fruits 13%. Peaks in litterfall nitrogen occurred in March and July–August, periods in which flowering and fruit formation, respectively, occur in the coffee plants.

Table 3 presents data for the amounts of nitrogen in litter fractions on the soil surface for three periods of the year. These data reflect not only the variations in litter biomass but also variations in nitrogen concentrations, which ranged from 2.1% N in December for all fractions to 1.4% N in March and June. Of this nitrogen store, shade tree leaves accounted for over 50% while coffee leaves made up less than 10%. Comparisons of litter N values for the soil and the corresponding fractions in litterfall show that coffee leaf litter decomposed faster than shade-tree leaf litter.

Litter decomposition rates (k) calculated as the ratio of annual litterfall vs the amount of soil litter for each fraction were 4.1 for shade-tree leaves, 10.0 for coffee leaves, 3.8 for twigs, and 20.5 for flowers and fruits. This assumes that organic matter has reached a steady state in the plantation[14]. Using the ratio of nitrogen transfer rate in litterfall vs the nitrogen store in soil litter fractions, one can obtain an estimate of the rate at which nitrogen was transferred from the litter compartment to the soil organic matter compartment. These values for the above litter fractions are 3.8, 8.4, 5.3 and 20.7, respectively. Leaves seem to have lost mass at a slightly faster rate than they lost nitrogen, while the flower and fruit fraction showed only a minor difference between the rates at which dry weight and nitrogen disappeared. Twigs on the other hand, apparently lost nitrogen at a faster rate than dry weight.

Relative as well as absolute increases of nitrogen in leaf litter have been reported and studied in detail for non-tropical systems (e.g.[3]), but little attention has been accorded to this process in tropical communities. Results obtained in our litter decomposition experiments clarify decomposition rate differences further. Table 4 presents nitrogen concentrations for each litter fraction as decomposition proceeded in litter bags. In both shade-tree leaves and coffee leaves, an increase in nitrogen concentration occurred with time and the increase accelerated when biomass losses of approximately 50% had occurred. Initial mean nitrogen concentrations in coffee leaves were higher than in shade-tree leaves, and the rates of weight loss were concomitantly higher. The increased nitrogen concentrations in the twig and flower and fruit fractions were lower than those of the leaf fractions. Nitrogen accumulation continued in all cases until 80% or more of the original litter mass had been lost. Although nitrogen concentrations increased, no net accumulation of nitrogen in the litter was observed, in contrast with some of the systems studied by Berg and Staaf[3]. The different phases of nitrogen and organic matter release in litter described by these authors were not distinct in our experiments; this may be because the processes occur at faster rates and apparently at higher critical values of nitrogen.

Table 5 shows average nitrogen concentrations in different plant compartments in the plantation as well as nitrogen stores in different parts of the coffee component. Coffee roots contained slightly over half of the total nitrogen

Table 3. Nitrogen (kg ha^{-1}) in litter on the soil of a coffee plantation under shade trees in Venezuela

Month	Leaves		Twigs	Flowers and fruits	NIF	Total
	Shade-tree	Coffee				
Dec	26.1	4.6	12.3	0.4	8.7	52.1
Mar	19.6	3.5	8.2	1.4	4.2	36.8
Jun	22.5	2.8	7.8	1.5	1.1	35.7

NIF, non identifiable.

Table 4. Nitrogen concentrations (% N) and remaining biomass (% of initial dry weight) of litter in mesh bags on the forest floor of a shaded Venezuelan coffee plantation

Month	Leaves				Twigs		Flowers and fruits	
	Shade-tree		Coffee					
	% N	% Biomass	% N	% Biomass	% N	% Biomass	% N	% Biomass
Jan	1.06	95	1.35	93	0.81	99	0.99	81
Feb	1.07	89	1.49	87	0.79	96	0.89	67
Mar	1.06	88	1.39	84	0.79	92	1.00	60
Apr	1.55	57	1.53	51	0.83	83	1.15	42
May	1.72	44	1.90	38	0.91	79	1.12	35
Jun	2.18	29	1.91	25	0.93	71	1.38	24
Jul	2.55	18	2.55	18	0.90	62	1.56	5
Aug	2.47	9	nd	0	0.95	42	nd	0
Sep	nd	nd	nd	0	nd	nd	nd	0
Oct	2.49	6	nd	0	1.08	39	nd	0
Nov	2.53	3	nd	0	1.11	38	nd	0
Dec	nd	0	nd	0	1.22	36	nd	0

nd, not determined.

store of the crop plants, branches about 25%, leaves about 16%, stems less than 10% and flowers and fruits at the time of sampling contained less than 5%. Leaves and fine roots, while storing less than 100 kg N ha^{-1}, are very dynamic compartments in this system because they turnover more often than do branches and stems.

Table 6 presents some soil characteristics. While organic carbon concentrations decreased with depth, total nitrogen showed higher values at intermediate and deeper horizons than at 0–20 cm depth. When concentrations were multiplied by bulk density to obtain nitrogen stores, this difference became even greater. Although we did not analyze the soil for fixed NH_4, it seems possible that the high nitrogen values for deeper layers may be due to a

Table 5. Nitrogen concentrations and stores in some phytomass compartments of a shaded coffee plantation in Venezuela

Plant	Part	%N	kg N ha^{-1}
Coffee	Leaves	1.51	61.7
	Flowers	1.03	0.9
	Fruits	1.10	21.3
	Branches	0.41	120
	Stems	0.53	38.1
	Roots	1.70	249
Erythrina sp.	Leaves	1.52	nd
	Branches	0.90	nd
Musa sp.	Leaves	1.54	nd
Inga sp.	Leaves	1.61	nd
	Branches	2.28	nd
Ficus sp.	Leaves	1.41	nd
	Branches	0.80	nd
Clethra sp.	Leaves	1.40	nd
	Branches	1.81	nd
Heliocarpus sp.	Leaves	2.19	nd
	Branches	0.95	nd

nd, not determined.

Table 6. Some soil characteristics and nitrogen stores in a Venezuelan coffee plantation

Soil depth (cm)	pH	Organic C%	Total N	
			%	10^3 kg N ha^{-1}
0–20	4.9	5.3	0.40	7.8
20–30	5.2	4.9	0.91	10.4
30–60	5.6	2.6	0.69	30.8

Table 7. Nitrogen concentrations in water in a coffee plantation in Venezuela during October and November

Water sample	NH$_4$–N	NO$_3$–N
	mg l^{-1}	
Bulk precipitation	2.23	0.06
Throughfall	1.23	0.06
Spring	0.82	0.28
Stream	0.77	0.42

proportion of the mineral nitrogen being retained in the interlayer of some clays.

Table 7 presents the nitrogen concentrations in bulk deposition, throughfall and stream water sampled during October and November. Bulk deposition was dominated by NH_4–N by a factor of 37 relative to NO_3–N, while in throughfall this ratio decreased to 21 due to a decrease in NH_4–N concentration after rain had passed through the canopy. In the stream water samples, the ratio of NH_4–N *vs* NO_3–N was about 2. Although the two month period of sampling is obviously too short to obtain a reliable estimate of inputs and losses, an order-of-magnitude value can be obtained from the data. By multiplying concentrations by monthly precipitations for the sampling period, an input of 3 kg N ha^{-1} can be calculated for October and November. Of this total, about 2 kg N ha^{-1} reached the ground with throughfall, the difference presumably being retained by the canopy and returned to the soil with litterfall. During the rainy season some losses of nitrogen could occur by leaching. Comparison of the nitrogen concentrations in rainfall, throughfall and streamflow seem to indicate that these losses were not large in the plantation studied.

The number of nitrifying bacteria (MPN) estimated in the soil samples taken in the plantation under shade were *ca.* 1100 bacteria g^{-1} and 1500 bacteria g^{-1} for Nitrosomonas and Nitrobacter, respectively. These numbers are very low when compared with data from agricultural soils but are comparable to values reported for undisturbed tropical forests where nitrification is almost negligible[11].

Our initial estimates for biological nitrogen fixation were below detection limits for the first three months of the year. This pathway of nitrogen cycling could nevertheless be of importance in the rainy season. Roskoski[55] has reported high inputs associated with root nodules of *Inga* sp. in Mexican coffee plantations.

An amount equivalent to 700 kg dry coffee berries were harvested from the plot. This corresponds to a total output of 20 kg N ha^{-1} yr^{-1}. Traditionally the berry pulp is returned to the field if coffee is processed near the plantations and this would represent a return to the soil litter of our plantation of 3 kg N ha^{-1} yr^{-1}. Coste[5] cites world average fertilizion inputs of 30 kg N ha^{-1} to maintain yields of 1000 kg dry beans ha^{-1}, but much higher N inputs have recently been reported for coffee plantations in Central America[4].

In coffee plantations under shade, the return of nitrogen by litterfall and its subsequent mineralization can be several times higher than output by harvest. If shade trees are nitrogen fixers or if they can exploit soil that is not available to the shallower rooted coffee plants, the outputs by harvest can be amply balanced by nitrogen coming to the forest floor in the form of shade-tree litter. Nitrogen is partially immobilized in decomposing litter, but this process does not lead to absolute accumulations as is the case in temperate forests. Litter decomposition proceeds very quickly in coffee plantations, so that for some litter fractions half of the nitrogen content is transferred to the soil in a matter of days. Shade-tree leaf litter in this study contributed over 80 kg N ha^{-1} yr^{-1} to the soil with a peak in litterfall in February–March. Flowering in coffee plants occurs at the end of March or beginning of April, so that the above increase in N input to the soil occurs at a time of high demand by the coffee plants. A second peak of nitrogen input from the shade litter occurred during the period of fruit filling in June–August, although this input comes mainly from the twig fraction which decomposes more slowly than the leaf fraction. This synchronization of inputs by

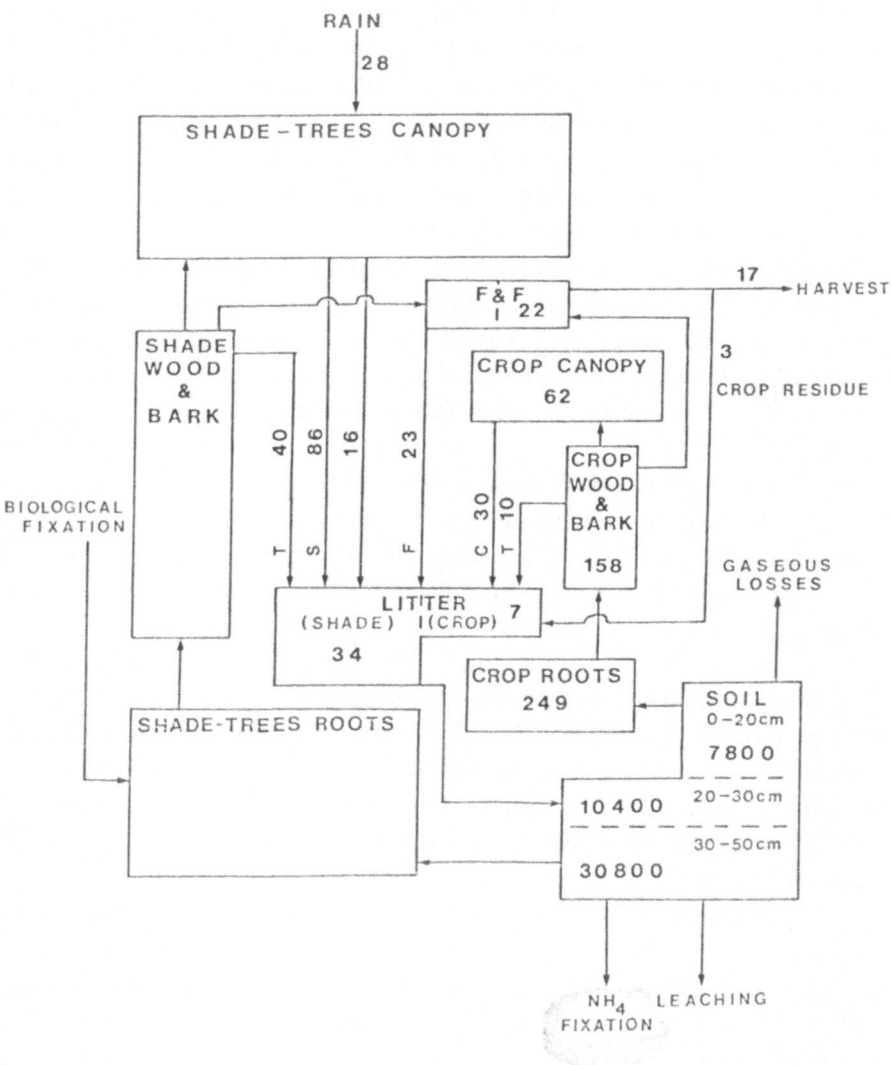

Fig. 1.   Nitrogen cycle in a coffee plantation under shade in Venezuela. Values for pool sizes = kg N ha^{-1}; flows = kg N ha^{-1} yr^{-1}. Litterfall inputs identified in diagram as T = twigs; S = shade-tree leaves; F&F = flowers and fruits; C = coffee leaves.

litterfall and coffee plant requirements is considered to be as important as the overall nitrogen subsidy supplied by the shade-trees. Aranguren[2] reported that 70% of the fine root biomass of coffee in the same plantation can be found in the upper 30 cm of the mineral soil. Additionally, during the wettest months, 9.3 g m^{-2} of fine living roots (expressed as dry matter) were collected in the soil litter quadrats intimately in contact with decomposing litter. This amount of roots is equivalent to that found in the first 5 cm of the A horizon and is thought to play an important role in the absorption of rapidly mineralized nutrients from litter.

Fig. 1 presents a summary of the nitrogen cycle in the coffee system studied. Although the nitrogen stores in the shade-tree compartments have not been studied, the flows from these to the litter pool and eventually to the coffee plants are indicated in the diagram. Arbitrarily, we have divided twig litterfall into shade-tree twigs ($40 \, kg \, N \, ha^{-1} \, yr^{-1}$) and coffee twig litterfall ($10 \, kg \, N \, ha^{-1} \, yr^{-1}$) for the purpose of this cycle. The amount indicated for bulk precipitation and throughfall inputs are only tentative.

Conclusions

In coffee plantations cultivated under the traditional system of shade-tree cover and little or no fertilization, the nitrogen output by harvest can be exceeded by nitrogen transfers from shade-tree leaves alone, while total inputs including all litter and nitrogen in throughfall can be over ten times higher than net nitrogen output by harvest. The synchronization of peaks in nitrogen inputs with periods of high demand by the coffee plants occurring during flowering and fruit formation is thought to play an important role in these traditional systems. To have a better understanding of the nitrogen cycle in coffee agro-ecosystems, more detailed studies are needed. Studies of nitrogen balance associated with the hydrological cycle, including denitrification and leaching losses are important in this respect. Better estimates of inputs by biological nitrogen fixation, gaseous losses and NH_4 fixation in clays are particularly necessary at this stage.

Acknowledgements We are deeply indebted to the late Dr Carlos Pérez for the opportunity of carrying out work in his plantation and for very stimulating discusions. One of the authors (J.A.) was partially funded by the Venezuelan CONICIT. We are also very grateful to Laura Martin and Elisa Orellana for laboratory assistance.

References

1 Ahmad N, Reid E D, Nkrumah M, Griffith S M and Gabriel L 1982 Crop utilization and fixation on added ammonium in soils of the West Indies. *In* Robertson G P, Herrera R and Rosswall T (eds.) Nitrogen cycling in Ecosystems of Latin America and the Caribbean. Plant and Soil 67, 167–186.

2 Aranguren J 1980 Contribución de la caida de hojarasca al ciclo de nutrientes en cultivos bajo árboles de sombra. M.Sc. Thesis Instituto Venezolano de Investigaciones Científicas, Caracas 285 p.

3 Berg B and Staaf H 1981 Leaching, accumulation and release of nitrogen in decomposing

forest litter. *In* Terrestrial Nitrogen Cycles. Clark F E and Rosswall T (Eds.). Ecol. Bull. Stockholm 33, 273–279.

4   Bornemisza E 1982 Nitrogen cycling in coffee plantations. *In* Robertson G P, Herrera R and Rosswall T (eds.) Nitrogen cycling in Ecosystems of Latin America and the Caribbean. Plant and Soil 67, 241–246.

5   Coste R 1975 El Café. Colección de Agricultura Tropical. Editorial Blume, Barcelona 240 p.

6   Delgado-Palacios G 1935 Contribución al estudio del café en Venezuela. Publ. Universidad Central de Venezuela, Caracas 63 p.

7   Franco C M and Inforzato R 1946 Sistema radicular do cafeeiro nos principais tipos do solo do Estado de Sao Paulo. Bragantia 6, 443–478.

8   Hardy F 1959 La relación carbono-nitrogeno en los suelos de cacao. Turrialba 9, 4–11.

9   Hardy R W F, Holstein R D, Lackson E K and Burns R C 1968 The acetylene-ethylene assay for N_2 fixation: laboratory and field evaluation. Pl. Physol. 43, 1185–1207.

10 Herrera R and Jordan C F 1981 Nitrogen cycle in a tropical Amazonian main forest: the caatinga of low mineral nutrient status. *In* Clark F E and Rosswall T (Eds.). Terrestrial Nitrogen Cycles. Ecol. Bull. Stockholm 33, 493–505.

11 Jordan C F, Todd R and Escalante G 1979 Nitrogen conservation in a tropical rain forest. Oecologia 39, 123–128.

12 Jordan C F, Caskey W, Escalante G, Herrera R, Montagnini F, Todd R and Uhl C 1982 The nitrogen cycle in a 'Tierra Firme' rain forest on oxisol in the Amazon Territory of Venezuela. *In* Robertson G P, Herrera R and Rosswall T (eds.) Nitrogen cycling in Ecosystems of Latin America and the Caribbean. Plant and Soil 67, 325–332.

13 Nômmik H 1981 Fixation and biological availability of ammonium in soil clay monerals. *In* Terrestrial Nitrogen Cycles. Clark F E and Rosswall T (Eds.). Ecol. Bull. Stockholm 33, 273–279.

14 Olson J S 1963 Energy storage and the balance of producers and decomposers in ecological systems. Ecology 44, 322–331.

15 Roskoski J P 1982 Nitrogen fixation in a Mexican coffee plantation. *In* Robertson G P, Herrera R and Rosswall T (eds.) Nitrogen cycling in Ecosystems of Latin America and the Caribbean. Plant and Soil 67, 283–291.

16 Rowe R, Todd R L and Waide J 1977 Microtechnique for most probable number analysis. Appl. Environ. Microbiol. 33, 675–680.

17 Suarez de Castro F 1953 Distribución de las raices de *Coffea arabica* en un suelo franco-limoso. Bol. Tecn. Federación Nacional de Cafeteros de Colombia 1, 5–28.

18 Suarez de Castro F and Rodríguez A 1955 Equilibrio de materia orgánica en plantaciones de café. Bol. Tecn. Federación Nacional de Cafeteros de Colombia 3, 5–28.

Plant and Soil 67, 259–269 (1982). 0032-079X/82/0672-0259$01.65.

© 1982 *Martinus Nijhoff/Dr W. Junk Publishers, The Hague.*

Nitrogen cycle of tropical perennial crops under shade trees

II. Cacao

Ciclo de nitrógeno en cultivos tropicales permanentes bajo árboles de sombra II. Cacao

J. ARANGUREN*, G. ESCALANTE and R. HERRERA

Centro de Ecología, Instituto Venezolano de Investigaciones Científicas (IVIC), Apartado 1827, Caracas 1010 A Venezuela

Key words Cacao Decomposition Litterfall N-cycling N_2fixation Shade trees Venezuela

Abstract A cacao (*Theobroma cacao*) plantation under shade trees was studied in northern Venezuela in order to estimate nitrogen stores in the soil and plantation trees and nitrogen fluxes associated with litterfall, decomposition and harvest. The cacao plants contained 302 kg N ha^{-1}, of which woody above-ground parts made up 50%. Fine roots and leaves contained *ca.* 60 kg N ha^{-1}. Litter on the soil surface was sampled quarterly and found to contain from 24 kg N ha^{-1} in November to 50 kg N ha^{-1} in May, with an annual average of 37 kg N ha^{-1}. Shade tree leaves made up 61% of the total nitrogen in the litter on the soil. Mineral soil stores of total nitrogen were 35×10^3 kg N ha^{-1}, 40% of which was found in the first 20 cm depth.

Litterfall rates were studied monthly for a year; 20.9×10^3 kg dry litter ha^{-1} fell during the year, the major contribution (50%) from shade-tree leaves. The rate of nitrogen transfer with litterfall was 321 kg N ha^{-1} yr^{-1}. Decomposition rates (k) for shade-tree leaves, cacao leaves, twigs, and flowers and fruits were 7.7, 9.5, 7.5 and 19.7 respectively. In litterbag decomposition experiments, a small increase in per cent N was observed as decomposition progressed for all fractions.

Rainfall collected during October and November contained nitrogen predominantly as NH_4, a fraction of which was retained by the canopy. Nitrogen output by harvest was *ca.* 45 kg N ha^{-1} yr^{-1}, with some 20 kg N ha^{-1} yr^{-1} returned to the field with pod shells after processing. The net harvest output can be amply compensated for by inputs of nitrogen in shade-tree leaf litter alone. Much of this shade-tree nitrogen may be from deeper soil horizons than those exploited by cacao and possibly from biological N_2-fixation by the shade trees.

Resumen Se estudio una plantación de cacao bajo sombra mixta en Venezuela con el fin de estimar las reservas de nitrógeno en el suelo y los cacaoteros asi como los flujos de materia orgánica y nitrógeno asociados a la caida de hojarasca, descomposición y extracción por cosecha durante un año. La fitomasa de los cacaoteros contenía 302 kg N ha^{-1} de los cuales el 50% estaba en las partes leñosas aéreas. Las raices finas y las hojas contenían *ca.* 60 kg N ha^{-1}. La hojarasca del suelo fué muestreada trimestralmente y se halló que contenía entre 24 kg N ha^{-1} en Mayo y 50 kg N ha^{-1} en Noviembre con un promedio anual de 37 kg N ha^{-1}. Las hojas de los árboles de sombra en este compartimiento contenían 23 kg N ha^{-1}. El suelo hasta 60 cm contenía 35×10^3 kg N ha^{-1}; *ca.* 40% de esta cantidad se encontró en los primeros 20 cm. La caida de hojarasca se estudió mensualmente; el total anual fué de 21×10^3 kg ha^{-1} a^{-1} materia seca con unas contribución de las hojas de sombra

* Present address: Departamento de Biología y Química, Instituto Universitario Pedagógico de Caracas, Venezuela.

mayor de 50%. La tasa de transferencia de nitrógeno en la caida de hojarasca fué de 321 kg N ha^{-1} a^{-1}. Las tasas de descomposición calculadas para los compartimientos hojas de árboles de sombra, hojas de cacaoteros, ramitas y flores mas frutos fueron de 7,7, 9,5, 7,5 y 19,7 respectivamente.

En experimentos de descomposición en bolsas de malla se encontraron pequenos aumentos en el % N en todas las fracciones de hojarasca mientras qué éstas se descomponían. La lluvia colectada en Octubre y Noviembre contenía nitrógeno predominantemente en forma amoniacal pero al pasar por el dosel la relación $NH_4 : NO_3$ se hizo mas estrecha. La salida de nitrógeno por cosecha se estimó en 45 kg N ha^{-1} a^{-1}, con un posterior retorno de ca. 20 kg N ha^{-1} a^{-1} en forma de residuos de cosecha una vez separados los granos de cacao. La salida neta de nitrógeno por cosecha puede ser ampliamente compensada por los ingresos de nitrógeno provenientes de la hojarasca de los árboles de sombra. La explotación de horizontes mas profundos del suelo y posiblemente la fijación biológica de nitrógeno en las especies de árboles leguminosos usados como sombra, pueden, explicar los rendimientos sostenidos sin fertilizantes.

Introduction

Cacao, like coffee, is commonly grown in Venezuela under shade trees. Chemical fertilization is used sparingly and although yields are only moderate, plantations maintain productivity for long periods. The role of shade trees and their litter in the nitrogen balance in cacao is still little known in tropical America. Boyer[3] reported litterfall of 8500 kg dry matter ha^{-1} yr^{-1} for a moderately shaded cacao plantation in Cameroon, corresponding to a nitrogen flux of 50 kg N ha^{-1} yr^{-1}. The rate of litter decomposition in this plantation was found to be similar to that of tropical forests; Boyer[3] reported a weight loss of 75% in the first 6 months for the plantation litter. Havord[7] found that cacao plantations under shade did not respond to nitrogen fertilization.

Hardy[6] estimated that 567 kg N ha^{-1} was stored in the phytomass of *Erythrina* sp. trees used as shade in cacao plantations in Trinidad. These trees were estimated to return 23 kg N ha^{-1} yr^{-1} in litterfall. Hardy[6] suggested that some of this return probably came from biological N_2-fixation in the roots of the *Erythrina* sp. trees; this was later supported by Cadima and Alvim[4] who found that cacao soils in Brazil were nitrogen enriched in the vicinity of *Erythrina* sp. Santana and Cabala[11] considered the transfer of nitrogen in senescent shade-tree root nodules to the soil to be an important nitrogen input to the cacao system.

In this paper we attempt to assess the role of both cacao and shade-tree litter in the nitrogen cycle of a cacao plantation in Venezuela. We concentrate in particular on the seasonality of nitrogen inputs to the soil *via* litterfall and subsequent decomposition. We also examine the nitrogen stores in different compartments of the cacao systems and some important fluxes between these compartments.

Study site

The cacao experimental area was located in 'Hacienda Monasterio' in Ocumare de la Costa in Northern Venezuela at 12 m elevation. Annual average temperature is 25°C and precipitation is 740 mm. Only 3% of this total precipitation falls between January and March. The onset of the rainy period occurs between April and June; *ca.* 29% of annual precipitation falls during this period. The main rainy season is from July to September with 43% of the total annual rainfall occurring during these three months. A second, less pronounced rainy period occurs between October and November and it is called 'nortes' because it is often accompanied by northern winds. The remaining 25% of annual rainfall occurs during the last quarter of the year.

A sample area of 7600 m² was selected in a 30 yr old plantation which had received no fertilization in the previous 7 yr. In this area grew 720 cacao (*Theobroma cacao*) trees var. 'Criollo Morado' under a dense canopy formed by 430 shade trees. The most common shade trees were *Castilloa elastica, Erythrina* sp., and *Artocarpus altilis*.

The soil of the area has been described in Cañizales[5] as a Psamment of recent alluvial origin. The terrain is well drained and almost flat. During the drier periods it is possible to irrigate the plantation by pumping water from a river nearby.

Methods

Methods not described fully below can be found in Aranguren *et al.*[2]

Cacao phytomass
Four typical plants from the plantation were selected for phytomass determinations carried out in the same manner as described for coffee by Aranguren *et al.*[2] For cacao, however, the entire root system was excavated and separated from the soil, a procedure that was facilitated by the light soil texture. Roots were separated into tap roots and lateral roots, the latter sorted by diameter classes.

Litter sampling
Litterfall was collected monthly starting in March by means of ten 1 m² trays. Litter on the soil was collected in February, May, August and November using the same methods described in Aranguren *et al.*[2]

Soil sampling
Soil samples were taken with an auger at 10 locations around the plot at 0–20, 20–40 and 40–60 cm depths. The samples were then processed as in Aranguren *et al.*[2]

Water sampling
During the months of October and November, at the end of the rainy season, water samples were collected for determinations of nitrogen in bulk precipitation, in throughfall, in river water and in water from the irrigation ditch in the plot. Samples were treated as in Aranguren *et al.*[2]

Litter decomposition
For estimating the decomposition rate of different litter fractions, 112 g total dry weight litter from cacao and shade trees were placed in 5 mm mesh bags[2]. A mixture of flowers and fruits (24 g dry weight) of all species which had flowered in February were included.

Chemical analyses
Laboratory procedures for chemical analyses have been outlined in Aranguren *et al.*[2]

Table 1. Monthly litterfall rates and nitrogen concentrations (%) N in a Venezuelan cacao plantation under shade-trees

Month	Litterfall $kg\,ha^{-1}\,mo^{-1}$	Per cent N			
		Leaves		Twigs	Flowers and fruits
		Shade-tree	Cacao		
Mar	2834	1.35	1.42	1.35	2.21
Apr	1961	1.32	1.21	1.28	1.73
May	1636	1.68	1.34	1.17	1.78
Jun	2124	1.80	1.47	1.48	1.85
Jul	1236	1.69	1.40	1.21	1.93
Aug	2782	1.93	1.58	1.00	1.86
Sep	1130	1.72	1.10	0.84	1.89
Oct	1702	1.64	1.30	1.11	1.93
Nov	1313	2.26	1.14	0.93	1.92
Dec	1060	1.30	1.00	0.81	1.67
Jan	1532	1.61	1.38	1.21	1.38
Feb	1539	1.39	0.94	1.18	1.64

Biological N_2-fixation and nitrifying bacteria

The same procedures outlined by Aranguren et al.[2] were used in the cacao plantation during the last month of sampling (February).

Results and discussion

Table 1 presents monthly litterfall rates and corresponding nitrogen concentrations for each litter fraction. Total annual litterfall was extraordinarily high as compared to the less densely shaded cacao plantation examined by Boyer[3]. The rate of litterfall shows large month-to-month variation with a maximum in March, a small peak at the onset of rains in May, and a larger peak at the height of the rainy season. A small peak corresponding to the 'nortes' rainfall peak was also observed.

Of the total annual litterfall, 50% was identified as shade-tree leaves, but their proportion in February reached 69 per cent of the monthly total. Cacao leaves comprised as little as 6 per cent of the total litterfall in March, and a maximum of 19 per cent in July. Amounts of twigs in litterfall followed those of shade-tree leaves, with peaks at the onset of rains and at the heights of the rainy season and 'nortes'. Flowers and fruits made up 15% of total litterfall on the average. The composition of litterfall in our plantation contrasts sharply with that reported by Boyer[3] for a cacao plantation in Cameroon, where cacao leaf litter was the most abundant fraction.

Shade-tree leaves and the flower and fruit fraction showed the highest N concentrations, with annual average concentrations of 1.64% N and 1.82% N,

Table 2. Nitrogen transfer rates (kg N ha^{-1} month^{-1} except total = kg N ha^{-1} yr^{-1}) in litterfall fractions in a Venezuelan cacao plantation under shade trees

Month	Leaves		Twigs	Flowers and fruits	Total
	Shade-tree	Cacao			
Mar	25.2	2.2	6.2	6.6	40.2
Apr	14.3	3.5	3.0	8.3	29.1
May	10.6	3.8	4.4	8.4	27.2
Jun	14.8	5.4	7.0	9.5	36.7
Jul	6.5	3.7	2.8	7.3	20.3
Aug	23.5	3.7	10.4	6.9	44.5
Sep	6.6	2.7	2.6	6.6	18.5
Oct	10.7	3.9	4.0	5.3	23.9
Nov	15.6	3.1	2.2	3.0	23.9
Dec	8.3	1.2	3.1	1.8	14.6
Jan	15.9	2.2	3.5	1.6	23.2
Feb	14.0	1.2	2.2	1.1	18.5
Total	166.0	36.6	51.4	66.4	320.6

Table 3. Nitrogen (kg N ha^{-1}) in litter on the mineral soil of a cacao plantation under shade trees in Venezuela

Month	Leaves		Twigs	Flowers and fruits	NIF	Total
	Shade-tree	Cacao				
Feb	23.0	3.3	6.4	3.2	2.7	38.6
May	33.3	4.7	6.9	4.0	0.8	49.7
Aug	22.3	3.3	6.7	1.9	1.0	35.2
Nov	11.4	4.3	5.6	1.6	1.2	24.0

NIF, non identifiable fraction.

respectively. These values are much lower than those reported by Santana and Cabala[11]; this is probably due to the differences in the species composition of the shade trees in the plantations studied. Additionally, some nitrogen in our litterfall samples could have been lost before analyses, however, since litter remained in the collection trays for up to one month periods. A shorter interval between litter harvests is advisable for future work in systems where decomposition proceeds very quickly. A peak in % N in shade-tree leaves was observed in November, a time when the % N in twigs was close to its minimum value. The reverse trend was observed during the main leaf shedding season (March–April). This observation suggests some retranslocation of nitrogen from leaves to twigs before senescent leaves are shed.

Table 2 shows monthly nitrogen transfer rates in the litterfall fraction. Concomitant with the very high rates of litterfall, large amounts of nitrogen ($321 \, kg \, N \, ha^{-1} \, yr^{-1}$) were transferred from the vegetation to the soil litter. Of this amount, shade tree leaves contributed 52%, cacao leaves 11%, twigs 16% and flowers and fruits the remaining 21%. A continuous although variable nitrogen input to the plantation floor was observed over all months, with maxima of over $40 \, kg \, N \, ha^{-1} \, month^{-1}$ during March and August. These peaks correspond to peaks in shade-tree leaf litterfall and to a lesser degree to peaks in twig litterfall. Cacao flowering and fruit formation also occur during these same periods.

Table 3 presents the amounts of nitrogen in different litter fractions on the plantation soil on four sampling dates. Total nitrogen in soil litter was lowest in November, with $24 \, kg \, N \, ha^{-1}$, and highest in May, with $50 \, kg \, N \, ha^{-1}$. These values are comparable to those cited by Boyer[3] but are very low in comparison with the nitrogen flux to this compartment via litterfall. Decomposition constants (k, Olson[10]) for shade-tree leaves, cacao leaves, twigs and flowers and fruits in terms of organic matter, were calculated by Aranguren[1] to be 7.7, 9.5, 7.5 and 19.7 respectively. These values are much higher than the decomposition rate for a plantation in Cameroon[3]. The decomposition constant (k) for the same litter fractions in terms of nitrogen fluxes and stores are 7.4, 9.4, 8.0 and 24.6, respectively. Thus, twigs and fruits and flowers apparently lose nitrogen at faster rates than they lose organic matter, while both shade-tree and cacao leaves lose N at slightly slower rates than organic matter. In the litter decomposition experiments in nylon bags (Table 4), all four fractions showed increases in

Table 4. Nitrogen concentration (% N) and remaining biomass (% initial dry weight) of litter in mesh bags on the forest floor of a shaded Venezuelan cacao plantation

Month	Leaves				Twigs		Flowers and fruits	
	Shade-tree		Cacao					
	% N	% IB	% N	% IB	% N	% IB	% N	% IB
Mar	1.40	90	1.26	97	0.73	93	0.97	92
Apr	1.49	70	1.38	60	0.85	83	1.09	42
May	1.57	58	1.48	37	0.88	68	1.38	22
Jun	1.72	45	1.48	5	0.87	53	1.55	9
Jul	1.85	33	nd	0	0.93	45	nd	0
Aug	1.85	21	nd	0	1.12	38	nd	0
Sep	1.83	14	nd	0	1.18	33	nd	0
Oct	1.92	13	nd	0	1.18	28	nd	0
Nov	1.97	3	nd	0	1.21	27	nd	0
Dec	nd	0	nd	0	1.25	23	nd	0
Jan	nd	0	nd	0	1.24	22	nd	0
Feb	nd	0	nd	0	1.27	19	nd	0

IB, initial biomass; nd, not determined.

nitrogen concentrations as decomposition proceeded. This same effect was observed for coffee litter by Aranguren et al.[2] but the ratios of initial $\%$ N to maximum $\%$ N were lower in all cases than in the cacao plantation. Nitrogen concentration increases were higher for the twigs and flower and fruit fractions in the cacao plantation than increases for the corresponding fractions for coffee litter[2]. On the other hand, only small relative increases in nitrogen concentrations were observed in cacao plantation shade-tree leaves and cacao leaf litter, while substantial increases for the respective fractions were found in the coffee plantation litter studied by Aranguren et al.[2]

Table 5 presents nitrogen concentrations and stores in the cacao plantation, including nitrogen concentrations in leaves and branches of the shade tree species present in the experimental plot. The highest $\%$ N in cacao plants was found in the fruits. A total store of 302 kg N ha^{-1} was estimated for the cacao plants. Branches, stems and roots contributed with 30%, 20% and 29% of this total, respectively. Cacao leaves contained 45 kg N ha^{-1}. Considering that 98% of the leaf phytomass is renewed annually[2], almost 20% of the nitrogen would be then retranslocated before leaves are shed. Turnover of fine root tissue could be another important pathway for internal nitrogen recycling in the plantation, but no studies that examine this flux exist.

The soil contained 34.2×10^3 kg N ha^{-1} to 60 cm depth; 41% of this amount was found in the first 20 cm (Table 6). In this same cacao plantation soil Aranguren[2] found 52% of the less-than 2 mm diameter cacao root phytomass. In these nitrogen-rich soils, only a fraction of 1% of total N mineralized per year is needed to supply the cacao roots with sufficient nitrogen for internal recycling and fruit production.

Table 7 presents the results of nitrogen determinations in water samples taken during October and November. As in the Venezuelan coffee plantation[2], NH_4–N was much higher than NO_3–N both in bulk precipitation and in throughfall. As rain passed through the canopy of the plantation, a slight decrease in NH_4–N concentration was observed, while NO_3–N remained unchanged. An upper limit of nitrogen input via precipitation can be calculated assuming that concentration in October and November are representative for the entire rainy season, and this value was 10.5 kg N ha^{-1} yr^{-1}. This is a very preliminary figure, however, since large variations can be expected both within and between years.

River and irrigation ditch water samples showed total inorganic nitrogen concentrations that were only slightly higher that those of rainfall, but the ratio of NH_4–N vs. NO_3–N was much lower.

The most probable number (MPN) of the nitrifying bacteria Nitrosomonas and Nitrobacter were 226 and 2435 cells g dry soil^{-1}, respectively, for the dry season when the first sampling was carried out. As in the coffee plantation studied by Aranguren et al.[2], these values are closer to those of climax forest ecosystems than to those of agricultural soils.

The initial estimates for biological N_2-fixation, also carried out during the dry season, were below the detection limits. This pathway of nitrogen input to the system could be of importance for cacao plantation under leguminous shade trees. Santana and Cabala[11] found nodule densities in *Erythrina* trees of *ca.* 2.0 g dry nodules m^{-2} in a cacao plantation in Brazil. No data is available for N_2-fixation by free living bacteria in cacao plantations, although some estimates for tropical forest in Amazonia, the site of origin of cacao, are significant[8,9].

The cacao plantation under study yielded an equivalent of 636 kg dry cacao almonds ha^{-1} yr^{-1}. This represents an output of 40 kg N ha^{-1} yr^{-1}. About half of this nitrogen output would be normally returned to the field in the fruit husks after processing. Husks analyzed contained 0.79% N. Cacao almonds are often fermented in tanks near the plantation; some additional return of nitrogen can occur when the clean-up waters are allowed to run into the fields.

Cacao plants are often pruned to eliminate unwanted shoots and diseased branches; firewood from the shade trees is also extracted from the plantations in some cases. In the plantation we studied neither practice was performed and therefore no estimates for these potential outputs are given.

Fig. 1 summarizes nitrogen stores and transfer rates in the plantation. An input of 329 kg N ha^{-1} yr^{-1} to the soil surface via throughfall and litterfall was estimated from our measurements. This nitrogen input is an order of magnitude higher than the net output by harvest.

Most of the nitrogen needed to sustain cacao production is probably absorbed from the upper soil or litter layers. Aranguren[1], for example, found that fine roots invading the litter layer reached an average of 72 kg dry matter ha^{-1} in the same plantation, though these roots were seasonal and almost disappeared during the dry season. Shade trees may take up nitrogen from deeper soil layers not exploited by cacao, and may thus enhance nitrogen availability for cacao by transferring this N to the soil surface via litterfall.

Flowering and fruiting in cacao occurs at different times during the year but the most important fruiting season is from May to November. During this period occurred one of the main peaks in litterfall. Nitrogen release from fallen litter was found to occur at very high rates.

Conclusions

In the cacao plantation studied, the nitrogen input to the soil by shade-tree leaf litter alone exceeded the net output by harvest of cacao almonds by a factor of more than six. This flux of nitrogen in the form of litterfall occurred continuously during the year but showed several peaks. Decomposition is very fast and while there is evidence of increasing nitrogen concentrations in all litter fractions, no net accumulation effect was observed.

Better estimates of the importance of both free-living bacteria and Rhizobium associated with legume shade trees are necessary. More complete studies of nitrogen leaching are needed, also. This possible nitrogen loss could be important

Table 5. Nitrogen concentrations and stores in some phytomass compartments of a shaded cacao plantation in Venezuela

Plant	Component	%N	kg N ha^{-1}
Cacao	Leaves	1.77	45.1
	Flowers	1.05	0.1
	Fruits	2.06	19.9
	Branches	0.91	89.9
	Stems	1.27	59.4
	Roots	1.12	88.0
Castilloa elastica	Leaves	2.98	nd
	Branches	1.75	nd
Erythrina sp.	Leaves	1.57	nd
	Branches	0.95	nd
Artocarpus altilis	Leaves	1.74	nd
	Branches	0.97	nd
Spondias lutea	Leaves	2.14	nd
	Branches	1.01	nd
Inga sp.	Leaves	1.82	nd
	Branches	1.93	nd

nd, not determined.

Table 6. Some soil characteristics and nitrogen stores in a Venezuelan cacao plantation

Soil depth (cm)	pH	Organic C (%)	Total N	
			(%)	(10^3 kg N ha^{-1})
0–20	7.4	1.50	0.62	13.9
20–40	7.1	0.29	0.47	9.7
40–60	6.1	0.21	0.18	10.6

Table 7. Nitrogen concentrations (mg N l^{-1}) in water fluxes in a cacao plantation in Venezuela. Averages for October and November

Source	NH$_4$–N	NO$_3$–N
Bulk precipitation	1.40	0.02
Throughfall	1.29	0.02
River	0.82	0.54
Irrigation ditch	1.30	0.41

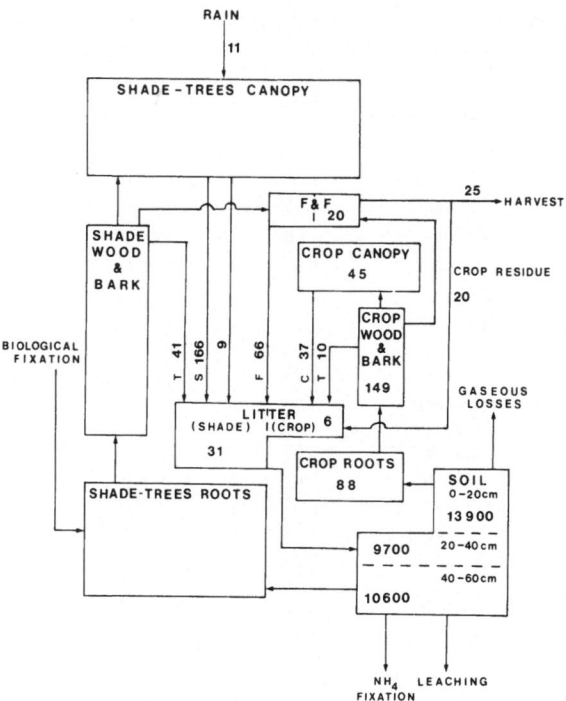

Fig. 1.   The nitrogen cycle of a cacao plantation under shade trees in Venezuela. Values for nitrogen stores = kg N ha^{-1}; for nitrogen flows = kg N ha^{-1} yr^{-1}. Litterfall inputs are identified as S = shade tree leaves, C = cacao leaves, T = twigs and F&F = flowers and fruits.

in light textured soils or fertilized plantations, but under the traditional cultivation system this pathway does not seem to be very important.

Acknowledgements   We wish to thank Fondo Nacional de Investigaciones Agropecuarias for the use of the experimental area in Hacienda Monasterio. One of the authors (J.A.) received research funds from the Venezuelan CONICIT. The laboratory assistance of Laura Martin and Elisa Orellana is gratefully acknowledged.

References

1   Aranguren J 1980 Contribución de la caida de hojarasca al ciclo de nutrientes en cultivos bajo árboles de sombra. M.Sc. Thesis. Caracas: Instituto Venezolano de Investigaciones Científicas. 285 p.
2   Aranguren J, Escalante G and Herrera R 1982 Nitrogen cycle of tropical perennial crops under shade trees. I Coffee. *In* Robertson G P, Herrera R and Rosswall T (Eds.). Nitrogen cycling in Ecosystems of Latin America and the Caribbean. Plant and Soil 67, 247–258.
3   Boyer J 1973 Cycles de matiére organique et des éléments mineraux dans une cacaoyére Camerounaise. – Café, Cacao, The 17, 3–23.
4   Cadima A and Alvim P T 1967 Influencia del árbol de sombra *Erythrina glauca* sobre algunos factores edafológicos relacionados con la producción del cacaotero. Turrialba 17, 330–336.
5   Cañizales R 1978 Estudio de suelos del Centro de Propagación de Cacao. Maracay, Venezuela: Instituto de Investigaciones Agrícolas, Ministerio de Agricultura y Cría. 35 p.

6   Hardy F 1959 La relación carbono : nitrógeno en los suelos de cacao. Turrialba 9, 4–11.

7   Havord G 1959 The nutrition and shade requirements of cacao. Turrialba 9, 138–148.

8   Herrera R and Jordan C F 1981 Nitrogen cycle in a tropical Amazonian rainforest: the caatinga of low mineral nutrient status. *In* Clark F E and Rosswall T (Eds.). Terrestrial Nitrogen Cycles. Ecol. Bull. Stockholm 33, 493–505.

9   Jordan C F, Caskey W, Escalante G, Herrera R, Montagnini F, Todd R and Uhl C 1982 The nitrogen cycle in a 'Tierra Firme' rain forest on oxisol in the Amazon Territory of Venezuela. *In* Robertson G P, Herrera R and Rosswall T (Eds.). Nitrogen cycling in Ecosystems of Latin America and the Caribbean. Plant and Soil 67, 325–332.

10 Olson J S 1963 Energy storage and the balance of producers and decomposers in ecological systems. Ecology 44, 322–331.

11 Santana M B and Cabala P F 1982 Observations on the dynamics of nitrogen in a cacao plantation. *In* Robertson, G P, Herrera R and Rosswall T (Eds.). Nitrogen cycling in Ecosystems of Latin America and the Caribbean. Plant and Soil 67, 271–281.

Plant and Soil 67, 271–281(1982).   0032-079X/82/0672-0271$01.65.
© 1982 Martinus Nijhoff/Dr W. Junk Publishers, The Hague.

Dynamics of nitrogen in a shaded cacao plantation

Dinamica del nitrógeno en un cacaotal bajo sombra

M. B. M. SANTANA and P. CABALA-ROSAND
Centro de Pesquisas do Cacau, CEPLAC, Ilhéus, Bahia, Brazil

Key words   Cacao Erythrina Leaching Mineralization N-cycling Nitrification Shade trees

Abstract   Studies of nitrogen mineralization and leaching were conducted in the cacao-growing region in the south of Bahia, Brazil, on plots fertilized with N, P and K and on plots without fertilizer in plantations 30–40 yrs old on CEPEC soil (Tropudalf) over a period of one year. Mini-lysimeters were installed at depths of 10, 20 and 40 cm and the leachate was collected weekly or after heavy rain. Net mineralization was measured in soil samples taken at depths of 0–5 and 5–15 cm and incubated for 30 days in plastic bags placed at the site of collection.

The degree of leaching was correlated with the amount of rainfall and, although it is difficult to quantify the losses per unit area, we estimate that these losses are minor. Ammonification and nitrification were both high during most of the year; nitrification was very rapid and was especially intensive on the fertilized area. Analyses of Erythrina and cacao litter show that these components make a considerable contribution to the nitrogen recycled in a cacao plantation. High concentrations of total nitrogen were detected in soil samples taken close to shade trees and, on average, the soil of shaded areas had more than 480 mg N kg soil^{-1} than soil of non-shaded areas. Removal of nitrogen in harvest can also be considerable. It is advisable to take nitrogen-cycle data into account when compiling tables of fertilizer recommendations.

Resumen   Se realizaron investigaciones sobre la mineralización y la lixiviación de nitrógeno en parcelas fertilizadas con N, P y K y en parcelas sin fertilizar en plantaciones de cacao de 30 años en la región sur de Bahia, Brasil. Los suelos del cacaotal eran CEPEC (Tropudalf), comunes en la zona. Las mediciones se realizaron durante un año. Se instalaron minilisímetros a 10, 20 y 40 cm de profundidad y se colectó el agua lixiviada semanalmente o después de intensas lluvias. La mineralización neta se midió en muestras de suelo tomadas a 0–5 y 5–15 cm de profundidad colocadas nuevamente en bolsas plásticas en el sitio de colecta. El grado de lixiviación se correlacionó con la cantidad de precipitación y aun cuando no es posible cuantificar las pérdidas por unidad de área, se estimó que estas pérdidas eran de menor cuantía. Tanto la tasa de amonificación como la de nitrificación fueron altas durante la mayor parte del año; la nitrificación fué particularmente intensa en el área fertilizada. Los análisis de la hojarasca fresca de Erythrina y de los cacaoteros mostraron que estos componentes contribuyen notablemente al ciclo del nitrógeno en la plantación de cacao. Se detectaron altas concentraciones de nitrógeno en muestras de suelo tomadas cerca de los árboles de sombra; en promedio los suelos de la zona sombreada contenían 480 mg N kg^{-1} suelo por encima del promedio de los suelos en plantaciones sin sombra. La cantidad de nitrógeno exportado por cosecha es notable. Se recomienda tomar en consideración la información procedente de los ciclos de nitrógeno para formular recomendaciones de fertilización.

Introduction

Despite the fact that in the humid tropics rainfall constitutes an important source of nitrogen, biological fixation and the application of fertilizers are the most important sources of supply of this element to most crops. Industrial fixation of nitrogen for fertilizers consumes large amounts of energy, and fertilizer inputs are expensive. This has limited the use of nitrogenous fertilizers in tropical areas and has awakened a growing interest in field investigations of the dynamics of nitrogen in the soil, mainly in relation to biological nitrogen fixation. The amounts of nitrogen fixed by this process exceed the amounts produced industrially[11].

Without underestimating the part played by free-living micro-organisms and the significant role of some rhizosphere associations as described by Döbereiner et al.[13], Dommergues et al.[14] and Yoshida and Arcanjas[35], the preponderant role of the bacteria of the genus Rhizobium stands out; in symbiosis with legumes they form the main natural process of capture of atmospheric nitrogen. This symbiosis benefits not only the legume: its effect can extend to other crops in the rotation or to plants growing in association with legumes. The first of these systems is most commonly employed in Brazil; usually a legume precedes the desired crop. In the state of Paraná, for example, cultivation of *Lupinus albus* prior to maize is equivalent to the application of 80 kg N per hectare[24].

In pastures or in perennial crops, association with legumes plays an important role in reducing needs for nitrogen fertilizers. Experimental evidence indicate that cacao under leguminous shade trees does not respond to applications of nitrogen[17]. On hydromorphic soils shading with *Erythrina glauca*, both improves soil drainage and enriches the surface layer of the soil[8].

These results indicate that investigations on cycling of nutrients, particularly of nitrogen, are important for cacao cultivation, enabling satisfactory levels of production to be obtained with management practices requiring the addition of only small quantities of fertilizers. The number of detailed investigations on this subject is in fact still very small; the outstanding one among them is the study conducted by Boyer[3] in a cacao plantation in the Cameroons.

Since its habitat in the Amazonian forests is normally under shade, cacao was until recently considered to be a shade-loving plant. However, investigations have clearly shown that the best yields are obtained at high intensities of light and fertilizers[5,10,23]. The interaction between shade, natural soil fertility and use of fertilizers is well represented by the scheme suggested by Alvim[1], in which the optimum production will result from the combination of sparse or no shade on soils of high fertility or those adequately fertilized.

Thus there has been increasing interest in the last few years in growing cacao trees at high light intensities with intensive use of fertilizers. For example, in the south of the state of Bahia, the main cacao-producing zone of Brazil, this new system of production now comprises about 50 per cent of existing cacao plantations[7].

Recently, however, the high cost of nitrogen fertilizers has started to restrict their use. One alternative to fertilizers involves using the tolerance of cacao trees to shade resourcefully. Although cacao production under shade is lower than full-light production, management costs are also lower because shaded plantations require fewer applications of fertilizers.

The present investigation is a preliminary contribution to the study of the cycling of nitrogen in cacao crops. We focus primarily on the removal of nitrogen in the harvested crop, the degree of nitrogen mineralization and leaching in the soil, and the nitrogen contribution of leguminous shade species via leaf litter.

Materials and methods

Leaching of nitrogen

Nitrogen leaching was measured in a plantation of 30–40 years old cacao trees, shaded by native tree species, and situated on Cepec soil (Tropudalf). Within this plantation, an area of 0.5 ha was subdivided into two equal plots, to one of which was applied 40, 90 and 60 kg ha^{-1} of N, P_2O_5 and K_2O respectively in the form of urea, triple superphosphate and potassium chloride. Subsequently, in each subplot four trenches 1.0 m square $\times$ 0.80 m deep were excavated, and mini-lysimeters were inserted into two of the walls for collection of leachates at depths of 0.10 m, 0.20 m and 0.40 m.

The mini-lysimeters[19,29] were constructed of rigid PVC pipes 7.5 cm diameter and 30 cm long. Apertures were made along one side and were covered with a nylon net, and the ends of the pipes were closed except for a drainage hole in one end which emptied leachate into a 5 l plastic container. Each container had 10 drops of toluene added prior to leachate collection to reduce the growth of micro-organisms.

The leachates were collected weekly, or after a period of rainfall heavy enough to fill the containers, their volumes were measured, and aliquots were taken for estimation of NH_4^+-N and NO_3^--N. Ammonium was evaluated by the calorimetric method of Weatherburn[33] and nitrate by the salicylic acid method described by Cataldo et al.[9] For nitrate analysis it was necessary to evaporate 9.8 ml of leachate to dryness in a beaker; a further 0.2 ml of the leachate was then added to dissolve the evaporate, and NO_3^--N determined in this volume. Throughout the duration of the experiment, rainfall data were recorded daily.

Mineralization

The degree of soil-N mineralization was measured by incubating samples in the field and in the laboratory. In the same area that the leaching studies were conducted, samples of soil were collected at depths of 0–5 and 5–15 cm. The samples were placed in polyethylene bags and incubated for 30 days at the same depths at which they were collected according to the method of Runge[27] and were analysed for NO_3^--N and NH_4^+-N, the results being compared with those of non-incubated samples collected at the same time and in the same manner.

For the determination of mineralization under laboratory conditions, soil samples collected at the same depths were incubated aerobically at 30°C for 28 days according to the method of Keeney and Bremner[21]. During the incubation period sub-samples were withdrawn at 7-day intervals for analysis of NO_3^--N and NH_4^+-N.

For samples incubated in the field, determinations were made of NH_4^+-N by the wet distillation process described by Bremner[4], and of NO_3^--N by the method of Cataldo et al.[9] For laboratory samples, NH_4^+-N and NO_3^--N were evaluated by methods described by Weatherburn[33] and Cataldo et al.[9], respectively.

Components of leaf litter and total N in the soil

From a 9-year-old cacao plantation on the same soil encompassing an area of 1.7 ha under the shade of *Erythrina glauca*, *Erythrina peoppigiana*, *Spondias lutea*, *Genipa americana* and *Bertholletia excelsa*, and from an adjacent area of unshaded cacao plantation, 0–20 cm deep soil samples were collected close to and distant from the shade trees. At the same points, collections were made of recently fallen leaves both of the cacao trees and of the shade trees, for determination of total Kjeldahl nitrogen. This sampling was carried out only once.

For the Erythrina trees, nodules were counted and total nitrogen was determined on samples of nodules and seeds.

Results and discussion

The amounts of NH_4^+–N and NO_3^-–N leached were proportional to amounts of rainfall (Fig. 1). On the fertilized area generally less nitrogen was leached than on the non-fertilized area. It is possible that the simultaneous application of nitrogen, phosphorus and potassium increased the development of the cacao roots, especially the rootlets, thus increasing nutrient-absorbing surfaces and in this way decreasing the amounts of nitrogen available for leaching. Overrein[25] measured leaching with lysimeters on a plot fertilized with ^{15}N-urea and observed limited leaching losses, attributing this to retention of these fertilizers in the surface organic layer owing to the urea itself or compounds resulting from its hydrolysis.

Although the methods used to measure leaching do not allow extrapolations

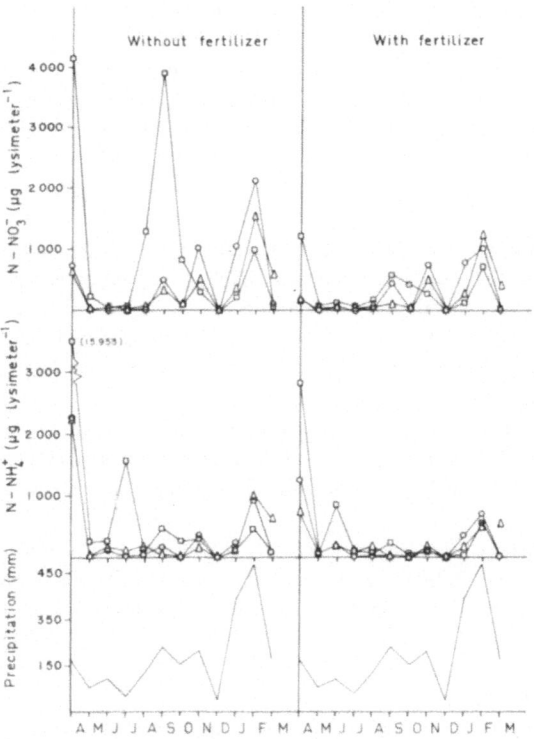

Fig. 1. NO_3–N and NH_4–N leached from a Cepec soil (Typic Tropudalf) at three depths in areas with and without fertilizer applications, and monthly rainfall averages.

to unit areas, results suggest that the nitrogen leaching losses do not seriously affect nitrogen availability to the cacao.

Fig. 2 shows the results of the mineralization experiments in the field. For both the incubated and not incubated (control) samples, the degree of nitrification was considerable; the ammonium nitrogen produced was rapidly transformed into nitrate. In general, ammonification was higher ($p < 0.01$) in the surface layer and also in the period from May to June. Nitrate production showed great monthly variability, with peaks in the periods May to September and February to March and a marked reduction in the period from October to December. The reduction in nitrification from October to December coincides with the period of most leaf fall and shoot production on the cacao trees. During this phase a higher intake of nutrients is to be expected, especially of nitrate-N[26]. A significant reduction in the concentration of NO_3–N in the soil results. In this period, the higher deposition of litter both from the cacao trees and from the shade trees could also bring about C/N ratios high enough to increase the demand for nitrogen by the micro-organisms present in the soil[16].

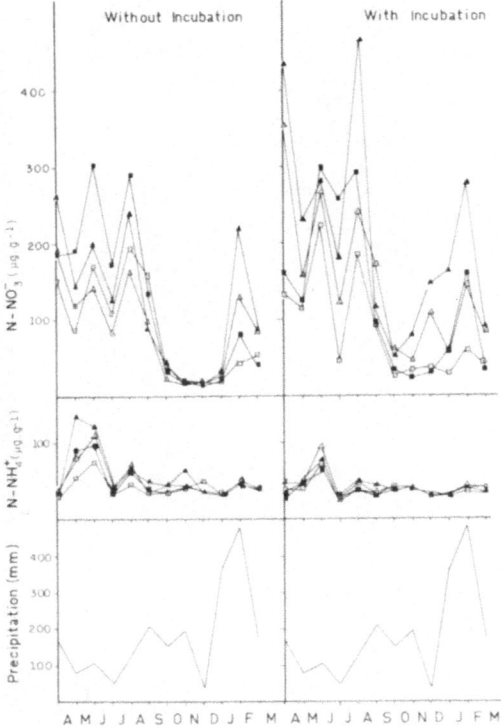

Fig. 2. NO_3–N and NH_4–N in samples of Cepec soil (Typic Tropudalf) collected at two depths and subjected or not subjected to incubation.

It was also observed that larger amounts of nitrate-N were encountered in the surface layer of the soil after the incubation period (Fig. 2). Meanwhile, the values for NH_4–N were low in both the incubated and non-incubated samples. The consumption of this form of nitrogen by the microorganisms[34] and the rapid conversion of ammonium into nitrate, explain this finding. Additionally, urea fertilization had a positive effect on nitrification ($p < 0.01$), possibly by stimulating the growth of nitrifying populations on account of an increase in pH[22].

The results for mineralization under laboratory conditions are in reasonable agreement with results for mineralization under field conditions[30], particularly with regard to the rates of ammonification and nitrification, and also to the low values for NH_4–N encountered in the fertilized samples.

Considering the high rates of mineralization during the greater part of the year, and also the quantity of leaves that fall from the cacao trees, which according to Alvim (personal communication) and Boyer[3] is *ca.* 5–8 × 10^3 kg dry matter ha^{-1} yr^{-1}, approximately 50–90 kg N ha^{-1} will be deposited annually in the surface

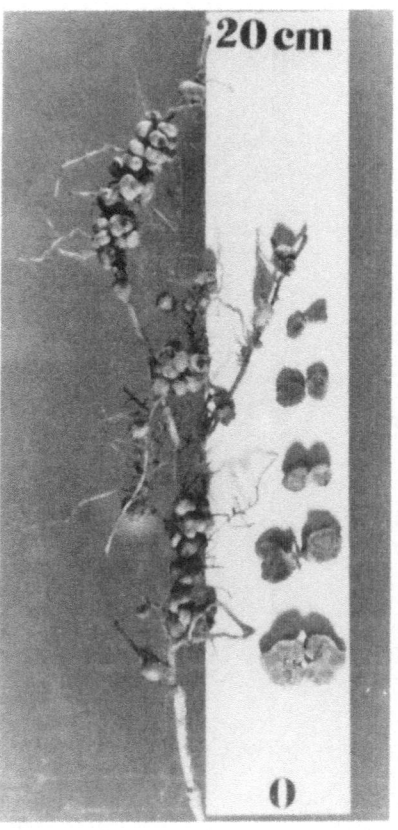

Fig. 3.   Details of the interior of Erythrina nodules and the extent of nodulation.

layer of the soil and will be available to the roots of the cacao trees *via* mineralization. Additionally, in plantations shaded with *Erythrina* spp., the annual fall of leaves, branches, flowers and seeds of this legume, with high nitrogen contents (Table 1), contributes even further to the amount of nitrogen available to the cacao tree.

The percentages of total N in our soils collected distant from (9 m) and close to (3 m) the shade trees show marked differences, especially for Erythrina (Table 2). These results are in agreement with those presented by Cadima and Alvim[8] for a

Table 1. Percent nitrogen in the nodules and in other residues of *Erythrina glauca* and the quantity of nodules found on roots. na = not available. Nodules were collected at 10 cm depth and values presented represent the means of 8 replicates per sample area

Plant part	Area sampled	Number/ m^2	Nodules g/m^2	% N
Recently fallen leaves	na	na	na	2.74
Fallen flowers	na	na	na	3.32
Seeds	na	na	na	6.15
Nodules	1	497	3.28	3.44
	2	239	1.34	3.48
	3	313	1.38	4.83
	4	100	0.71	3.82
	5	101	0.93	4.04
	6	191	3.41	4.13

Table 2. Percent nitrogen in recently fallen leaves and in soil samples (0–20 cm depth) collected from a non-shaded area and from a shaded area at sites close to (3 m) and distant from (9 m) shade tree stems

Species	Percent N		Fallen leaves
	Soil		
	3 m	9 m	
Erythrina glauca and *E. poeppigiana*	0.322	0.245	2.74
Genipa americana	0.207	0.197	2.14
Spondias lutea	0.210	0.193	1.61
Others*	0.190	0.192	0.92
Average	0.232	0.207	1.85
Unshaded	0.171		
Cacao			0.91

* *Bertholletia excelsa* and an unidentified species of Bombacaceae.

20 + yr old stand of cacao shaded by Erythrina. Total N in soil samples collected at increasing distances from the base of Erythrina show a gradual decline which starts at a distance of 4.5 m (Fig. 4).

The average total-N content of soil in the shaded areas was 485 mg N kg^{-1} more than soil in non-shaded areas (Table 2). A similar but smaller difference (240–250 mg N kg soil^{-1}) was found by Santana and Morais[28] in a comparison of a cacao plantation shaded with *Erythrina vs* another shaded with bananas.

Erythrina has a deep root system with lateral roots spreading to 8–10 m in the adult phase[8]. Thus, when Erythrina are planted at the spacings commonly employed (24 × 24 m), a large part of the planted area is under the influence of both the canopy and the root system of the shade trees. Erythrina is leafless during July–September. Alvim and Alvim[2] concluded from phenological studies that *E. glauca* is extremely sensitive to photoperiodic and/or thermoperiodic stimuli, and that leaf fall is probably induced by short days. The flowering period of these trees also occurs between July and September[2]. During this period litter is very abundant; fallen leaves, branches, flowers and seeds can reach a total of 2000 kg dry matter ha^{-1}, double that estimated by Hardy[15] for the annual fall of leaves and flowers of this species. Within about two months these residues are almost completely humified, and can contribute 30–60 kg N ha^{-1} to the soil-N

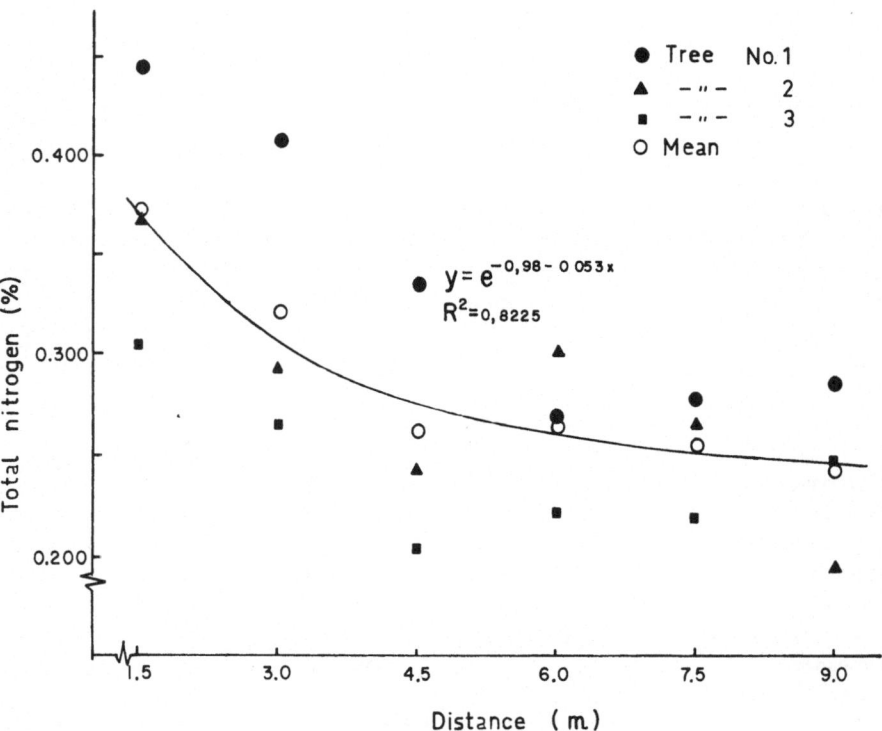

Fig. 4. Contents of nitrogen in soil at increasing distances from trunks of Erythrina.

Table 3. Contents of nitrogen in dry beans, husks and whole fruits of cacao

Plant part	N	Authors
Beans	22.0	Santana and Cabala (unpubl.)
kg N · 1000 kg dry beans⁻¹)	20.0	Zeller (Dierendonck[12])
	24.0	Urquhart[32]
	20.0	Kanapathy[20]
	20.4	Thong and Ng[31]
Husks	10.6	Thong and Ng[31]
(kg N · 1000 kg dry beans)	12.0	Santana and Cabala (unpubl.)
Fruits	16.0	Humphries[18]
(kg N · 1000 kg dry fruits)	15.6*	Kanapathy[20]

* Average from husks and beans

pool. Also following leaf fall, Rhizobium becomes inactive because of the lack of photosynthate and in consequence nodule senescence occurs. Nodules are very abundant in Erythrina (Fig. 3 and Table 1), and on decomposition transfer nitrogen to the soil.

Rainfall also constitutes an important source of nitrogen. According to Boyer[3], a cacao tree under moderate shade may receive $17 \text{ kg N ha}^{-1} \text{ yr}^{-1}$.

The main loss of nitrogen from a cacao plantation is probably through harvesting, and this aspect has been considered recently in view of the need for its replacement by fertilizers[6]. Analysis of the seeds and husks of cacao fruits collected in our study area revealed average values of 2.2% N and 1.2% N respectively, values similar to those reported by Humphries[18], Dierendonck[12], Urquhart[32], Kanapathy[20] and Thong and Ng[31] (Table 3). Because cacao husks generally remain in the plantation, the removal of nitrogen through harvesting is relatively low. An annual production of 1000 kg dry cocoa beans ha⁻¹ removes no more than $20–24 \text{ kg N ha}^{-1}$.

The content of N in the branches and trunk of the cacao tree does not exceed *ca.* 3.5 kg N per ha[3]. Recently, in detailed investigations of the content of nutrients in the various parts of the cacao tree at different stages of development, Thong and Ng[31] showed that in the productive phase, with annual yields of 1000 kg dry beans ha⁻¹, about 469 kg N ha⁻¹ is taken up and that of this amount only 6% is removed with the yield.

Conclusions

1. There is strong evidence that in cacao trees under Erythrina shade, the entry of N into the system through litter plus input *via* rainfall can be greater than those removed even by harvests of 1500 kg dry beans ha⁻¹.

2. Erythrina increases the N-content of soil and this enrichment extends to a considerable distance from the trunks of these trees.

3. In the soil under study, the losses of N through leaching were small and apparently did not limit the availability of this nutrient to the cacao trees.

4. Mineralization and production of NO_3–N is fairly intensive, and occurs almost throughout the year.

5. The balance of gains and losses of N in cacao plantations needs to be studied carefully with the aim of improving fertilizer recommendations.

References

1   Alvim P de T 1977 Cacao. *In* Alvim P de T and Kozlowsky T T (Eds.). Ecophysiology of Tropical Crops, pp 279–313. Academic Press, New York.

2   Alvim P de T and Alvim R 1978 Relation of climate to growth periodicity in tropical trees. *In* Tropical Trees as Living Systems. Tomlinson, P B and Zimmerman M H (Eds.). Cambridge, Cambridge University Press, pp 445–464.

3   Boyer J 1973 Cicles de la matière organique et des elements mineraux dans une cacaoyère camerounaise. Café, Cacao, Thé 17, 3–24.

4   Bremner J M 1965 Inorganic forms of nitrogen. *In* Black C A (Ed.). Methods of Soil Analysis, pp 1179–1237. American Society of Agronomy, Madison, Wisconsin.

5   Cabala-Rosand F P, Pires de Prado E, Ruy de Miranda E and Leondy de Santana C J 1971 Effect of shade removal and manuring on the production of the cacao tree in Bahia. Rev. Theobroma (Brasil) 1, 43–57. (In Portugese, English summary.)

6   Cabala-Rosand P, Miranda E R, Santana M B M and Santana C J L 1975 Nutritional requirements and the use of fertilizer in cacao. Ilhéus, Ba., Brasil. CEPLAC/CEPEC Boletim Técnico, no 30. 59 p. (In Portugese.)

7   Cabala-Rosand P, Santana M B M and Santana C J L de 1980 Aspectos da adubação fosfatada em alguns cultivos do Nordeste do Brasil. *In* Reunião Grupo de Trabalho sobre Adubação Fosfatada no Brasil. Brasilia, Embrapa, Outubro 1980. Centro de Pesquisas do Cacao, CEPLAC. 52 p.

8   Cadima-Zevallos A and Alvim P de T 1967 Influencia del arbol de sombra *Erythrina glauca* sobre algunos factores edafológicos relacionados con la produción del cacaotero. Turrialba (Costa Rica) 17, 330–336.

9   Cataldo D A, Haroon M, Schrader L E and Youngs v L 1975 Rapid colorimetric determination of nitrate in plant tissue by nitration of salicylic acid. Commun. Soil Sci. Plant Anal. 6, 71–80.

10 Cunninghan R K and Arnold P W 1962 The shade and fertilizer requirements of cocoa (*Theobroma cacao*) in Ghana. J. Sci. Food Agric. 13, 213–221.

11 Delwiche C C 1970 The nitrogen cycle. Sci. Am. 222, 137–146.

12 Dierendonck F J E van 1959 The manuring of coffee, cocoa, tea and tobacco. Centre d'Etude de l'Azote, Genève. 205 p.

13 Döbereiner J, Day J M and Dart P J 1973 Rhizosphere associations between grasses and nitrogen fixing bactéria. Effects of O_2 on nitrogenase activity in the rhizosphere of *Paspalum notatum*. Soil Biol. Biochem. 5, 157–160.

14 Dommergues Y, Balandreau J, Rinaudo G and Weinhard P 1973 Non-symbiotic nitrogen fixation in the rhizosphere of rice, maize and different tropical grasses. Soil Biol. Biochem. 5, 83–89.

15 Hardy F 1961 Suelos de cacao: aspectos pedológicos. *In* Hardy F (Ed.). Manual de Cacao, pp 119–134. Instituto Interamericano de Ciências Agrícolas, Turrialba, Costa Rica.

16 Harmsen G W and van Schreven D A 1955 Mineralization of organic nitrogen in soil. Adv. Agron. 7, 299–390.

17 Havord G 1959 The nutrition and shade requirements of cacao. Turrialba, Costa Rica 9, 138–148.

18 Humphries E C 1940 Growth rate and mineral intake by the pod. In Port of Spain, Trinidad, ICTA. A Report on Cacao Research, 1939. pp 43–46. St. Augustine, Trinidad.

19 Jordan C F 1968 A simple, tension-free lysimeter. Soil Sci. 105, 81–86.

20 Kanapathy K 1976 Guide to Fertilizer Use in Penisular Malaysia. Ministry of Agriculture and Rural Development of Malaysia, Kuala Lumpur. 160 p.

21 Keeney D R and Bremner J M 1966 Comparison and evolution of laboratory methods obtaining an index of soil nitrogen availability. Agron. J. 58, 498–503.

22 Morril L G and Dawson J E 1967 Patterns observed for the oxidation of ammonium to nitrate by soil organisms. Soil Sci. Soc. Am. Proc. 31, 757–760.

23 Murray D B 1954 A shade and fertilizer experiment with cacao III. In St. Augustine, Trinidad and Tobago. ICTA. A Report on Cacao Research 1953, pp 30–37. St. Augustine. Trinidad.

24 Muzilli O 1980 The use of phosphate fertilizers in Paraná State. In Reunião Grupo de Trabalho sobre a Adubação fosfatada no Brasil. Outubro 1980, Brasília, Embrapa. Fundação Instituto Agronômico do Paraná. (In Portuguese.)

25 Overrein L N 1969 Lysimeter studies on tracer nitrogen in forest soil: 2. Comparative losses of nitrogen through leaching and volatilization after the addition of urea – ammonium – and nitrate-N[15]. Soil Sci. 107, 149–159.

26 Rodriguez R M, Carvajal J F, Machicado M and Jimenez E 1963 Requerimentos nutricionales del cacaotero durante un ciclo anual. Rev. Cacao (Costa Rica) 8(4), 1–7.

27 Runge M 1971 Investigation of the content and the production of mineral nitrogen in soils. In Integrated Experimental Ecology, pp 191–202. Berlin-Heidelberg-New York, Springer-Verlag.

28 Santana M B M and Morais F I 1977 The effect of shading with 'eritrina' (Erythrina fusca) on the nitrogen levels of soils planted with cocoa. In Conference on Limitations and Potentials for Biological Nitrogen Fixation in the Tropics, Brasília, Brasil, Julho 1977. Proceedings, Abstracts p. 345.

29 Santos O M and Medeiros A G 1980 Redistribution and leaching of copper fungicide in a cacao plantation after aerial and ground spraying. Rev. Theobroma (Brasil) 10, 69–78. (In Portugese, English abstract.)

30 Santos O and Santana M B M 1980 Nitrogen mineralization in soils used for cacao cultivation in the Bahia: I CEPEC Modal, Itabuna Modal and Água Sumida soils. Rev. Theobroma 12, in press. (In Portugese, English abstract.)

31 Thong K C and Ng W L 1978 Growth and nutrient composition of monocrop cocoa plants and coconuts, Kuala Lumpur, Malaysia, 1978. Kuala Lumpur, Malaysia, 25 p. (Mimeographed.)

32 Urquart D A 1963 Cacao. Traduçaõ do inglês por Juvenal Valério. Turrialba, Costa Rica, IICA. 322 p.

33 Weatherburn M W 1967 Phenol-hypochlorite reaction for determination of ammonia. Anal. Chem. 39, 971–974.

34 Weeraratna C S 1979 Pattern of nitrogen release during decomposition of some green manures in a tropical alluvial soil. Plant and Soil 53, 287–294.

35 Yoshida T and Arcanjas R R 1973 The fixation of atmospheric nitrogen in the rice rhizosphere. Soil Biol. Biochem. 5, 153–156.

Plant and Soil 67, 283–291 (1982).   0032-079X/82/0672-0283$01.35.   SU-25
© 1982 *Martinus Nijhoff/Dr W. Junk Publishers, The Hague*

Nitrogen fixation in a Mexican coffee plantation

Fijación de nitrógeno en una plantacion de café en México

J. P. ROSKOSKI*
Instituto Nacional de Investigaciones Sobre, Recursos Bióticos, Apartado Postal 63, Xalapa, Veracruz, México.

Key words Banana trees Coffee *Inga jinicul* *Inga vera* Mexico N-cycling N$_2$-fixation Orange trees

Abstract Fertilizer studies in Mexico indicate that coffee production can be stimulated by added nitrogen. One traditional method of coffee cultivation employs leguminous trees for shade, but these species may also play an important role in coffee production by biologically fixing nitrogen.

The presence and importance of nitrogen fixation was evaluated in four systems: coffee only, coffee plus the leguminous shade tree *Inga jinicuil* Schletchter, coffee plus the leguminous tree *Inga vera* H.B. and K., and coffee plus banana and orange trees. In all systems coffee leaves with epiphylls, wood litter, soil, roots, and root nodules were assayed for nitrogen fixing activity with the acetylene reduction technique.

All components of these systems exhibited activity except roots. Total apparent fixation was highest in the *Inga jinicuil* site, and equivalent to > 40 kg N ha^{-1} yr^{-1} assuming a 3:1 C$_2$H$_2$:N$_2$ ratio. The activity was primarily associated with *Inga jinicuil* nodules. Apparent fixation in the other three sites was less than 1 kg N ha^{-1} yr^{-1}. Nitrogen fixed in the *I. jinicuil* site was 53% of the average amount of fertilizer nitrogen applied annually, suggesting that fixation by non-crop legumes can be an important nitrogen source for coffee agro-ecosystems.

Resumen En estudios de fertilización en México se ha encontrado que la produccion de café puede ser estimulada por la adición de nitrógeno. Uno de los métodos tradicionales de cultivo de café es el de proveer al cultivo de sombra por medio de árboles de la familia Leguminosae los cuales pueden jugar un papel importante en la producción del café a través de la fijación biológica de nitrógeno.

La presencia e importancia de fijación de nitrógeno se evaluó en cuatro sistemas: café solo, café con *Inga junicuil* Schletcher, café con *Inga vera* H.B. & K. y café con bananos y naranjos. En todos los sistemas las hojas de café con epifilos, restos leñoso en el mantillo, suelo, raices y nódulos fueron estudiados por el método de reducción de acetileno para fijación de nitrógeno. Todos estos componentes presentaron actividad fijadora excepto las raices. El total de fijación aparente fué mayor en el sitio con *I. jinicuil*, equivalente a > 40 kg N ha^{-1} ão^{-1} asumiendo una relación de 3:1 para C$_2$H$_2$:N$_2$. La actividad estaba relacionada con los nódulos de *I. jinicuil*. La fijación aparente en los otros tres sitios fue menos de 1 kg N ha^{-1} año^{-1}. La fijación de nitrógeno para el sitio con *I. jinicuil* equivale al 53% de la cantidad promedio de fertilizante nitrogenado aplicado anualmente, lo cual puede tomarse como indicativo de que la fijación de nitrógeno por leguminosas adicionales al cultivo puede ser una fuente importante de nitrógeno para el café.

* Present address: Plant Sciences Department, University of Arizona, Tucson, Arizona 85721, USA.

Introduction

Coffee is currently cultivated in thirteen of the thirty-two states of the Republic of Mexico[7]. This large geographical area includes highly diverse topography, soils, and climate.

Consequently, coffee cultivation practices differ widely. However, one practice employed throughout the region is nitrogen fertilization, since in all areas nitrogen is essential for high coffee production[7]. Another cultivation technique common to some areas is the use of leguminous trees for shade. However, these species may also affect coffee production by adding nitrogen to the systems via symbiotic fixation.

Because little information about this potential nitrogen source exists, I initiated a comparative study to examine nitrogen fixation in coffee sites with different degrees and types of shade. Specific objectives were to determine in what components of each site nitrogen fixation occurs and to quantify fixation in each site.

Methods

This study formed part of a larger ecosystem study conducted by researchers from the Instituto Nacional de Investigaciones sobre Recursos Bioticos and took place in a commercial coffee plantation near the city of Xalapa, Veracruz (19°27' N, 96°57' W, 1225 m elevation). The climate of the area is semi-hot and humid with warm summers and cool winters[5]. Annual mean temperature is 19 ($\pm$ 2)°C, and annual precipitation averages 1758 ($\pm$ 193) mm.

Four sites that represent the four planting regimes common in the area were chosen for study. The sites differed primarily in the degree and type of shade present. Site 1 contained coffee and herbaceous ground cover; site 2 had coffee and the leguminous shade tree *Inga jinicuil* Schletchter, site 3 contained coffee plus *Inga vera* H.B. & K., another leguminous shade tree; and site 4 contained coffee plus banana (*Musa sapientum* L.) and orange (*Citrus sinensis* Osb.). Coffee cover and density were similar for all four sites. Shade species cover ranged from 0 in site 1 to over 40,000 m^2 ha^{-1} for site 4[8].

Nitrogen fixing activity was evaluated for coffee leaves with epiphylls, for wood litter, for soil to a depth of 15 cm, for fine root samples from coffee, shade trees, and herbs in all four sites, and for root nodules from site 2. Field sampling methods varied for each of the above substrates. All samples were assayed for nitrogen fixing activity using the acetylene reduction technique[6]. A 3:1 ratio for $C_2H_4 : N_2$ fixed was assumed for all fixation estimates.

Once a month from February 1978 through February 1979, twenty coffee leaves with visible epiphyllic growth were randomly collected from each site and assayed for nitrogen fixing activity as described in Roskoski[16]. In addition, the number of coffee leaves with epiphylls in each site was determined in May 1978 and February 1980. For these estimates, three 25 m^2 plots were chosen randomly in each site. Each contained four coffee plants. Each plant was then divided into four equal longitudinal sections, and the total number of coffee leaves and the number of coffee leaves with epiphylls were counted for one randomly-chosen section. These values, together with the epiphyll fixation data, were used to calculate fixation by coffee leaf epiphylls on a ha^{-1} yr^{-1} basis.

Soil cores (3.5 cm diameter × 15.0 cm depth) were collected every two months from May 1978 through May 1980 at various distances from randomly chosen coffee trees in each site, and assayed for nitrogen fixing activity.

Nodule biomass for *I. jinicuil* and changes in nodular acetylene reduction activity were monitored monthly from March 1979 through October 1980. Once each month ten nodule samples were randomly collected every three hours over a thirty-three hour period and assayed for nitrogen fixing activity[17].

Results and discussion

Nitrogen fixation in rhizospheres and wood

No rhizosphere samples from any of the sites exhibited a capacity to fix nitrogen. The few wood litter samples that were assayed fixed less than 1 nm N g dry wood^{-1} day^{-1}. However, since site 1 lacks shade trees and all the dead wood in the three shaded sites is periodically removed and sold in the local markets for firewood, the amount of wood litter in all four sites was small. Consequently nitrogen input from fixation in wood litter is insignificant in all systems.

Nitrogen fixation by coffee leaf epiphylls

Nitrogen-fixing leaf epiphylls were found in all sites and N-fixation varied monthly (Table 1). The monthly variation in nitrogen fixation may be related to phenological changes in the coffee plant. The highest rates of nitrogen fixation occurred in January and February, when the number of mature epiphyll-covered leaves was also greatest. In late February, coffee leaf-replacement began and fixation decreased as young, uncolonized leaves formed an increasingly larger portion of the leaf population. Leaf-replacement was complete by May, and the new leaves rapidly colonized by epiphylls. However, fixation remained low from June through September, perhaps because heavy summer rains that contain nitrogenous compounds[14] supplied sufficient nitrogen to the phylloplane during

Table 1. Monthly N_2 fixation by coffee epiphylls. Data shown is for site 2; similar rates and patterns were observed for other sites. Values are extrapolated from single monthly measurements of fixation by active leaves (20 replicates). After Roskoski[16]

Month	g N fixed ha^{-1}
January	0.37
February	0.40
March	0.14
April	0.06
May	0.02
June	0.02
July	0.01
August	0.01
September	0.02
October	0.04
November	0.08
December	0.22

this period to reduce the need for nitrogen fixation. In addtition, summer rains may dilute the concentration of carbohydrates on leaf surfaces[18], thus reducing the primary energy source for nitrogen fixation. Increased fixation in the fall and winter may be caused by the stimulatory effects of nutrients that leach from aging leaves[10, 12].

Significant differences were found among the sites for the amount of nitrogen fixed by epiphylls, which varied from 0.7 to 1.4 g N ha^{-1} yr^{-1}, with higher amounts in the shaded sites due to a greater number of leaves with epiphylls[16]. However, these low fixation values suggest that nitrogen fixation by coffee epiphylls is not an important source of nitrogen for any of these sites.

Nitrogen fixation in soil

No inter-site differences were detected in the total amounts of nitrogen fixed in soil samples from these sites. However, fixation did change significantly through time (Fig. 1). During the first nine months of the study rates of nitrogen fixation decreased. After January 1979, fixation began to increase and continued to do so until September. A decrease in activity in November was then followed by a dramatic increase in January 1980. The effects of the late-November fertilizer application on this increase are presently under investigation.

Calculated fixation for the soil in these four sites, about 0.5 kg ha^{-1} yr^{-1}, is less than that reported for other areas[11, 19, 20]. Relatively high nitrogen levels in the soil (50 ppm NH$_4$ and 0.5% total N (Jimenez pers. comm.)), may have depressed soil nitrogen fixation[15] in these sites.

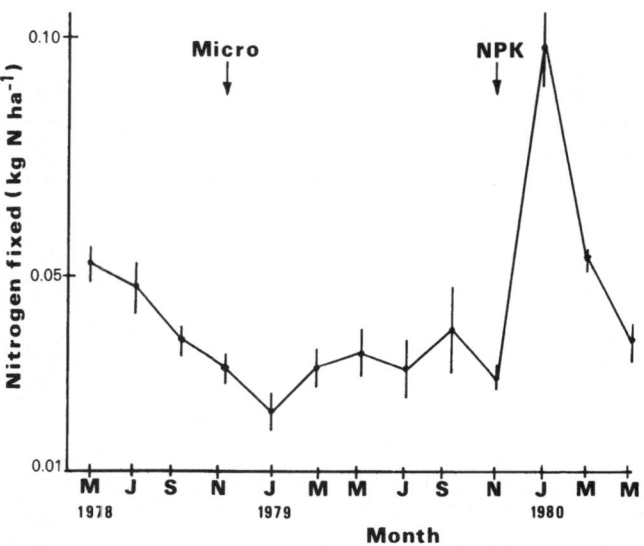

Fig. 1.   Seasonal nitrogen fixation in soils (to 15 cm) of 4 coffee plantations with different shade managements. Values plotted are means (± standard errors, n = 4 plantations). Arrows indicate times of NPK and micronutrient (Mo, Bo, Co, Cu, Zn, Mn) fertilizer applications.

Nitrogen fixation by nodules

Examination of soil cores from the *I. jinicuil* and *I. vera* sites established that nodules occurred only on *I. jinicuil* roots. Furthermore, intensive sampling around one coffee tree in the *I. jinicuil* site suggested that nodules of this species are distributed in clumps around the trunks of coffee trees. This information led to a systematic sampling of the *I. jinicuil* site, which verified that fine roots and nodules of *I. jinicuil* are concentrated around the coffee trunks within or just below the litter (L + F + H) layer. Average nodule biomass in the *I. jinicuil* site was 70.7 kg dry nodules ha^{-1}. Cultivation practices routinely employed in the plantation may be responsible for this distribution pattern. Prior to the initiation of the rainy season in June, the leaf litter under each coffee tree is scraped aside to a radius of approximately 50 cm and N–P–K fertilizer is applied. Since nodules require high levels of phosphorus for growth and nitrogen fixation[2,4], temporarily high levels of available phosphorus under coffee trees could promote rooting and nodulation of *I. jinicuil* after fertilization.

Analyses of soil samples collected at different distances from coffee trunks showed that nitrogen and phosphorus levels decreased with distance from the trunk (Table 2). Although nitrogen depresses nodulation and nitrogen fixation[1,3,13], these data suggest that at the level of fertilizer used in this plantation the stimulatory effect of phosphorus overrides the inhibitory effect of nitrogen.

Observed variations in monthly rates of nodule N-fixation (Fig. 2) appear related to phenological changes in *I. jinicuil*. Flowering and leaf fall occur during the spring dry season (March through May), and N-fixation declined following flowering in both May 1979 and April 1980. This may be due to decreased levels of photosynthate at this time. Pods developed during June and July 1979 and fixation increased during this period to a yearly maximum in July. After the pods dropped in August fixation decreased. From September through December a second dry season occurs. As soil moisture dropped during this period (Jimenez, pers. comm.), *I. jinicuil* again shed its leaves and fixation declined. By January the *I. jinicuil* canopy was again closed and nodular activity had begun to increase. The increase continued until the onset of the first yearly dry season in March. From March to September the activity pattern for 1980 was similar to that for

Table 2. Nodule biomass and soil chemistry 30, 60, and 90 cm from coffee stems. Soil values (van Kessel and Roskosi[21]) are means of 3 replicates; nodule values (Roskoski[17]) are means of 25–100 replicates. Values in parentheses are standard errors

Distance from stem (cm)	Nodules (g m^{-2})	N (%)	Available-P (mg P kg dry soil^{-1})	pH
0–30	86.50 (19.04)	0.16 (0.02)	64 (2)	4.8 (0.1)
30–60	15.74 (5.54)	0.15 (0.02)	54 (3)	5.2 (0.1)
60–90	6.77 (4.00)	0.12 (0.01)	44 (3)	5.8 (0.1)

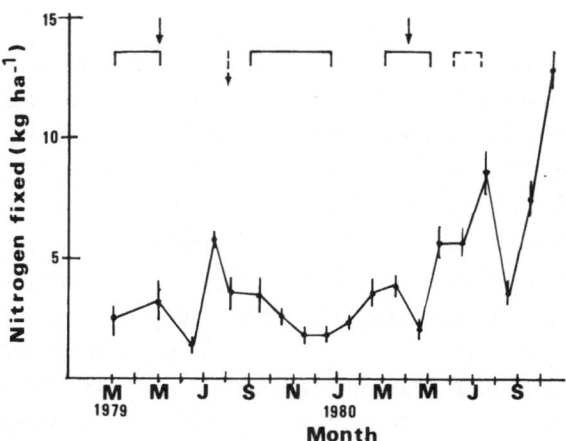

Fig. 2.   Seasonal nitrogen fixation by nodules of *Inga jinicuil* in a coffee plantation. Values plotted
are means (± standard erros). Arrows indicate times of flowering (solid line) or pod-drop (broken
line); horizontal brackets indicate periods of leaf fall (solid line) or defoliation (broken line).

1979. However, a severe insect defoliation in June and July 1980 removed all
foliage from the trees and the immature pods abscissed. Consequently, no fruits
were produced in 1980. New leaf production began in September at a time when
I. jinicuil leaves are normally being shed. Nitrogen fixation during this period
rather than decreasing as in 1979, increased dramatically to the highest levels
observed to date.

Total nitrogen fixation

Total N-fixation varied from 0.4 to 47 kg N ha^{-1} yr^{-1} (Table 1). Epiphyll
fixation differed among sites but was extremely low. The amount of nitrogen fixed
in the soil was similar for all sites and averaged about 0.5 kg N ha^{-1} yr^{-1}.
Nodular fixation by *I. jinicuil* was over 40 kg ha^{-1} yr^{-1}, and accounted for the
major differences in fixation among sites.

The importance of the large amounts of nitrogen fixed by *I. jinicuil* nodules will
depend on the amount of nitrogen the plantation receives from other sources and
whether the fixed nitrogen is available to the coffee plants.

From 1976 to 1980 the *I. jinicuil* site received between 45 and 157 kg of N
fertilizer ha^{-1} yr^{-1}, equivalent to an average input of 88 kg N ha^{-1} yr^{-1}.
Nitrogen inputs from *I. jinicuil* fixation could therefore equal 30 to 100% of the
amount of nitrogen applied *via* fertilizer or an average of 53%. At the same time,
nitrogen inputs from precipitation were about 10 kg N ha^{-1} yr^{-1}, less than 25%
of fixation inputs in this system. Nitrogen fixation by *I. jinicuil* then can represent

Table 3. Nitrogen fixation (kg N ha^{-1} yr^{-1}) in four coffee sites with different types of shade management

N-fixation source	Shade source			
No shade	I. jinicuil	I. vera	Citrus	
Epiphylls	0.0007	0.0014	0.0013	0.0011
Soil	0.410	0.550	0.550	0.470
Nodules	0	46.8	0	0
Total	0.411	47.3	0.551	0.471

a significant nitrogen input to the coffee ecosystem. Furthermore, the potential for even higher inputs exists: *I. jinicuil* has not been bred to maximize fixation, and levels of soil N in the study site were relatively high because of years of nitrogen fertilization. Since selective breeding can generally increase a species capacity to fix nitrogen and the presence of nitrogen can inhibit nodulation and nitrogen fixation[1,3,13], nitrogen fixation by *I. jinicuil* in this site may be an underestimate of the nitrogen fixing potential of this species.

Although the majority of the nitrogen fixed by *I. jinicuil* is initially incorporated into its own biomass, much of it becomes available for uptake by coffee trees and herbs after litterfall and mineralization. In addition, since *I. jinicuil* supplies part of its own nitrogen requirements *via* fixation, more fertilizer nitrogen is available to coffee trees in areas with *I. jinicuil* than in sites with non-nodulated shade trees. In fact, the *I. jinicuil* site in this study has higher coffee yields than an adjacent area with the non-nodulated *I. vera* trees[9], despite the *I. jinicuil* site's older[8] and subsequently lower-yielding (Instituto Mexicana del Cate) coffee plants. However, whether this yield difference can be ascribed to the added nitrogen from fixation has yet to be determined (Table 3).

Conclusions

Nitrogen fixation occurred in soil and in samples of epiphyll-bearing coffee leaves in four study sites that differed in shade management practices, but N-fixation occurred at low levels. In contrast, fixation by *I. jinicuil* nodules in a shaded site was over 40 kg N ha^{-1} yr^{-1}. Fixation in all three system-components varied through time, suggesting that long-term studies are required to achieve reliable estimates for annual N-fixation. The nitrogen input from nodular fixation by *I. jinicuil*, when compared to other nitrogen inputs associated with management activities, can represent a major source of nitrogen for the coffee ecosystem. The amount of nitrogen fixed by *I. jinicuil* can apparently be affected by simple management techniques such as phosphorus fertilization. However, more experimental field studies are needed before the nitrogen-fixing potential of *I. jinicuil* and other tree legumes in intercropping systems can be realized.

Acknowledgements I thank R. Alarcon, G. Castilleja, J. Calderon, R. Hernandez, M. Koterba, J. Montano, and especially C. van Kessel for help with lab and field work. Financial support was provided by a Rockefeller Foundation Postdoctoral Fellowship in Environmental Affairs and the Instituto Nacional de Investigaciones sobre Recursos Bioticos.

References

1   Dart P J 1974 The infection process. *In* Quispel A (Ed.). The Biology of Nitrogen Fixation, pp 381–429. Amsterdam: North Holland Pub. Co.

2   DeMooy D J, Pesek J and Spaldon E 1973 Mineral nutrition in soybeans: Improvement, production, and uses. Agronomy 16, 267–352.

3   Dixon R O D 1969 Rhizobia, with particular reference to relationship with host plants. Annu. Rev. Microbiol. 23, 137–158.

4   Edwards D G 1977 Nutritional factors limiting nitrogen fixed by rhizobia. *In* Ayanaba A and Dart P J (Eds). Biological Nitrogen Fixation in Farming Systems of the Tropics, pp 189–204. N.Y.: John Wiley and Sons.

5   Garcia E 1970 Los climas del estado de Veracruz. Anales Inst. Biol. Univ. Nac. Mex. Ser. Bot. 41, 3–42.

6   Hardy R W F, Holsten R D, Jackson E K and Burns R C 1968 The acetylene-ethylene assay for N_2 fixation: laboratory and field evaluation. Plant Physiol. 43, 1185–1207.

7   Instituto Mexicano del Cafe 1974 Tecnologia Cafetalera Mexicana. Mexico City: Imprenta Venecia S.A. 194 p.

8   Jimenez E 1979 Ecological study of the coffee agro-ecosystem. I. Structure of a coffee plantation in Coatepec, Ver., Mexico. Biotica 4, 1–12.

9   Jimenez E and Martinez P 1979 Ecological studies of the coffee agro-ecosystem. II. Organic matter production in different types of coffee plantations. Biotica 4, 109–126.

10 Jones K 1970 Nitrogen fixing bacteria in the canopy of conifers in a temperate forest. *In* Dickinson C H and Preece T F (Eds). Microbiology of Aerial Plant Structures, pp 41–66. N.Y.: Academic Press.

11 Koch B L and Oya J 1974 Non-symbiotic nitrogen fixation in some Hawaiian pasture soils. Soil Biol. Biochem. 6, 363–367.

12 Last F T and Deighton F C 1965 The non-parasitic micro-flora on the surfaces of living leaves. Trans. Br. Mycol. Soc. 48, 83–99.

13 Lie T A 1974 Environmental effects on nodulation and symbiotic nitrogen fixation. *In* Quispel, A. (Ed.). The Biology of Nitrogen Fixation, pp 555–582. Amsterdam: North Holland Pub. Co.

14 McColl J G 1970 Properties of some natural waters in a tropical wet forest of Costa Rica. BioScience 20, 1096–1100.

15 Postgate J 1974 Prerequisites for biological nitrogen fixation in free-living heterotrophic bacteria. *In* Quispel A (Ed.). The Biology of Nitrogen Fixation. pp 663–686. Amsterdam: North Holland Pub. Co.

16 Roskoski J P 1980 N_2 fixation (C_2H_2 reduction) by epiphylls on coffee, *Coffea arabica*. Microbial Ecol. 6, 349–355. ʹ

17 Roskoski J P 1981 Nodulation and N_2 fixation by *Inga jinicuil*, a woody legume in coffee plantations. I. Measurements of nodule biomass and field C_2H_2 reduction rates. Plant and Soil 59, 201–206.

18 Ruinen J 1974 Nitrogen fixation in the phyllosphere. *In* Quispel A (Ed.). The Biology of Nitrogen Fixation, pp 86–121. Amsterdam: North Holland Pub. Co.

19 Steyn P L and Delwiche C C 1970 Nitrogen fixation by non-symbiotic microorganisms in some California soils. Environ. Sci. Tech. 4, 1122–1128.

20 Todd R L, Meyers and Waide J B 1978 Nitrogen fixation in a deciduous forest in the southeastern United States. Ecol. Bull. 26, 172–177.

21   van Kessel C and Roskoski J P 1981 Nodulation and N_2 fixation by *Inga jinicuil*, a woody legume in coffee plantations. II. Effect of soil nutrients on nodulation and N_2 fixation. Plant and Soil 59, 207–216.

Plant and Soil 67, 293–304 (1982).   0032-079X/82/0673-0293$01.65.   SU-26
© 1982 *Martinus Nijhoff/Dr W. Junk Publishers, The Hague.*

Nitrogen cycling in South American savannas

El ciclo del nitrógeno en las sabanas sudaméricanas

J. PEREIRA
Empresa Bras. de Pesq. Agrop. (*EMBRAPA*), *Centro de Pesq. Agrop. dos Cerrados* (*CPAC*), *Planaltina, D.F., Brazil*

Key words Burning Denitrification N-cycling N$_2$-fixation Nitrification Oxisol Rhizobium Savanna South America Ultisol

Abstract Savannas cover about 300 million hectares of South America. The soils are mainly oxisols and ultisols and their natural fertility is very low with high acidity. The natural vegetation varies in density and in the amount of biomass produced annually, which can be equal to that produced by forests in the region. Among the nitrogen-fixing micro-organisms, the only ones well-studied are Rhizobium bacteria. In managing the biomass in these areas, it is important to consider biological nitrogen-fixation as a possible source of nitrogen to replace that removed in crops. Nitrification and denitrification in these soils are intense but not well studied. The rainfall distribution during the growing season seems to have a considerable influence of the nitrogen supply to the soils. A considerable loss of nitrogen occurs in this environment when vast areas are burned annually.

Resumen Las sabanas ocupan alrededor de 300 millones de hectáreas de Sudamérica. Los suelos son básicamente oxisoles y ultisoles de muy baja fertilidad y alta acidez. La vegetación natural varía en densidad y en la cantidad de biomasa producida anualmente, la cual puede llegar a ser igual a la producida por bosques de la región.

Entre los microorganismos fijadores de nitrógeno, los únicos bien estudiados son las bacterias del género Rhizobium. En el manejo de la biomasa de estas áreas, es importante considerar la fijación del nitrógeno, como una fuente posible que reemplace al que fué exportado en las cosechas.

La nitrificación y la denitrificación en estos suelos, es intensa pero no bien estudiada. La distribución de lluvias durante la estación de crecimiento parece tener una influencia considerable en la provisión de nitrógeno de los suelos. Se registran considerables pérdidas de nitrógeno en este ambiente, cuando amplias áreas son quemadas anualmente.

Introduction

Savannas cover 300×10^6 ha of South America, with 188×10^6 ha in Brazil, 30×10^6 ha in Venezuela, 28×10^6 ha in Colombia, 23×10^6 ha in Bolivia, and smaller areas in other regions. Soils are mainly oxisols and ultisols. In northern Brazil they are well-drained in contrast with other northern South American savannas[17], they are very permeable and of low water retaining capacity, their texture varies from clayey to sandy, and they are very deep. Their natural fertility is very low, being poor in calcium, magnesium, phosphorus, potassium, sulphur and zinc[30], with high acidity and with exchangeable aluminum at levels toxic to plants. Topographical relief varies from gently undulating to flat. The natural vegetation reflects the edaphic gradient[29]. Savanna soils are generally conducive to mechanized agriculture and served by an extensive network of roads; consequently they are rapidly coming under agricultural production.

The climate of tropical South American savanna regions is characterized by an 8-month rainy season, 900–1600 mm of annual precipitation, and temperatures of 22–26°C. Equatorial savannas have a 6–9 month rainy season, 3000–4000 mm of rain annually, and temperatures of 26–30°C[6]. Potential evapotranspiration varies throughout from 950–1050 mm in the rainy season[7].

In spite of the low nutrient status of these areas, annual biomass production varies considerably; in some cases it is comparable to that of forests. Peres et al.[34] determined from litter production studies that 8.52×10^3 kg dry matter ha^{-1} yr^{-1} are produced on hillsides in the cerradão (savanna containing trees) and 2.78×10^3 kg dry matter ha^{-1} yr^{-1} on flat hilltops in the herbaceous savannas within the CPAC (Centro de Pesq. Agrop. des Cerrados, DF, Brazil) grounds. This suggests that the quantities of nitrogen cycling in these systems are higher than usually expected. The potential natural supply of nitrogen for the formation of biomass in supposedly low due to the low soil fertility noted above, low microbial activity, a low cation exchange capacity, and the absence of 2:1 clays. Thus it would be valuable to describe and quantify the nitrogen cycle in undisturbed savannas. However, no such in-depth study has yet been carried out.

Sources of nitrogen for plants in the savannas include atmospheric N_2, naturally-occurring soil-N, and fertilizer nitrogen. Various legumes are naturally abundant and are effective N-fixers. In view of population pressures in savanna regions and the need to maximize food production, new areas are being brought under cultivation at an accelerating rate, massive fertilization is common, and new plant species are being introduced. These changes may considerably affect the nitrogen balance of ecosystems in these regions, and with the production of more biomass in less time, it becomes necessary to use soil management methods which recycle the nitrogen and other nutrients to make them available for subsequent crops in the most economical way possible. The need to understand nitrogen cycling in these ecosystems is pressing.

Biological nitrogen fixation

Asymbiotic fixation
About 80% of the earth's atmosphere is N_2 and there are various living organisms able to assimilate it, subsequently making it available for plant growth. The intermediate level of organic matter in savanna soils support heterotrophic micro-organisms. Although heterotrophic nitrogen-fixers do not generally fix large quantities of nitrogen, the aerobic species could be important in savanna in view of their organic matter and other requirements[1]. Since *Azotobacter chroococcum, Beijerinckia* spp. and *Derxia* spp. are not tolerant of low calcium concentrations and high exchangeable aluminum, savanna soils do not favor their growth; once limed and fertilized, however, conditions are ideal

for these organisms. Beijerinckia may be the most important heterotrophic nitrogen-fixer in savanna soils due to its tolerance for low calcium and high aluminum. The same species and genera of heterotrophic N-fixers may occur under different climatic conditions because so few species of these heterotrophs have been found. These species include *A. paspali* and *Derxia grumosa*, which have never been found outside the tropics and sub-tropics[16].

Asymbiotic nitrogen-fixing bacteria occur most frequently in the rhizosphere, where they depend on plant photosynthesis and the subsequent supply of carbohydrates to the roots[11]. Recent studies have shown that significant levels of nitrogen can be fixed in the rhizosphere of many tropical grasses and cereals by bacteria able to utilize energy in root exudates. In Brazil, *Paspalum notatum*, a grass present in some sandy savannas, stimulates nitrogen fixation by *Azotobacter paspali*, established permanently on root surfaces with a thin layer of mucilage[14].

Sugar cane is presently cultivated in the savannas, and these systems can fix about 67 kg N ha^{-1} yr^{-1} *via* rhizosphere N-fixation by micro-organisms such as Beijerinckia[14]. Sorghum, maize, rice and various forage grasses have also been shown to stimulate nitrogen fixation in their rhizospheres. *Azospirillum lipoferum* and *A. brasiliense* have been suggested to be the main nitrogen fixers in these grasses[5,12,13,15,31]. On the other hand, Jensen[25] showed that Beijerinckia grows copiously on the leaves of numerous tropical plants, from which nitrogen may be transferred to the soil.

Although most savanna soils are well drained and aggregated, anaerobic heterotrophic nitrogen-fixers could be important. Greenwood (1963, cited by Allisson[1]) states that these bacteria may be found within soil aggregates. The most numerous genera include Clostridium, Klebsiella, Enterobacter, Escherichia, Spirillum, Bacillus and Polymyxa[26]. Among the most important anaerobic N-fixers in South American savanna soils are Clostridium, although Clostridium are probably less important than aerobic Azotobacter.

Autotrophic micro-organisms can also fix nitrogen. Best studied are the blue-green algae *Anabaena variabilus*, *Nostoc punctiforme*, and other Cyanophyceae, and also the bacteria Colecothrix and Fizerella. Some of these may contribute to nitrogen-fixation in the savannas.

Knowledge of nitrogen-fixation by free-living micro-organisms in savannas is based more on qualitative observations than on analytical data. According to Nye and Greenland[32], 39 kg N ha^{-1} yr^{-1} may be fixed by free-living nitrogen fixers in African savannas. Conditions in South American savannas are similar to those in Africa.

Symbiotic nitrogen fixation

i) Legumes South America has the largest reserve of Leguminosae species in the world, and the largest concentrations of legumes occur in the savannas. This suggests that symbiotic nitrogen fixation is very important in the equilibrium ecosystem. However, many of these legumes seem to be inactive due to poor soil conditions[33].

Data are not available for evaluating potential N-fixation by native legumes, but about 3×10^9 kg yr^{-1} of soybeans are being produced in the Brazilian cerrados and symbiotic nitrogen-fixation contributes 150×10^6 kg N to these systems. Scientists are working with considerable success to ensure that these soils, which do not contain *Rhizobium japonicum*, are inoculated with more efficient Rhizobium strains. Yields greater than 1.5×10^3 kg fresh grain (13% H_2O) ha^{-1} have been achieved without nitrogen fertilizer but with highly efficient strains of *R. japonicum*[38].

Lack of specificity for Rhizobium strains is more common in tropical than in temperate legumes[10]. Thus it is more difficult to introduce efficient Rhizobium strains, particularly in forage legumes. However, many of these legumes may form an efficient symbiosis with non-specific Rhizobium strains. This has been observed with some green manure plants. In competition studies of 24 green manure legumes that host Rhizobium, I have observed that *Crotalaria juncea, C. paulina, Styzolobium atterrimum, Cajanus cajan* and *Canavalia ensiformis* produce more dry weight, achieve better nodulation, and fix more nitrogen than the 19 others (Pereira, unpublished). The cowpea group of Rhizobia, found widely in South American savannas, is an exception to the general lack of specificity. Certain legumes, for example some *Stylosanthes guianensis* and *Centrosema pubescens*, are relatively specific for members of this Rhizobia group, although greater efficiency might be achieved by the introduction of improved strains[10]. In all cases of symbiotic nitrogen-fixation in savannas, the potential for fixation probably is much greater than the quantity actually fixed. Correcting pH, phosphorus and molybdenum deficiences of savanna soils often improves symbiotic fixation[20].

Large plantations of non-leguminous trees are being established in Brazilian savanna areas as part of a programme for alternative energy sources. According to Döbereiner[9] the possibility of forestation with leguminous trees has been ignored. Various species such as acacias and other Mimosoidae which show specifity for Rhizobium strains could be inoculated in nurseries.

High levels of NO_3^- in the soil inhibit nitrogen fixation, but this would be unlikely to occur in South American savannas due to intense leaching. Suhet and Ritchey[37] attribute that to leaching of nitrate in oxisols from Brazilia. Much emphasis has been given to legume-grass associations with a view to not only an overall increase of protein in pastures, but also to a transfer of nitrogen from legumes to grasses. Additionally, rotating legumes with other crops benefits subsequent crops with nitrogen[38]. Problems with soybean nodulation were encountered with newly cleared cerrado in the most propitious soybean variety for the region (IAC-2), but experiments at CPAC[18] showed that higher levels of inoculant (1 kg strain 965 inoculant for every 40 kg seed) led to satisfactory nodulation.

Andrew[2] reviewed the influence of soil nutrition on nitrogen fixation in the tropics and concluded that levels of nitrogen fixed could be increased by improving Rhizobium strains and physical and chemical soil conditions, particularly the level of essential elements in the soil. Correction of pH eliminates the inhibitory effect of exchangeable aluminum, and fertilization can increase the availability of nutrients which directly benefit the symbiosis such as Ca, Mg, P, K, S, Mo, B and Zn. More research and improved management techniques could be developed to increase symbiotic nitrogen-fixation.

ii) Non-legumes Bond[3] described a series of non-leguminous genera able to nodulate and fix nitrogen. Of these, Coriaria, Myrica and Discaria originate in Latin America and probably in savanna. These dicotyledonous angiosperms form a symbiosis with an N-fixing endophytic actinomycete[35]. Casuarina, although introduced into Brazil, forms a symbiosis in fertilized soils. Bond[3] states that Cycadaceae fix much nitrogen by means of nodules; cycads are found in more humic savannas on the edge of the Amazon rainforest. Introduced Alnus and Myrica grow at pH 4 and grow well in waterlogged soils. Symbiotic nitrogen fixation by non-legumes has not been observed in savannas, however.

Non-biological immobilization of nitrogen

Apart from nitrogen immobilized in roots and micro-organisms, reactions of ammonia and nitrites with lignin in the organic matter and the fixation of ammonium in clays as well as ammonium aluminum phosphates could also immobilize nitrogen. Workers at CPAC[18] have observed high maize yields up to four years after N-fertilization of cerrado soils containing moderate levels of organic matter (Table 1). The immediate source of this nitrogen is not yet known. Fixation by mineral clays is not important because amounts of clay are very small and mainly in the 1 : 1 form in savanna soils.

Table 1. Corn (Var. Cargill 111) yields on an oxisol soil (2.7% organic matter) cultivated 3 years previously with corn fertilized with different levels of nitrogen (EMBRAPA/CPAC[18])

Applied nitrogen (kg N ha^{-1})				Yield (10^3 kg ha^{-1})
1975–76	1976–77	1977–78	1978–79	1978–79
100	100	30	0	2.63
80	100	30	0	2.23
200	0	0	0	2.71
0	0	30	0	1.90
0	0	0	0	1.94

Nitrification

Nitrification may be the most evident phase of the nitrogen cycle in savannas due to its high activity, the result of good soil aeration and favorable temperatures. Nitrification is carried out by autotrophic micro-organisms such as Nitrosomonas and Nitrobacter and by heterotrophs such as *Clostridium butyricum, Aspergillus flavus, Aspergillus niger* and others. The most frequently occurring nitrifier under natural conditions is *Aspergillus niger*. Autotrophic nitrifiers may not be important in savannas due to oligotrophic conditions and high acidity. However, once the soil is fertilized, environmental conditions seem very favorable for nitrification. Heterotrophic nitrifiers are activated much more rapidly than autotrophs. Whereas the former have a cycle of about one hour, autotrophic cycles are *ca.* 11 hours.

With increasing soil pH, the populations of Nitrosomonas and Nitrobacter increase, reducing the stability of mineral nitrogen as NO_3^- leaches more readily than NH_4^+ and as nitrate and nitrite react with lignin to form immobile nitrogen species. Chemical nitrification apparently occurs, but the phenomenon is not well understood.

Denitrification

Denitrification may be rare in savanna soils due to high aeration and acidity; according to Bremner and Shaw[4], cited in Allisson[1], the appropriate conditions for denitrification in the field are low O_2 and neutral to slightly alkaline pH. These conditions may be met at the microsite level, however.

Many micro-organisms are able to use NO_3^- for protein and other substances, and also for electron transport in which it is reduced to nitrite[22]. In these two cases nitrogen is rarely lost from the soil.

The main denitrifying genera in the cerrados are likely to be Pseudomonas, Achromobacter, Bacillus, Micrococcus, and some species of Thiobacillus[22,40].

Nitrogen volatilization

It is possible that low levels of the unstable NO or ammonium nitrite occur. Large losses of nitrogen occur in these regions of South America when vast areas are annually burned.

Nitrogen movement in the soil

Savanna soils range from fine clays to gross sands, and are deep, but clays are mainly localized in areas with well defined wet seasons. Both types of soil have good infiltration properties and consequently nitrate leaches easily. This is particularly important where there is much rainfall at low intensity. Nitrogen losses by erosion are important where high-intensity rainfall occurs.

Differences in the compaction of the soil surface affect infiltration and thus leaching. Vegetation cover and root development also affect leaching. More numerous and deeper roots physically facilitate leaching, although the root surface-area available for nitrogen absorption is often greater.

Ion concentrations in the soil solution also have a great influence on leaching; this is particularly important in cultivated savannas. The greater the solution ion concentration, the greater the loss of ions by leaching. Kinjo and Pratt[27] showed that subsoils of oxisols in Brazil possess a low NO_3^--adsorbing capacity, and Kinjo and Pratt[28] showed that negative adsorption of nitrate in these soils occurred with higher concentrations of SO_4^- and $H_2PO_4^-$; with increasing concentrations of Cl^-, SO_4^- and $H_2PO_4^-$ the amounts of NO_3^- adsorbed decreased proportionally. Veloso[39] showed that in soils of Sao Paulo, Goiás and the Federal District of Brazil, adsorption of NO_3^- increased with pH. This indicates that savanna soils, whether cultivated or not, have a considerable nitrate-leaching potential. However, the depth of the soil and the even distribution of the rainfall diminish the likelihood of the NO_3^- reaching the water table at levels which could cause pollution. Coating the nitrogen fertilizer with sulfate can reduce losses of nitrogen *via* leaching and increase recovery by the plants (Table 2).

Nitrogen fertilization and crop yields

Grove[23] concluded from comparisons of results from fertilizer experiments and various oxisols, ultisols and other tropical soils that 80–120 kg N-fertilizer ha^{-1} results in yields of maize equivalent to 95% of those obtainable with 200 kg N-fertilizer ha^{-1} (Fig. 1). Using the methods of Standford and Hunter[36], Grove[23] showed that the nitrogen level in the dry matter of the above-ground

Table 2. Grain yield and nitrogen recovery of maize with various sources of fertilizer-N (from Fox *et al.*[19]). SCU = sulfur-coated urea, and lime + N_{min} = lime-coated ammonium nitrate. Fresh grain = 13% H_2O

Location	Fertilizer-N (kg N ha^{-1})	Yield (10^3 kg fresh grain ha^{-1})	N-content of above-ground dry matter (kg N ha^{-1})	Number of crops
Puerto Rico				
Urea–N	67	4.6	103	5
SCU–N	67	3.8	86	5
Brazil				
Urea–N	140	6.0	99	3
SCU–N	140	5.4	90	3
Lime + N_{min}	140	5.6	90	3

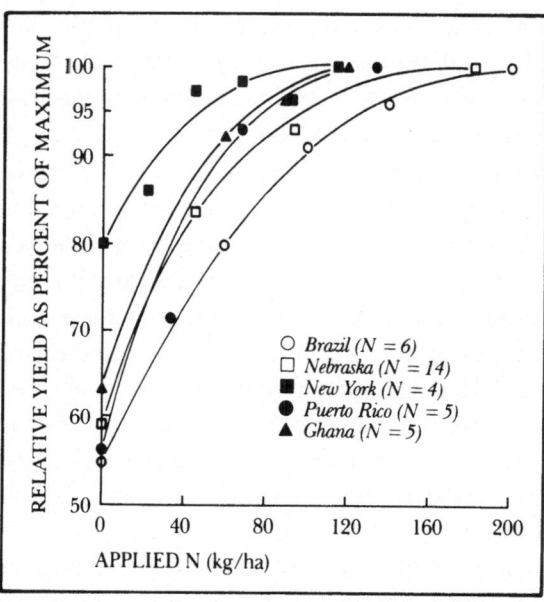

Fig. 1.   Average effect of fertilizer-N on relative maize yield. Note position origin on y-axis. (From Grove[23]).

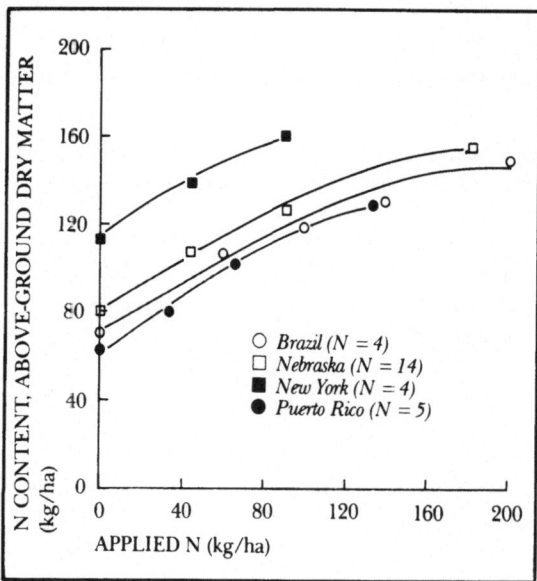

Fig. 2.   Average effect of fertilizer-N on the N-content of above-ground dry matter in maize. (From Grove[23]).

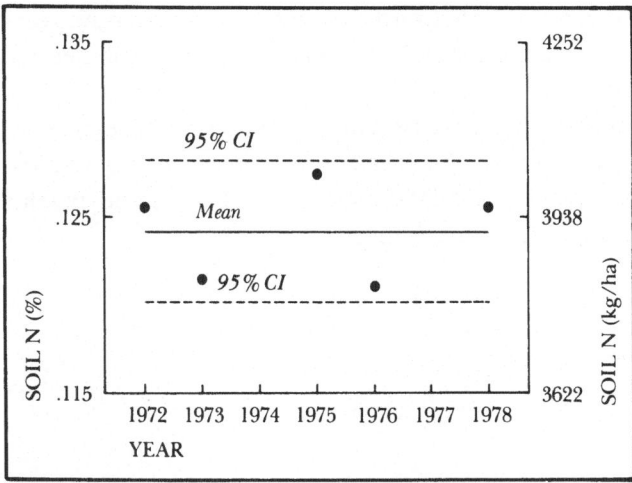

Fig. 3. Soil nitrogen (to 20 cm) in check plots for six crop years in Brazil. CI = confidence interval. Note positions of y-axis origins. (From Grove[23]).

vegetation increased 0.56 kg N for every kg N-fertilizer that was applied between 0–100 kg fertilizer-N ha^{-1} (Fig. 2). Production was similar at all the sites. The concentrations of organic nitrogen in unfertilized plots in the Brazilian cerrado he studied did not change significantly from 1972–1978, although 384 kg N ha^{-1} were removed from the sites at harvests. Apparently nitrogen mineralization was low (Fig. 3).

Rainfall during the growing season seems to influence nitrogen availability in the soil. Freitas[21] found a significant correlation between increased yield due to nitrogen-fertilizer and December rainfall for October-planted maize on a Matão, São Paulo, oxisol ($r = 0.89$, $p < .01$, n = 6, for 50 kg fertilizer-N ha^{-1}, and $r = 0.89$, $p < .01$, n = 6 for 100 kg fertilizer-N ha^{-1}). At less than 200 mm December precipitation there was little response to added-nitrogen, whereas in the years with 293 mm and 364 mm precipitation the responses to nitrogen were very high (Table 3). Yields inversely related to rainfall ($r = -0.94$, $p < .01$, n = 4)

Table 3. Relation between the production increment of corn due to nitrogen fertilization and the rainfall for December in a cerrado of Matão, SP (Freitas[21])

	1962–63	1963–64	1964–65	1965–66	1966–67	1967–68
Yield without N (10^3 kg grain ha^{-1})	2.53	3.05	3.31	3.12	3.80	5.14
Production increment (%)						
50 kg fertilizer-N ha^{-1}	73	5	64	24	17	(1.9)
100 kg fertilizer-N ha^{-1}	98	11	87	29	30	3.0
Rainfall in December (mm)	293	126	364	206	198	191

were observed by workers at CPAC[18] in Brasilia oxisols over 4 consecutive years in maize plots with no nitrogen fertilizer. Suhet and Ritchey[37] attributed the relationship to NO_3^- leaching.

Symbiotic nitrogen fixation reduces the need for fertilizer. For forage legumes, however, little is known about the legume species or varieties that may be most efficient in this association. Workers at CPAC[18] and at CIAT[24] are studying this with some of the more promising forage legumes.

Acknowledgements   I thank my colleague R. Bradley for suggestions and translation.

References

1   Allisson F E 1973 Soil Organic Matter and its Role in Crop Production, pp 181–276. Amsterdam: Elsevier.

2   Andrew C S 1962 Influence of nutrition on nitrogen-fixation and nitrogen in the tropics with particular reference to pastures: *In* A Review of Nitrogen in the Tropics with Particular Reference to Pastures. Commonwealth Agricultural Bureaux of Australia. Bulletin 46, 130–146.

3   Bond G 1967 Fixation of nitrogen by higher plants other than legumes. Annu. Rev. Plant Physiol. 18, 107–126.

4   Bremner J M and Shaw K 1958 Denitrification in soils. II. Factors affecting denitrification. J. Agric. Sci. 51, 40–52.

5   Bülow J F W van and Döbereiner J 1975 Potential for nitrogen-fixation in maize genotypes in Brazil. Proc. Nat. Acad. Sci. USA 72, 2389–2393.

6   Camargo A P de 1969 Problema climático inexistente. Coopercotia 26 (232), 21–25.

7   Cochrane T T 1978 Evaluation de los ecosistemas de sabana de America Tropical para la produccion de ganado de carne; un estudio en marcha. *In* Tergas L E and Sanchez P A (Eds). Produccion de Pastos en Suelos Acidos de los Tropicos, pp 3–16. Cali, Colômbia: CIAT.

8   Dantas M C and Drosdowicz A 1972 Influencia dos adubos minerais na atividade celulolítica em solos de Cerrado. Rev. Microbiol. 3, 25–34.

9   Döbereiner J 1967 Efeito da inoculacao de sementeiras da Sabia (*Mimosa caesalpnifolia*) no estabelecimento e desenvolvimento das mudas no campo. Pesq. Agrop. Bras. 2, 301–305.

10   Döbereiner J 1978 Potential for nitrogen-fixation in tropical legumes and grasses. *In* Döbereiner R H, Burris A H, Hollaender A A, Franco C A, Neyra C A and Scot D B (Eds). Limitations and Potentials for Biological Nitrogen-Fixation in the Tropics, pp 13–24. New York: Plenum Press.

11   Döbereiner J and Campello A B 1971 Non-symbiotic nitrogen-fixing bacteria in tropical soils. Plant and Soil Spec. Vol. 457–470.

12   Döbereiner J and Day J M 1975 Nitrogen fixation in the rhizosphere of tropical grasses. *In* Stewart WDT (Ed.). Nitrogen Fixation by Freeliving Micro-organisms, pp 29–36. Cambridge: Cambridge University Press.

13   Döbereiner J and Day J M 1976 Associative symbioses in tropical grasses: characterization of micro-organisms and dinitrogen-fixing sites. *In* Newton W E and Nyman C J (Eds). Proceedings of the First International Symposium on N_2 Fixation, pp 518–538. Pullman, Washington: Washington State University Press.

14   Döbereiner J, Day J M and Dart P J 1972 Nitrogen activity and oxygen sensitivity of *Paspalum notatum*. J. Gen. Microbiol. 71, 103–116.

15   Döbereiner J, Marriel I E and Nery N 1976 Ecological distribution of *Spirillum lipoferum* Beijerinckia. Can. J. Microbiol. 22, 1464–1473.

16 Drosdowicz A 1976 Equilíbrio dos solos de cerrados. *In* Ferri M G (Ed.). Simpósio Sobre o Cerrado, 4, pp 233–458. Brasilia: Belo Horizonte, Itatiaia Editora.

17 Eiten G 1972 The cerrado vegetation of Brazil. Bot. Rev. 38, 201–341.

18 EMBRAPA/CPAC (Empresa Brasileira de Pesquia Agropecuaria/Centro de Pesquisa Agropecuaria dos Cerrados) 1980 Relatorio Tecnico 1978–79. Brasilia, D.F.: CPAC.

19 Fox R H, Talleyrand H and Bouldin D R 1974 Nitrogen fertilization of corn and sorghum grown in ultisol and oxisol in Puerto Rico. Agron. J. 66, 534–540.

20 Franca G E de and Carvalho M M 1970 Ensaio exploratorio de fertilizaçao de cinco leguminosas tropicais em solo de cerrado. – Pesq. Agropec. Bras. 5, 147–153.

21 Freitas L M M 1980 Alternativas de uso do cerrado. *In* Marchetti D (Ed.). Simpósio Sobre o Cerrado, Uso e Manejo, 5, pp 279–316. Brasilia: Editerra Editora.

22 Graham P 1972 El ciclo del nitrogeno. *In* Medina H (Ed.). El Uso del Nitrógeno en el Trópico, pp 119–140. Medellin, Colombia: Centro de Publicaciones de la Universidad Nacional (Sede de Medellin).

23 Grove T L 1979 Nitrogen fertility in oxisols and ultisols of the humid tropics. Cornell Univ. Internat. Agric. Bull. 36, 1–28.

24 Halliday J 1979 Respuestas en el campo de luguminosas forrageiras tropicales en la inoculación com Rhizobium. *In* Tergas L E and Sanchez P A (Eds). Próduccion de Pastos en Suelos Acidos de los Trópicos, pp 135–150. Cali, Colombia: CIAT.

25 Jensen H L 1965 Non symbiotic nitrogen fixation. *In* Bartholomew W V and Clark F E (Eds). Soil Nitrogen, pp 436–480. Madison, Wisconsin: American Society of Agronomy.

26 Jurgensen M F and Davey C B 1970 Non symbiotic nitrogen fixing micro-organisms in acid soils and the rhizosphere. Soils Fertil. 30, 435–446.

27 Kinjo T and Pratt P F 1971 Nitrate adsorption I In some acid soils of Mexico and South America. Soil Sci. Soc. Am. Proc. 35, 722–725.

28 Kinjo T and Pratt P F 1971 Nitrate adsorption II In competition with chloride, sulfate and phosphate. Soil Sci. Soc. Am. Proc. 35, 725–728.

29 Lopes A D and Cox F R 1977 Cerrado vegetation: an edaphic gradient. Agron. J. 9, 828–831.

30 McClung A C and Freitas L M M de 1959 Sulfur deficiency in soils from Brazilian Camps. Ecology 40, 315–317.

31 Neyra C A, Döbereiner J, Leilande L and Knowles R 1977 Denitrification by N_2-fixing *Spirillum lipoferum*. Can. J. Microbiol. 23, 300–305.

32 Nye P H and Greenland D J 1960 The Soil under Shifting Cultivation, Tech. Comm. No. 51. Farnham Royal, England: Commonwealth Agricultural Bureaux. 156 p.

33 Pereira J 1972 Efeito de Fontes e Doses de Fosforo, na Adubacao e Cultura da Soja em um Solo de Campo-Cerrado, Vicosa. Minas Gerais, Brasil: Univ. Fed. Vicosa (Tese). 70 p.

34 Peres J R R, Suhet A R and Vargas M 1978 Produção de liteira e sua decomposição microbiana no Cerradão e Cerrado. Anais do IX Congresso Brasileiro de Microbiologia, Belo Horizonte, M.G., Brasil, 69.49 (Abstract).

35 Rozo E V de 1972 Fijación biologica de nitrogen in suelos equatoriais. *In* Medina H (Ed.). El Uso del Nitrogeno en el Trópico, pp 93–118. Bogotá: Sociedad Colombiana de Suelos.

36 Stanford G and Hunter A S 1973 Nitrogen requirements of winter wheat (*Triticum aestivum*, L. var. 'Blueboy' and 'Redcoat'). Agron. J. 65, 442–447.

37 Suhet A R and Ritchey K D 1981 Nitrate leaching in cerrado soil. Brasilia, D.F.: Centro de Pesquisa Agropecuária dos Cerrados (Unpublished manuscript).

38 Vaz C A, Lobato E, Pereira G and Pereira J 1980 Reciclagem da matéria Organica na Agricultura Brasileira, Informe Especial No. 1. Brasíleira: Secretaria Nacional de Produçao Agropecuária. 37 p.

39 Veloso A C Y 1975 Adsorçao de nitrato em latossolos sob vegetaçao de cerrado. Turrialba 25, 404–409.

40 Woldendorp J W, Dilz K and Kolenbrander G J 1966 The fate of fertilizer nitrogen on permanent grassland soil. *In* Nitrogen and Grasslands, pp 53–76. Wageningen: PUDOC.

Plant and Soil 67, 305–314 (1982).   0032-079X/82/0673-0305$01.50.   SU-27
© 1982 Martinus Nijhoff/Dr W. Junk Publishers, The Hague.

Nitrogen balance in the Trachypogon grasslands of Central Venezuela

Balance de nitrógeno en las sabanas de Trachypogon del centro de Venezuela

E. MEDINA
Centro de Ecología, I.V.I.C., Apartado 1827, Caracas 1010 A, Venezuela

Key words Burning Internal N-recycling N-cycling N-inputs N-losses Savanna *Trachypogon* sp. Venezuela

Abstract   The nitrogen balance of a Trachypogon grassland in Calabozo, Venezuela, is calculated for average conditions using biomass accumulation, nitrogen content, and turnover rates of organic matter. Burning Trachypogon grasslands results in losses of 8.5 kg N ha^{-1} yr^{-1}, while rainfall inputs average 2.6 kg N ha^{-1} yr^{-1}. Uptake of N by vegetation is 14.8 kg N ha^{-1} yr^{-1}, but the total N required to build new tissue during a growing season is about 30 kg N ha^{-1} yr^{-1}, so that about 50% of the nitrogen in the vegetation is recycled internally. Nitrogen losses *via* fire are probably balanced by biological N$_2$-fixation, but no data are available for N-fixation in these savannas. The calculations presented in this paper are based on few data and more measurements are needed to develop a conclusive picture of the N-balance of Trachypogon grasslands.

Resumen   El balance de nitrógeno de una sabana de Trachypogon en Calabozo, Venezuela fué calculado para las condiciones medias utilizando la acumulación de biomasa, su contenido de nitrógeno y las tasas de producción de materia orgánica. Como resultado de las quemas de las sabanas de Trachypogon se pierden 8,5 kg N ha^{-1} año^{-1}, mientras que las entradas por precipitación son de 2,6 kg N ha^{-1} año^{-1}. La absorción por la vegetación es de 14,8 kg N ha^{-1} año^{-1} mientras que el N total requerido para formar nuevos tejidos durante el período de crecimiento es de ca. 30 kg N ha^{-1} año^{-1}. Asi se puede estimar que cerca del 50% del nitrógeno requerido por las plantas es reciclado internamente. Las pérdidas de nitrógeno por la quema son probablemente compensadas por la fijación biológica de nitrógeno pero no existen datos para estas sabanas. Los cálculos presentados en este trabajo se basan en pocos datos y se requerirían mas mediciones para establecer un balance de nitrógeno mas definitivo.

Introduction

The Trachypogon savanna is one of the most important vegetation types in Central and Eastern Venezuela[18], covering more than 100,000 km^2. This savanna is dominated by several species of the genus Trachypogon (*T. vestitus, T. plumosus, T. ligularis*), and it has been described as a park savanna with scattered isolated trees, including among the most frequent but often dwarf *Curatella americana, Bowdichia virgilioides, Byrsonima crassifolia, Roupala complicata, Palicourea rigida, Platycarpum orinocense,* and *Caraipa guianensis*[4,15,18].

The climate is seasonal with a humid season of 6–8 months. Annual rainfall throughout the savanna region varies from 900 to 2200 mm, with the greater precipitation along the southern border of the main region[9,13].

Trachypogon savannas are relatively infertile and are often burned during the dry season to stimulate forage production for cattle. Productivity of above-ground biomass increases after burning[9,20], and new forage has a higher nutritional value and is more palatable to cattle[12].

The burning of grasslands results in losses of nitrogen and sulphur through volatilization, and in the redistribution of other nutrients in ash by prevailing winds. This burning has raised several important ecological questions regarding the Trachypogon savannas: a) is fire progressively increasing nutrient deficiency in savanna soils by causing losses in organic matter, N and S; and b) is productivity of native vegetation limited by low N-availability, or are other nutrients limiting?

There are no definitive answers to these questions as yet. It appears that annual burning volatilizes less than 1% of the total N-inventory in the Trachypogon grassland[12], but this loss is not compensated for by inputs through rainfall. Sulfur losses through fire, however, seem to be balanced by rainfall inputs[8].

Fertilizer experiments have shown that above-ground production of native vegetation increases with both N and P additions[12,19]. On the other hand, retranslocation from above-ground biomass to the roots in native grassland is higher for P (about 80%) than for N (about 60%). This retranslocation efficiency is markedly reduced in plots receiving additional P, while it remains unchanged in plots receiving additional N[12]. Similar results were reported by Norman[16] for Themeda grasslands in Northern Australia.

Comprehensive studies of the nitrogen cycle in South American savannas are lacking. In this paper I attempt to summarize available information regarding nitrogen relationships in the Trachypogon grasslands of the Venezuelan Llanos in order to highlight the many gaps in existing knowledge.

Nitrogen cycling in the Trachypogon savanna

Soil nitrogen

Typical soils of the Trachypogon savannas range from clay and clay loam to sandy clay (frequently sandy in the south), and have low organic matter, N, and P contents[4,15]. A hard-pan is often found 30–120 cm below the soil surface though it may appear at the soil surface in eroded areas[15].

Soil analyses of a typical profile near Calabozo in Central Venezuela appear in Table 1. Values are for a depth of 0–40 cm because frequently more than 90% of the root biomass is found above this depth[4,6,9,21]. Samples were collected at the beginning of the dry season, and thus available N and P values may not represent steady state values during the previous period of maximum growth. Often in regions with a pronounced seasonality in plant growth, NO_3–N accumulates when vegetation growth has ceased if there is still sufficient water in the soil to allow continued organic matter decomposition.

Most soil-N is in the organic form, NO_3–N represents only 1.4% of the total N to a depth of 40 cm. At the end of the growing period total N in the soil was

4.42×10^3 kg N ha^{-1}, available NO_3–N was 60 kg N ha^{-1}, and available P was 17 kg P ha^{-1}. Losses of available N at the beginning of the next rainy season are probably minimal because the rainfall-evaporation relationships are such that soil is water-saturated only towards the middle of the rainy season. At that time, vegetation is growing vigorously and can absorb most of the N that is released through mineralization of organic matter.

Precipitation inputs and leaching losses

The amount of water which drains below the root zone can be calculated as the difference between rainfall and evapotranspiration when the soil is saturated. In Calabozo, the soil profile is saturated from June to September, although this can vary with rainfall amount and distribution in a particular year. Using rainfall and transpiration data for a burned grassland plot, San José and Medina [20] estimated drainage losses of 109 mm of water. The ratio of drainage rainfall in this case was

Table 1. Soil characteristics to 40 cm depth in a Trachypogon grassland near Calabozo, Edo. Guárico, Venezuela. Values are means ($\pm$ s.d.) of 8 samples. Vegetation is dominated by *Trachypogon vestitus* and *Axonopus canescens*. Available-P = acid-extractable-P

Component	Value
Texture	
Sand (%)	44.9 (3.6)
Silt (%)	24.9 (2.6)
Clay (%)	25.2 (2.3)
Bulk density (g cm^{-3})	1.6
pH	4.9 (0.08)
Organic matter (%)	1.23 (0.20)
(kg ha^{-1})	78720
Total-N (%)	0.069 (0.025)
(kg ha^{-1})	4420
NO_3–N (mg N kg soil^{-1})	9.6 (3.5)
(kg N ha^{-1})	60
Available-P (mg P kg soil^{-1})	2.3 (1.0)
(kg P ha^{-1})	17

Table 2. Nutrient inputs (kg ha^{-1} yr^{-1}) in bulk precipitation in savanna stations in Venezuela (Escobar [8]). Mineral N = NH_4^+–N + NO_3^-–N

Station	Mineral-N	P	K
Estación Biológica de los Llanos	1.3	4.6	3.8
Estación Experimental de los Llanos	1.8	3.4	3.9
San Fernando de Apure	4.7	4.2	4.4

Table 3. Vegetative biomass (10^3 kg ha^{-1}) in Trachypogon savannas in Venezuela. Values in parentheses are standard deviations of the means

Location and dominant species	Above-ground	Below-ground	Reference
Puerto Ayacucho *(Trachypogon vestitus Paspalum carinatum, Bulbostylis capillaris)*	3.37 (0.20)	3.87 (0.76)	9
Puerto Ayacucho, 'tree-less savanna' *(T. plumosus, Axonopus pulcher)*	1.86	nd	6
Puerto Ayacucho, 'sandy savanna' *(T. vestitus, A. canescens)*	1.35	nd	6
Puerto Ayacucho, 'ripio savanna' *(Trachypogon* spp., *Axonopus* spp., *Mesosetum* spp.)	0.94	nd	6
Calabozo *(T. plumosus, T. vestitus, A. canescens)* after fire	3.00 ± (0.69)*	2.32**	4, 5, 12
protected > 4 years	11.35 ± (0.46)***	2.70	21
Calabozo (soil > 1 m deep) *(T. plumosus)*	5.70	2.27	21
Calabozo (shallow soil) *(T. ligularis)*	1.98	2.29	21

* $n = 10$
** $n = 3$
*** $n = 4$
nd, not determined

6%. Blydenstein[4] used meteorological data to estimate drainage losses of 158 mm H_2O, which is a ratio of drainage rainfall of 14%. Judging from the annual variation in depth of the water table (about 2 m; Blydenstein[4]) the intensity of drainage is approximately 100–150 mm H_2O yr^{-1}, or between 7–11% of the average rainfall in Calabozo (1300 mm, Monasterio[14]).

Nitrogen concentrations in drainage waters have not been measured in Calabozo, but because mineralization of organic matter takes place mainly during the rainy season, when demand from vegetation is at its peak, concentrations are probably very low. Assuming that mineral N in drainage waters is 0.5 mg/l, and that annual drainage is 125 mm H_2O, I estimate drainage losses through deep drainage to be *ca.* 0.6 kg N ha^{-1} yr^{-1}.

Mineral input through rainfall has not been frequently measured in Central

Venezuela. The only data available to me (Table 2) indicate an average input for 3 stations of 2.6 kg N ha^{-1} yr^{-1}.

Biomass, nitrogen, and N-translocation in vegetation

Vegetation biomass in the Trachypogon savannas varies widely depending on location, availability of nutrients, and temporal amount and distribution of rainfall (Table 3). In the Trachypogon grasslands of Central Venezuela maximum above-ground biomass (about 10×10^3 kg ha^{-1}) is reached after approximately 4 years in the absence of burning[4,21]. This biomass may be that at which production and decomposition is balanced in the absence of perturbation. Trachypogon savannas, however, are regularly burned, usually annually, and one is not likely to find such accumulations over extended areas. 3.0×10 kg ha^{-1} of above-ground biomass is probably representative for areas regularly burned with 2.4×10^3 kg ha^{-1} of below-ground biomass (as measured near Calabozo). These values obviously are subject to annual variations.

Nitrogen concentrations in biomass varies strongly with age due to retranslocation from old to new tissue and from above-ground to below-ground organs[12,24], and also due to the accumulation of structural carbohydrate and minerals. Measurements of N-concentrations in above-ground dead and green material give approximate values for retranslocation (though not corrected for structural carbohydrate accumulation).

In these savannas, N-concentrations in living tissue are generally low and similar among different grass species; concentrations for several dominant species show overlapping standard deviations (Table 4). Although retranslocation factors measured in these savannas differ (Table 4), this may be due to the way the factors were calculated: in Calabozo, the average concentration of the green material during the middle of the growing season was compared with the concentration of dead above-ground biomass during the

Table 4. Mean nitrogen concentration and retranslocation factors ($\pm$ s.d.) in vegetation of Trachypogon savannas

Location and dominant spp.	Green tissue (% N)	Dead tissue (% N)	Living roots (% N)	Retranslocation factor (%)
Calabozo (*Trachypogon vestitus, Axonopus canescens*)[12]	0.79 (0.11)	0.28 (0.04)	0.62	64 (6)
Puerto Ayacucho (*T. vestitus, Paspalum carinatum*)[9]	0.70 (0.08)	0.48 (0.03)	nd	32 (6)

nd, not determined.

Table 5. Organic matter and nitrogen contents of soil and vegetation in Trachypogon savannas

	Organic matter (10^3 kg ha^{-1})	Nitrogen (kg ha^{-1})
Vegetation		
Above ground	3.00	23.7
Below ground	2.40	14.6
	5.40	38.3
Soil (0–40 cm)	78.7	4420
% in vegetation	7%	0.9%

Table 6. Nitrogen inputs and losses in several grassland savannas. Per cent burned is per cent of above-ground biomass lost

Location	Above ground biomass (10^3 kg ha^{-1})	N content (% N)	(kg N ha^{-1})	Rain-fall (kg N ha^{-1} yr^{-1})	Burned (%)	Fire loss (kg N ha^{-1} yr^{-1})	N fixation (kg N ha^{-1} yr^{-1})	Reference
Nigeria	3.04	0.35	10.6	4.0–5.0	83.	8.8	3.0–9.0	10, 11
Cote d'Ivoire Lamto	3.79	0.28	10.5	5.0	100.	9.9	9.0	1, 23
Katherine, Australia	1.07	0.45	4.4	1.5	nd	4.5–5.6	nd	16, 17, 25
Calabozo Venezuela (average)	3.00	0.28	8.5	2.6	74.	8.5	nd	

nd, not determined.

middle of the dry season, whereas in Puerto Ayacucho, the N concentrations of green and dead material at each sampling date were compared. Because differentiation between the green and dead compartments is difficult, some old plant material in which translocation was still taking place may have been included as dead material in the Puerto Ayacucho estimates. The nitrogen retranslocation factor of 64% for Calabozo may be the more representative for the entire savanna.

The estimated proportion of the total system nitrogen contained in the vegetation is low (0.9%, Table 5), and actual values may be as high as 1.5% depending on biomass accumulation.

Detailed measurements of production and decomposition in the Trachypogon savannas near Puerto Ayacucho in southern Venezuela indicate a root biomass turnover rate of 43% per year[9]. Turnover rate for the above-ground biomass is 100% for a burned plot and 74% for an unburned plot.

The nitrogen balance in a Trachypogon grassland

Ung the foregoing data, I have developed a provisional nitrogen balance for a Trachypogon grassland (Fig. 1). In a grassland burned during the dry season, 8.5 kg N ha^{-1} yr^{-1} are lost from the system by burning. This loss is not compensated for by rainfall input, ca. 2.6 kg N ha^{-1} yr^{-1}. In a grassland which is not burned, 6.3 kg N ha^{-1} yr^{-1} are re-incorporated to the soil through decomposition and biomass leaching, while 2.2 kg N ha^{-1} yr^{-1} remain in the undecomposed above-ground organic matter. In Trachypogon grasslands that are burned annually, there is a nitrogen deficit. Only in grasslands with a very low vegetation biomass accumulation (about 0.9×10^3 kg ha^{-1} yr^{-1}), such as those reported by Bulla and Lourido[6] (cf Table 3), does rainfall input compensate for N-losses *via* fire.

Most of the remaining fluxes in Fig. 1 have been calculated from production

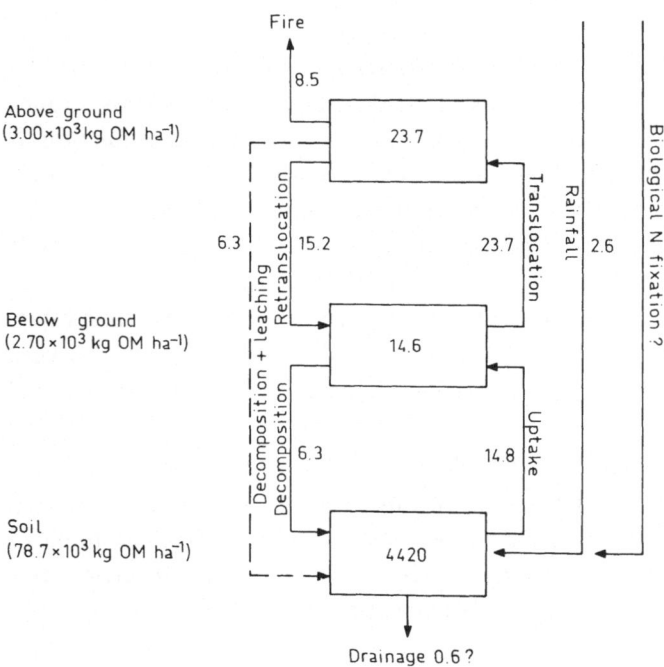

Fig. 1. Nitrogen balance in an average Trachypogon grassland in Central Venezuela. Fluxes = kg N ha^{-1} yr^{-1}, pools = kg N ha^{-1}. OM = organic matter, dashed line = flux from unburned savanna.

and decomposition data and therefore require confirmation by independent methods. Though these flux estimates may lack precision, order of magnitude estimates are useful for future studies. For example, $30 \, kg \, N \, ha^{-1} \, yr^{-1}$ are needed for organic matter synthesis, but only $14.8 \, kg \, N \, ha^{-1} \, yr^{-1}$ need to be taken up from the soil since $15.2 \, kg \, N \, ha^{-1} \, yr^{-1}$ apparently circulate internally in the vegetation.

Assuming that these savannas are in a long term steady-state with respect to soil nitrogen, which would be expected if the systems are adapted to frequent burning, the missing nitrogen is probably recovered by other processes, most probably biological N-fixation. The capacity for savanna soils to fix N_2 has been documented by several authors. In Brazil, Döbereiner[7] found fixation rates of 7.8–$23.9 \, mg \, N \, m^{-2} \, day^{-1}$ (or approximately 14–$43 \, kg \, N \, ha^{-1} \, yr^{-1}$ assuming a 180-day growing season) in *Paspalum notatum* grasslands. If these rates are typical for Trachypogon savannas, the highest losses resulting from burning even protected savannas could be recovered in one growing season. Balandreau and Villemin[1] report asymbiotic fixation rates up to $9.0 \, kg \, N \, ha^{-1} \, yr^{-1}$, while Isichei[10] found that in Nigerian savannas blue-green algal crusts may contribute 3.3–$9.2 \, kg \, N \, ha^{-1} \, yr^{-1}$. In grasslands dominated by *Trachypogon vestitus* on sandy clay soils of Calabozo, blue-green algae are abundant during the rainy season, but no data on their fixation activity are available. Although there is a wide variety of legumes in the Trachypogon grasslands, and most of them are nodulated[2], their contribution to the total N balance of the grassland may not be significant because they constitute only a small percentage of total vegetation biomass[13].

A comparison of the nitrogen balance in Trachypogon grasslands with other savanna areas in Africa and Australia shows that the amount of N lost through fire is correlated with the amount of above-ground biomass accumulation, N content at time of burning and % of organic matter burned (Table 6). Values of N losses that I calculated are very similar to those reported for Nigeria and Cote d'Ivoire, but are higher than those of Northern Australia grasslands, mainly because the latter had very low biomass at the time of burning.

Rainfall-N input in Central Venezuela appears considerably lower than values reported for African savannas[3,22]. However, only a few stations have been studied in Venezuela, and longer series of measurements are required before final conclusions can be reached.

In none of the savannas compared are N losses through fire compensated for by rainfall inputs during the year (a possible exception may be the Lamto savanna studied by Villecourt and Roose[22], who found a very high input of organic N in rainfall). Therefore, it appears that biological N fixation must play a fundamental role in maintaining the N balance in savanna ecosystems.

Acknowledgements I thank Drs. Sandra Brown (University of Illinois) and Ariel Lugo (Institute of Tropical Forestry, Puerto Rico) for their many suggestions for improvements of this paper.

References

1 Balandreau J and Villemin G 1973 Fixation biologique de l'azote moleculaire en savanna de Lamto (basse Cote d'Ivoire), resultats preliminaires. Rev. Ecol. Biol. Sol 10, 25–33.
2 Barrios S and González V 1971 Rhizobial symbiosis on Venezuelan savannas. Plant and Soil 34, 707–719.
3 Bate G 1980 Nitrogen cycling in savanna ecosystems. In Clark F and Rosswall T (Eds.). Terrestrial Nitrogen Cycles. Ecol. Bull. Stockholm 33, 463–475.
4 Blydenstein J 1962 La sabana de Trachypogon del Alto Llano. Bol. Soc. Ven. Cienc. Nat. 23, 139–206.
5 Blydenstein J 1963 Cambios de la vegetación después de la protección contra el fuego. Bol. Soc. Ven. Cienc. Nat. 23, 233–238.
6 Bulla L and Lourido L 1980 Production, decomposition and diversity in three savannas of the Amazonas Territory (Venezuela). In Furtado I (Ed.). Tropical Ecology and Development. pp 73–77. Kuala Lumpur.
7 Döbereiner J 1977 Potential for nitrogen fixation in tropical legumes and grasses. In Döbereiner J, Burris R and Hollaender A (Eds.). Limitations and Potentials for Biological Nitrogen Fixation in the Tropics, pp 13–24. New York: Plenum Press.
8 Escobar A 1977 Estudio de las sabanas inundables de Paspalum fasciculatum. Thesis Magister Scientiarum. Caracas: IVIC.
9 Guinand E and Sánchez P 1979 Productividad primaria, fenología y composición florística de un tipo de sabana situada en el Territorio Federal Amazonas. Thesis. Caracas: Escuela de Biología, UCV.
10 Isichei A O 1981 Nitrogen fixation by blue-green algal soil crusts in Nigerian savannas. In Rosswall T (Ed.). Nitrogen cycling in West African Ecosystems, pp 191–198. Stockholm: Royal Swedish Academy of Sciences.
11 Isichei A O and Sanford W W 1981 Nitrogen loss by burning from Nigerian grasslands ecosystems. In Rosswall T (Ed.). Nitrogen Cycling in West African Ecosystems, pp 325–331. Stockholm: Royal Swedish Academy of Sciences.
12 Medina E, Mendoza A and Montes R 1978 Nutrient balance and organic matter production in the Trachypogon savannas of Venezuela. Trop. Agric. Trinidad 55, 243–253.
13 Medina E and Sarmiento G 1979 Tropical grazing-land ecosystems of Venezuela. In Tropical Grazing-lands. pp 612–619. Paris: UNESCO.
14 Monasterio M 1968 Observations sur les rythmes annuels de la savanne des 'Llanos' de Venezuela. Thesis. Montpellier.
15 Monasterio M and Sarmiento G 1968 Análisis ecológico y fitosociológico de la sabana de la Estación Biológica de los Llanos. Bol. Soc. Ven. Cienc. Nat. 27, 477–524.
16 Norman M 1962 Response of native pasture to nitrogen and phosphate fertilizer at Katherine, N.T. Aust. J. Exp. Agric. Anim. Husb. 2, 27–34.
17 Norman M and Wetselaar R 1960 Losses of nitrogen on burning native pasture at Katherine, N.T. J. Aust. Inst. Agric. Sci. 26, 272–273.
18 Ramia M 1967 Tipos de sabana en los Llanos de Venezuela. Bol. Soc. Ven. Cienc. Nat. 37, 264–288.
19 San José J J and García-Miragaya J 1981 Factores ecológicos operacionales en la producción de materia organica de las sabanas de Trachypogon. Bol. Soc. Ven. Cienc. Nat. 36, 347–374.
20 San José J J and Medina E 1975 Effect of fire on organic matter production and water balance in a tropical savanna. In Golley F B and Medina E (Eds.). Tropical Ecological Systems. Ecological Studies 11, 251–264.
21 San José J J and Medina E 1976 Organic matter production in the Trachypogon savannas of Venezuela. Trop. Ecol. 17, 113–124.
22 Villecourt P and Roose E 1978 Charge en azote et en éléments minéraux majeurs des eaux de

pluie, de pluviolessivage et de drainage dans la savanne de Lamto (Cote d'Ivoire). Rev. Biol. Ecol. Sol 15, 1–20.

23   Villecourt P, Schmidt W and Cesar J 1979 Recherche sur la composition chimique (N, P, K) de la strate herbaceé de la savane de Lamto (Cote d'Ivoire). Rev. Ecol. Biol. Sol 16, 9–15.

24   Weinmann H 1942 On the autumal re-migration of Nitrogen and Phosphorus in *Trachypogon plumosus*. J. South Afr. Bot. 6, 179–196.

25   Wetselaar R and Hutton S T 1963 The ionic composition of rain water at Katherine, N.T. and its part in the cycling of plant nutrients. Aust. J. Agric. Res. 14, 319–329.

Plant and Soil 67, 315–323 (1982).   0032-079X/82/0673-0315$01.35.   SU-28
© 1982 *Martinus Nijhoff/Dr W. Junk Publishers, The Hague.*

Net primary productivity and nitrogen and carbon distribution in two xerophytic communities of central-west Argentina

Productividad primaria y la distribución de nitrógeno y carbono en dos comunidades xerofíticas de la Argentina centro occidental

R. H. BRAUN WILKE

National Council of Scientific and Technical Research (CONICET), Argentine Institute for Research on Arid Lands (IADIZA), Casilla de Correo 507, Mendoza (5500), Argentina

Key words Argentina Carbon Energy Legumes Litter Mineral soil N-cycling *Prosopis flexuosa* Xerophytic forests

Abstract   Net primary productivity and the nitrogen, carbon, and energy contents of the leaf, aerial wood and root components of the five most important woody dominants in two xerophytic forests in central-west Argentina were measured. Nitrogen and carbon contents of litter and mineral soil beneath individual plant canopies were also studied.

The woody dominants in the 8-yr old 'chaco' woodland in Chamical, La Rioja, covered a greater proportion of total community area but had less aerial biomass than the 5 woody dominants of the 50-yr-old open *Prosopis flexuosa* woodland in Ñacuñán, Mendoza. Marked differences in net primary production among species of the two communities were also noted (29–115 kg aerial biomass ha^{-1} yr^{-1} in the Chamical *vs* 51–524 kg ha^{-1} yr^{-1} in the Ñacuñán woodland). Nitrogen in vegetation varied by species, and within species, varied by season and plant component. In general, leaf-N was higher in legumes in summer than in non-legumes in summer, and for most species higher in summer than in winter. Differences in %N in other plant components and in per cent C among species and seasons were less consistent. In both communities, soil N and C were higher and more variable with depth under individual plant canopies than in non-vegetated areas, and differences among species were apparent.

Resumen   Se midieron la productividad primaria neta y los contenidos de carbono, nitrógeno y energía en los componentes foliares, leñosos y radicales de cinco de las plantas dominantes en dos bosques xerofíticos de la Argentina centro occidental. También se midió el contenido de carbono y de nitrógeno en el mantillo y el suelo mineral bajo el dosel de plantas individuales.

Las dominantes leñosas en la vegetación de bosque chaqueño de ocho años en Chamical, La Rioja cubrían una proporción mayor del área total pero tenían una menor biomasa que las cinco dominantes leñosas de la vegetación abierta de cinco años de *Prosopis flexuosa* en Ñacuñán, Mendoza. Se notaron también marcadas diferencias en las productividades primarias netas de ambas comunidades con valores entre 29–115 kg ha^{-1} año^{-1} en biomasa aérea para Chamical y de 51–524 kg ha^{-1} año^{-1} en biomasa aérea para Ñacuñán. El nitrógeno de la vegetación varía según las especies y dentro de las mismas asi como estacionalmente y entre los componentes. En general el nitrógeno foliar era mayor en las leguminosas que en las no leguminosas durante el verano y para la mayoría de las especies era mayor en verano que en el invierno. Las diferencias en el % N en otros componentes y en el % C entre las especies y estaciones no mostraron tendencias claras. En ambas comunidades el N y el C del suelo era mas variable con la profundidad bajo el dosel de plantas individuales que en areas no vegetadas; se hallaron también diferencias entre las especies.

Introduction

Research directed to a better understanding of the role of woody components in ecosystems of arid zones of Argentina is limited and recent. In this paper two such systems investigated by IADIZA in the central west of Argentina are described with regard to net primary production as well as the spatial distribution of nitrogen[1,2,3].

The two study areas, representative of extensive regions in Argentina, are presently devoted mainly to cattle breeding, though at one time the forests were exploited for wood. However, land is frequently being reclaimed for agricultural use; extensive dry farming cultures are being established in the chacoan forest, and in the Mendoza 'monte' shrubland intensive cultures that are irrigated with river and groundwater occur.

Study sites

The western xerophytic chacoan forest is located at Chamical, La Rioja Province (30° 30' S, 66° 15' W at 467 m elevation). The forest was cleared eight years before the present study and is presently grazed. The landscape is an undulating piedmont plain with alluvial sandy soil derived from granite. The desert-like climate is subtropical warm and monzoonic, with 350 mm of rain per year, more than half of which falls in summer. Yearly mean temperature is 19°C; mean maximum is 27°C, mean minimum is 11°C, and freezing temperatures are not common.

Table 1. Density and per cent cover of 5 most important woody species in an 8-yrs old Chamical, La Rioja province Chaco forest and a 50 yrs old Ñacuñán, Mendoza province forest of the Monte formation

Community, species	Density (ind ha^{-1})	Cover m^2 ind^{-1}	% cover
Chamical			
Aspidosperma quebracho blanco	508	0.44	2.26
*Prosopis nigra**	490	1.37	6.72
*Cercidium praecox**	141	2.40	3.38
*Acacia caven**	272	0.70	1.93
Larrea divaricata	3530	0.75	26.40
Ñacuñán			
*P. flexuosa**	220	1.23	2.71
*Geoffroea decorticans**	440	0.16	0.68
L. divaricata	1200	0.18	2.16
Atriplex lampa	1000	0.08	0.76
Lycium tenuispinosum	2900	0.04	1.20

* Legumes
Per cent cover = m^2 100 ha^{-1}.

Table 2. Net primary production (kg dry matter ha^{-1} yr^{-1}) for the 5 most important woody species in the communities described in Table 1 legend

Community, species	Aerial			Roots	Total
	Leaves	Wood	Total		
Chamical					
Aspidosperma quebracho blanco	5.3	36.8	42.1	25.4	67.5
*P. nigra**	3.5	33.6	37.1	21.9	59.0
*C. praecox**	9.5	105.0	115.0	34.2	149.0
*Acacia caven**	9.8	55.0	64.8	34.8	99.6
Larrea divaricata	8.8	20.5	29.3	12.1	41.4
Ñacuñán					
*P. flexuosa**	165	359	524	na	na
*G. decorticans**	52	108	160	na	na
L. divaricata	126	214	340	na	na
Atriplex lampa	108	68	176	na	na
Lycium tenuispinosum	11	40	51	na	na

* Legumes
na, not available

Table 3. Maximum living biomass (10^3 kg dry matter ha^{-1}) for the 5 most important species in the communities described in Table 1 legend

Community, species	Aerial			Roots	Total
	Leaves	Wood	Total		
Chamical					
Aspidosperma quebracho blanco	0.024	0.125	0.149	0.145	0.294
*P. nigra**	0.018	0.157	0.175	0.110	0.284
*C. praecox**	0.031	0.306	0.337	0.106	0.443
*Acacia caven**	0.049	0.267	0.316	0.187	0.504
Larrea divaricata	0.727	2.73	3.46	1.020	4.48
Ñacuñán					
*P. flexuosa**	0.157	5.84	5.99	na	na
*G. decorticans**	0.044	1.21	1.25	na	na
L. divaricata	0.173	1.47	1.65	na	na
Atriplex lampa	0.127	0.77	0.88	na	na
Lycium tenuispinosum	0.024	0.27	0.29	na	na

* Legumes
na, not available

Aspidosperma quebracho blanco and *Prosopis nigra* are the dominant woody plants in the area; also important are *Acacia caven* and *Cercidium praecox* (Table 1). Cactii (*Opuntia quimilo, Cereus validus*) are scarce, except where there is heavy grazing. Panicoid grasses (*Setaria, Gouinia, Digitaria, Eragrostis,* and *Trichloris* spp.) make up 37% of the herbaceous layer.

The second system studied, a xerophytic open forest of the western 'monte' formation, is located at Ñacuñán, Mendoza Province (34° S, 68° W, at 572 m elevation). The forest was 50 years old when sampled and is also on an undulating plain, with a deep sandy loam soil that is permeable, of eolic origin. The climate is desert like and monzoonic; 260 mm of rain fall per year, mainly in summer. Yearly mean temperature is 15°C; mean maximum is 24°C and mean minimum is 6.6°, and freezing temperatures occur during eight months of the year.

Prosopis flexuosa is the only tree species present. *Geoffroea decorticans* is the next most important woody species by biomass (Table 2). *Atamisquea emarginata, Condalia microphylla, Larrea* spp., *Atriplex lampa,* and *Lycium* spp. are also important. Only 17% of the herbaceous layer consists of panicoid grasses; present are *Aristida, Digitaria, Neobouteloua, Pappophorum, Setaria, Sporobolus* and *Trichloris* spp.

Methods

All values in this study are based on samples from the 5 most important woody species in each system. Net production was estimated by annual phytomass differences. Phytomasses were calculated by means of allometric relationships between dry weights and dimensions of plants of each population under consideration. Corresponding correlations for adjustment curves were high. Litter was harvested annually and dried at 70°C.

By means of conventional analyses, total nitrogen and carbon for leaves, stems, and (for the chacoan forest) roots as well as N and C in litter and soil was determined. The caloric values of plant material were also determined, and samples for all analyses were obtained at two times during the year.

Results and discussion

The vegetation of the chacoan ecosystem of Chamical represents a stage in the secondary succession of a true forest; four of the five most important species are arboreal (Table 1). In contrast, although the older Prosopis system in Ñacuñán has a greater plant density and biomass, only one tree species is present and plant cover in general is considerably less important. Legumes are important in both communities, especially in the Chamical forest, where they account for 12 per cent of total cover. One important species (*Larrea divaricata*) is common to both systems.

Net primary productivity is apparently several times greater among the 5 community dominants in the Ñacuñán community than among those in the younger Chamical forest (Table 2). The biomass of the dominants is also higher in the older community (Table 3).

The nitrogen contents of the plant biomass vary among species and within species vary by season and plant part (Tables 4 and 5). In all cases but *Lycium*

Table 4. Total nitrogen (%N) during the winter and summer in 5 woody dominants in the communities described in Table 1 legend

Community, species	Leaves		Aerial wood		Roots	
	Summer	Winter	Summer	Winter	Summer	Winter
Chamical						
Aspidosperma quebracho blanco	1.61	1.69	0.98	1.38	1.68	1.40
*P. nigra**	2.94	1.26	1.45	1.20	1.68	0.95
*C. praecox**	4.20	1.06	1.54	0.98	2.1	1.12
*Acacia caven**	3.24	**	2.80	1.54	1.68	1.96
Larrea divaricata	2.31	1.14	2.03	1.38	2.94	1.46
Ñacuñán						
*P. flexuosa**	2.11	**	1.15	1.36	na	na
*G. decorticans**	2.51	**	1.03	1.19	na	na
L. divaricata	1.98	1.20	1.17	1.22	na	na
Atriplex lampa	2.11	0.81	1.13	0.54	na	na
Lycium tenuispinosum	2.84	**	0.91	0.88	na	na

* Legumes
** species without leaves in the winter
na, not available

Table 5. Total nitrogen (kg N ha^{-1}) in the woody dominants of the communities described in Table 1 legend. Values are for summer-collected samples

Community, species	Aerial			Roots	Total
	Leaves	Wood	Total		
Chamical					
Aspidosperma quebracho blanco	0.5	1.6	2.1	2.9	4.9
*P. nigra**	0.6	2.8	3.4	2.2	5.6
*C. praecox**	1.7	6.3	8.0	3.0	11.1
*Acacia caven**	1.9	9.0	10.9	3.7	14.7
Larrea divaricata	17.0	55.9	76.9	30.4	103.0
Ñacuñán					
*P. flexuosa**	3.3	67.3	70.6	na	na
*G. decorticans**	1.1	12.5	13.6	na	na
L. divaricata	3.4	17.2	20.6	na	na
Atriplex lampa	2.7	8.7	11.4	na	na
Lycium tenuispinosum	0.7	2.5	3.2	na	na

* Legumes
na, not available

tenuispinosum in the Ñacuñán community, per cent N in summer leaves of legumes is greater than in leaves of other species. In general, %N tends to decrease

from summer to winter, although this is not always the case (*e.g. Aspidosperma* sp.
leaves and wood, and wood of several Ñacuñán species).

Carbon and energy contents of the community dominants (Table 6) also varied
with species, plant part and season, though general trends were not apparent.

Nitrogen and carbon contents of the litter layer in the Chamical community
(Table 7) varied by canopy species but not apparently by season. Litter layer
nitrogen was highest under *Aspidosperma* sp. canopies (0.45 *vs* 0.05–0.06 g N
m^{-2}) mainly because of a greater litter biomass under this species.

Table 6. Total carbon (%C) and (inside parentheses) energy (kcal g^{-1}) contents of 5 woody dominants
in the communities described in Table 1 legend in winter and summer

Community, species	Leaves		Aerial wood		Roots	
	Summer	Winter	Summer	Winter	Summer	Winter
Chamical						
Aspidosperma quebracho blanco	44 (4.9)	32 (4.8)	35 (4.1)	30 (4.2)	42 (4.8)	36 (4.7)
P. nigra*	33 (4.8)	48 (4.8)	35 (4.5)	42 (4.7)	36 (4.7)	44 (4.3)
C. praecox*	37 (5.0)	41 (4.3)	36 (4.8)	38 (4.6)	27 (4.7)	41 (4.4)
Acacia caven*	34 (4.5)	**	29 (4.5)	42 (3.8)	36 (4.2)	49 (4.5)
Larrea divaricata	37 (4.4)	34 (4.9)	33 (4.6)	38 (4.0)	33 (5.4)	55 (4.3)
Ñacuñán						
P. flexuosa	46 (4.9)	**	42 (4.4)	33 (4.7)	na	na
G. decorticans	42 (4.5)	**	38 (4.5)	32 (4.4)	na	na
L. divaricata	39 (4.7)	35 (4.9)	39 (4.6)	43 (4.6)	na	na
Atriplex lampa	29 (3.6)	28 (3.7)	41 (4.6)	39 (4.4)	na	na
Lycium tenuispinosum	25 (3.3)	**	42 (4.9)	44 (4.5)	na	na

* Legumes
** species without leaves in the winter

Table 7. Total carbon (%C) and total nitrogen (%N) in the winter and summer litter layers underneath
individual plant canopies of 5 woody dominants in the Chamical community described in Table 1
legend. N-values expressed as g N m^{-2} refer to N-distributions beneath individual plants; kg N ha^{-1}
values refer to community-wide means. %C and %N values are means of 3 individual plants per
species

Species	Summer		Winter			
	%C	%N	%C	%N	g N m^{-2}	kg N ha^{-1}
Aspidosperma quebracho						
blanco	28	1.17	28	1.44	0.45	0.10
P. nigra*	22	1.55	26	1.64	0.06	0.04
C. praecox*	20	1.54	23	1.41	0.05	0.02
Acacia caven*	24	1.62	25	1.56	0.06	0.01
Larrea divaricata	26	1.51	27	1.61	0.06	0.14

* Legumes

Soil nitrogen was higher in the nonvegetated areas as of the Chamical community than in those of the Ñacuñán, though the reverse was true for per cent carbon (Table 8). There was little difference in per cent N or C with soil depth in these areas in either system.

In general, soil nitrogen and carbon were substantially greater under the canopies of individual plants than in the open, and in many cases these differences were more pronounced directly under the canopy than at the edge of the canopy (Table 9). Trends with soil depth were also more obvious under canopies, with both per cent N and C usually decreasing between 5 and 45 cm depth and often between 45 and 90 cm. The sharpest decrease with depth occurred under *Acacia caven* in the Chamical community; this species has a shallow, predominantly horizontal root system. No consistent differences in soil per cent N and C were detected between winter and summer values in either community (data not shown).

Conclusions

1. Marked differences in biomass nitrogen and net primary production exist between two xerophytic forest communities of different ages, among species within these communities, and among plant parts within each species. Legumes appear to have a generally higher leaf-N content, but legume *vs* non-legume differences in wood and root per cent N are not apparent.

2. Litter accumulation under individual canopies leads to marked differences in soil N and C contents among species and especially between canopy and non-canopy (bare) areas. Fertility islands result.

3. Highest soil N and C contents occur in the upper soil profile. The sharpest decrease with depth occurred beneath the shallow-rooted *Acacia caven* canopy.

Acknowledgements This research was performed by members of the Argentine Institute for Research on Arid Lands (IADIZA), within the framework of the Organization of American States (OAS) Special Project on Arid Zones.

Table 8. Total soil nitrogen (%N) and carbon (%C) in open areas of communities described in Table 1 legend

Community	Depth (cm)	%N	%C
Chamical	5	0.08	0.33
	45	0.07	0.43
	90	0.05	0.33
Ñacuñán	5	0.04	0.25
	45	0.05	0.22
	90	0.05	0.25

Table 9. Total soil nitrogen (%N) and carbon (%C) under and at the edges of the canopies of individual dominants in the communities described in Table 1 legend. Values are means of samples taken from under 3 plants of each species

Community, species	Soil depth (cm)	%N		%C	
		Under	Edge	Under	Edge
Chamical					
Aspidosperma quebracho blanco	5	0.12	0.10	0.77	0.52
	45	0.10	0.08	0.37	0.47
	90	0.08	0.07	0.25	0.44
*P. nigra**	5	0.11	0.12	0.71	0.52
	45	0.08	0.07	0.38	0.47
	90	0.08	0.06	0.31	0.44
*C. praecox**	5	0.12	0.12	0.77	0.74
	45	0.10	0.08	0.33	0.49
	90	0.10	0.08	0.32	0.35
*Acacia caven**	5	0.17	0.10	1.02	0.98
	45	0.13	0.09	0.40	0.52
	90	0.10	0.05	0.32	0.49
Larrea divaricata	5	0.10	0.11	0.36	0.60
	45	0.09	0.09	0.35	0.42
	90	0.07	0.07	0.27	0.40
Ñacuñán					
P. flexuosa	5	na	0.06	na	0.27
	45	na	0.05	na	0.17
	90	na	0.04	na	0.25
*G. decorticans**	5	na	0.07	na	0.54
	45	na	0.05	na	0.44
	90	na	0.08	na	0.15
L. divaricata	5	na	0.06	na	0.34
	45	na	0.06	na	0.25
	90	na	0.05	na	0.20
Atriplex lampa	5	na	0.04	na	0.18
	45	na	0.05	na	0.15
	90	na	0.06	na	0.15
Lycium tenuispinosum	5	na	0.04	na	0.34
	45	na	0.04	na	0.45
	90	na	0.04	na	0.22

* Legumes
na, not available

References

1   Braun Wilke R H, Candia R J, Leiva R, Paez M N, Stasi C R and Wuilloud C F 1978 Primary aboveground net productivity in the *Prosopis flexuosa* community of Ñacuñán (Mendoza, Argentina). Deserta 5, 7–43 (In Spanish, English summary).

2   Braun Wilke R H, Lasso R H, Cordero A, Ramacciotti J, Leiva R A, Medero R N and Paez M
    N 1980 Distribution and balance of phytomass and nutrients in two communities resulting from
    clearing of Los Llanos, La Rioja Province, Argentina. Deserta 6, 59–90 (In Spanish, English
    summary).
3   Braun Wilke R H and Candia R J 1980 Caloric values and nitrogen and carbon contents in plant
    species of the *Prosopis flexuosa* community of Ñacuñán, Mendoza (Argentina). Deserta 6, 91–99
    (In Spanish, English summary).

Plant and Soil 67, 325–332 (1982). 0032-079X/82/0673-0325$01.20.   SU-29

The nitrogen cycle in a 'Terra Firme' rainforest on oxisol in the Amazon territory of Venezuela

Ciclo de nitrógeno de un bosque pluvial de Tierra Firme sobre oxisol en el Territorio Amazonas de Venezuela

C. JORDAN, W. CASKEY,
Institute of Ecology, University of Georgia, Athens, GA 30602, USA

G. ESCALANTE, R. HERRERA,
Centre de Ecologia, Instituto Venezolana de Investigaciones Científicas (IVIC), Caracas, Venezuela

F. MONTAGNINI, R. TODD and C. UHL
Institute of Ecology, University of Georgia, Athens, GA 30602, USA

Key words Amazonas Denitrification N-conservation mechanisms N-cycling N_2-fixation Oxisol Rainforest Venezuela

Abstract Standing stocks and fluxes of nitrogen, including nitrogen fixation and denitrification, were measured in a tropical rainforest on Oxisol in the Amazon Territory of Venezuela. The standing stock of nitrogen was comparable to that of temperate forests, but was higher than that in an adjacent forest on Spodosol. Fluxes were higher than in forests in the temperate zone, but lower than in another tropical forest on more fertile soil. Even though nitrogen was abundant, this does not mean that nitrogen could not be limiting to agriculture if the forest is cleared and the land cultivated. The nitrogen fixing and nitrogen conserving mechanisms are dependent upon the structure of the undisturbed forest, and destruction of the forest would eventually decrease the input of nitrogen to the soil.

Resumen Los contenidos y flujos de nitrógeno, incluyendo fijación de nitrógeno y denitrificación, se midieron en un bosque lluvioso tropical que crece sobre oxisoles en el Territorio Amazonas de Venezuela. El contenido de nitrógeno era comparable con el de bosques templados, pero era más elevado que en un bosque adyacente sobre spodosoles. Los flujos eran más elevados que en bosques de zonas templadas, pero más bajos que en otro bosque tropical de suelos más fértiles. A pesar de que el nitrógeno era abundante, esto no significa que este elemento no podría ser un factor limitante para la agricultura si el bosque se corta, y se cultiva la tierra. Los mecanismos de fijación y conservación de nitrógeno dependen de la estructura del bosque no perturbado, y la destrucción del bosque eventualmente disminuiría el suministro de nitrógeno al suelo.

Introduction

The San Carlos project is an ecological study of a tropical rain forest in the Amazon Territory of Venezuela[10, 21]. Because it has been hypothesized that nutrient element shortages may limit production in the central Amazon region, the project has focused to a large extent on nutrient cycling, including physiological and morphological adaptations to low nutrient conditions.

We have published papers dealing with calcium, potassium, magnesium and phosphorus cycling in the forest at San Carlos[14,19,27]. These reports have emphasized the nutrient-conserving mechanisms of vegetation living on soil with a low nutrient content. Among the nutrient-conserving mechanisms described are: superficial roots which are very efficient in taking up nutrients released by decomposing organic matter[16,27], mycorrhizae which also increase the efficiency of nutrient uptake[10], rapid turnover of small roots[13], sclerophyllous leaves which are resistant to insect attack and to leaching[24], secondary plant compounds which discourage herbivory[26], epiphylls which adsorb cations and fix nitrogen[19], thick bark which insulates the phloem from leaching by stem flow[12], and large pore spaces in the soils which result in rapid drainage and decreased opportunity for nutrient leaching[8].

Jordan et al.[18] have shown how low pH and high tannin content of the root humus layer at San Carlos is effective in conserving nitrogen through depressing populations of denitrifying bacteria.

Herrera and Jordan[9] have studied the nitrogen cycle in a 'caatinga' forest on Spodosol sands as part of the San Carlos project. The water table in the San Carlos caatingas often reaches the soil surface during heavy storms, but the level subsides as soon as the rain ceases. The caatinga forest appears to be one of the most nitrogen-poor forests in the tropics[9]. The standing crop of nitrogen in the ecosystem is less than one half of that of other tropical forests. In this report we describe the standing stock and fluxes of nitrogen in a 'terra firme' forest on Oxisols, just a few hundred meters away from the caatinga site. The terra firme forest contrasts with the caatinga forest in that the former rarely if ever floods, is underlain by a different type of soil, and has a distinct vegetation type.

Study site

The study area is near the confluence of the Casiquiare River and the Rio Negro in the Amazon Territory of Venezuela ($1°56'$ N, $67°03'$ W, 119 m altitude). The climate is typically equatorial with a $26°C$ annual mean temperature, and average rainfall is about 3500 mm. No month receives less than 100 mm of rain on the average, but there is considerable monthly and yearly variability in rainfall. River level follows a modal pattern with the peak in July–August about 6 m higher than the low in January–February[4]. The topography consists of gently rolling hills to 50 m above the surrounding terrain. The hills are composed of plinthitic material (laterite), which contains abundant ferric concretions near its contact with overlying sand.

The sand horizon forms an irregular cover over the clay, and may have formed in place or may be sediment washed in from the sandstone remnants of the Guiana Shield arising to the north and east. With the exception of the variable sand cover, the soil is similar to the Xanthic Ferrasol soil type which is mapped as the dominant soil in the central Amazon Basin[5]. Ferrasols are equivalent to Oxisols in the U.S. soil taxonomy system and to Latasols in the Brazilian system[22]. The soils are locally referred to as 'terra firme'. In the depressions between the terra firme hills there are acid Spodosols with well-developed B horizons; these support vegetation similar to that of the Brazilian Amazon caatinga or campina and to that of the heath forests of Borneo.

The study reported here was carried out on an undisturbed site 4 km east of the village of San

Carlos de Rio Negro, Venezuela. Terra firme hills are common close to San Carlos, but Spodosols probably are the most common soil type in the Rio Negro-Casiquiare region of Venezuela[8]. The terra firme site supports a weakly seasonal but predominantly evergreen, very complex equatorial-tropical lowland forest. It is 25–40 m in height.

Methods

Biomass of all major components of the terra firme forest have been determined by Jordan and Uhl[17]. During the biomass determination, subsamples were taken from the leaves, bark, heartwood, sapwood, twigs, and branches of 42 individuals of 28 species. Root samples were taken from 18 soil pits by Stark and Spratt[28]. Samples were dried, digested, and analyzed for total nitrogen by standard methods[1]. Forty samples of soil and surface mat also were taken and analyzed for total nitrogen. Because more than 90% of the roots occurred above the sand/clay interface at an average depth of 16 cm, mass of soil was calculated to this depth.

Leaf-fall was measured for one year. Total leaf-fall was collected in 42 0.12-m^2 litter traps, dried, and weighed. Subsamples were analyzed for total nitrogen.

Measurements of ammonium and nitrate nitrogen in precipitation, dry-fall, and throughfall began in 1975. Throughfall was measured for one year only. Precipitation is still being measured and data through 1979 are included here. Methods of collection and analysis have been given by Jordan[12] and Jordan et al.[19] Nitrogen in subsurface runoff was determined by multiplying ammonium and nitrate nitrogen concentrations in collections from zero-tension lysimeters (soil water collectors) located below the root zone times total subsurface runoff determined from the water balance.

Nitrogen fixation was estimated by the acetylene reduction technique[6] using a $3:1$ $C_2H_4:N_2$ fixation ratio. Samples were taken from leaves of 18 species (half of the leaves with and half without epiphyllic lichens), bark of four species with and without lichens, three root-mat samples, and three soil samples in March 1977, July 1977, and February 1978. Fixation rates multiplied by the mass of the respective compartments gave total fixation rates.

Denitrification potentials were estimated from measurements of denitrifying activity in soil slurries during Phase I of denitrification[23]. Triplicate samples of the humus layer and mineral soil were taken in January and July 1980. Total potential denitrification rates were determined by multiplying measured rates times the mass of the respective compartments.

Results and discussion

Results of this study are compared to the results of the caatinga study at San Carlos as well as to one other tropical study and three temperate studies in Table I. Several striking patterns emerge from the data. First, the mass of nitrogen in the leaves and stems of the terra firme forest at San Carlos is more similar to that of the Ivory Coast forest than it is to the caatinga forest. Because the aboveground biomasses of the three tropical forests are within the same range (300–380 $\times$ 10^3 kg dry matter ha^{-1}), the overall concentration of nitrogen in the caatinga forest is less than in the other two.

A large difference in concentration of nitrogen also occurred in the leaves and stems of the temperate forests shown in Table 1. The Douglas fir forest, with a biomass of 720 $\times$ 10^3 kg ha^{-1} had aboveground nitrogen stocks of 538 kg N ha^{-1}, while an angiosperm forest at Hubbard Brook, with only one-fourth the biomass, had 351 kg N ha^{-1}. Sollins et al.[25] speculated that the nitrogen concentration in the Douglas fir forest is lower than that in the angiosperm forest

Table 1. Nitrogen cycling in the terra firme forest compared to other tropical and temperate forests. Pools are in kg N ha^{-1} and fluxes in kg N ha^{-1} yr^{-1}. Brackets indicate summed compartments

	Amazon Rain Forest Terra Firme[a,g]	Amazon Rain Forest Caatinga on Podsol[b]	Seasonal Forest Banco 1 Ivory Coast[c]	Hardwood Forest Coweeta North Carolina[d]	Hardwood Forest Hubbard Brook New Hampshire[e]	Douglas Fir Andrew Forest Oregon[f]
Pools						
Leaves	143	72	⎱1150	⎱995	⎱351	144
Stems and branches and bark	941	264	⎰	⎰	⎰	394
Roots	586	834	–	–	181	197
Litter and superficial humus	406	132	–	140	1100	798
Soil	3507	785	6500	6917	3626	3397
Total	5583	2087	7650	8052	5258	4930
Inputs						
NH$_4$–N in precipitation	11.3	⎱21.0	21.2	2.7	⎱6.5	⎱2.0
NO$_3$–N in precipitation	0.2	⎰		3.6	⎰	⎰
N-fixation	16.2	>35	–	12.0	14.2	2.8
Outputs						
NH$_4$–N leached	8.4	⎱9	⎱21.2	0.06	⎱4.0	⎱1.5
NO$_3$–N leached	5.7	⎰	⎰	0.1	⎰	⎰
Denitrification	2.9	–	–	10–18	–	–
Balance (input-output)	+8.9			~0	+16.7	+4.3
Internal fluxes						
N in leaf-fall	61.3	⎱24	⎱170	⎱33	⎱54.2	⎱10.8
NH$_4$–N in throughfall	25.0	9	80	4	9.3	3.4
NO$_3$–N in throughfall	0.3					
Total fluxes	86.6	33	250	37	63.5	14.2

[a] This study.
[b] Herrera and Jordan[9].

[e] Bormann et al.[3]
[f] Sollins et al.[25]
[g] Jordan et al.[19]

at Hubbard Brook, New Hampshire, because there is a higher nitrogen content in the rain at Hubbard Brook than in Oregon. While this may be the case, one might expect that if nitrogen were limiting in the Douglas fir forests, the trees would take it up as quickly as it entered the system and thereby minimize nitrogen leaching losses. With low N-leaching losses for several decades, differences in input should not affect the overall nitrogen concentration because biomass should increase proportionately. In any event, rainfall cannot explain the differences in nitrogen concentration between the terra firme forest and the caatinga forest because both receive the same precipitation inputs.

Another explanation for the difference in overall nitrogen concentration between the Douglas fir forest and the angiosperm forest at Hubbard Brook is that the angiosperm trees dominating at Hubbard Brook contain more ray parenchyma (living tissue with high nitrogen content) in woody tissues than do gymnosperm trees dominating in Oregon[25]. This may partly explain the terra firme-caatinga difference as well. Jordan and Herrera[14] have hypothesized that the structure and function of the caatinga forest is more similar to gymnosperm forests of the temperate zone than it is to angiosperm forests. Although they did not compare parenychma tissues, the caatinga forest may be similar to gymnosperm forests in this respect also.

Despite the low standing crop of nitrogen in the caatinga forest, nitrogen in the roots was higher than in any other forest shown in Table 1. This is due to the relatively large biomass of roots in the caatinga forest which, in turn, may result from the nutrient scarcity in the soil[9].

Another striking pattern that emerges from Table 1 is the relatively high input of nitrogen in precipitation in the tropical ecosystems examined. This is due in part to a greater amount of precipitation in tropical ecosystems than in the temperate systems cited, but higher concentrations of nitrogen, principally in the ammonium form, also play a role.

Rates of nitrogen fixation are also higher in the tropical ecosystems. Nitrogen fixation probably is due in large part to the blue-green algae in both the free living form and in lichens[20,29]. These algae and lichens live on surfaces of leaf, bark, and soil. Since they are active the entire year in the wet tropics, total annual nitrogen fixed is greater in these regions.

Comparisons of potential denitrification among these systems are difficult to make because of the general lack of data.

An important trend shown in Table 1 is that total biological fluxes of nitrogen into, through, and out of the terre firme forest are greater than for any other forest listed except for the forest in Ivory Coast. In other publications about the nutrient cycles of the Amazon forest ecosystems we have emphasized that the Amazonian forest is very conservative of nutrient elements. For example, in the report of nitrogen cycling in the caatinga forest we emphasized the conservation mechanism of N-transfer from the leaf into the twig before the leaf is abscised. However, in the terra firme forest conservation of nitrogen does not seem to be so important, judging from relative rates of internal cycles and nitrogen-loss rates. The reason for this may be that in contrast to other nutrients, the supply of nitrogen is not dependent on the nutrient-holding and supplying capacity of the soil. Rather, although nitrogen-leaching is high, the nitrogen-fixing capacity of the superficial organisms and the legumes in the forest, plus the nitrogen in the rainfall, appear to be sufficient to meet the needs of the undisturbed forest.

This does not mean that nitrogen will not be a problem if the forest is cut and converted to some other use such as agriculture or pasture. The nitrogen-fixing organisms are dependent upon the undisturbed forest for survival. We found virtually no nitrogen fixation in the soils and plants of a slash and burn plot on Oxisol and adjacent to the terra firme forest (Montagnini and Escalante, unpubl.). Also, in the undisturbed forest, surfaces such as leaves, bark, roots, and the soil organic matter are important scavengers of nitrogen from the rainwater which passes through the system[19,27]. Many of these mechanisms are destroyed when the forest is cleared.

Conclusions

In contrast to other nutrient element supplies in the terra firme forest of the Amazon Basin, the supply of nitrogen does not appear to be highly critical. However, since nitrogen-fixing organisms and nitrogen-scavenging mechanisms are parts of the undisturbed forest N-cycle that are sensitive to disturbance, conversion of the forest to field or pasture could result in nitrogen shortages.

Acknowledgements The San Carlos project is a multidisciplinary, multinational project directed through the Ecology Center, Instituto Venezolano de Investigaciones Cientificas, Caracas, Venezuela, with European participation coordinated through the Max Planck Institute for Limnology, Plön, Germany, and North American participation coordinated through the Institute of Ecology, University of Georgia, Athens, Georgia. Financial support comes in part from CONICIT de Venezuela, US National Science Foundation, OAS, and Unesco.

References

1   Allen S E, Grinshaw H M, Parkinson J A and Quarmby C 1974 Chemical Analysis of Ecological Materials. Blackwell, Oxford.
2   Bernhard-Reversat F 1975 Recherches sur l'ecosysteme de la foret subequatoriale de basse Cote-D'Ivoire VI. Les cycles des macro-elements. La Terre et la Vie 29, 229–254.
3   Bormann F H, Likens G E and Melillo J M 1977 Nitrogen budget for an aggrading northern hardwood forest ecosystem. Science 196, 981–983.
4   Brunig E E, Herrera R, Heuveldop J, Jordan C, Klinge H and Medina E 1977 The international Amazon project. *In* Transactions of the International MAB-IUFRO Workshop on Tropical Rain Forest Ecosystems Research, pp 104–120. Brunig E F (Eds.). University of Hamburg, Hamburg-Reinbek.
5   FAO-UNESCO 1971 Soil map of the world. Vol. IV. South America. Food and Agricultural Organization, Rome.
6   Hardy R W F, Holsten R D, Jackson E K and Burns R C 1968 The acetylene-ethylene assay for N_2 fixation: laboratory and field evaluation. Plant Physiol. 43, 1185–1207.

7   Henderson G S, Swank W T, Waide J B and Grier C C 1978 Nutrient budgets of Appalachian and Cascade region watersheds: a comparison. For. Sci. 24, 385–397.

8   Herrera R 1979 Nutrient distribution and cycling in an Amazon caatinga forest on spodosols in southern Venezuela. Ph.D. thesis, Dept. of Soil Science, Univ. of Reading.

9   Herrera R F and Jordan C F 1981 Nitrogen cycle in a tropical rain forest of Amazonia: the case of low mineral nutrient status in the Amazon caatinga. In Terrestrial Nitrogen Cycles. Processes, Ecosystem strategies and Management Practices. Clark F E and Rosswall T (Eds.). Ecol. Bull. Stockholm 33, 493–505.

10 Herrera R, Jordan C F, Klinge H and Medina E 1978 Amazon ecosystems: their structure and functioning with particular emphasis on nutrients. Interciencia 3, 223–232.

11 Herrera R, Merida T, Stark N and Jordan C F 1978 Direct phosphorus transfer from leaf litter to roots. Naturwissenschaften 65, 208.

12 Jordan C F 1978 Stem flow and nutrient transfer in a tropical rain forest. Oikos 31, 257–263.

13 Jordan C F and Escalante G 1980 Root productivity in an Amazonian rain forest. Ecology 61, 14–18.

14 Jordan C F and Herrera R 1981 Tropical rain forests: are nutrients really critical? Am. Nat. 117, 167–180.

15 Jordan C F and Heuveldop J 1981 The water budget of an Amazonian rain forest. Acta Amazonica 11, 87–92.

16 Jordan C F and Stark N 1978 Retencion de nutrientes en la estera de raices de un bosque pluvial Amazonico. Acta Cientifica Venezolana 29, 263–267.

17 Jordan C F and Uhl C 1978 Biomass of a 'terra firme' forest of the Amazon Basin. Oecol. Plant. 13, 387–400.

18 Jordan C F, Todd R L and Escalante G 1979 Nitrogen conservation in a tropical rain forest. Oecologia 39, 123–128.

19 Jordan C F, Golley F, Hall J and Hall J 1980 Nutrient scavenging of rainfall by the canopy of an Amazonian rain forest. Biotropica, 12, 61–66.

20 Mague T H 1977 Ecological aspects of dinitrogen fixation by bluegreen algae. In A Treatise on Dinitrogen Fixation, pp 85–140. Hardy R W F and Gibson A H (Eds.). Wiley, New York.

21 Medina E, Herrera R, Jordan C and Klinge H 1977 The Amazon project of the Venezuelan Institute for Scientific Research. Nat. Resour. 13, 4–6.

22 Sanchez P A 1976 Properties and Management of Soils in the Tropics. Wiley, New York. 618 p.

23 Smith M S, Firestone M K and Tiedje J M 1978 The acetylene inhibition method for short-term measurement of soil denitrification and its evaluation using nitrogen-13. Soil Sci. Soc. Am. J. 42, 611–615.

24 Sobrado M A and Medina E 1980 General morphology, anatomical structure, and nutrient content of sclerophyllous leaves of the 'bana' vegetation of Amazonas. Oecologia 45, 341–345.

25 Sollins P, Grier C C, McCorison F M, Cromack K and Fogel R 1980 The internal element cycles of an old-growth Douglas fir ecosystem in western Oregon. Ecol. Monogr. 50, 261–285.

26 Sprick E G 1979 Composicion mineral y contenido de fenoles foliares de especies leñosas de tres bosques contrastantes de la region Amazonica. Tesis de grado Licenciado, Univ. Central de Venezuela. Facultad de Ciencias, Escuela de Biologia. Caracas, Venezuela. 242 p.

27 Stark N and Jordan C F 1978 Nutrient retention by the root mat of an Amazonian rain forest. Ecology 59, 434–437.

28 Stark N and Spratt M 1977 Root biomass and nutrient storage in rain forest oxisols near San Carlos de Rio Negro. Trop. Ecol. 18, 1–9.

29 Stewart W D P, Sampaio M J, Isichei A O and Sylvester-Bradley, R 1978 Nitrogen fixation by soil algae of temperate and tropical soils. In Limitations and Potentials for Biological Nitrogen Fixation in the Tropics. pp 41–63. Döbereiner J, Burris R H and Hollander A (Eds.). Plenum, New York.

30 Swank W T and Douglass J E 1975 Nutrient flux in undisturbed and manipulated forest

ecosystems in the southern Appalachian mountains. Publication no. 117 de l'Association Internationale des Sciences Hydrologiques Symposium de Tokyo, Dec. 1975, pp 445–446.

31 Todd R L, Meyer R D and Waide J B 1978 Nitrogen fixation in a deciduous forest in the southeastern United States. *In* Environmental Role of Nitrogen-Fixing Blue-green Algae and Asymbiotic Bacteria. Granhall U (Ed.). Ecol. Bull. Stockholm 26, 172–177.

Plant and Soil 67, 333–342 (1982). 0032-079X/82/0673-0333$01.50. SU-30

Nitrogen cycling in the seasonally dry forest zone of Belize, Central America

Ciclo de nitrógeno en la zona de bosque seco estacional de Belize, América Central

J. T. ARNASON
Biology Department, University of Ottawa, Ottawa, Ontario, Canada

and J. D. H. LAMBERT
Biology Department, Carleton University, Ottawa, Ontario, Canada

Key words Agriculture Belize Cohune palm Dry forest Fallow Hardwood Litter N-cycling

Abstract Two forest associations, cohune palm (Cohune Ridge) and mixed tropical hardwood (High Bush), were assessed on the basis of nutrient movement and storage for their suitability for agriculture. Continuous monitoring of soil nitrogen and leaf litterfall over a one-year period provided information on soil building processes in the forest fallow. Destructive cuts revealed the storage of 690 kg N ha^{-1} in the standing biomass of the Cohune forest versus 203 kg N ha^{-1} in the High Bush. Litter biomass was exceptionally high in the Cohune Ridge (497 kg ha^{-1} dry matter) as compared to the High Bush (65 kg ha^{-1} dry matter) and other tropical forests. This is probably because of a low rate of decomposition in the Cohune Ridge palm forest. A substantial reserve of nitrogen is present in both forests' fallows, and this can in part be harvested by the small farmer for crop production.

Resumen Dos asociaciones boscosas (Palma Cohune y Matorral Alto) se estudiaron en relación a sus reservas y flujos de nutrimentos y sus posibilidades de aprovechamiento agrícola. Se siguió la evolución del nitrógeno en el suelo y en la caida de hojarasca durante un año obteniéndose asi información sobre los procesos pedogenéticos en el barbecho del bosque.

   A través de muestreos destructivos se encontró que en la biomasa del bosque de Palma Cohune habían 690 kg N ha^{-1} y en el Matorral Alto solo 203 kg N ha^{-1}. La biomasa de hojarasca era excepcionalmente alta en Palma Cohune alcanzando un valor de 497 kg materia seca ha^{-1}; en el bosque de Matorral Alto la hojarasca era de 65 kg materia seca ha^{-1}. Esto probablemente se deba a la baja velocidad de descomposición en el caso de la palma.

   Las reservas sustanciales encontradas en ambos barbechos para el nitrógeno podrían ser parcialmente utilizadas para la producción agrícola campesina.

Introduction

   Although a considerable amount of information on nutrient cycling and storage in tropical forests has been collected[11, 12], data for the lowland forests of the Yucatan peninsula are limited. Our study site at Indian Church, Belize (17°45′ N, 88°40′ W) is typical of the secondary forests of the region. According to Beard's[1] classification, the site falls within the semi-evergreen seasonal forest zone. Precipitation is 1480 mm annually and a distinct dry season occurs from January to May. Soils of the region are typical Central American rendolls of the Yaxa group[13], which are underlain by moderately hard amorphous limestone. In a previous study[7], we identified six distinct plant associations by ordination of vegetation transect data. Two associations are presently used by local Maya farmers for shifting or 'milpa' agriculture and by nearby Mennonite farmers for mechanized agriculture.

As part of a Canada-Belize agriculture project, we have assessed nutrient cycles and nutrient stores in these two associations. Our analysis of one of these associations, a mixed tropical hardwood forest known locally as 'High Bush', has been reported previously[8]. The second association is dominated by the cohune palm (*Orbignya cohune*) and is known locally as 'Cohune Ridge'. Cohune Ridge and High Bush forests often grade into each other. The Cohune forest usually occurs on level lowland sites with deep clay soils while High Bush is more prevalent on rocky upland sites.

Nitrogen cycling in the Cohune forest forms the basis of this report and a summary of our data for the High Bush site is used for an unusual comparison of a monocotyledonous and a dicotyledonous forest. Assessment of the major nitrogen pools and fluxes is critical to the understanding and improvement of traditional agro-ecosystems[11,12], as well as for the efficient management of these areas as forestry resources[6].

Materials and methods

Vegetation was sampled in line transects of ten 10×10 m plots. Within each quadrat the presence of all tree species was recorded as well as each tree's basal area (ba) at breast height. Relative frequency, density and dominance were summed to yield an importance value to 300[14]. Nomenclature follows Standley and Steyermark[16].

Destructive sampling of the vegetation was carried out on three representative 10×10 m plots. All trees (≥ 80 cm^2 ba), saplings (< 80 cm^2 ba and ≥ 150 cm high) and seedlings (saplings < 150 cm high) were harvested and separated into leaves, wood greater than 2.5 cm diameter, and wood less than 2.5 cm diameter. Material was weighed on site and five representative samples of each part were collected for each plant harvested. Standing litter (FF layer) was collected from 15 randomly placed 1×1 m plots in February 1978. Litterfall was collected biweekly from four 1×1 m screens placed at random in the forest. All samples were dried at 60°C for biomass determinations and then ground in a Wiley mill and analysed for total nitrogen in triplicate using a semi-micro Kjeldahl method[15].

Five 12-cm cores from each of four forest sites were collected monthly, composited by site, air dried, and ground to pass a 2-mm seive. Soil profile samples were collected below the FF layer from 0–12, 12–24, and 24–36 cm depths in a similar manner in February 1978. Total soil nitrogen was determined for the replicates per site composite as described for plant material. Organic matter was estimated by removing carbonates with HCl before combustion at 450°C for 4 hours. For calculation of C:N ratios, percentage C was assumed to be 45% of organic matter.

Stand age was estimated on the basis of historical information and tree ring analyses.

Results and discussion

The High Bush site, *ca.* 40 years old, was a mixture of tropical hardwoods dominated by *Guazuma ulmifolia*. The closed canopy was about 20 m high except

Table 1. Tree composition of High Bush forest. Importance is the sum of relative frequency, density and dominance. Values are based on those trees with basal area $\geq 80\ cm^2$

Species	Importance	Stems ha^{-1}
Guazama ulmifolia	76	190
Trichilia havannesis	40	100
Spondias mombin	32	35
Enterolobium cyclocarpum*	26	5
Cupania belizenis	22	55
Trichilia hirta	22	55
Sapindus saponaria	19	40
Parmentiera aculeata	13	30
Hirtella paniculata	6	10
Anona cherimola	6	10
Pithecolobium sp.*	6	5
Cedrela mexicana*	4	5
Dendropanax arboreus	4	5
Piper amalago	3	5
Lonchocarpus guatemalensis	3	5
Bursera simaruba	3	5
Stemmandenia donnell-smithii	3	5
Simaruba glauca	3	5
Citrus sp.	2	5
Guarea giabra	2	5
Erythoxylon affine	2	5
Trophis racemosa	2	5
Total		590

* Emergent trees (trees above closed canopy)

for occasional emergent trees such as *Enterolobium cyclocarpum*. During the dry season, several species such as *Cedrela mexicana, Spondias mombin* and *Enterolobium cyclocarpum* shed their leaves. The Cohunge Ridge association (Table 2) was especially striking because of a cathedral quality created by the tall palms (30–35 m) and a dense canopy that created a cool and dim atmosphere even at midday. Ground cover and lianas were sparse compared to the high bush, and enormous piles of litter (to 1.5 m) had accumulated at the foot of the large palms. The age of this site was *ca.* 65 years.

The destructive cuts revealed that in the Cohune Ridge forest, *Orbignya cohune* stored the largest amounts of above-ground live-tree biomass nitrogen in both the sapling layer and the tree layer (76% and 84% of total live-tree biomass-N respectively, Tables 3 and 4). This was mainly by virtue of its large biomass, since the percentage N of *O. cohune* (the only monocot in the forest sample) was fairly low compared to that of the dicot species, especially in the leaves and branches. The ability to produce a large amount of photosynthate with a minimum amount

Table 2. Tree composition for Cohune Ridge Forest. See Table 1 legend for further explanation

Species	Importance	Stems ha^{-1}
Orbignya cohune[a]*	100	120
Spondias mombin	44	90
Sabal mauritiiformis[a]	34	80
Cupania belizensis	20	50
Bursera simaruba	18	40
Matayba sp.	16	30
Swartzia cubensis	10	20
Trichilia havannesis	10	20
Guarea giabra	8	20
Hirtella paniculata	7	20
Pseudolmedia sp.	5	10
Unknown sp.	5	10
Trophis racemosa	5	10
Unknown sp.	5	10
Coccoloba belizensis	5	10
Protium copal	5	10
Nectandra sp.	5	10
Total		540

* Emergent trees
[a] monocots (Palmae)

of nitrogen (a high nitrogen use efficiency) may be a factor in the success of this species as has been suggested for C_4 plants[2].

In the sapling layer several small palms contributed to the biomass. These include *Chrysophyla argentia*, *Sabal mauritiformis*, *Bactris major*, and *Chamaedorea sp.* Here again percent N was significantly lower in palms than in the remaining dicots (1.35 *vs* 1.77%N in leaves, respectively; $p = 0.1$). Percent-N in leaves of all species tested was moderate ($< 3\%$N); legumes are uncommon and were, in fact, not present in the sample.

The total amounts of above-ground biomass and nitrogen in the woody vegetation and litter of the Cohune Ridge forest (Table 5) are comparable to values reported for other mature tropical forests throughout the world[11,12]. Totals for the High Bush forest (Table 5) are lower, probably because of the lower age of the forest. The overall concentration of nitrogen in the above-ground woody vegetation was comparable (0.38%N for the Cohune Ridge and 0.36%N for the High Bush). An extraordinarily high amount of litter (43.6×10^3 kg ha^{-1}) was present in the Cohune Ridge forest, a value that exceeds others reported for tropical forests: 1.16×10^3 kg ha^{-1} in Trinidad (Cornforth,[3]), 13.0×10^{-3} kg ha^{-1} in Guatemala[4] and $2.91 \times 10^3 - 14.1 \times 10^3$ kg ha^{-1} in upland forests of Panama[5]. One exception is a mangrove forest in Panama with 102×10^3 kg dry

Table 3. Live-tree above-ground biomass (10^3 kg dry matter ha^{-1}) and nitrogen stores (kg N ha^{-1}) in the Cohune Ridge association in February, 1978. Values do not include those for saplings

Species	Wood				Leaves		Fruit		Flowers		Total	
	≥ 2.5 cm diam.		< 2.5 cm diam.									
	Biomass	N	Biomass	N	Biomass	N	Biomass	N	Biomass	N	Biomass	N
Orbignya cohune[a]	77.9	187	50.0	192	7.05	113	2.11	11.1	0.456	8.18	138	511
Guazuma ulmifolia	13.9	20.0	0.991	5.01	0.116	1.95					15.0	27.0
Trophis racemosa	6.88	21.3	0.680	6.75	0.090	2.01					7.65	30.1
Simaruba glauca	3.57	8.91	0.330	2.25	0.081	1.19					3.98	12.4
Unknown sp.	2.06	3.25	0.467	6.18	0.052	1.43					2.58	10.9
Swietenia macrophylla	0.870	0.88	0.377	2.01	0.253	3.25					1.50	6.14
Coccoloba belizensis	1.22	6.89	0.281	3.78	0.100	1.62					1.60	12.3
Spondias mombin	0.183	0.83	0.018	0.00	0.002	0.05					0.20	0.88

[a] monocots (Palmae)

Table 4. Biomass (kg dry matter ha^{-1}) and nitrogen contents (kg N ha^{-1}) of saplings (trees < 80 cm^2 basal area) in Cohune Ridge forest. Note different biomass units from those in Table 3

	Wood		Leaves		Total	
	Biomass	N	Biomass	N	Biomass	N
Orbignya cohune[a]	5660	17.9	2490	34.2	8150	52.1
Sabal mauritiiformis[a]	288	1.28	150	0.991	438	2.27
Protium copal	2.2	0.027	1.8	0.027	4.0	0.054
Chamaedorea sp.[a]	1.47	0.020	0.73	0.011	2.20	0.031
Alibertia sp.	164	0.969	14.5	0.258	179	1.23
Plumeria sp.	14.6	0.087	0.30	0.005	14.9	0.092
Piper amalago	4.4	0.059	0.50	0.013	4.9	0.072
Cryosophyla argentia[a]	260	2.64	57.0	1.11	317	3.75
Hirtella paniculata	13	0.107	0.67	0.014	13.7	0.121
Pouteria magdelina	27	0.122	4.30	0.045	31.3	0.167
Bactris major[a]	120	0.938	48.0	0.625	168	1.56
Guarea giabra	257	1.08	3.7	0.983	261	1.16
Spondias mombin	178	0.620	7.0	0.095	185	0.715
Cupania belizensis	159	1.28	1.73	0.031	161	1.31
Coccoloba belizensis	523	4.05	2.90	0.052	526	4.10

[a] monocots (Palmae)

Table 5. Biomass (10^3 kg dry matter ha^{-1}) and nitrogen contents (kg N ha^{-1}) of above-ground woody vegetation and litter in the High Bush and Cohune Ridge forests

		Cohune Ridge		High Bush	
		Biomass	N	Biomass	N
Live vegetation					
Trees	Leaves	8.64	125	1.08	21.8
	Wood < 2.5 cm diameter	53.1	218	5.92	50.1
	Wood ≥ 2.5 cm	107	249	36.6	53.7
	Flowers	0.456	8.16	<0.001	<0.001
	Fruit	2.11	11.1	<0.001	<0.001
Saplings	Leaves	2.78	37.5	1.89	22.2
	Wood/Branches	7.67	31.2	9.41	30.3
Seedlings	Leaves	0.212	2.58	0.140	2.4
	Branches	0.003	0.02	0.160	1.7
Lianas	Leaves	0.022	0.38	0.050	2.0
	Wood/Branches	1.19	7.23	0.360	18.8
Total		183	690	56.0	203
Litter		43.6	480	7.18	64.8
Total		226	1170	63.2	269

litter ha^{-1}. The amount of litter biomass in the High Bush forest is comparable to or lower than these values.

The large accumulation of litter on the forest floor in the Cohune Ridge is probably due to a slow rate of decomposition rather than a large litter influx, as the annual influx of litter is lower in the Cohune Ridge than in the High Bush (Table 6). A litter-N turnover factor can be estimated from the relationship A = kL, where A is the annual litter fall, k is the rate factor for decomposition, and L is the litter on the forest floor[10].

The low k values for Cohune Ridge litter suggest a very low litter-N turnover rate in the Cohune Ridge. Ewel[4] reached a similar conclusion in Guatemala, where *Orbignya cohune* had the slowest decomposition rate of any species tested. Lignification, suberization or the presence of toxic secondary metabolites may retard decomposition.

Although the concentration of nitrogen remains relatively constant in the litter

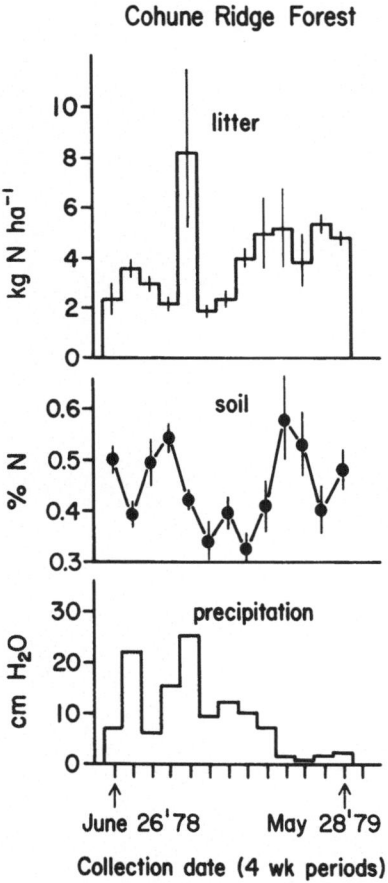

Fig. 1.   Leaf litterfall, soil total-nitrogen, and precipitation levels in the Cohune Ridge forest for May 1978 to April 1979. Vertical bars denote standard errors.

Table 6. Biomass and N-contents of litterfall and standing-litter pool and calculated turnover rate (k) for the litter pool in High Bush and Cohune Ridge forests. k = (litterfall/litter-pool)

Forest	Litterfall		Litter biomass		k	
	Biomass (10^3 kg ha^{-1} yr^{-1})	N (kg ha^{-1} yr^{-1})	Biomass (10^3 kg ha^{-1})	N (kg ha^{-1})	Biomass (10^3 kg yr^{-1})	N (kg yr^{-1})
High Bush	12.6	156	7180	64.8	1.8	2.4
Cohune Ridge	4.26	42.9	43600	480	0.10	0.09

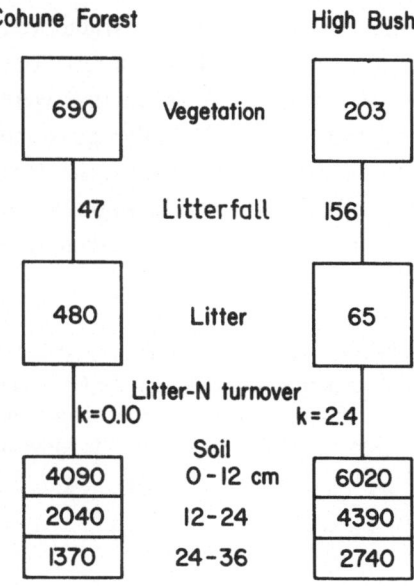

Fig. 2. Nitrogen in above-ground vegetation, litter, and soil pools (kg N ha^{-1}), nitrogen in litterfall (kg N ha^{-1} yr^{-1}) and estimated rate of litter-N turnover (kg N yr^{-1}) in the Cohune Ridge and High Bush forests.

fall, monthly collections indicated highest inputs of litter nitrogen to the forest floor during the dry season and early wet season (Fig. 1). The extraordinary peak in October is atypical and represents *ca.* 900 kg of litter ha^{-1} that fell during hurricane Greta, which passed 50 miles from the site. Total mineral-soil nitrogen levels declined during the same period that leaf litter inputs were minimal (November–January, Fig. 1). Because of the slow decomposition of the litter and large reserves on the forest floor, the effect of cool winter temperatures on microbial action is a more plausible reason than low litterfall for the low mineral-soil nitrogen levels.

Soil nutrient profiles (Table 7), particularly the upper portions[9], can provide an estimate of soil nitrogen reserves. Levels of organic matter are high in both sites, leading to high C:N ratios.

Table 7. Total-N, organic matter (OM), and C:N ratios with soil depth in Cohune Ridge forest. Values are means ($\pm$ standard errors) of 4 sites (Cohune Ridge) or 1 site (High Bush)

Depth (cm)	Cohune Ridge					High Bush		
	%N		%OM		C:N		%N	%OM	C:N
0–12	0.313	(0.01)	14.1	(0.7)	20.3	(0.5)	0.46	20.2	19.8
12–24	0.144	(0.006)	9.0	(0.2)	28.1	(0.9)	0.31	15.0	21.8
24–36	0.090	(0.005)	8.0	(0.4)	40.0	(1.1)	0.18	11.8	29.5

Nitrogen cycling in the Cohune Ridge palm forest appears to be characterized by the slow decomposition of litter and a subsequent buildup of forest floor biomass. In other ways the pools and transfer rates for nitrogen that were examined resemble those for other tropical forests (Fig. 2). Although some biomass nitrogen is volatilized during burning in the traditional agriculture cycle, much can be returned to the soil (especially that in charred wood which decomposes rapidly during wet weather[11]) and can represent a substantial reserve of nitrogen for crop growth.

Acknowledgements   We thank Alejandro Torres and Feliz Uck for assistance with the field work, and Janet Gale for the laboratory analyses. Financial support was provided by the Canadian International Development Agency and the Belize Ministry of Agriculture. Travel to this symposium was made possible by the generosity of the faculty of Science and the Institute for International Cooperation, University of Ottawa.

References

1   Beard J S 1955 The classification of tropical American vegetation types. Ecology 36, 89–100.
2   Brown R H 1978 A difference in nitrogen use efficiency in C_3 and C_4 plants and its implication in adaptation and evolution. Crop Sci. 18, 93–98.
3   Cornforth J S 1970 Leaf litterfall in a tropical forest. J. Appl. Ecol. 7, 603–608.
4   Ewel J 1969 Dynamics of litter accumulation under forest succession in Eastern Guatemala lowlands. MS Thesis. University of Florida, Gainsville, Florida.
5   Golley F B, McGinnis J T, Clements R G, Child G I and Duever M J 1969 The structure of tropical forests in Panama and Columbia. BioScience 19, 693–696.
6   Keeney D 1980 Prediction of soil nitrogen availability in forest ecosystems. For. Sci. 26, 159–171.
7   Lambert J D H and Arnason T 1978 Distribution of vegetation on Maya ruins and its relation to ancient land use at Lamanai, Belize. Turrialba 28, 33–41.
8   Lambert J D H and Arnason T 1980 Leaf litter and changing nutrient levels in a seasonally dry tropical hardwood forest. Plant and Soil 55, 429–443.
9   Lawson G W, Armstrong-Mensah K D and Hall J B 1970 A catena in tropical moist semi-deciduous forest near Kade, Ghana. J. Ecol. 58, 371–398.
10 Nye P H 1961 Organic matter and nutrient cycles under moist tropical forest. Plant and Soil 8, 333–346.
11 Nye P H and Greenland D J 1960 The Soil under Shifting Cultivation. Commonwealth Bureau of Soils. Technical Communication 51. Harpenden, England: Commonwealth Agricultural Bureau.
12 Sanchez P A 1976 Properties and Management of Soils in the Tropics. John Wiley and Sons, New York.
13 Wright A C S, Romney D H, Arbuckle R H and Vial V E 1979 Land in British Honduras. Colonial Research Publication 24. London.
14 Curtis J 1959 Vegetation of Wisconsin. Madison, Wisconsin: University of Wisconsin Press. 651 p.
15 McKeague J A 1978 Manual on Soil Sampling and Methods of Analysis. 2nd Edition. Ottawa: Canadian Society of Soil Science. 212 p.
16 Standley P C and Steyermark J 1958 Forest and flora of Guatemala Fieldiana Bot. 24:1.

Plant and Soil 67, 343–353 (1982).   0032-079X/82/0673-0343$01.65.
© 1982 *Martinus Nijhoff/Dr W. Junk Publishers, The Hague.*

Seasonal dynamics of nitrogen cycling for a Prosopis woodland in the Sonoran Desert

Dinamica estacional del ciclo de nitrógeno de un bosque de Prosopis en el desierto Sonorense

P. W. RUNDEL, E. T. NILSEN, M. R. SHARIFI,
Dept. of Ecology and Evolutionary Biology, University of California, Irvine, California 92717, USA

R. A. VIRGINIA, W. M. JARRELL,
Dept. of Soil and Environmental Sciences, University of California, Riverside, California 92521, USA

D. H. KOHL and G. B. SHEARER
Dept. of Biology, Washington University, St. Louis, Missouri 63130, USA

Key words   Desert woodland   N-cycling N_2fixation   *Prosopis glandulosa*   Sonoran Desert

Abstract   Prosopis woodlands in the Sonoran Desert have levels of above-ground biomass and productivity much higher than those predicted for desert plant communities with such low levels of precipitation. A stand of *P. glandulosa* near the Salton Sea, California, has 13,000 kg ha^{-1} above-ground biomass and a productivity of 3700 kg ha^{-1} yr^{-1}. Such a high level of productivity is possible because Prosopis is decoupled from the normal limiting factors of water and nitrogen availability. Soil nitrogen contents for the upper 60 cm of soil beneath Prosopis canopies have 1020 g m^{-2} total nitrogen, 25 per cent of which is in the form of nitrate. Such accumulations of nitrogen may be the result of active symbiotic nitrogen fixation. Early estimates suggest that about 25–30 kg N ha^{-1} yr^{-1} is fixed in these stands. Since Prosopis covers only 34% of the ground surface and its water resources are not limiting, much higher levels of nitrogen fixation and productivity may be possible in managed stands at greater densities.

Resumen   Los bosques de Prosopis en el desierto Sonorense tienen niveles de producción de biomasa (parte aérea) y productividad mucho mayores que las predecibles para comunidades de plantas de desierto con muy bajos niveles de precipitación. Los bosques freatofiticos de *P. glandulosa* cerca del Mar de Salton, California, producen 13 000 kg ha^{-1} de biomasa aérea con una productividad 3 700 kg ha^{-1} año^{-1}. Tan alto nivel de productividad es posible porque Prosopis no es afectado por los factores que limitan la aprovechabilidad de agua y nitrógeno.
  Los primeros 60 cm del perfil del suelo bajo el dosel de Prosopis contienen 1 020 g m^{-2} de nitrógeno total, el 25% existe en la forma de nitrato. Tales acumulaciones de nitrógeno pueden ser el resultado de la fijación simbiótica activa. Los primeros valores estimados sugieren que son fijados entre 25–30 kg N ha^{-1} año^{-1} en estos bosques. Puesto que Prosopis cubre solamente 34% de esta área y sus recursos de agua no son limitantes, puede ser posible la obtención de mayores niveles de fijación de nitrógeno y productividad de los cultivos si se manejan con mayores densidades.

Introduction

Productivity in desert and semi-arid plant communities is generally very low. These low levels of production result from the limiting factors of high water stress and limited nitrogen availability[22]. There are exceptions to this general pattern, however, particularly in desert and semi-arid ecosystems dominated by woody legumes. The genus Prosopis, the mesquites, forms the dominant coverage on millions of square kilometers of arid woodlands in North and South America. There are numerous anecdotal and semiquantitative accounts of high levels of productivity in Prosopis woodlands, but in the past there had been no good quantitative data published.

In recent studies[15] we have measured biomass and productivity rates in a stand of *Prosopis glandulosa* var. *torreyana* in the northwestern Sonoran Desert. Despite a very low mean annual precipitation of only 70 mm yr^{-1} at our study site near the Salton Sea in southern California, we have found a standing above-ground biomass of 13,000 kg ha^{-1} with a net production of 3700 kg ha^{-1} yr^{-1}. We have found even higher values for biomass and productivity in a second stand near Catavina in Baja California Norte, Mexico[10]. These levels of productivity far exceed those of other desert ecosystems, which generally range from 150–1000 kg ha^{-1} yr^{-1}.

Our hypothesis has been that Prosopis is able to reach such remarkable levels of productivity because it is decoupled from the normal limiting factors of water and nitrogen availability. It is well established that deep roots of Prosopis tap permanent water tables at great depths, making water available throughout the year[11]. Water alone, nevertheless, will not ensure high productivity without readily available nitrogen for growth. Such nitrogen is normally not present in desert soils[21]. Symbiotic nitrogen fixation, however, could provide a source for such nitrogen[2]. In this paper we report on our studies of the seasonal and long-term dynamics of nitrogen cycling in *Prosopis glandulosa* stands. We describe pool sizes of nitrogen in individual ecosystem compartments as well as the magnitude of fluxes between pools.

Materials and methods

Field studies were carried out at Harper's Well near the southern margin of the Salton Sea in the Sonoran Desert of California. Here *Prosopis glandulosa* dominates an extensive woodland community. Total Prosopis cover is 33.9% within the stand, with all other perennials providing a total of 4% additional coverage. The stand elevation is − 30 m, with a permanent ground water supply at a depth of 5 m in the soil. Biomass and productivity measurements necessary to calculate total nitrogen levels in above-ground biomass were determined using a dimensional analysis technique modified from that of Whittaker and Marks[23]. Individual tissue components were analyzed at roughly four-week intervals for their nitrogen contents by micro-Kjeldahl techniques. Soils to 60 cm depth at 30 cm intervals were analyzed for total nitrogen as well as organic nitrogen, nitrate-nitrogen and ammonium-nitrogen.

Results and discussion

Nitrogen pools

Soil nitrogen pools under the canopies of *Prosopis glandulosa* are remarkably high. The mean total nitrogen content of the upper 60 cm of soil under these canopies in our study site was 1020 g N m^{-2} in 1980 (Table 1). By comparison,

Table 1. Nitrogen (g N m^{-2}) in the upper 60 cm of soil under (canopy) and not under (non-canopy) individual Prosopis in Sonoran desert Prosopis stands

N-fraction	Canopy	Non-canopy	Total
Organic-N	762	103	294
NO$_3^-$-N	253	55	112
NH$_4^+$-N	5	2	3
Total	1020	160	409

open areas between canopies of individual Prosopis contained only 160 g N m^{-2}, while adjacent areas outside of the stand had only 45 g N m^{-2}. Very little nitrogen was present at greater depths in the soil. Even more surprising than the high levels of total soil nitrogen under Prosopis canopies was the proportion of nitrate in this total. Nitrate comprised 25% of the total soil nitrogen pool, with concentrations as high as 1000 mg NO$_3$-N kg dry soil^{-1}. This concentration of nitrate far surpasses that of even most agricultural soils. Ammonium-nitrogen concentrations were quite low in comparison to NO$_3$-N concentrations, but also high in relation to NH$_4$-N concentrations in agricultural soil.

The total above-ground nitrogen content of *Prosopis glandulosa* at Harper's Well was 54.6 g N m^{-2} of canopy area. We lack good data for below-ground tissues. During July 1980 (the period of maximum leaf biomass) nearly 60% of above-ground biomass nitrogen was contained in woody branches and trunks (Fig. 1). Leaves, reproductive tissues, and new stems comprised the remainder with 19.9, 11.9 and 2.1% of the total, respectively. On a weight basis, individual tissues varied greatly in % N. In mid-March 1980, at the peak of the spring

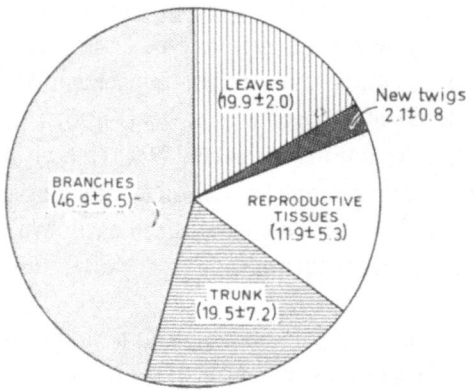

Fig. 1. Above-ground nitrogen distribution in *Prosopis glandulosa*, in July 1980. Values in parentheses represent per cent distribution ± standard errors.

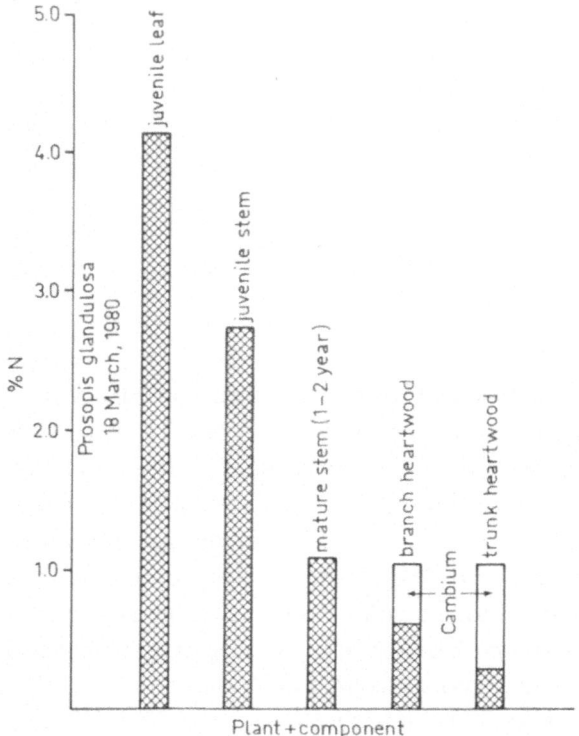

Fig. 2.   Pcr cent N (dry weight basis) in *Prosopis glandulosa* tissues in March 1980.

growth period, juvenile leaves contained over 4% N and new stems 2.7% N (Fig. 2). The previous year's stems had about 1.1% N while older woody tissues had less than 0.7% N. Nitrogen concentrations of new tissues changed rapidly, however, during this early period of growth. New leaves contained more than 5% N when they first formed, but the N-content dropped rapidly as the leaves matured over the next two months (Fig. 3). There was very little decline in leaf nitrogen concentration from late April to January when the majority of leaves were abscissed. A very similar seasonal progression of nitrogen concentrations occurred with new stems, although the levels of nitrogen were lower (Fig. 3). For leaves, our data suggest that the rapid early decline in leaf per cent N was largely a function of dilution as the leaves expanded and became heavier on an area basis. Leaf specific weights (mg cm^{-2}) of Prosopis leaflets increased until early May (Fig. 4), but nitrogen specific weights (mg N cm^{-2}) did not vary greatly after mid-March.

Monthly data on the seasonal dynamics of above-ground biomass components and the nitrogen concentration of these components allowed us to calculate the mean daily rate of nitrogen accumulation for each sample period. These data indicate a very high rate of nitrogen accumulation in leaves early in

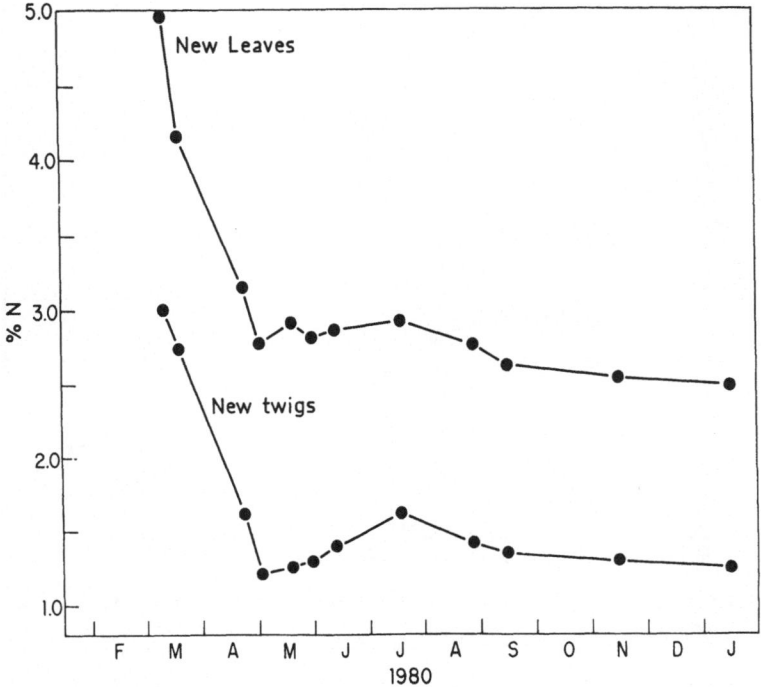

Fig. 3. Change in per cent N in *Prosopis glandulosa* leaves and stems with age.

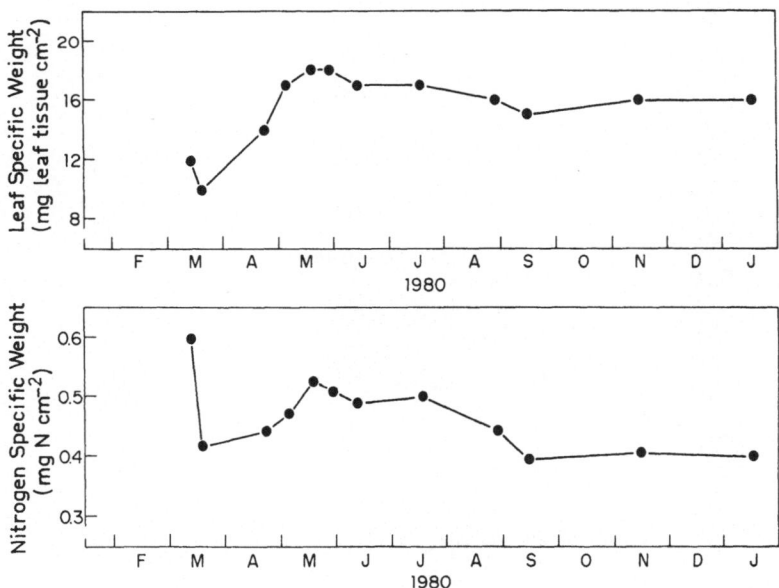

Fig. 4. Changes in leaf specific weight (mg dry leaf tissue cm^{-2} leaf area) and N concentration (mg N cm^{-2} leaf area) in *Prosopis glandulosa* with age.

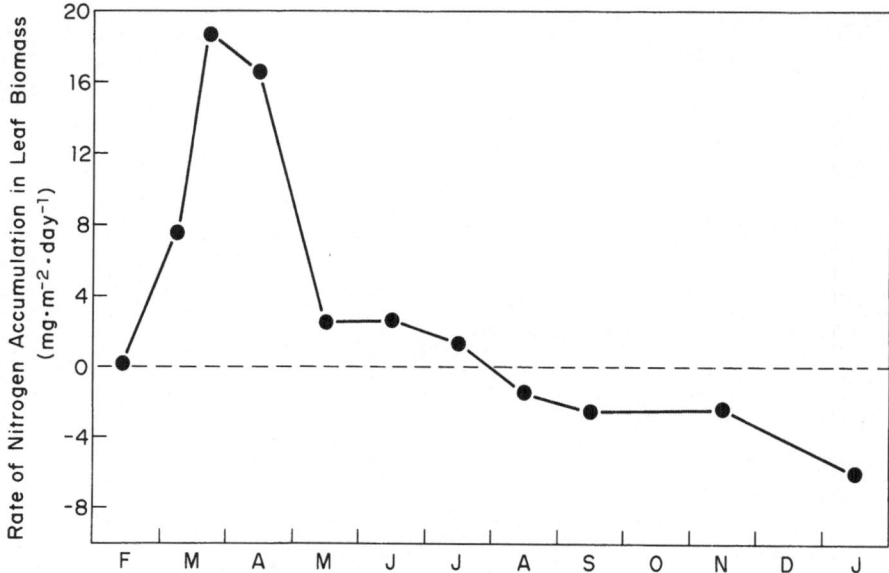

Fig. 5.   Rate of nitrogen accumulation in leaf biomass of a *Prosopis glandulosa* community in the Sonoran Desert, California.

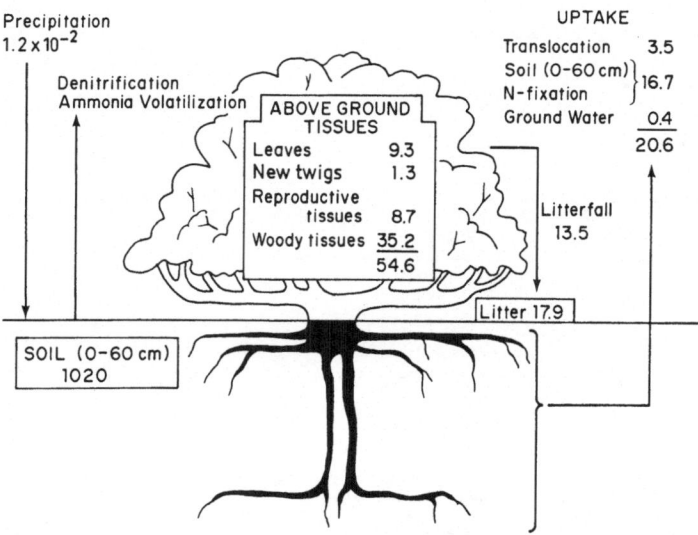

Fig. 6.   Nitrogen cycling beneath the *Prosopis glandulosa* canopy. Units for N pools are g N m^{-2}, and for fluxes, g N m^{-2} yr^{-1}. Uptake is based on above-ground biomass only.

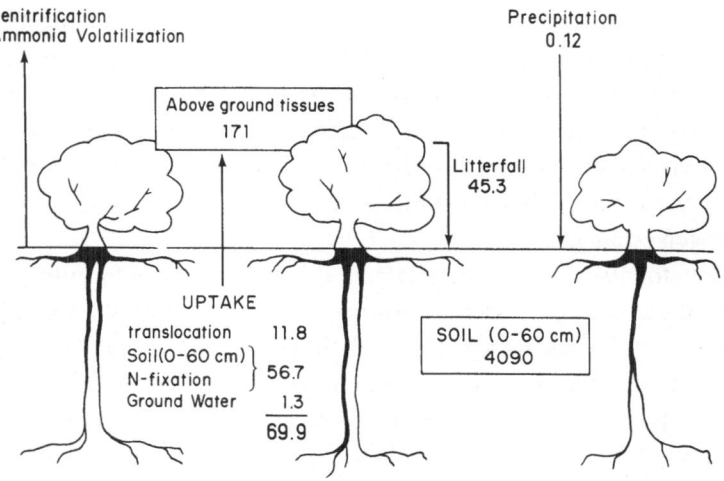

Fig. 7. Nitrogen cycling in a *Prosopis glandulosa* community in the Sonoran Desert, California. Biomass-N pools (kg N ha^{-1}) and fluxes (kg N ha^{-1} yr^{-1}) are for *Prosopis glandulosa*, and uptake is based on above-ground biomass only.

the growing season (during March), and then a rapid decline that continued through April. Very little net accumulation of nitrogen in leaves took place through the early summer, and by mid-summer there was a net loss of nitrogen from the above-ground biomass as leaf abscission began to occur (Fig. 5).

Nitrogen fluxes

Most external inputs and outputs from our system are relatively easy to account for since they are so small. For the inputs, we have considered precipitation input, animal inputs, and nitrogen fixation. With a mean annual precipitation of only 70 mm, we calculate that precipitation inputs of nitrogen are only about 1.2×10^{-2} g N m^{-2} yr^{-1} (assuming a mean rainwater-N concentration of 0.17 mg Nl^{-1}; Schlesinger and Hasey[13]). Animal inputs through fecal accumulation by vertebrates is primarily an internal recycling factor in our Prosopis stand, but some very small net input of nitrogen may come from birds or larger carnivores which feed outside the stand. Nitrogen fixation is a very important input, however. While our preliminary data suggest that free-living fixation by bacteria, bluegreen algae and lichens is very small, we present data below to suggest that symbiotic fixation by nodules of Prosopis must be very great.

Outputs of nitrogen from our ecosystem could occur through leaching, ammonia volatilization, denitrification and erosion. Although we have not directly measured the first two of these fluxes, they are clearly negligible factors because of the very low levels of precipitation, low soil ammonia concentrations, and a soil pH of 7–7.5, which should inhibit ammonia volatilization[6]. Our measurements of denitrification using the acetylene inhibition method[13] suggest that about 0.05 g N m^{-2} of canopy cover is lost following a 50 mm artificial rainfall, but this amount is again insignificant in the overall nitrogen budget[20]. Nitrogen lost by erosion and surface runoff is difficult to quantify. In the majority of years, as in 1980, there is no significant nitrogen loss by erosion. During exceptional years, however, sheet flow flooding across the stand can be considerable; it clearly has occurred in the past. While such events may only occur a few times a century, they are an important consideration in long-term models of nitrogen cycling.

Using our data for 1980 we have calculated a simple flow model for nitrogen cycling beneath the canopy of Prosopis at Harper's Well (Fig. 6). Of the 54.6 g N m^{-2} of above-ground biomass nitrogen beneath the canopy, 20.3 g N m^{-2} was in new leaves and twigs, reproductive tissues and the new increment of woody tissues. The quantity of nitrogen reabsorbed before leaf tissue loss by abscission is a minimum estimate for nitrogen which could be translocated from stored tissues into the new nitrogen pool (11.8 kg N ha^{-1} yr^{-1}). From measurements of seasonal transpiration fluxes through the foliage of the Prosopis canopies[10], we can calculate that only about 0.04 g m^{-2} of nitrogen uptake could occur from groundwater (where NO_3–N is present at 1 mg N l^{-1} or less). The remaining 16.7 g N m^{-2} taken up must come from a combination of uptake from surface roots and from symbiotic nitrogen fixation. This, of course, is a minimum estimate since we do not know the quantity of nitrogen accumulated in roots.

On a system-wide basis the net nitrogen uptake by above-ground Prosopis biomass is 69.8 kg N ha^{-1} yr^{-1} (Fig. 7). Of this total, 56.7 g m^{-2} comes from the combination of surface root uptake and symbiotic nitrogen fixation.

Nitrogen fixation

What evidence is there that symbiotic nitrogen fixation is taking place in these stands? First, nodules have been found on a young Prosopis individual growing in a moist wash in early spring. However, no nodules have been found in surface soils under larger trees. Second, measurements of the natural ^{15}N abundance of soil and Prosopis tissues indicate an input of symbiotically-fixed nitrogen[8,19]. Since the $^{15}N/^{14}N$ ratio of soil nitrogen usually exceeds that of the atmosphere, plants that fix N_2 have $^{15}N/^{14}N$ ratios below those of soil nitrogen and associated non-nitrogen-fixing plants[1,7,16]. With appropriate sampling techniques, the magnitude of this difference should be proportional to the amount of nitrogen fixed[4]. While our studies of ^{15}N abundance are still in a preliminary stage, our data indicate that nitrogen fixation is actively occurring in *Prosopis glandulosa*. These data suggest that possibly as much as about 50% of total nitrogen uptake may come from fixation.

Our second line of evidence that fixation is important comes from an analysis of the long-term dynamics of nitrogen cycles in our Prosopis stand. Our study site at Harper's Well is located on the old bottom deposits from Lake Cahuila, which covered the present Salton Sea basin up until very recent times. Archeological data suggest that the lake dried *ca.* 400–500 yrs BP. Thus we have a maximum stand age for our calculations. If we take the present canopy soil nitrogen (1020 g $N m^{-2}$) and subtract the background soil nitrogen (45 g $N m^{-2}$) from adjacent non-Prosopis sites, we calculate a differential nitrogen accumulation under Prosopis canopies of 975 g $N m^{-2}$. Assuming a 500 year stand age, the mean annual nitrogen accumulation has been *ca.* 1.95 g $N m^{-2}$ of canopy for this period. On a stand basis this is slightly over 6 kg $N ha^{-1} yr^{-1}$. Since fixation is the only significant system input of nitrogen, these values represent a minimum figure for mean annual levels of fixation. Expected erosional losses at irregular intervals over the 500 year period would likely increase this maximum level considerably. Since above-ground Prosopis biomass is only 50–75 yrs old (by ring counts), the nitrogen found beneath the Prosopis individuals has probably accumulated over a much briefer time period than the maximum of 500 yrs.

The actual mean level of nitrogen fixation in our stand is undoubtedly higher than our minimum estimates. Erosional losses of nitrogen in irregular surface runoff in the stand certainly occur at irregular intervals. If such losses were of sufficient magnitude to cycle soil nitrogen on a 100 year rather than a 500 year cycle, then stand nitrogen fixation could be about 30 kg $N ha^{-1} yr^{-1}$. Another estimate of nitrogen fixation can be based on our preliminary figures from ^{15}N abundance measurements that approximately 50% of the total uptake comes from fixation. This suggests that at least 23–36 kg $N ha^{-1} yr^{-1}$ is fixed on a stand basis; including the below-ground nitrogen accumulation in root production may increase this estimate to > 40 kg $N ha^{-1} yr^{-1}$.

These estimated rates of nitrogen fixation in our Prosopis stand are quite similar to those reported for other natural communities in regions with much higher levels of precipitation[12,18]. Since the high levels of nitrate in the upper 60 cm of soil probably inhibit nodule formation[5], it is not surprising that we have not found nodules in the surface root zone. We hypothesize that nodulation is currently restricted in our stand to the capillary fringe above the ground water table.

Conclusions

The high levels of production and nitrogen fixation we have measured in Prosopis suggest that managed woodlands may provide economically significant sources of wood fuels and forage for sheep or goats. The low total coverage of Prosopis in natural stands is almost certainly related to problems of seedling establishment[9,17] and not resource availability. So long as water resources do not become limiting, our study suggests that stand levels of nitrogen fixation as high as 150 kg $N ha^{-1} yr^{-1}$ might be possible with plantation growth of selected genotypes of *Prosopis glandulosa*[3]. Such a level would be close to that reached in agricultural production of alfalfa or soybeans. Plantation productivity of Prosopis could be as high as 15,000 kg $ha^{-1} yr^{-1}$, equally split between new woody tissues and high-quality forage[10].

References

1   Delwiche C C, Zinke P J, Johnson C M and Virginia R A 1979 Nitrogen isotope distribution as a presumptive indicator of nitrogen fixation. Bot. Gaz. 140 (suppl.), 65–69.

2   Eskew D L and Ting I P 1978 Nitrogen fixation by legumes and blue green algal-lichen crusts in a Colorado desert environment. (unpublished).

3   Felker P 1979 Mesquite: an all purpose leguminous arid land tree. pp 89–132. *In* Ritchie G A (Ed.). New Agricultural Crops. Colorado: Westview Press.

4   Fried M and Broeshart H 1975 An independent measurement of the amount of nitrogen fixed by a legume crop. Plant and Soil 43, 707–711.

5   Gibson A H 1976 Recovery and compensation by nodulated legumes to environmental stress. pp 385–403. *In* Nutman P S (Ed.). Symbiotic Nitrogen Fixation in Plants. Cambridge: Cambridge University Press.

6   Klubek B, Eberhardt P J and Skujins J 1978 Ammonia volatilization from Great Basin desert soils. pp 107–129. *In* West N E and Skujins J (Eds.). Nitrogen in Desert Ecosystems. New York: Dowden, Hutchinson, and Ross, Inc.

7   Kohl D H, Schearer G and Harper J E 1980 Estimates of N_2 fixation based on differences in the natural abundance of ^{15}N in nodulating and non-nodulating isolines of soybeans. Plant Physiol. 66, 61–65.

8   Kohl D H, Bryan B A, Schearer G and Skeeters J L 1981 Natural abundance of ^{15}N of *Prosopis* as an index of N_2-fixation in desert ecosystems. Bull. Ecol. Soc. Am. 62, 133–134.

9   Mooney H A, Gulmon S L, Rundel P W and Ehleringer J 1980 Further observations on the water relations of *Prosopis tamarugo* of the northern Atakama desert. Oecologia 44, 177–180.

10 Nilsen E T, Rundel P W and Sharifi M R 1982 Productivity in native stands of *Prosopis glandulosa*, mesquite, in the Sonoran desert of southern California and some management implications. California Riparian Environment Symposium. Sept. 17–19., Davis, CA.

11 Phillips W S 1963 Depth of roots in soil. Ecology 44, 424.

12 Phillips D A 1980 Efficiency of symbiotic nitrogen fixation in legumes. Annu. Rev. Plant Physiol. 31, 29–49.

13 Ryden J C, Lund L J, Letey J and Focht D D 1979 Direct measurement of denitrification loss from soils. II. Development and application of field methods. Soil Sci. Soc. Am. J. 43, 110–118.

14 Schlesinger W H and Hasey M M 1980 The nutrient content of precipitations, dry fallout, and intercepted aerosols in the chaparral of southern California. Am. Mid. Nat. 103, 114–122.

15 Sharifi M R, Nilsen E T and Rundel P W 1982 Biomass and net primary production of *Prosopis glandulosa* (Fabaceae) in the Sonoran desert of southern California. Am. J. Bot. (in press).

16 Shearer G and Kohl D H 1978 ^{15}N abundance in N_2-fixing and non-N_2-fixing plants. pp 605–622. *In* Frigerio A (Ed.). Recent Developments in Mass Spectrometry in Biochemistry and Medicine, Vol. 1. New York: Plenum Press.

17 Simpson B B 1977 Mesquite-its biology in two desert ecosystems. Stroudsburg, Pa: Dowden, Hutchinson, and Ross, Inc.

18 Stewart W D P 1977 Present day nitrogen fixing plants. Ambio 6, 166–173.

19 Virginia R A, Jarrell W M, Kohl D H and Shearer G B 1981 Symbiotic nitrogen fixation in a *Prosopis* (Leguminosae)-dominated desert ecosystem. 483 p. *In* Gibson A H and Newton W E (Eds.). Current Perspectives in Nitrogen Fixation. Canberra: Aust. Acad. Science.

20 Virginia R A, Jarrell W M and Franco-Vizcaino E 1982 Direct measurement of denitrification in a *Prosopis* (mesquite)-dominated Sonoran desert ecosystem. Oecologia (in press).

21 West N E and Skujins J (Eds.) 1978 Nitrogen in Desert Ecosystems. Stroudsburg, Pa: Dowden, Hutchinson, and Ross, Inc. 307 p.

22 Whittaker R H 1975 Communities and Ecosystems. New York: MacMillan. 385 p.

23 Whittaker R H and Marks P L 1975 Methods of assessing terrestrial productivity. pp 55–118. *In* Lieth H and Whittaker R H (Eds.). Primary Productivity in the Biosphere. New York: Springer-Verlag.

Plant and Soil 67, 355–358 (1982). 0032-079X/82/0673-0355$00.60. SU-32
© 1982 *Martinus Nijhoff/Dr W. Junk Publishers, The Hague.*

Litterfall and nitrogen turnover in an Amazonian blackwater inundation forest

Caida de horajasca y ciclo de nitrógeno en un bosque de innundación en la región amazonica de aguas negras

U. IRMLER

Abt. Angewandte Ökologie/Küstenforschung, Zoolog. Inst., Universität, D-2300 Kiel, Olshausenstrasse 40–60, F. R. Germany

Key words Amazonas Arthropods Decomposition Inundation forests Litterfall N-cycling

Abstract In 1976/77 energy flow and nutrient cycling in an Amazonian blackwater inundation forest were studied. The major part of the litter biomass turnover occurred during the emersion phase. 95% decomposition rate for nitrogen was measured with 15 mm mesh litter bags and was 4.7 years. Over 30 per cent of the annual leaf-fall was decomposed by soil-dwelling arthropods.

Resumen Durante los años 1976 y 1977 se estudiaron los flujos de energía y nutrimentos en un bosque de innundación en la región de aguas negras del Amazonas. La mayor parte de la caida de hojarasca ocurre durante la fase de aguas altas mientras que la mayor descomposición ocurrió durante la fase de aguas bajas. La tasa de descomposición para el nitrógeno en bolsas de 15 mm de malla, fue de 4.7 años para el 95%. Cerca del 30% de la caida anual de hojas fue descompuesta por artrópodos del suelo.

Introduction

The valleys of the Amazonian rivers are seasonally inundated. Water levels and inundation periods (which average about 6 months) change gradually within the inundation forest depending on the distance from the river bank. During the year of study inundation at the investigation site occurred from April to July. Most of the inundated areas are forested and these forests are affected by the chemical and physical conditions of the adjacent rivers. In 1976/77, we studied energy flow and nutrient cycling in an Amazonian blackwater inundation forest situated at Tarumã Mirím River about 15 km north of Manaus, Brazil. Litterfall, litter decomposition and energy flow and nutrient turnover in a soil-dwelling cockroach population were quantified. I present here the results concerning litterfall, litter decomposition, and N-cycling in this system.

Methods

Litterfall was measured using 1-m² open-top samplers. During the submersion phase, samplers were placed both at water surface and at ground levels in order to measure both litterfall and the net proportion of fresh litter that is lost *via* water flow.

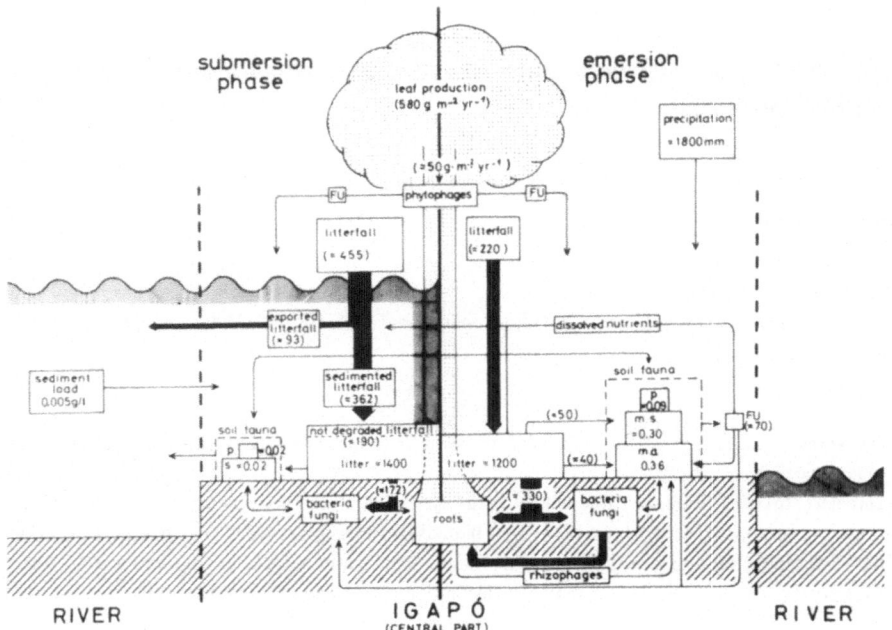

Fig. 1. Standing-crop biomass and production in the emersion and submersion phases of an Amazonian blackwater inundation forest. Values in parentheses refer to dry matter production, values without parentheses refer to standing biomass; p = predators, s = saprophagous fauna, ms = macro saprophagous fauna, ma = microarthropods. (From Irmler[2]).

Leaf litter decomposition was studied directly by monitoring weight and N loss from freshly-fallen leaves placed in 15-mm mesh bags and incubated on the forest floor for 100 days of the emersion phase[4]. Decomposition rates were also studied indirectly by feeding experiments with cockroaches[3]. Nitrogen was determined by Kjeldahl methods.

Results and discussion

Energy flow and nutrient turnover were investigated for both the submersion phase (SP) and the emersion phase (EP). Fig. 1 presents standing biomass, litterfall, and production data for each phase. The major part of the litterfall occurred during the inundation phase[1], although the major part of the litter turnover occurred during the emersion phase. 455 g dry litter m^{-2} fell during the submersion phase; of this, 93 g was lost *via* river flow and 172 g was decomposed during the submersion phase. Litterfall during the emersion phase was 220 g dry litter m^{-2}. The saprophagous fauna consumed about 90 g dry leaf litter m^{-2} during the emersion phase, or over 30% of the annual leaffall (not total litterfall) available to it; the remainder was presumably decomposed by bacteria or fungi[2].

Nitrogen was distributed in and transferred between the different components of the inundation forest ecosystem as listed in Tables 1 and 2. The input of nitrogen to the system *via* rainwater and water of the inundations was extremely

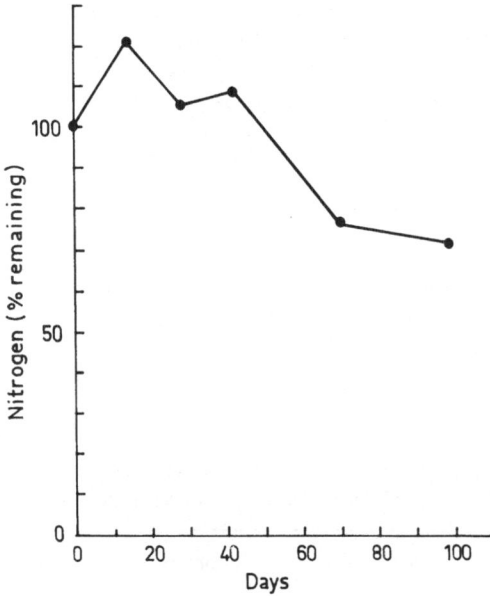

Fig. 2.   Release of nitrogen from fresh leaf litter in 15 mm mesh bags during incubation on the forest floor for 100 days of the emersion phase in the blackwater inundation forest.

low. A major part of the system's total nitrogen content is in the standing vegetation, not investigated in this study. 13 g N m^{-2} was in the standing litter. Leaf-production resulted in a net apparent N-uptake of 11.25 g N m^{-2} yr^{-1}, of which 3.57 g N m^{-2} yr^{-1} was translocated back to the twigs before leaffall. Litterfall transferred 9.67 g N m^{-2} yr^{-1} to the forest floor. 1.31 g N m^{-2} yr^{-1} was lost by leaf export *via* river current during the inundation phase. The remaining 8.36 g N m^{-2} yr^{-1} in the litter was presumably mineralized. 95% decomposition for N was 4.7 years; Olson's[5] decomposition rate (k) was 0.64.

Fig. 2 shows the loss of litter-N from mesh bags during 100 days of the emersion phase. The N-concentration of the litter was between 1.2 and 1.5% N initially, and over 100 days, decreased to 1.0% N.

Investigations with cockroaches showed that litter was assimilated when it contained 1.4–1.5% N but not when it contained 1.0% N. This suggests that cockroaches feed mainly on fresh rather than older litter. Older leaf litter is probably assimilated only with dead animal biomass (*ca.* 14%N). The N-concentration of the cockroach feces was 1.5% N when cockroaches were fed animal biomass plus old litter (1.0% N), or about the same N-content as fresh litter, indicating that cockroaches may effectively redistribute nitrogen in the litter.

Acknowledgements This research was a cooperative project of Max-Planck-Institut für Limnologie, Abt. Tropenökologie, Plön, Germany, and Instituto Nacional de Pesquisas de Amazonia (INPA) Manaus/Amazonas, Brazil.

Table 1. Nitrogen in litter and soil invertebrates of the blackwater inundation forest, and in river water at time of peak flooding. Biomass values are on dry-weight basis

Component	Mass g m^{-2}	Nitrogen %	Nitrogen g N m^{-2}
Litter	1300	1.0	13.0
Soil invertebrates	0.75[b]	13.9	0.11
River water	3 × 10^6	40 × 10^{-6}[a]	1.2

[a] 0.40 mg N l^{-1}.
[b] From Irmler[2].

Table 2. Nitrogen fluxes in various parts of the blackwater inundation forest. Exported litter is the difference between submerged and nonsubmerged litter traps during the submersion phase

	Mass g m^{-2} yr^{-1}	N-content %	N-flux g N m^{-2} yr^{-1}
Bulk precipitation	2.2 × 10^6[a]	44 × 10^{-6}[b]	1.0
Leaf production	580[c]	1.94	11.3
Litterfall	675	1.44	9.72
Exported litter	93	1.44	1.31
Soil-invertebrate feces	90[c]	1.52	1.37

[a] 2200 mm.
[b] 0.44 mg Kjeldahl-N l^{-1}.
[c] From Irmler[2].
[d] Emersion phase values are 1.8 × 10^6 g H$_2$O m^{-2} yr^{-1}, and 0.8 g N m^{-2} yr^{-1}.

References

1   Adis J, Furch K and Irmler U 1979 Litter production of a Central-Amazonian inundation forest. Trop. Ecol. 20, 236–245.
2   Irmler U 1979 Considerations on structure and function of the 'Central-Amazonian inundation forest ecosystem' with particular emphasis on selected soil animals. Oecologia 43, 1–18.
3   Irmler U and Furch K 1979 Production, energy and nutrient turnover of the cockroach *Epilampra irmleri* Rocha e Silva and Aguiar in a Central-Amazonian inundation forest. Amazoniana 6, 497–520.
4   Irmler U and Furch K 1980 Weight, energy and nutrient changes during the decomposition of leaves in the emersion phase of Central-Amazonian inundation forests. Pedobiologia 20, 118–130.
5   Olson J S 1963 Energy storage and the balance of producers and decomposers in ecological systems. Ecology 44, 322–331.

Plant and Soil 67, 359–365 (1982). 0032-079X/82/0673-0359$01.05.   SU-33

Nitrogen in litterfall and precipitation and its release during litter decomposition in the Chilean piedmont matorral

El nitrógeno en la precipitación y la caida de hojarasca y su liberación durante la descomposición en el matorral premontano de chile

R. E. CISTERNAS* and L. R. YATES

Laboratorio de Ecología, Instituto de Ciencias Biológicas, Pontificia Universidad Católica de Chile. Casilla 114-D, Santiago, Chile

Key words   Chile *Lithraea caustica* Litterfall N-cycling Piedmont-matorral *Quillaja saponaria*

Abstract   Parts of the nitrogen cycle involving two dominants (*Lithraea caustica* and *Quillaja saponaria*) in the Chilean piedmont matorral have been studied over a 15-month period. Analyses showed that $8.2 \, \text{kg N ha}^{-1} \, \text{yr}^{-1}$ entered the system in rainfall and dry deposition, though impaction of N-containing compounds on vegetation (not measured) may elevate this value. *L. caustica*, by virtue of its greater percent cover, contributed more leaf litter than did *Q. saponaria* to the system (1089, *vs* $737 \, \text{kg dry matter ha}^{-1} \, \text{yr}^{-1}$, respectively), although on an individual basis *Q. saponaria* produced more litter (640, *vs* $350 \, \text{g dry leaf litter m}^{-2} \, \text{yr}^{-1}$ r *L. caustica*). This plus the greater nitrogen release of *L. caustica* leaf litter during decomposition (2.61, *vs* $0.60 \, \text{g N kg dry litter}^{-1} \, \text{yr}^{-1}$ for *Q. saponaria*) and *Q. saponaria*'s higher N-content of dropped leaves (0.54, *vs* 0.37% N for *L. caustica*) may indicate a more external cycling of nitrogen in *Q. saponaria* relative to that in *L. caustica*. These two species may therefore represent two different strategies of individual nitrogen cycling, external and internal.

Resumen   En el matorral premontano de Chile se estudiaron algunas partes del ciclo de nitrógeno de dos de sus dominantes (*Lithraea caustica* y *Quillaria saponaria*) durante 15 meses. Los análisis mostraron que los ingresos de nitrógeno por lluvia y deposición seca fueron de $8,5 \, \text{kg N ha}^{-1} \, \text{año}^{-1}$; este valor podría ser mayor si se considera el ingreso adicional por impactación de compuestos nitrogenados sobre la vegetación (no medido). Debido a su mayor cobertura porcentual, *L. caustica* contribuyó con mas hojarasca que *Q. saponaria* obteniendose valores de 1089 y $737 \, \text{kg ha}^{-1}$ materia seca año^{-1} respectivamente. Sin embargo, al considerar el area bajo los individuos, *Q. saponaria* produjo mas hojarasca que *L. caustica* (640 *vs* $350 \, \text{g}$ materia seca m^{-2} año^{-1}. Durante la descomposición de la hojarasca, *L. caustica* liberó $2,61 \, \text{g N kg}^{-1}$ hojarasca seca año mientras que *Q. saponaria* liberó $0,60 \, \text{g N kg}^{-1}$ hojarasca seca año^{-1}.

Introduction

The matorral is a degraded evergreen schlerophyll forest that is the most common vegetation formation in the central zone of Chile. Its structure is very similar to ecosystems of other mediterranean regions of the world[5]. The piedmont matorral located in the foothills of the Cordillera de los Andes near Santiago, Chile, is characterized by almost the same species as those in the plains

* Present address: Departamento de Biologia y Química, Universidad de Talca, Talca, Chile.

matorral, though individual shrubs in the piedmont matorral are smaller and species densities differ.

Although the soils of this ecosystem have very low nitrogen concentrations (0.2% N), the standing plant biomass is very high. This suggests the presence of mechanisms which compensate for this apparent soil-nitrogen deficiency. To our knowledge there are a few studies which consider the movement of nutrients in this type of ecosystem (e.g.[6, 7, 8]), but comprehensive studies that consider mobilization of nutrients among the different parts of these plant communities are lacking. The aim of this study is to investigate these mechanisms by quantifying nitrogen fluxes among different compartments of the ecosystem during an annual cycle. We have focused our attention on N-fluxes affected by the two dominant species, Lithraea caustica and Quillaja saponaria. The fluxes and N-pools studied include 1) nitrogen inputs to the system through bulk precipitation, 2) nitrogen transferred to the soil from the canopy in litterfall and throughfall, 3) nitrogen released from decomposing leaf-litter, and 4) nitrogen in different plant compartments such as green leaves, bark, and roots.

Study site and methods

The study was conducted in a Cordillera de los Andes piedmont area at 1000 m elevation located 10 kilometers northeast of downtown Santiago. The climate of the area is Mediterranean, with cool moist winters and hot dry summers. The matorral vegetation is dominated by L. caustica and Q. saponaria, and Trevoa trinervis, Baccharis sp., Kageneckia oblonga, and Colliguaya odorifera are the most conspicuous subordinate species. The canopy of L. caustica covers three times more area (31.1% cover) than Q. saponaria (11.5% cover). About 25% of the soil of the system is exposed, and 32% of the study site is covered by subordinate shrub species.

Within a sampling site of approximately 1 ha, 10 L. caustica and 10 Q. saponaria trees were selected at random for nitrogen determinations. Precipitation and throughfall were recorded daily in 9 plastic pluviometers 20 cm in diameter. Stemflow was monitored using a plastic gutter made from 1.3 cm plastic tubing split in half and coiled around the main stem of 4 individuals of each dominant species. All solutions were collected separately in 10 l plastic jars, removed after each rainfall and stored for later inorganic nitrogen determinations. One ml of 40% formaldehyde l^{-1} of sample solution was added to all samples and analyses were performed within 30 days of collection.

Sixteen 0.5 m^2 litter-collecting trays made of 2 mm fine plastic mesh suspended on wooden frames were set out the canopies of eight randomly selected trees of each of the two dominant species. Sampling took place from January 1979 to April 1980. Throughout the 15 months of the experiment, the leaf litter that fell into the trays was collected at monthly intervals, dried at 70°C and stored for later Kjeldahl N-analyses[1]. Samples were stored for no more than 4 wks before analysis.

L. *caustica* and *Q. saponaria* leaf litter decomposition rates were estimated separately by placing 10 g of litter (fresh weight) into each of seventy 2-mm nylon mesh bags that were then incubated under the canopy of the species from which it was derived. Every month from February 1979 until March 1980, five bags were collected from under the canopy of each species and dry weights and nitrogen contents determined.

Seasonal samples of green leaves, bark and roots of both species were also collected for nitrogen determinations. Green leaves of different ages were separated to a determined nitrogen concentration trends during the species life-cycles.

Results

Nitrogen in precipitation and throughfall

Nitrogen enters the system both as wet deposition in rainfall and as dry deposition in atmospheric dust which sediments on plant structures and soil. 8.2 kg N ha^{-1} yr^{-1} entered the piedmont matorral in bulk precipitation. 43% of this nitrogen was intercepted by the two dominant tree species, a portion hit other shrub species, and the remainder fell on the ground.

For both species, nitrogen transferred in throughfall was considerably higher than that mobilized in stemflow. System-wide, 2.0 and 0.7 kg N ha^{-1} yr^{-1} were washed from *L. caustica* and *Q. saponaria*, canopies, respectively (Fig. 1).

Leaf-litter nitrogen

As do many evergreen mediterranean species, both *L. caustica* and *Q. saponaria* shed a portion of their leaves during the dry season. About 60% of both species' litterfall took place during the summer months.

Overall, *L. caustica* contributed 1.8 times more leaf litter to the system than did *Q. saponaria* (Table 1). However, since leaf-litter nitrogen concentrations were 1.25 times higher in *Q. saponaria* (Table 2), the litterfall nitrogen contributed by each species to the soil is not substantially different (*ca.* 4.0 kg N ha^{-1} yr^{-1}, Fig. 1).

Nitrogen released during leaf-litter decomposition

Annual changes in *Q. saponaria* and *L. caustica* leaf-litter nitrogen during decomposition in mesh litterbags is shown in Fig. 2. During decomposition there was 1) a rapid initial decrease in leaf-litter nitrogen, 2) a prolonged period of slow but steady accumulation of nitrogen, and 3) a final rapid nitrogen fall. Very similar trends were observed for both species leaf-litter, although monthly values of nitrogen and the slopes of the different stages were higher in *Q. saponaria* than in *L. caustica*. The final stage of nitrogen release was simultaneous with the vegetative growth period of both species.

On a system-wide basis N-mineralization in leaf-litter was very similar for both

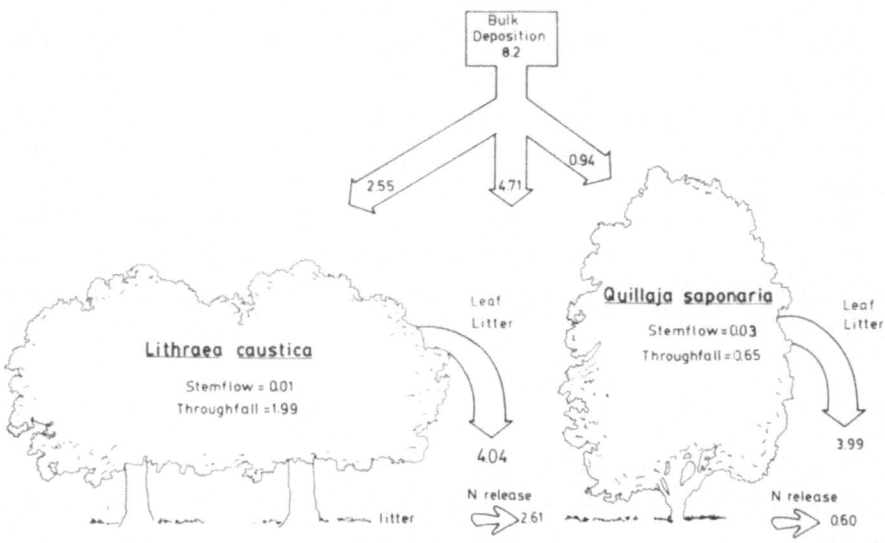

Fig. 1. Aspects of the nitrogen cycle beneath the dominant species *Quillaja saponaria* and *Lithraea caustica* in the Chilean piedmont matorral. Values are in kg N ha^{-1} yr^{-1} and take into account the respective % covers of each species.

Table 1. Leaf and leaf-litter production, nitrogen released from litter bags, and N-uptake (leaf production × N-content of 1 yr-old green leaves) by the two dominants of the piedmont matorral

	Production		N-release		Plant nitrogen uptake	
	Leaf-litter (kg dm ha^{-1} yr^{-1})	Leaves* (kg dm ha^{-1} yr^{-1})	Indi-vidual (g N m^{-2} yr^{-1})	System-wide (kg N ha^{-1} yr^{-1})	Indi-vidual (g N m^{-2} yr^{-1})	System-wide (kg N ha^{-1} yr^{-1})
L. caustica	1039	1066	1.09	3.39	3.64	11.7
Q. saponaria	736	748	1.23	1.41	7.15	7.25

* Adapted from Mooney[5].
dm, dry matter.
Individual refers to fluxes beneath individual plants.

species (Table 1). *L. caustica* litter lost 0.78 kg Kjeldahl-N ha^{-1} yr^{-1} while *Q. saponaria* lost 0.81 kg.

Nitrogen in different plant compartments

The partial nitrogen concentrations of different plant structures are summarized in Table 2. Generally, *L. caustica* green leaves live twice as long as *Q. saponaria* green leaves, and show a higher mean nitrogen concentration. A

Table 2. Mean nitrogen concentrations ($\pm$ standard deviations) in plant components of the two dominant species in the Chilean piedmont matorral

	Q. saponaria	L. caustica
Green leaves		
1.0-yr-old	0.97 $\pm$ 0.18	1.10 $\pm$ 0.23
2.0-yr-old	0.65*	0.75*
Leaf-litter	0.54 $\pm$ 0.06	0.37 $\pm$ 0.03
Bark	0.57	0.43
Roots	1.30	0.53

*From Mooney[5].

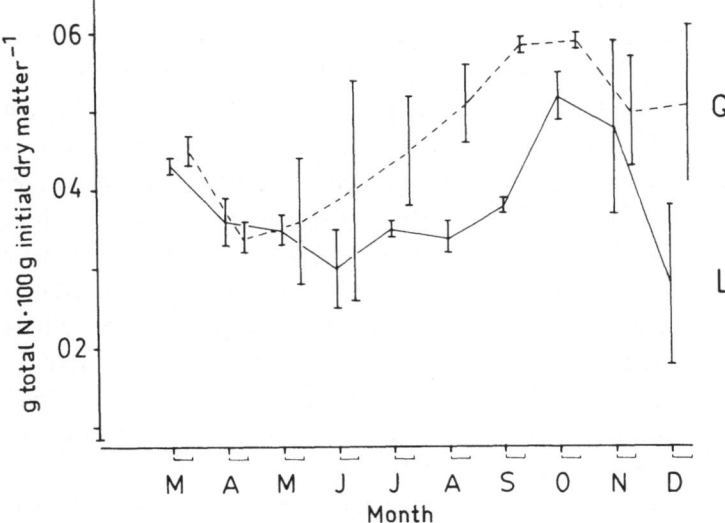

Fig. 2. Nitrogen content (g total-N $\cdot$ 100 g initial dry matter^{-1}) of *Q. saponaria* (Q) and *L. caustica* (L) leaf-litter during decomposition in litterbags under the respective canopies. Note that y-axis origin $\neq$ 0.

green-leaf nitrogen to decrease with age is observed in both species. Other plant structures of *Q. saponaria* had higher nitrogen concentrations than corresponding parts of *L. caustica*.

Discussion

The total amount of nitrogen entering the system as wet and dry deposition, 8.2 kg N ha^{-1} yr^{-1}, is very significant because it is over twice as much as the amount of nitrogen released through decomposition under either dominant, 3.32 kg ha^{-1} yr^{-1}. This unusually high deposition input of nitrogen may be because of the proximity of Santiago. Santiago has a high degree of air pollution[2] and prevailing winds blow toward the experimental site. It would be interesting to elucidate the actual role of the city as a source of additional nitrogen to surrounding ecosystems.

The initial rapid decrease in the nitrogen content of decomposing leaf-litter is likely due to soluble nitrogen's leaching from dead leaves. The accumulation observed during the second stage can be explained by 1) nitrogen deposition from throughfall, 2) nitrogen fixation by free-living bacteria that use leaf-litter as an energy source, or 3) fungal translocation of nitrogen to the litter from the soil. It seems more likely that the increase is principally due to N-fixation or fungal translocation since the nitrogen in rainfall and throughfall is not great enough to explain this increase alone. The release of nitrogen noted in the third stage of decomposition (Fig. 2), which occurs during the growing season, could be due to the combined actions of denitrifying bacteria, decomposition mesofauna, NH_3 volatilization, and leaching.

Yates et al.[9] found that Q. saponaria litter decomposed faster than did L. caustica litter, which may indicate that the former is the more available substrate for the microflora and mesofauna. In fact, Montenegro (unpublished) has found that Q. saponaria leaves have no tricomes on their surface, their cuticle is thin, and they have a low lignin content. L. caustica leaves in contrast have tricomes, relatively thick cuticles, and high lignin concentrations[4]. L. caustica leaf-litter would therefore be expected to decompose more slowly[3] and consequently release nitrogen at a lower rate than Q. saponaria leaf-litter, as is apparently the case (Table 1). Q. saponaria's higher leaf-litter nitrogen content and greater accumulation of nitrogen during decomposition may also contribute to Q. saponaria's faster decomposition. The net result is that a more intensive movement of available nitrogen to the soil under the canopy of Q. saponaria may occur. This nitrogen could then be utilized by the plant or be absorbed by the roots of neighbors.

The different decomposition rates of these species[9] and their different green-leaves and leaf-litter nitrogen concentrations (Table 2) may indicate different strategies for mobilizing nitrogen L. caustica may favor internal cycling of nitrogen, minimizing nitrogen release to the environment, while Q. saponaria may favor external cycling. The conservative strategy of L. caustica may give it a competitive advantage over concomitant shrub species of the matorral and may have resulted in its relatively high density. It could be advantageous for other shrub species to live near Q. saponaria due to its release of nitrogen to the environment.

Conclusions

1. Nitrogen inputs from bulk deposition were $8.2 \, kg \, N \, ha^{-1} \, yr^{-1}$. The proximity of Santiago could be the cause of these high deposition inputs.

2. L. caustica and Q. saponaria respectively contributed to the entire system 2.0 and $0.7 \, kg \, N \, ha^{-1} \, yr^{-1}$ via throughfall and stemflow, and 4.04 and $3.99 \, kg \, N \, ha^{-1} \, yr^{-1}$ via litterfall. Of this litterfall N, 2.61 and $0.60 \, kg \, N \, ha^{-1}$ were released from mesh litterbags over the subsequent year, 60% during summer months.

3. One-yr-old *Q. saponaria* leaves contained 0.97% N, whereas fresh leaf litter (with leaves generally greater than 2 yrs old) contained 0.54% N. Corresponding N-contents for *L. caustica* leaves and litter were 1.10% N and 0.37% N, respectively. Based on these data, equivalent throughfall and stemflow fluxes, and decomposition rates, we postulate that *L. caustica* and *Q. saponaria* may have different nitrogen-economy strategies. The former may favor internal cycling of nitrogen and the latter maximize external cycling.

Acknowledgements This study was funded by DIUC (Dirección de Investigación de la Pontificia Universidad Católica de Chile) under contract 44/79.

References

1   Gilbert O and Bocock K L 1960 Changes in leaf-litter when placed on the surface of soils with contrasting humus types. II. Changes in the nitrogen content of oak and ash leaf-litter. J. Soil Sci. 11, 10–19.

2   INTEC (Instituto de Investigaciones Tecnológicas) 1978 A Report to the Health Ministry: A Program to Control Atmospheric Pollution in the Metropolitan Area of Santiago: INTEC. 120 p. (Mimeo, in Spanish.)

3   Meentemeyer V 1978 An approach to the biometeorology of decomposer organisms. Intl. J. Biometeror. 22, 94–102.

4   Montenegro G, Riveros F and Alcalde C 1980 Morphological structure and water balance of four Chilean shrub species. Flora 170, 554–564.

5   Mooney H 1977 A Study of Convergent Evolution: Scrub Ecosystems of California and Chile. Stroudsburg, Pennsylvania: Dowden, Hutchinson and Ross. 224 p.

6   Rapp M and Lossaint P 1981 Some aspects of mineral cycling in the Garrigue of Southern France. *In* di Castri F, Goodall D W and Specht R L (Eds.). Ecosystems of the World, Vol. 11, Mediterranean-type shrublands.

7   Rundel P E and Neel J W 1978 Nitrogen fixation by *Trevos trinervis* (Rhamnaceae) in the Chilean matorral. Flora 167, 127–132.

8   Shaver G R 1981 Mineral nutrient and nonstructural carbon utilization. *In* Miller P C (Ed.). Resource Use by Chaparral and Matorral. A Comparison of Vegetation Function in Two Mediterranean Type Ecosystems, pp 237–257. N.Y.: Springer-Verlag.

9   Yates L R, Cisternas R and Cadiz D 1979 Leaf-litter production and decomposition in the piedmont matorral of Central Chile. Trabajo presentado en la XXII Reunión Anual de la Sociedad de Biología de Chile, Valdivia, Chile (unpublished).

Plant and Soil 67, 367–376 (1982).   0032-079X/82/0673-0367$01.50.   SU-34
© 1982 Martinus Nijhoff/Dr W. Junk Publishers, The Hague.

Regional gains and losses of nitrogen in the Amazon basin

Ganancias y pérdidas regionales de nitrógeno en la cuenca Amazónica

E. SALATI,
Centro de Energia Nuclear na Agricultura (CENA), CP 96, Piracicaba, Brazil,

R. SYLVESTER-BRADLEY,
Centro Internacional de Agricultura Tropical (CIAT), Aptdo. Aereo 6713, Cali, Colombia,

and R. L. VICTORIA
CENA, CP 96, Piracicaba, Brazil

Key words   Amazonas Entisol Hydrologic losses N-cycling N$_2$-fixation Oxisol Streamflow
Regional budget Ultisol

Abstract   In order to better understand the relative importance of different ecosystems and nitrogen cycling processes within the Amazon basin to the nitrogen economy of this region, we constructed a generalized nitrogen budget for the region based on data for hydrologic losses of nitrogen and nitrogen fixation in Amazon forests. Data included information available for nitrogen in water entering and leaving both the entire basin and watersheds on oxisol and ultisol soils near Manaus, Brazil, in addition to biological nitrogen fixation in forests on ultisol, oxisol and entisol ('varzea') soils in Central Amazonia.

Available data indicate that 4–6 kg N ha^{-1} yr^{-1} are lost via the River Amazonas, and that a similar amount enters in rainfall. Root-associated biological nitrogen fixation contributes ca. 2 kg N ha^{-1} yr^{-1} to forests on oxisols, 20 kg N ha^{-1} yr^{-1} to forests on utisols, and 200 kg N ha^{-1} yr^{-1} to forests on fertile varzea soils. There is 5–10 fold more NH$_4$$^+$–N than NO$_3$–N in rain and stream water entering and leaving the waterbasin near Manaus.

Calculations based on these data plus certain assumptions yield the following regional nitrogen balance estimate: inputs through bulk deposition of 36×10^8 kg N yr^{-1} and through biological nitrogen fixation of 120×10^8 kg N yr^{-1}, and outputs via the River Amazonas of 36×10^8 kg N yr^{-1} and via denitrification and volatilization (by difference) of 120×10^8 kg N yr^{-1}.

Resumen   Con el fin de comprender la importancia relativa de los ecosistemas y los procesos del ciclo de nitrógeno dentro de la cuenca Amazónica, hemos establecido un balance generalizado para este elemento basado en los datos de pérdidas de nitrógeno por las aguas y la fijación biológica en los bosques Amazónicos. Se incluye en el modelo la información disponible para las aguas que entran y salen de la cuenca entera asi como para una cuenca experimental sobre oxisoles y ultisoles cerca de Manaus, Brasil. También se incluyen datos de fijación biológica de nitrógeno en suelos diferentes de la Amazonia Central.

Entre 4–6 kg N ha^{-1} año^{-1} salen por el rio Amazonas y cantidades semejantes entran al sistema por precipitación. La fijación biológica de nitrógeno en las raices contribuye con ca. 2 kg N ha^{-1} año^{-1} en bosques sobre oxisoles, 20 kg N ha^{-1} año^{-1} en bosques sobre ultisoles y unos 200 kg N ha^{-1} año^{-1} en bosques sobre entisoles mas fértiles en la 'varzea'. La relación NH$_4$–N vs NO$_3$–N está alrededor de 5–10 tanto para las aguas de lluvia como para las de los rios cerca de Manaus.

Para la cuenca del Amazonas se obtuvo, basándose en los datos arriba mencionados y en algunas premises, el siguiente balance regional. Entradas por precipitación, 36×10^8 kg N año^{-1}; fijación biológica de nitrógeno, 120×10^8 kg N año^{-1}. Las salidas por el rio Amazonas, 36×10^8 kg N año^{-1} y por desnitrificación y volatilización (por diferencia), 120×10^8 kg N año^{-1}.

Introduction

Nitrogen cycles in tightly-managed, homogeneous ecosystems such as sugar cane, rice and corn are difficult to quantify, but even more complex are nitrogen cycles in tropical forests, characterized by a very heterogeneous flora and fauna. Soil variability and climatic diversity within the Amazon region result in ecosystems that range in vegetation type from campina savanna to tropical rain forest. Depending upon the degree of detail considered, a number of geobotanic associations can be found within the region, and a nitrogen balance could be defined for each. By integrating these budgets, a regional N-balance for the entire Amazon basin could be calculated. Unfortunately, however, such detailed information is not available, and measurements of nitrogen fluxes and pools which exist are mostly preliminary estimates of either the nitrogen gained or lost via a particular mechanism or the amounts of nitrogen in certain system components. Exceptions include several more-complete studies that have been carried out in specific ecosystems in Venezuela (e.g.[3,4]).

In this study, data on three aspects of the nitrogen balance of the Amazon basin are considered in order to derive a tentative estimate of the nitrogen balance of the whole basin: 1) nitrogen concentrations and flow rates of the rivers Solimões and Amazonas, 2) NO_3^--N and NH_4^+-N in precipitation and river water in Central Amazonia near Manaus, Brazil, and 3) inputs of nitrogen via biological nitrogen fixation in three widespread Central Amazonian ecosystems.

This study is intended to help identify suitable sites and topics for further investigations. We emphasize the urgent need for filling some of the gaps in our present knowledge, particularly since increasing population pressures in the region may soon result in natural conditions being irreparably altered by human activities.

Hydrological losses of nitrogen from the Amazon basin

We have made an estimate of the nitrogen lost from the Amazon basin via river flow using Amazon River water fluxes at Obidos[8] and mineral-N and organic-N concentrations[2,10]. These data are summarized in Tables 1 and 2. To extrapolate these data to the whole Amazon basin it was necessary to assume that a) the concentrations of the various nitrogen compounds do not vary along the River Amazonas after Manaus, and b) the flow rate of the River Amazonas expressed on a land area basis ($m^3 H_2O ha^{-1}$ of watershed) is the same for both the river above Obidos and the river between Obidos and the Atlantic.

Data for monthly river discharge and nitrogen concentrations show that NO_3^--N concentrations are related to rates of river flow; in our composite water year, high discharges are coincident with low NO_3^--N concentrations (Fig. 1). This trend is also evident for organic nitrogen, although we have no explanation for the August organic-N peak.

Table 1. Estimates of NO_3–N discharge from the Amazon basin above Obidos, Brazil, based on Oltman's[8] water discharge rate and various NO_3–N concentration measurements. NO_3–N concentrations were measured at either (a) Obidos or (b) Manaus, Brazil

Month	H_2O-discharge $10^9 m^3$ H_2O month^{-1}	NO_3–N Oltman[8,a] mg N l^{-1}	10^6 kg N	Gibbs[2] mg N l^{-1}	10^6 kg N	Schmidt[10,b] mg N l^{-1}	10^6 kg N
January	295	nd	nd	nd	nd	0.065	19.2
February	349	nd	nd	nd	nd	0.064	22.3
March	455	nd	nd	nd	nd	0.043	19.6
April	557	nd	nd	nd	nd	0.035	19.5
May	643	nd	nd	nd	nd	0.023	14.8
June	622	nd	nd	nd	nd	0.018	11.2
July	549	0.1	54.9	nd	nd	0.028	15.4
August	442	0.1	44.2	nd	nd	0.024	10.6
September	311	nd	nd	nd	nd	0.040	12.4
October	241	nd	nd	nd	nd	0.084	20.2
November	220	nd	nd	nd	nd	0.076	16.7
December	254	nd	nd	nd	nd	0.071	18.0
Monthly $\bar{x}$	nd	nd	nd	0.2	82.5	nd	16.7
Annual total	nd	nd	nd	nd	990.	nd	200.

nd, not determined.

Almost 90 per cent of the nitrogen in the River Amazonas was in the organic form throughout the 1969–70 water year[10] (Fig. 2). Most of this organic nitrogen was in solution rather than suspended, although the proportion varied during the year. Gibbs[2] and Schmidt[10] concluded that soluble NH_4^+–N is unimportant relative to soluble organic-N, as is the relative contribution of NO_2^-–N.

Based on our composite water year, the Amazon River apparently discharges about 234×10^6 kg N month^{-1} (Table 2), with a maximum in April (378×10^6 kg N) and a minimum in October (130×10^6 kg N). The total annual loss of nitrogen from the Amazon basin, then, is 3.1×10^9 kg N, which is equivalent to approximately 4–6 kg N ha^{-1} yr^{-1}. This is a minimum value since calculations are based on the lowest values in the literature (see Table 1).

NO_3^- and NH_4^+ flows in a Central amazonian basin

In 1979, water balance studies were initiated in a Central Amazonian waterbasin located approximately 60 km north of Manaus. The basin is about 2200 ha and supports 'terra firme' (non-flooded) rainforest growing mainly on oxisol soils, although ultisols are present along the stream beds. Over the past two-years, precipitation has been measured with 11 pluviometers, stream flow

Table 2. Total-N discharged from the Amazon basin based on water discharge (Table 1) and total-N concentrations (Schmidt[10]) of the Amazon River at Obidos, Brazil. Discharge past the mouth of the river assumes that the watersheds below Obidos lose the same amount of N ha^{-1} as the watersheds upriver

Month	Total-N (Obidos	Mouth (corrected)	
	10^6 kg N	10^6 kg N	kg N ha^{-1}
January	238	267	0.445
February	292	315	0.525
March	270	301	0.502
April	332	370	0.617
May	229	255	0.425
June	286	318	0.530
July	235	262	0.436
August	339	378	0.629
September	134	150	0.249
October	117	131	0.218
November	154	172	0.287
December	189	212	0.353
Monthly $\bar{x}$	235	261	0.435
Annual total	2815	3130	5.22

has been monitored at two sites, and conventional meteorological measurements have been made at a climatological station that includes a $20 \, m^2$ evaporation pan.

In 1980, rainwater, throughfall and streamwater were sampled at weekly intervals and analyzed for various elements, including NH_4^+–N and NO_3^-–N. Table 3 presents preliminary data for the balance of NH_4^+–N and NO_3^-–N in May and June of 1980. Nitrogen in bulk precipitation included 0.02 mg NH_4^+–N l^{-1} and 0.005 mg NO_3^-–N l^{-1} for both months. Total mineral nitrogen input to the waterbasin in bulk precipitation was calculated to be 49.0 and 44.2 kg N for May and June, respectively, or 0.022 and 0.020 kg N ha^{-1}.

Streamwater NH_4^+–N concentrations for May and June were 0.043 and 0.025 mg NH_4^+–N l^{-1}, respectively. NO_3^-–N concentrations were lower, at 0.005 mg NO_3^-–N l^{-1} for each month. Total outflow was 1.1×10^6 and $0.98 \times 10^6 \, m^3$ H_2O for May and June, respectively, so that total mineral-N output from the waterbasin was 37.8 and 20.2 kg N for these months, or 0.017 and 0.009 kg N ha^{-1}, respectively.

Output rates of NO_3^-–N (ca. 0.0005 kg NO_3^-–N ha^{-1} month^{-1} for both months) are much lower than those reported for the same months by Schmidt[10], who had found approximately 0.025 kg NO_3^-–N ha^{-1} month^{-1} draining from the entire Amazon basin.

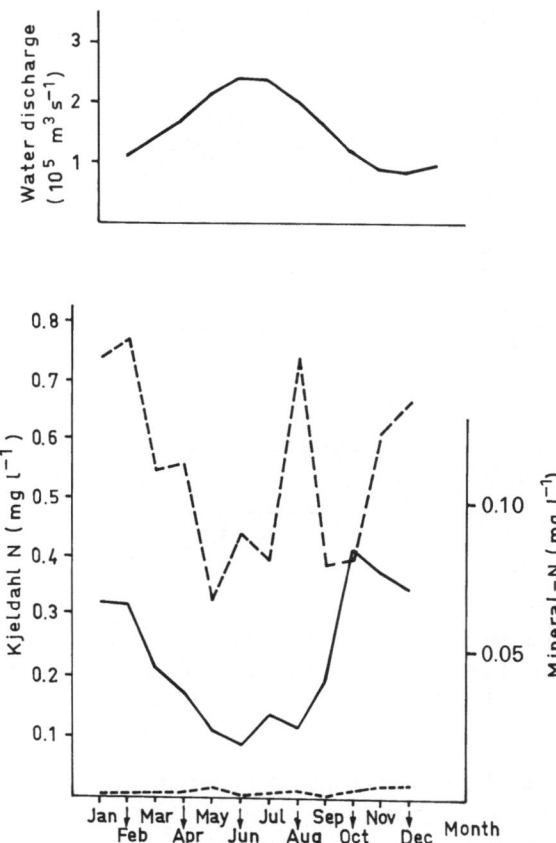

Fig. 1.   Amazon River water flow past Obidos, Brazil (Oltman[8]) and concentrations of organic N
(– – – –), $NO_3{}^-$–N (————), and $NO_2{}^-$–N (- - - - - -) (Schmidt[10]) by month. Note different scales for
Kjeldahl-N and mineral-N. Kjeldahl-N includes suspended and dissolved organic-N plus $NH_4{}^+$–N.

Table 3. Mineral-N inputs *via* bulk precipitation and outputs via streamflow from an experimental
waterbasin near Manaus, Brazil, during May and June, 1980. Watershed area is 2200 ha^{-1}

Month	H_2O-flux	NH_4–N		NO_3–N		Total mineral-N
		mg N l^{-1}	kg N ha^{-1}	mg N l^{-1}	kg N ha^{-1}	kg N ha^{-1}
Bulk precipitation						
May	133[a]	0.02	0.0210	0.005	0.0015	0.0225
June	133[a]	0.02	0.0187	0.005	0.0014	0.0201
Stream water						
May	1.1[b]	0.042	0.0166	0.005	0.0005	0.0171
June	0.98[b]	0.025	0.0087	0.005	0.0005	0.0092

[a] mm precipitation.
[b] 10^6 m^3 H_2O.

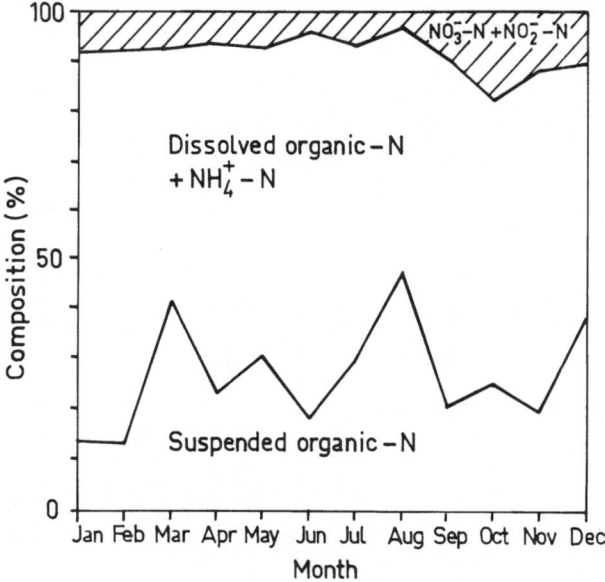

Fig. 2.   Relative composition of total-nitrogen discharged past Obidos, Brazil, in the Amazon River. Values are based on water discharge and nitrogen concentrations presented in Fig. 1. Absolute rates are presented in Table 2. (After Oltman[8]).

Nitrogen inputs *via* biological nitrogen fixation

Nitrogen fixation has been studied in various systems in the vicinity of Manaus by acetylene reduction, nodule distribution and natural isotope variation methods. These studies have been concentrated on three soil types which are significant in the region for different reasons. The 'varzea' soil is a fertile alluvial soil which occurs along the banks of whitewater ('agua branca') rivers such as the Solimões and the Amazonas. The natural forest vegetation on this soil has nearly all been cleared and the land is used for relatively-intensive production of vegetables, milk, sorva, jute, and other crops, even though it is flooded for three or four months of the year. Many introduced and native grasses, legumes, and other plants that grow here are adapted to flooding; some can float.

The second soil is white, sandy, and covered with a layer of semidecomposed material and live roots; this layer can be up to one meter thick. Water in streams draining these areas is the color of dark tea due to high concentrations of organic compounds, and this drainage forms blackwater rivers such as the Rio Negro. The campina and campinarana vegetation in these areas appear xerophytic, although this is in fact an adaptation to the oligotrophic conditions. The vegetation is low in stature, is dominated by only a few species (some of which are legumes), and tree trunks, branches and leaf surfaces are covered by epiphytes and lichens. Sites containing this soil tend to be small and separated by other types of forest, and a certain degree of endemism has been found[9]. The soil is of no use for agriculture, but such sites are being destroyed because the sand is needed for road construction.

The third class of soils, and by far the most important class in the Manaus area in terms of area occupied and problems presented for agricultural development, is oxisols and ultisols supporting rainforest vegetation. Rainforests are more diverse than forests of campina vegetation, and legumes are one of the five most abundant families[1]. Rainforests also have a high biomass[5] (ca. 470×10^3 kg dry matter ha^{-1}) and are being increasingly cleared to grow a wide range of cash crops such as sugar cane, fodder, rubber, cocoa, and fruit, as well as subsistence crops such as rice, cowpeas and cassava.

Recent N-fixation studies have shown that:

1) Although there is a high proportion of legumes in the primary rainforest growing on oxisols, the legumes are only sparsely nodulated[6,12]. The highest rates of acetylene reduction that have been detected in the root mat extrapolate to 1.5 kg N-fixed ha^{-1} yr^{-1} (Table 5).

2) There is no difference between the δ^{15}N‰ ratios of legumes and non-legumes in primary rain forests on oxisols, but there is a significant difference between leaves of legumes and non-legumes in campina vegetation (Table 4). Nodules were not found on roots of legume plants in campina vegetation, but this may be because they were very deep (Sylvester-Bradley, unpublished).

3) Epiphytes, lichens and *Stigonema parriforme*, a bluegreen algae growing on

Table 4. δ^{15}N‰ in leaves of legume and non-legume trees growing on two soil types in Central Amazônia, Brazil, relative to ^{15}N/^{14}N of atmospheric N_2 (Victoria *et al.*, in press). Reserva Ducke is an experimental area near Manaus

Plant	Reserva Ducke (clayey soil)	Campina (sandy soil)
Legumes		
Macucu (*Aldinalatifolia* sp.)	na	− 1.9
Faveira (*Pithecolobium* sp.)	+ 3.9	na
Angelim (*Pithecolobium raceniorum*)	+ 7.8	na
Muirapiranga (*Caesalpinoidea*)	+ 6.5	− 1.6
Visgueiro (*Pankia* sp.)	+ 10.0	na
Copaiba roxa (*Caesalpinoidae*)	+ 8.1	na
Tento (*Papilionidae*)	na	− 2.2
Non-legumes		
Mata-mata (Lecitidaceae)	+ 5.8	na
Ucuuba branca (*Virola surinasuensis*)	+ 5.9	na
Casca doce (*Glycoxylon thophiluru*)	na	− 7.0
Andiroba (*Carapa guianensis*)	+ 7.0	na
Pitomba de macaco (Sapotaceae)	na	− 7.0

na, not available.

Table 5. Nitrogen-fixation (C_2H_2 reduction assuming $3:1$ $C_2H_4:N_2$ ratio) in various components of Central Amazonian systems. Brackets indicate uncertainty. Data are from sources cited in text

Component	kg N ha^{-1} yr^{-1}
Roots	
in primary forest on clayey soil	1.50
in secondary forest on clayey soil (no cultivation)	2.45
in secondary forest on sandy soil	20
in primary forest on alluvial soil	243
in 'improved' pasture on oxisol	(0–1)
in pasture on alluvial ('varzea') soil	(>1)
Epiphytes	
on campina vegetation	(5–20)
on primary forest on clayey soil	(0–5)
Aquatic Macrophytes	(100)
Termites	(0–5)

the soil surface in campina vegetation show acetylene-reducing activity (Table 5) but this is difficult to quantify on a system-wide basis (Leite dos Anjos and Sylvester-Bradley, unpublished).

4) Legumes in rainforests on ultisols are nodulated[6] and their roots reduce acetylene. The highest rates detected extrapolate to 20 kg N-fixed ha^{-1} yr^{-1} (Table 5)[12].

5) Legumes in primary forests on varzea soil are very well nodulated. The highest rates of acetylene reduction detected extrapolate to 243 kg N-fixed ha^{-1} yr^{-1} (Table 5)[12].

6) Acetylene-reduction activity in roots of aquatic plants (Sylvester-Bradley, unpublished), and in termites[11,13] also occurs. Acetylene reduction by termites is extremely sensitive to disturbance.

7) Inoculation of cowpeas (*Vigna* sp.) and winged beans (*Phaseolus* sp.) grown in oxisols does not increase their nodulation, but inoculated plants fertilized with phosphorus and potassium grow much better and are well-nodulated (Oliveira and Sylvester-Bradley, in press).

8) Burning existing vegetation before planting is not as effective for increasing plant growth as is fertilizing with phosphorus and potassium[7].

These nitrogen fixation data, although limited in scope, have some interesting implications. First, the large differences in the levels of root-associated nitrogen fixation in equilibrium vegetation on oxisols, ultisols and entisols imply that there are also large differences in losses of nitrogen from these systems. Second, the limiting factors for growth and nitrogen fixation in each soil type are presumably different. For example, varzea soils contain over 30 mg P kg^{-1} dry soil whereas the oxisols and ultisols contain around 2 mg. Thus the potential for nitrogen fixation in the varzea soils may be much greater than that in oxisols and ultisols because phosphorus is probably not limiting in the former. Subsequently, farmers may have many more problems with biological nitrogen fixation as a source of nitrogen for crops on ultisols and oxisols than for crops on varzea soils. The tenfold-greater nitrogen fixation detected in ultisols over that in oxisols is presumably due to greater leaching losses of nitrogen from ultisols, though nitrogen fixation in oxisols may be low due to a combination of high soil-nitrogen levels (8000 kg N ha^{-1})[5], low phosphorus, and possibly toxic substances (produced by the vegetation) which inhibit nitrogen fixation. Ultisols are considered by local farmers to be easier to cultivate than oxisols, and they support better nodulation in pot experiments, but ultisols tend to occur on steeper grades and therefore are easily eroded. Consequently, oxisols are preferred for agricultural development.

A third interesting aspect of these N-fixation findings is the greater nitrogen fixation associated with epiphytes in campina vegetation than in primary rainforest vegetation. This implies that the importance of epiphytes should be assessed in different kinds of tropical rainforests so that the epiphytic contribution of nitrogen to the system relative to other nitrogen inputs can be determined.

General considerations

Based on the data presented here and information presented by others (C. Jordan, R. Herrera and E. Sanhueza, personal communications), a balance of nitrogen for the Amazon basin was established. The balance is based on three arguments:

1) the overall flow of nitrogen past the mouth of the River Amazonas is 6 kg N watershed-ha^{-1} yr^{-1};

2) the input of nitrogen *via* bulk precipitation is 6 kg ha^{-1} yr^{-1};

3) that sites on oxisols occupy 50% of the Amazon basin and fix 2 kg N ha^{-1} yr^{-1}, that sites on ultisols occupy 45% of the basin and fix 20 kg N ha^{-1} yr^{-1}, and that sites on soils occupying the remaining 5% of the basin fix 200 kg N ha^{-1} yr^{-1}. Thus, oxisol sites in the region (3×10^8 ha) fix 6×10^8 kg N yr^{-1}, ultisol sites (2.7×10^8 ha) fix 54×10^8 kg N yr^{-1}, and varzea and other fertile-soil sites (0.3×10^8 ha) fix 60×10^8 kg N yr^{-1}, for a basin-wide average of 20 kg N-fixed ha^{-1} yr^{-1}.

Accordingly, losses of nitrogen by denitrification and volatilization should be of a similar magnitude (*ca.* 20 kg N ha^{-1} yr^{-1}), unless the system is accumulating nitrogen, or unless nitrogen is removed by crop harvests and not replaced by fertilizer-N or increased N-fixation.

The overall nitrogen balance for the Amazon basin therefore includes inputs of 36×10^8 kg mineral-N yr^{-1} from bulk precipitation, inputs of 120×10^8 kg N yr^{-1} from biological nitrogen fixation, hydrological outputs of 36×10^8 kg N yr^{-1} *via* the R. Amazonas, and outputs of 120 $\times$ 10^8 kg N yr^{-1} by denitrification and volatilization (by difference).

Acknowledgements The authors wish to acknowledge support from the Instituto Nacional de Pesquisas de Amazonia (INPA), Financiadora de Estudos e Projetos (FINEP) and United Nations Project BRA/72/010.

References

1 Black G A, Dobzhansky Th and Pavan C 1950 Some attempts to estimate species diversity and population density of trees in Amazonian forests. Bot. Gaz. 3, 413–425.
2 Gibbs R J 1972 Water chemistry of the Amazon River. Geoch. Cosmoch. Acta 36, 1061–1066.
3 Herrera R and Jordan C F 1981 Nitrogen cycle in a tropical Amazonian rain forest: The caatinga of low mineral nutrient status. *In* Clark F E and Rosswall T (Eds.). Terrestrial Nitrogen Cycles. Ecol. Bull. Stockholm 33, 493–505.
4 Jordan C F, Caskey W, Escalante G, Herrera R, Montagnini F, Todd R and Uhl C 1982 The nitrogen cycle in a 'Terra Firme' rain forest on oxisol in the Amazon territory of Venezuela. *In* Robertson G P, Herrera R and Rosswall T (Eds.). Nitrogen Cycling in Latin American and Caribbean Ecosystems. Plant and Soil 67, 325-332.
5 Klinge H 1976 Bilanzierung von Hauptnährstoffen im Ökosystem tropischer Regenwald (Manaus)-vorläufige Daten. Biogeografia 7, 59–77.
6 Norris D O 1969 Observations on the nodulation status of rain forest leguminous species in Amazonia and Guyana. Trop. Agric. Trinidad 46, 145–151.
7 de Oliveira L A and Sylvester-Bradley R 1982 Effect of different Central Amazonian soils on growth, nodulation and occurrence of N$_2$-fixing *Azospirillum* spp. in roots of some crop plants. Turrialba. (In press).
8 Oltman R E 1967 Reconnaissance investigations of the discharge and water quality of the Amazon. Atas do Simposio sobre a Biota Amazônica 3, 163–185.
9 France G T 1978 The origin and evolution of the Amazon flora. Interciencia 3, 207–222.
10 Schmidt G W 1972 Amounts òf suspended solids and dissolved substances in the middle reaches of the Amazon over the course of one year (August, 1969–July, 1970). Amazoniana (Kiel, Germany) 3, 208–223.
11 Sylvester-Bradley R, Bandeira A G and de Oliveira L A 1978 Fixação de nitrogênio (redução de acetileno) em cupins (Insecta: Isoptera) da Amazônia Central. Acta Amazonica 9, 621–627.
12 Sylvester-Bradley R, de Oliveira L A, de Podestá Filho J A and St John T V 1980 Nodulation of legumes, nitrogenase activity of roots, and occurrence of nitrogen-fixing *Azospirillum* spp. in representative soils of Central Amazonia. Agro Ecosystems 6, 249–266.
13 Sylvester-Bradley R, de Oliveira L A and Bandeira A G 1982 Nitrogen Fixation in Nasutitermes in Central Amazonia. *In* Jaisson, Michener and Howse (Eds.). Proceedings of the International Symposium on Social Insects in the Tropics, 9–13 November, 1980, Cocoyoc, Mexico. Paris: University of Paris Press. (In press).
14 Victoria R L, Stewart J. W B, Matsui E and Salati E 1979 Variação natural de ^{15}N em plantas e solos da Região Amazônica. Abstract No. 173. XVII Brazilian Congress of Soil Science, Manaus 8–13 July 1979, p. 63. Manaus: Sociedade Brasileirade Ciência do Solo.

Plant and Soil 67, 377–387 (1982).   0032-079X/82/0673-0377$01.65.   SU-35
© 1982 Martinus Nijhoff/Dr W. Junk Publishers, The Hague.

Eutrophy in Lake Aculeo, Chile

Eutrofia de la laguna de Aculeo en Chile

S. CABRERA S.
Department of Cell Biology and Genetics, University of Chile, PO Box 6556, Correo 7, Santiago, Chile

and V. MONTECINO B.
Department of Biology, University of Chile, PO Box 653, Santiago, Chile

Key words   Chile   Energy flow   Eutrophication   Lake Aculeo   N-cycling   Primary production

Abstract   We examined energy flow and nitrogen turnover in a highly eutrophic lake in Chile. Carbon fixation varied seasonally between 25–450 mg C m^{-2} h^{-1} in the photic zone, and *ca.* 6.25×10^6 kg C were fixed in the lake during 1980. Nitrogen turnover in the phytomass was *ca.* 1.04×10^6 kg N yr^{-1} based on an arbitrary C:N ratio of 6.0.

Resumen   En un lago muy eutrofizado en Chile se estudiaron el flujo de energía y la tasa de recambio de nitrógeno. La fijación de carbono mostró variaciones estacionales entre 25 y 450 mg C m^{-2} h^{-1} en la zona fótica; cerca de $6,25 \times 10^6$ kg C fueron fijados durante 1980. El recambio de nitrógeno, calculado asumiendo una relación arbitraria C:N = 6,0, fue de $1,04 \times 10^6$ kg N año.

Introduction

Studies of primary productivity are fundamental to an understanding of both the flow of energy in ecosystems and the state of eutrophy in lakes and lagoons. In accordance with Margalef[5], eutrophy is taken here to mean the production capacity of a system.

When we come across a body of water whose oxygen curve displays severe O_2 depletion at depth, detect sporadic mass mortality of fish and observe that the green colour of the water makes it rather unpleasant for swimming and water sports in general, we can be certain that we have encountered an eutrophied lake. Such is the case with Lake Aculeo in Chile.

The object of this report is to attempt to place approximate values on the amounts of nitrogen cycled by this body of water during one year, based on values for carbon fixation by phytoplanktonic algae. We hope that the information thus acquired will serve as a basis for decisions regarding the management of this lagoon.

Site description

Lake Aculeo is situated 50 km southeast of Santiago (33° 50′ 30″ S, 70° 54′ 24″ W, 360 m elevation); it has a maximum depth of 7 m and a maximum length of 5.8 km, is alluvial in origin, and the local climate is Mediterranean. The lagoon is surrounded by the hills of the Coastal Cordillera (to 2000 m elevation), and by agricultural land (Fig. 1), and is fed or drained by the same stream depending on the level of water in the lake.

The transparency of the water as determined with a Secchi disc is to at least 0.25 m. Especially in periods with no wind, clinograde distributions of oxygen from the surface to the bottom are observed together with massive blooms (48–520 × 10^6 cells 1^{-1}) of the cyanophyte *Microcystis aeruginosa* Kuetz. (I. Vila, personal communication.)

Methods

Periodic measurements of primary productivity were made *in situ* from the surface to the bottom at site 5 (Fig. 2) during one annual cycle. Depth measurements were carried out with a Simrad EY echo-sounder; 14 transects were traced to cover the lake completely. The profiles were sketched onto aerial photographs (scale 1 : 25000) taken by the Chilean Air Force Aerial Photogrammetric Service. In order to obtain the volumes of water comprised at each depth, a hypsographic curve was traced (Fig. 3).

The values used for the total incident radiation represent a mean of 16 years of measurements[1] taken at the Quelentaro station close to the lake. A continuous daily recording actinograph was used. To determine the photosynthetically active radiation (400–700 nm) and its penetration into the water, an LI-190S quantum sensor was used in the open air and an LI-192S underwater (LI Corporation), connected to a transportable Minigor RE 501 Goerz data logger. Radiation penetration into the water was measured only on those days on which experiments were carried out.

Water temperature was measured with a thermometer graduated in tenths of a degree and enclosed in a transparent Van Dorn bottle. Concentrations of dissolved oxygen were determined by Winkler's method as modified by Golterman[2]. A portable Beckmann pH meter was used to monitor pH, alkalinity and available inorganic carbon; alkalinity was measured by titration and available inorganic carbon was calculated according to Margalef[5].

Primary productivity was quantified by the carbon-14 technique[8,11]. Samples were obtained with an opaque Van Dorn bottle between 100% and 3% of photosynthetically active radiation. Incubation took place for periods of 2.5 hours between 1100 and 1500 hrs. Between 0.5 and 2 μCi of NaH^{14}CO$_3$ from a stock solution were added to each bottle, and the activity of the solution was subsequently determined by fixing 2.5 ml of water from the bottle with 0.5 ml of CO$_2$ sequestering agent (Amersham Corporation). The scintillation liquid was principally PPO, POPOP, Triton and toluene, with final counts (to 2% error) carried out in a liquid with a dioxane base. Integrations of productivity values for the total depth of water per day were carried out by planimetry using graphs of photosynthetically active radiation.

Concentration of chlorophyll-a (chl-a) were determined according to the technique recommended by SCOR-UNESCO (1964), with optical densities measured by a Schimadzu UV-190 scanning spectrophotometer.

To determine the horizontal homogeneity in the lake, 8 sites were sampled (Fig. 2). The samples were taken with a water-depth integrating bottle[9] between the surface and a depth of 3 m, the extent of the photic layer at that time. The concentration of chl-a was determined in these samples.

Fig. 1. Aerial view (1 : 100,000) of Lake Aculeo (Aerial Photogrammetric Service, Chilean Air Force).

Once the productivity of each stratum of water was determined, it was then multiplied by the total volume of each stratum obtained from the hypsographic curve. The calculation of annual productivity was carried out by plotting the values for total incident radiation (daily means for each month) against the integrated values for the primary productivity of the lake, interpolating the values for those months in which no *in situ* determinations were made.

Results

The morphometric parameters (Table 1) and the bathymetric chart (Fig. 2) show an elliptic lake with an average depth of 4.4 m. The volumes of water per stratum at maximum water level are given in Table 2.

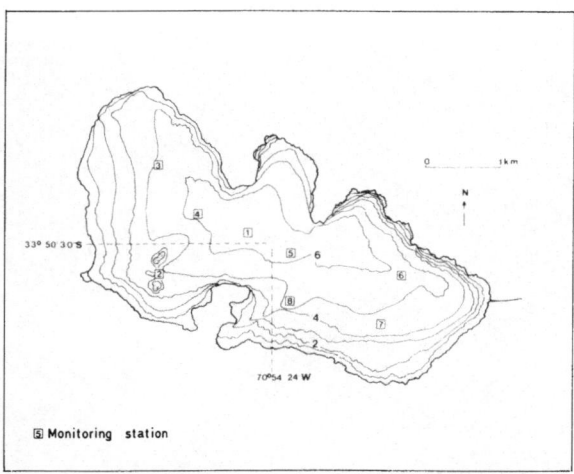

Fig. 2.   Bathymetric chart of Lake Aculeo.

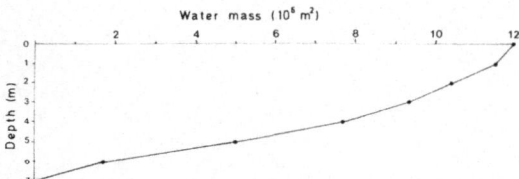

Fig. 3.   Hypsographic curve of Lake Aculeo.

Table 1. Morphometric parameters for Lake Aculeo

Parameter	Value
Area (km^2)	11.99
Mean depth (m)	4.4
Volume (10^6 m^3)	5.16
Maximum length (km)	5.85
Maximum width (km)	2.88
Mean width (km)	2.07
Maximum depth (m)	7.0
Perimeter (km)	18.3
Development of the shoreline	1.48
Altitude (m)	360

The annual curve of total incident radiation from the daily means for each month is unimodal, with a maximum in January of 603 langley and·a minimum in June of 154 langley (Table 3). The penetration of photosynthetically active radiation throughout the year fluctuated between 1.88 m and 5.3 m in January and November respectively.

Concentrations oi chl-a varied between 487 mg m^{-3} at 0.6 m depth in January 1981 and 28.4 mg m^{-3} at 1.0 m depth in September 1980. Primary productivity fluctuated between 578 $\pm$ 24.8 mg C m^{-3} h^{-1} at 0.0 m depth in November 1978 and 3.8 $\pm$ 0.9 mg C m^{-3} h^{-1} at 2.5 m depth in November 1980. Primary productivity per square metre fluctuated between 25.3 mg C m^{-2} h^{-1} in June 1980 and 451 mg C m^{-2} h^{-1} in November 1978 (Table 4). The integrated values for chl-a in the photic layer of the lake fluctuated between 0.8 and 4.4 $\times$ 10^3 kg (Table 4) in June 1980 and January 1981 respectively. The integrated values for primary productivity as a total for the photic layer fluctuated between

Table 2. Water volumes by stratum in Lake Aculeo

Depth (m)	Volume (10^6 m^3)
0–1	11.77
1–2	10.96
2–3	9.83
3–4	8.53
4–5	6.35
5–6	3.33
6–7	0.86

Table 3. Total incident radiation by month (daily means) for Lake Aculeo. Values are 16-yr means (Arata[1])

Month	Langley Value
January	603
February	524
March	426
April	313
May	196
June	154
July	177
August	251
September	351
October	434
November	540
December	592

Table 4. Seasonal primary productivity ($\pm$ s.d.) and chlorophyll-a (chl-a) levels in Lake Aculeo

Month	Depth	Primary productivity		Chlorophyll-a	
		$(mg\,C\,m^{-3}\,h^{-1})$	$(mg\,C\,m^{-2}\,h^{-1})$	$(mg\,chl\text{-}a\,m^{-3})$	$(10^3\,kg)$
Nov 1978	0.0	578 $\pm$ 24.8		84.5	
	0.5	213 $\pm$ 9.7		85.9	
	2.0	75 $\pm$ 0.3		67.2	
	0–2		451		2.2
Jan 1980	0.0	14.3 $\pm$ 3.8		56.8	
	0.6	178.8 $\pm$ 12.5		93.4	
	1.2	91.9 $\pm$ 25.0		95.2	
	2.5	28.9 $\pm$ 7.0		90.7	
	0–2.5		232		3.1
Jun 1980	0.0	82.7 $\pm$ 29.3		84.3	
	0.5	9.2 $\pm$ 1.8		65.6	
	1.0	0.0		73.6	
	0–1.0		25.3		0.8
Sep 1980	0.0	47.8 $\pm$ 8.8		31.0	
	1.0	24.1 $\pm$ 3.3		28.4	
	1.5	24.4 $\pm$ 0.9		33.5	
	2.0	18.2 $\pm$ 0.2		36.3	
	3.0	24.4 $\pm$ 0.9		39.9	
	0–3.0		126		1.4
Nov 1980	0.0	218.8 $\pm$ 52.9		468.7	
	0.2	170.1 $\pm$ 1.2		42.2	
	1.0	74.3 $\pm$ 1.3		75.3	
	1.5	23.7 $\pm$ 1.1		55.9	
	2.5	3.8 $\pm$ 0.9		53.9	
	0–2.5		179		2.4
Jan 1981	0.1	329.2 $\pm$ 41.6		329.0	
	0.6	487.4 $\pm$ 46.9		487.0	
	1.1	65.4 $\pm$ 52.9		65.4	
	0.1–1.1		403		4.4

$0.16 \cdot 10^6$ kg C month^{-1} in June 1980 and $1.74 \cdot 10^6$ kg C month^{-1} in January 1981 (Table 5).

A correlation (r $=$ 0.99, p $<$ 0.001) was found between the quantity of carbon fixed per day and the concentration of chl-a in the photic layer. The average concentration of chl-a at 8 sites in the lake (Fig. 2) was 95.15 $\pm$ 11.37 mg m^{-3} (Table 6). The inferred values for monthly carbon fixation in the photic layer appear in Table 7. The annual total is 6250 $\times$ 10^3 kg C yr^{-1} fixed by the phytoplankton in the illuminated layer of the lagoon.

Table 5. Carbon fixation integrated by month in Lake Aculeo. Total is for depths specified

Month	Depth (m)	10^3 mg C stratum^{-1} hr^{-1}	10^6 kg C month^{-1}
Nov 1978	0–1	3.18	
	1–2	0.988	
	2–3	0.344	
	Total		1.85
Jan 1980	0–1	1.27	
	1–2	0.737	
	2–3	0.247	
	3–4	0.057	
	Total		1.00
Jun 1980	0–1	0.541	
	1–2	0.000	
	Total		0.157
Sep 1980	0–1	0.425	
	1–2	0.233	
	2–3	0.209	
	3–4	0.183	
	Total		0.368
Nov 1980	0–1	1.53	
	1–2	0.438	
	2–3	0.046	
	Total		0.837
Jan 1981	0–1	3.63	
	1–2	0.367	
	Total		1.74

Table 6. Mean chlorophyll-a (chl-a) values ($\pm$ standard deviations) in the 0–3 m depth of water at 8 sites in Lake Aculeo, 16 December 1980. *In situ* primary productivity (see text) was measured at site 5

Site	mg chl-a m^{-3}
1	93.7 (1.5)
2	96.0 (2.1)
3	84.0 (0.2)
4	91.8 (4.8)
5	88.7 (0.1)
6	87.8 (0.4)
7	120.8 (4.2)
8	98.6 (2.8)

Table 7. Inferred monthly carbon fixation values for the lake Aculeo.

Month	C-fixation (10^3 kg C month^{-1})
January	1002
February	680
March	450
April	300
May	200
June	157
July	160
August	220
September	368
October	540
November	837
December	1340

Discussion

Analysis of the horizontal homogeneity in Lake Aculeo on the basis of integrated samples of the depth of water in the photic layer shows that the values for chl-a fluctuate between 83.9 and 120.8 mg/m^{-3}, seven sites being within an acceptable range of homogeneity. Only site 7 deviates from the mean and it would appear that this value is biased.

If we consider that the quantity of chl-a of the phytoplankton bears a direct relationship to primary productivity, extrapolating the values encountered at site 5 (where the incubations were carried out) to the total for the lake may carry a maximum error between 5.7% (sites 3 and 5 differ in chl-a) and 26% (sites 7 and 5).

The lake receives water during winter and spring from the mountain belt that surrounds it to nearly 320° (Fig. 1); the remainder of its circumference opens on to a valley used basically for agriculture. The rate of renewal in the summer is very slow, since it receives water from only a small stream which empties into the north-eastern margin and has practically no effect on its hydrology. The input is approximately 0.3 m^3 s^{-1}. The number of seasonal tributaries increases during the winter and involves all the lake margins. During this period the water backs up through the north-eastern sector.

At two points in the lake there are water offtakes for irrigation (by means of pumps); these, plus the very high evaporation during the summer (the water reaches 30.4°C), are the main ways in which water is lost during this period.

According to the hypsographic curve, the bottom of the lake is level, except for two small islands situated on the south-western flank. The margins are nearly all gentle slopes. The morphology of the basin probably favours the recycling of nutrients because the bottom is shallow and the distance between the photosynthetic layer and the sediments is constant over the major part of its area.

The only agents of mixing appear to be the wind, which is moderate during the greater part of the afternoon, and pleasure boats during the spring and summer. The lake thus has all the characteristics that favour a high production capacity and a homogeneity of horizontal distributions of the phytoplankton.

The maximum values for primary productivity, which attain 451 mg C m^{-2} h^{-1} with concentrations of chl-a above 28 mg m^{-3} and maxima of the order of 480 mg m^{-3}, indicate that this lake is indeed eutrophic: Wetzel[12] suggests that all bodies of water with chl-a values between 10 and 500 mg m^{-3}, with averages of 1000 mg C m^{-2} day^{-1} for primary productivity, with ranges of total nitrogen between 500 and 11000 µg l^{-1}, and with Cyanophyceae, Bacillariophyceae, Chlorophyceae and Euglenophyceae predominant in the phytoplankton, should be considered eutrophic. Sancha et al.[7], in fact, found that organic-N values for the lake fluctuate between 1700 and 25800 µg N l^{-1}, which indicates the existence of a hypereutrophic system according to Wetzel[12]. It will be necessary to update these values.

C:N ratios in lakes have been studied by a large number of investigators working on primary productivity. Vollenweider[11] has noted that 'the use of conversion factors used in biology is only a crude approximation. These are influenced by biological, environmental and historic factors'. Redfield et al.[6] stated that the quantity of nitrogen in the phytoplankton can vary between 4 and 20 parts per unit of carbon, and that this depends on the accuracy of the analysis and the external conditions of the sample. Margalef[4] gave a C:N ratio of 7.14.

Nakanishi[3] studied the relationship between the quotient of carbon and nitrogen versus the concentration of chl-a and found it to be approximately 10.5 to 11 for values of chl-a between 10 and 15 mg m^{-3}, and 5 for values of chl-a between 0.5 and 7.0 mg m^{-3}.

The quantity of nitrogen cycled through primary producers in the photic layer of Lake Aculeo during one annual cycle, then, can be estimated on the basis of an arbitrary C:N ratio of 6.0. This means that if 6.25×10^6 kg C yr^{-1} are fixed, the quantity of nitrogen involved is ca. 1.04×10^6 kg N yr^{-1}. This value may be an underestimate since, as we have shown, the estimated value for carbon carries a 20–25% error.

Several options for managing the lake exist. One possibility using the production potential of this lake is to extract the phytoplankton for industrial use, either for fertilizer or to enrich the diets of domestic animals. This would decrease the biomass of algae in the lake and would form a pathway for nutrients to leave the hydrographic basin via export of crops, livestock and poultry products.

If the lake is preserved in its current state of eutrophy, certain steps should be taken to reduce deleterious effects. These include, for example, that of encouraging the oxygenation of the deepest layers during periods of anoxia, either by direct aeration of the bottom of the lake or by constructing artificial cascades with recirculated water. Those familiar with the lake environs will be aware of the absence of wastewater treatment in the watershed, so that wastes reach the lake untreated either via surface or groundwater. This is one of the main sources of nitrates and phosphates for the system. Fertilizers used on adjacent cropland also tend to augment the nutrients in the lake.

In order to avoid these problems it would be necessary to intensify regulatory controls on sanitation and on fertilizer and pesticide use or to construct a canal around the lake to intercept these residues so that they could be processed in a treatment plant or discharged into a drainage basin of low productivity. These measures would be expensive to implement, but greater economic losses may result if steps are not taken soon to control the eutrophication process.

Conclusions

1. Lake Aculeo is highly eutrophic and its waters are very fertile.

2. Owing to physical, biological, and environmental characteristics, the lake is extremely homogeneous in the horizontal distribution of its phytoplankton.

3. Carbon fixation in the entire lake was estimated to be $6.25 \times 10^6 \, kg \, C \, yr^{-1}$ and nitrogen turnover to be *ca.* $1.04 \times 10^6 \, kg \, N \, yr^{-1}$. The lake shows a strong tendency towards hypereutrophy.

Acknowledgements Thanks are due to Professor Nibaldo Bahamonde N. of the Chilean National Museum of Natural History, in his capacity as National Coordinator of the MAB-5 programme in Chile, for revising this manuscript and making valuable suggestions. Thanks also go to Dr. Javier García de Cortazar for his collaboration in this project and for carrying out the morphometric determinations on the lake.

This research was supported by the Department of Scientific Development and International Cooperation, University of Chile (Project B 1083-801), and the MAB-5 UNESCO Chile Programme SC RP 551-381 ('Biological potential of the dams and lakes of central Chile').

References

1 Arata A 1980 Datos para Proyectos de Energía Solar, Departamento de Publicaciones de la Universidad Técnica Federico Santa María. Valparaíso, Chile, 90 p.
2 Golterman H L 1969 Methods for Chemical Analysis of Freshwater. IBP Handbook, Vol. 8 Blackwell. Sci. Publ., Oxford.
3 Nakanishi M 1976 Seasonal variations of chlorophyll 'a' amounts, photosynthesis and production rates of macro and microphytoplankton in Shiozu Bay, Lake Biwa. Physiol. Ecol. Japan 17, 535–549.
4 Margalef R 1974 Ecología. Ed. Omega S A, Barcelona. 950 p.
5 Margalef R 1976 Limnología de los embalses españoles. Dirección General de Obras Hidraúlicas, Min. de Obras Públicas. Madrid. 422 p.
6 Redfield A C, Ketchum B H and Richards F A 1963 The influence of organisms on the composition of sea water. *In* The Sea. Ed. M A Hill. Vol. 2, 27–79. Interscience, New York.

7   Sancha A M, Castillo G and Thiers R 1977 Estudio limnológico de la Laguna de Aculeo.
     Factibilidad de uso del agua en riego y recreación. Ingeniería Sanitaria, Universidad de Chile.
     Publicación 1–44.
8   Schindler D W 1966 A liquid scintillation method for measuring carbon-14 uptake in
     photosynthesis. Nature London 211, 844–845.
9   Schröder R 1969 Ein summierender Wasserschöpfer. Arch. Hydrobiol. 66, 241–243.
10   SCOR-UNESCO 1964 Report of SCOR-UNESCO Working Group 17 on 'Determination of
     photosynthetic pigments', Sydney. Mimeographed.
11   Vollenweider R 1974 Primary Production in Aquatic Environments. IBP Handbook vol. 12.
     Blackwell Sci. Publ., Oxford.
12   Wetzel R 1975 Limnology. W. B Saunders Company, Philadelphia. 743 p.

Plant and Soil 67, 389–394 (1982).   0032-079X/82/0673-0389$00.90.
© 1982 *Martinus Nijhoff/Dr W. Junk Publishers, The Hague*.

Report of work group shifting cultivation and traditional agriculture

S. R. GLIESSMAN (rapporteur), J. D. H. LAMBERT (chairman), J. T. ARNASON and P. A. SANCHEZ

Introduction

In many parts of the American tropics shifting cultivation has been the traditional form of agriculture for centuries. Such a system has provided the basis for survival for a majority of the population. Traditional agriculturalists have developed relatively stable, efficient cropping systems which reflect an empirical knowledge of ecological theory. Farmers have utilized a strategy of sustained yield with a minimum of imports from outside the system, unlike the modern technological approach of maximizing yields on the short term by maximizing outside imports. The intensive study of the structure and function of traditional agroecosystems, especially the aspects of N-cycling, can provide information of immediate application in the improvement of such systems.

Studies of N-cycling in natural ecosystems, based upon an understanding of the balance of inputs by outputs, as well as the complexity of mechanisms which function to keep N within the system once it is there, have been outlined in this workshop for several sites in the American tropics. Such information offers guidelines for which to study agricultural systems from the ecosystem point of view. Outputs, or production, can be understood and managed in light of the complexity of mechanisms which are tied to the internal stability of N in the system. The potential applications are great for knowledge concerning different means of increasing local N-inputs, as well as maintaining N within the system.

Present knowledge

During the course of the workshop N-cycling and distribution data in traditional agroecosystems were presented for 4 areas: Indian Church, Belize; Yurimaguas, Peru; San Carlos, Venezuela; and Tabasco, Mexico. Systems varied from shifting cultivation to permanent traditional agriculture, and covered a range of habitats within the tropics of Latin America (Table 1). Good descriptive information is available for the Belize and Mexico sites, and less so for the other two sites.

Information on N-cycling within these systems is most complete in terms of the distribution of N in the soil and plant biomass components of the agroecosystems (see presentations of Lambert, Arnason, and Gliessman). Data on the mechanisms of N-fluxes in these systems are extremely limited (Table 2). It can be seen that the most significant inputs of N come from the slash, weeds, and crop residues, although the importance of fixation with associated legume grain or cover crops are also noted. The large volatilization of N with burning is evident also, but all other mechanisms of N-flux are still in great need of study.

In Table 3, the available data are combined in a summary of N-budgets for the systems presented. It is important to note the net loss of N in shifting cultivation systems where fire is used as a management practice. This reaffirms the theory that a fallow period is necessary to allow nutrient replenishment. On the other hand, in the permanent-yield agroecosystems, N-balances are more favorable. In two cases, the overall budgets demonstrate a remarkable N-addition. A brief explanation for each system follows:

System 1: A shifting cultivation system employing fire, with N losses through volatilization and other processes about equalling N-outputs for the crop. Soil N is maintained.

System 2: A corn/bean intercrop system, traditional in Tabasco, demonstrating an overall positive N-balance, probably attributable to fixation by the bean component. Soil N is slightly increased.

System 3: A corn system without the use of fire, aided by considerable N-input from a legume cover-crop. Soil N is increased.

Table 1. General site characteristics. Nitrogen data for shifting agriculture and traditional agro-ecosystems

	Belize (Lambert and Arnason)	San Carlos Venezuela (Herrera and Jordan)	Tabasco Mexico (Gliessman)	Yurimaguas Peru (Sanchez)
Mean temp. °C	28°	26°	25°	26°
Annual rainfall	1700 mm	3600 mm	4000 mm	2100 mm
Months < 100 mm	4	0	2	2
Natural veg.	Semi Dec. seasonal forest	Evergreen rain forest	Evergreen rain forest	Evergreen rain forest
Soil type	Mollisol	Oxisol	Ultisol (Tropept?)	Ultisol
Veg. biomass 10^3 kg/ha	~150	500	600	500(?)
Biomass N kg/ha	550	1500	N.D.	1600
Soil total N (1 m deep)	N.D.	4000	N.D.	7800
Surface soil (0–15 cm)				
C%	7.0	2.8	2.9	1.5
N%	0.35	0.3	0.30	0.15
P (avail) ppm	15	2	5.0	1.0
K meq/100 g	1.9	0.003	1.2	0.1
pH	7.2	4.0	5.5	4.0
Land use: (%)				
Cultivated land	15	N.D.	30	15
Intensive grazing	0	N.D.	0	0
Extensive grazing	0	N.D.	50	15
Unused	85	N.D.	20	70
Natural vegetation complex				
Annual grasses	0	N.D.	0	0
Perennial grasses	25%	N.D.	20%	15
Herbaceous legumes	0	N.D.	10%	N.D.
Other herbs.	0	N.D.	30%	N.D.
Trees or shrubs	75%	N.D.	40%	85

System 4: A rice shifting cultivation system employing fire, which promoted the loss of considerable N. This system showed the lowest N-output with the crop. Soil N dropped slightly. Such a system would require a fallow period before replanting. Introducing Azolla with its N-fixing bacteria would help to maintain N levels.

System 5: A streamside system in permanently moist, periodically flooded soil. A remarkable increase in biomass N, as well as soil N is observable, due to a combination of sediment inputs from flooding, weed N capture, and possibly blue-green algae N-fixation. One of the most easily sustainable systems in terms of N.

Table 2. System N-fluxes

	Belize Corn	San Carlos Venezuela Yuca/etc.	Tobasco Mexico Various systems	Peru	
				Shifting	Permanent (3 crops a year) rice-corn soybeans
Inputs kg/ha/yr					
Fixation:					
Rhizobium	N.D.	N.D.	150 (with legumes)	N.D.	50
Bacteria	N.D.	N.D.	N.D.	N.D.	N.D.
Algae	N.D.	N.D.	N.D.	N.D.	N.D.
Rhizosphere	N.D.	N.D.	N.D.	N.D.	N.D.
Free living	N.D.	N.D.	N.D.	N.D.	N.D.
Atmospheric:					
Wet	N.D.	N.D.	N.D.	N.D.	N.D.
Dry	N.D.	N.D.	N.D.	N.D.	N.D.
Fertilizer:					
Chemical	0	0	0	0	180
Organic	268	2000	170	150	20
Total	268	2000+	320	150	250
Outputs kg/ha/yr					
Harvest	70	N.D.	45 (17–70)	40	160
NH_3 Volatization		N.D.	N.D.	N.A.	
Leaching		N.D.	N.D.	N.A.	~70
Denitrification	67	N.D.	N.D.	75	
Erosion		N.D.	N.D.	N.A.	0
Burning		N.D.	120 (rice)	N.A.	20
Grazing		N.D.	N.D.	N.A.	0
Total	137	N.D.	~100 (30–140)	115	250

N.D. – not determined.
N.A. – not available.

System 6: A yuca system where nitrogen balance is maintained by large crop biomass inputs and the use of a legume cover crop. Soil N increased.

System 7: A conuco shifting agriculture system in Venezuela, employing fire.

System 8: A shifting agriculture system in Peru, employing fire initially.

System 9: Same as above but continuous cropping (3 crops per year – rice, corn, soybeans) with fertilizer.

Table 3. Summary nitrogen budgets in several agro-ecosystem (dry weight)

System	Location	Initial biomass (N) A	Fert. (N) added A'	Final biomass (N) B	Harvest (N) removed C	Other (N) losses D	Balance B−A+ (C+D)	Initial soil N	Final soil N	Crop yield t/ha
1) Corn (burned)	Belize	268	0	131	70	67 (calc)	0	.3 (5.04)	.3 (5.46)	N.A.
2) Corn/Beans (not burned)	Tabasco	138.6	0	109.4	69.4	N.D.	+ 40.2	.24 (5.04)	.26 (6.51)	3.0 t/ha corn 0.5 beans
3) Corn/leg. cover (not burned)	Tabasco	260.2	0	192.2	49.2	N.D.	− 18.8	.24 (5.04)	.31 (4.62)	2.6 t/ha
4) Rice (burned)	Tabasco	192.8	0	35.2	16.8	120.5 (fire)	− 20.3	.24 (3.78)	.22 (5.04)	1.6 t/ha
5) Malanga (not burned)	Tabasco	35.0	0	71.8	31.4	N.D.	+ 68.2	.18 (4.41)	.24 (5.04)	8–10 t/ha
6) Yuca/leg. cover crop (not burned)	Tabasco	199.2	0	142.0	60.7	N.D.	+ 3.5	.21	.24	15 t/ha
7) Conuco (burned) (3 yr cycle)	Venezuela	1200.0	0	600	N.D.	N.D.	N.D.	N.D.	N.D.	8 t/ha
8) Yurimaguas 1 yr (burned)	Peru	150	0	40	40	75	− 5	0.15% N	0.12% N	20 t/ha
9) Yurimaguas cont. cropping	Peru	150	150	300	250	100	+ 350	0.15% N	0.12% N	N.A.

% soil N below, total weight N in the upper 15 cm in parentheses (t/ha)

Systems 1, 4, 7 and 8 are of the shifting cultivation type.
Systems 2, 3, 5 and 6 are more permanent sustained yield traditional agriculture type.
N.D. – not determined.
N.A. – not available.

General observations

In all systems, except the malanga, total biomass N in the system at the end of the study period was reduced. Soil N was probably increased in all cases by inputs from surface litter decomposition. The need to maintain and increase the biomass component of the systems, coupled with legume N-fixation inputs, is an integral part of the management strategies for sustained yields. However, it must be stressed that N is only one component of these systems, and must be studied as one of a great number of factors which together or separately may be limiting to production.

Areas lacking information

Information is lacking for the majority of traditional agroecosystems where sustainable yields and local N-inputs have been managed for long periods of time. We do know that there is a wealth of empirical knowledge that has not been exploited. Because of the rapid cultural changes currently taking place in tropical areas, it is feared that most of these traditional agroecosystems will be irreversibly altered or even destroyed before we will have the opportunity to study them. A combined ethnobotanical/agroecological study is proposed to determine how the agricultural and cultural components for their immediate study have evolved under the ecological and economic conditions which the small-scale farmer must function.

Priority areas for research

+ + +1) Immediate initiation of combined ethnobotanical/agroecological studies of the different traditional agroecosystems which demonstrate long-term yield stability and independence from outside inputs of N. Focus for research in these systems: Crop species and varieties; Crop calendar; Practices; N-fluxes and balances, especially the means of increasing local imports, increasing recycling efficiency, and reduction of N-losses. Information is required for all nutrients, not just nitrogen.

+ + +2) Studies of traditional intercropping and multiple cropping systems, especially those that include legumes, in order to quantify capacities for biomass accumulation, N-inputs, and yield potential on a sustained basis. By understanding the mechanisms by which N moves through such systems, design improvements can be tested and incorporated, especially for the small farmer.

+ +3) Identify specific crop associations with specific soil types to obtain reasonably high yields without rapidly depleting nutrient reserves or increasing the need to use fertilizers.

+ + +4) Breeding programs to increase yields in local varieties but still retain hardiness and pest resistance, both in the field and under storage.

+ +5) Study non-crop components of the traditional agroecosystems for the capacity to accumulate certain nutrients, especially N, microclimate modification and biological control.

+ +6) Compare the value of mulching non-harvestable biomass in the agroecosystem as compared to incorporation, removal or burning.

+ + +7) Extensive trials of local non-crop legumes, either associated with the crop or between crop periods, for N improvement, biomass input, weed control and alternative fallows for more rapid regeneration of soil productivity in shifting cultivation.

+ + +8) Develop a system for evaluating crop inputs in a manner that integrates the ecological, agronomic and economic aspects of agroecosystem management, the focus to be more on the mechanisms of production stability rather than the return for capital invested. This allows the best aspects of traditional agroecosystems to be maintained and improved. Where necessary and permissable, in both economic and ecological terms, new technologies which complement these systems can be incorporated. What might appear to be small, insignificant nutrient losses from the system today might eventually be limiting in the future.

Future areas of cooperation

1) There is a need to stimulate more communication between agronomists and ecologists. Workshops such as this are ideal situations for this interchange. Joint research should be an obvious followup.

2) Development of educational programs in (agricultural) schools which focus on the local, traditional systems of agricultural production, stressing the positive, relatively low-cost management benefits. At the same time this would help create a better attitude towards traditional agriculture, and by studying it, it can be improved. In the long term, better attitudes towards traditional agriculture affect the small farmers also, aiding in reducing the flow of people out of rural areas into the cities.

3) Stimulate and convince major research organizations to increase research and field trials which employ traditional varieties and practices, ideally on the farms under the conditions available for the small farmer.

Plant and Soil 67, 395–397 (1982).   0032-079X/82/0673-0395$00.45.   SU-37
© 1982 *Martinus Nijhoff/Dr W. Junk Publishers, The Hague*.

Report of the work group on sugarcane

A. P. RUSCHEL (chairman), F. R. TAMAS (rapporteur), N. AHMAD, H. H. MENDOZA-PALACIOS and S. VALDIVIA-VEGA

Nitrogen balance in sugarcane

We have considered the major aspects of N-cycling in sugarcane crops of Latin America. A general nitrogen balance for this crop is difficult to construct because of widely differing agronomic practices and growing conditions and also a lack of knowledge concerning certain processes. Agronomic complications include a) a crop-growth period that varies from 10 to 22 months, b) yields that range from 50 to 120×10^3 kg ha^{-1} crop^{-1}, c) burning at harvest that may disturb soil organic-N accumulation by destroying potential litter but that is not always practiced, and d) variable recycling practices for post-harvest waste products (such as filter cake mud or cachaza, bagasse, vinhaza, or begacillo). Important parts of the N-cycle in sugarcane for which our understanding is limited, mainly for lack of accurate methods, include a) the origin of the nitrogen in the plant, *i.e.* whether the nitrogen is derived from soil, fertilizer, or the roots of the preceding crop, b) nitrogen in throughfall, which has been observed but not yet adequately measured, and c) the rate at which nitrogen is immobilized after the first harvest.

Nevertheless, sugarcane is a crop which has been widely studied throughout the area and therefore we can make some justified generalizations about its N-balance. For example, in general about 30 percent of the fertilizer-N applied to sugarcane is recovered by the crop. Additionally, the average N-contents for the various commercially-grown sugarcane varieties are known and therefore it is possible to calculate how much nitrogen is incorporated in the various plant parts.

Table 1. Selected characteristics of sugarcane agroecosystems in Latin America and the Caribbean

Feature	General	Peru
Climate		
Mean annual temperature (°C)	15–30	**
Precipitation (mm yr^{-1})	1000–2500	
Growth period (months crop^{-1})	12–22*	
Rainy season (months yr^{-1})	4–6	12**
Soil		
C (%)	<1.0	1.2
N	<0.07	<0.09
P (%)	<0.002	0.0003–0.0012
K (%)	<0.01	0.06–0.15
pH	4–7	7.4–8.0
Biomass N-distribution (kg N ha^{-1})		
Stalks	70.	
Leaves	30.	
Roots	20.	

* Lower value for ratoon crops.
** Peru crops grown on irrigated desert soils.

Table 2. The N-balance in sugarcane cropping systems of Latin America and the Caribbean. All values kg ha^{-1} yr^{-1}, unless noted

Component	Brasil	Ecuador	Dominican Republic	Peru	Trinidad	Range
Inputs						
N-fixation						
Rhizosphere	15–25	na	na	na	na	na
N-fertilizer						
Commercial	60–100*	150	200	200	80	60–200
Manure**	5**	5–10	5–10	5–10	5–10	5–10
Bulk deposition	*ca.* 5	na	5–10	na	na	5–10
Total inputs	100	na	na	282	na	na
Outputs						
Harvest	50–60	50–60	50–60	150	50–60	50–150
NH$_3$-volatilization	na	na	na	na	na	na
Leaching	na	na	na	20	na	na
Burning	na	na	na	45	na	*ca.* 30
Forage	na	na	na	20	na	na
Total outputs	100	100	100	230	100	100–230
Within system						
Fertilizer recovery (%)	50	50	50	70	50	50–70

* Higher value is for ratoon crops.
** Filter-cake mud except Brazil (vinhaça).
na, data not available.

Table 3. Priorities for N-cycle research in Latin American and Caribbean sugarcane cropping systems

Process	Priority
Inputs	
N$_2$-fixation:	
by Rhizobia (intercropping systems)	+
by bacteria	
in rhizosphere	+ + +
free-living	+ +
by algae	O
Atmospheric deposition	
wet	+
dry	O
Fertilizer	
commercial	+ +
manure and organic matter	+ + +

Outputs

Harvest	+ +
NH_3-volatilization	+
Leaching	+ +
Denitrification	+ + +
Erosion	+
Burning	+ + +
Forage	+

Within system

Source of plant-N	+ + +

We utilized these data, and data for other parts of the cycle that seem well understood, to construct generalized budgets for this cropping system in several Latin American regions. A description of these regions is presented in Table 1, and Table 2 presents the nitrogen budgets.

Research priorities

We have also drawn up a list of nitrogen-cycle research priorities for sugarcane cropping systems (Table 3). Highest priority is given to those aspects of the cycle that could lead to an increased N-supply with a minimal input of energy. These include 1) recycling nitrogen in crop residue and factory waste that would otherwise be lost by burning, 2) identifying the sources of plant nitrogen to allow more efficient utilization of fertilizer and soil nitrogen, 3) biological N_2-fixation, particularly N_2-fixation in the rhizosphere and the role played by mycorrhiza, and 4) denitrification of nitrogen that might otherwise be available for plant uptake.

Plant and Soil 67, 399–402 (1982).   0032-079X/82/0673-0399$00.60.   SU-38
© 1982 *Martinus Nijhoff/Dr W. Junk Publishers, The Hague*.

Report of the work group on cereal and grain crops

A. A. FRANCO (rapporteur), J. R. JARDIM FREIRE (chairman), J. ARRIVETS, J. GÓMEZ CARRIÓN, L. FRIONI, D. J. GREENWOOD, Y. Z. ISHAC, M. A. LÁZZARI, P. L. LIBARDI, L. LONGERI, M. P. SALEMA, M. N. VERSTEEG and R. L. VICTORIA

Introduction

We have attempted to construct a general nitrogen budget for arable grain crops. There are serious gaps in our knowledge of the nitrogen cycle of grain cropping systems, and for many values we have had to rely upon rough estimates made from the best available information. In particular, we have seldom been able to separate losses of nitrogen via denitrification from those via immobilization and ammonia volatilization. Even so, we have prepared tables of nitrogen inputs and outputs for wheat, corn, beans, peas and soybeans grown in climates ranging from temperate to tropical.

Nitrogen balance for grain crops

Harvest yields depend on growing conditions and vary enormously; consequently we have included a description of the conditions under which each set of data were obtained (Table 1). The individual yields for each crop in Table 1 assume that the effects of other nutrient deficiencies are minimal. We have also included rooting depth in the table as this can have a decisive influence on the amount of nitrogen that can be extracted from the soil. In addition, estimates of the total amounts of organic-N within the rooting depth are also included because these amounts are always very large compared with the nitrogen removed by the crop, and a small change in organic-N levels can have a large influence on the ability of the soil to meet crop requirements.

Nitrogen fluxes in these systems are presented in Table 2. These figures are for a single monoculture harvest of each crop, although we recognize that some crops may rely on nitrogen fixed by a previous crop and that polyculture cropping systems predominate in some areas. We have not considered cropping simultaneously with two crops because too little data is available for these types of systems.

Grain yields in these systems vary from 1000 to 10,000 kg ha^{-1}, annual rainfall varies from 3 to 1200 mm, and annual mean temperature varies from 10 to 22°C. Soil pH varies from 5.5 to 7.5, and some of the soils are rich in available P and K while others are very deficient (Table 1).

Despite this variation, however, several general patterns emerge. First, almost every crop has roots that penetrate far below the depth of cultivation. It therefore appears that most of these crops can absorb nitrogen at depths of 60 to 100 cm. The amount of organic-N at this depth varies from 2000 to 12,000 kg N ha^{-1}, enormous when compared to any input or output.

Although it appears that the information is lacking regarding portions of the input-output budget (Table 2), nitrogen fixed by bluegreen algae and by rhizosphere and free-living organisms seldom exceeds 20 kg N ha^{-1} yr^{-1}. Little inorganic-N is generally added by rainfall. Most of the nitrogen in legumes is supplied by rhizobia, and most of the nitrogen in cereal grain, by fertilizer.

The main output of nitrogen is that removed by harvest of the grain, and this removal amounts to 45 and 165 kg N ha^{-1} crop^{-1}. Leaching losses are on occasion considerable. In some situations losses of nitrogen from erosion and from denitrification can also be very large.

The difference between total nitrogen inputs and outputs is the amount of nitrogen gained by the soil. For legumes in these systems, inputs rarely exceed outputs by more than 50 kg ha^{-1} yr^{-1}. This implies that legume grains may generally leave only small amounts of available-N for subsequent crops. The balances for non-legumes, on the other hand, often show that N-removal is greater than total inputs, so that crops may essentially be 'mining' the soil for nitrogen. In other words, nitrogen is obtained at the expense of soil organic matter.

Table 1. Selected characteristics of cereal and grain cropping systems described further in Table 2

Feature	Temperate				Sub-tropical	Tropical			Semi-arid/temperate		Arid/sub-tropical		
	Wheat (England)	Pea (England)	Soy-bean (Chile)	Wheat (Chile)	Corn (Madagascar)	Bean (Brazil)	Soy-bean (Brazil)	Corn (Brazil)	Wheat (Argentina)	Corn (Argentina)	Bean (Peru)	Soy-bean (Peru)	Corn (Peru)
Climate													
$\bar{x}$ annual temperature (°C)	10.	12.	10.	10.	17.5	22.	22.	22.	14.7	16.5	18.	18.	18.
Precipitation (mm yr^{-1})	700.	700.	1200.	1200.	1200.	1200.	1200.	1200.	600.	780.	3.	3.	3.
Rainy season (days yr^{-1})	365.	365.	180.	180.	180.	160.	160.	160.	20.	30.	0.	0.	0.
Growing season (days yr^{-1})	200.	90.	120.	200.	150.	100.	120.	140.	160.	160.	110.	130.	160.
Soil (to 20 cm)													
C (%)	2.0	2.0	2.5	2.5	2.0	2.0	2.0	2.0	1.2	1.6	0.25	0.25	0.25
N$_{total}$ (%)	0.2	0.2	0.2	0.2	0.1	0.1	0.1	0.1	0.1	0.1	0.02	0.02	0.02
(kg N ha^{-1})	12000.	9000.	6000.	6000.	7000.	4000.	4500.	4500.	4000.	6000.	2000.	2000.	2.000
N$_{inorganic}$ (kg N ha^{-1})	40.	30.	na	na	na	na	na	na	36.	40.	na	na	na
P (available)* (mg P kg^{-1})	30.	30.	5.	5.	5.	8.	8.	8.	12.	10.	2.	2.	2.
K (mg K kg^{-1})	200.	200.	100+	100+	40.	80.	80.	80.	100+	100+	100.+	100.+	100.+
pH (in H$_2$O)	6.5	6.5	6.0	6.0	5.5	5.5	5.5	5.5	7.0	6.5	7.5	7.5	7.5
Yield (kg ha^{-1} crop^{-1})	10000.	3000.	3000.	4000.	5000.	1000.	2000.	4000.	4000.	4500.	2000.	3000.	6000.
Rooting depth (cm)	100.	60.	na	na	60.	40.	50.	50.	60.	80.	60.	60.	60.

* By Olsen's method.
na, not available.

Table 2. N-balance for grain and cereal cropping systems described in Table 1. Units are in kg N ha^{-1} yr^{-1}

Process	Temperate				Sub-tropical	Tropical			Semi-arid/temperate		Arid/sub-tropical		
	Wheat (England)	Pea (England)	Soy-bean (Chile)	Wheat (Chile)	Corn (Madagascar)	Bean (Brazil)	Soy-bean (Brazil)	Corn (Brazil)	Wheat (Argentina)	Corn (Argentina)	Bean (Peru)	Soy-bean (Peru)	Corn (Peru)
Inputs													
N-fixation													
by rhizobia	0.	130.	200.	0.	0.	35.	130.	0.	0.	0.	75.	140.	0.
by bluegreen algae	na	na	0.	0.	na	0.	0.	0.	na	na	na	na	na
in rhizosphere	15.	15.	0.	na	na	0.	0.	10.	na	20.	na	na	na
by free-living bacteria	na	na	0.	na	na	0.	0.	na	5.	na	na	na	na
Wet Deposition	30.	30.	0.	na	na	5.	5.	5.	na	0.	25.	25.	40.
Irrigation	na	na	0.	na	na	na	0.	0.	100.	0.	na	na	na
Fertilizer-N	220.	20.	0.	100.	135.	40.	0.	100.	na	0.	na	na	160
Total In	265.	195.	200.	100.+	135.+	80.	135.	115.+	105.+	20.+	100.+	165.+	200.+
Outputs													
Harvest	150.	150.	150.	80.	80.	45.	120.	60.	100.	90.	80.	165.	120.
Leaching	40.	na	na	low	50.	6.	5.	6.	na	na	10.	35.	80.
NH$_3$-volatilization	na	na	na	na	na	na	na	na	na	na	0.	0.	0.
Denitrification	na	na	na	high	na	na	na	na	high	high	0.	0.	0.
Erosion	na	0.	low	high	low	0.	0.	0.	0.	0.	0.	0.	0.
Burning	50.	0.	0.	25.	0.	0.	0.	0.	0.	0.	0.	0.	0.
Grazing	0.	0.	0.	0.	0.	0.	0.	0.	0.	15.	0.	0.	0.
Total Out	240.+	150.+	150.+	105.+	130.+	51.+	125.+	66.+	100.+	105.+	90.	200.	200.

na, not available.

Research priorities

We have ranked in Table 3 the N-cycle processes most in need of attention in these cropping systems. Of highest priority is research into mineralization-immobilization processes, but emphasis should also be placed on research into the use of fertilizer-N by these crops and losses via leaching and denitrification. Nitrogen-fixation by Rhizobia and by rhizosphere bacteria are additional areas that deserve priority.

Conclusions

1. Nitrogen balance studies are of considerable importance for predicting and understanding changes in the crop nitrogen cycle that take place as a result of different agronomic practices.

2. Too few detailed N-balance studies have been made of arable crops in Latin America.

3. Much useful information can be obtained by carrying out studies at different degrees of detail. Ideally, ^{15}N should be used for these studies, but we recognize that such experiments can be carried out at only a few sites because of economic limitations. We believe that considerable progress can nevertheless be made by using conventional methods of analysis to measure simultaneously at various sites major inputs and outputs of nitrogen and changes in soil organic-N (to the depth of rooting) over a number of years.

Table 3. Priorities for nitrogen-cycle research in cereal and grain cropping systems. $\bigcirc$ = lowest and $+ + + +$ = highest priorities

Process	Priorities
Inputs	
N$_2$-fixation	
by Rhizobium	$+ +$
by algae	$\bigcirc$
by bacteria	
in rhizosphere	$+ +$
free-living	$\bigcirc$
Atmospheric deposition	
wet	$+$
dry	$\bigcirc$
Fertilizer	
commercial	$+ + +$
manure	$+$
Outputs	
Harvest	$\bigcirc$
NH$_3$-volatilization	$+$
Leaching	$+ + +$
Denitrification	$+ + +$
Erosion	$+$
Burning	$\bigcirc$
Grazing	$\bigcirc$
Within-system	
Mineralization-immobilization	$+ + + +$

Plant and Soil 67, 403–407 (1982). 0032-079X/82/0673-0403$00.70.
© 1982 *Martinus Nijhoff/Dr W. Junk Publishers, The Hague.*

Report of the work group on coffee and cacao plantations

J. P. ROSKOSKI (rapporteur), E. BORNEMISZA (chairman), J. ARANGUREN, G. ESCALANTE
and M. B. M. SANTANA

Introduction

Coffee and cacao are two of the most important perennial agricultural crops in Latin America. Plantations of these crops occur in all Latin American countries except for those in the southern cone of South America. In addition, Brazil is the largest coffee exporter in the world.

Available data indicate that both crops are cultivated at two technological levels. One is the traditional cultivation regime, characterized by minimal agro-chemical inputs. The second employs modern plantation techniques. Both crops are grown under a variety of environmental conditions with marked differences in soil, climate, and topography. These environmental differences lead to variations in plantation structure, productivity, and consequently management practices. However, nitrogen fertilizers are used in most coffee plantations since agronomic data indicate that added-nitrogen is essential for adequate production, and their use in cacao plantations is increasing as the introduction of hybrids intensifies. At the same time, the rising cost of nitrogen fertilizer, which today accounts for approximately 10% of the production costs for coffee, coupled with the increased use of fertilizer as modern agricultural technology expands, makes an understanding of the nitrogen dynamics of these two systems essential for efficient management.

Despite the importance of nitrogen to these two systems, data available for nitrogen cycling in coffee and cacao plantations are scarce. In this report we have attempted to construct nitrogen budgets for these two systems using the existing data base.

Nitrogen balances in coffee and cacao plantations

We have attempted to set ranges for different parameters of coffee and cacao nitrogen cycles for both the traditional and the modern regimes. Table 1 presents general climatic, vegetation cover, and soil characteristics for these systems.

Soil values were reported to a depth of 30 cm because coffee and cacao absorb the majority of their nutrients from this portion of the soil profile. The values for the two species represent the range of values from Mexican, Venezuelan, Costa Rican, and other sites.

Two soils were generally considered as representative of the types that are used for cultivation of these two crops; an andosol with a specific gravity of $0.8 \, \text{g cm}^{-3}$ and an ultisol with a specific gravity of $1.2 \, \text{g cm}^{-3}$. These two soil types were used to calculate per ha amounts of total nitrogen for the two types of plantations using the percent nitrogen data indicated. We estimate that the amount of inorganic nitrogen in these two types of soils is low, in the order of 5–$100 \, \text{kg N ha}^{-1}$. The only exception that might be important is the case of a soil which fixes ammonium.

The distributions of biomass nitrogen within each system are presented in Table 2. These values are mainly based on data from study sites in Venezuela, Mexico, and Brazil. The weed biomass component is low in well-managed coffee plantations. From the Venezuelan coffee project we have taken values of 10–$15 \, \text{kg}$ dry matter ha^{-1} as an average value. This was multiplied by a nitrogen content of 2.5% to arrive at the nitrogen in this compartment. In cacao plantations the intense shade normally leads to very little, if any, herbaceous cover. Consequently, the value reported for this compartment is 0.

Under traditional cultivation systems the coffee plant density is about 1200 trees ha^{-1}, while intensive cultivation leads to densities of about 6000 trees ha^{-1} (*e.g.* the Venezuelan sites). The average mass of a coffee tree was calculated using data from the Venezuelan study. The values for the various components were used to calculate coffee biomass for both the low and high density plantations.

Table 1. Selected characteristics of coffee and cacao plantations in Latin America

Aspect	Coffee	Cacao
Climate		
Average annual temperature (°C)	18–25	22–26
Annual precipitation (mm)	1000–3000	600–3000
Rainy season (days yr^{-1})	180–270	200–365
Growing season (days yr^{-1})	365	365
Vegetation (% cover)		
Weeds	0.004–0.04	0.0
Crop	10–30	20–60
Shade trees	70–90	40–80
Soil (to 30 cm)		
C (%)	0.8–4.5	0.8–3.0
N_{total} (%)	0.02–1.00	0.05–0.5
(kg ha^{-1})	7200–24 000	1200–12 000
$N_{inorganic}$ (kg ha^{-1})	5–100	5–100
P (available) (%)	0.0005–0.01	0.0001–0.003
K (available) (%)	0.008–0.060	0.002–0.02
pH (in H_2O)	4.5–7.0	4.2–6.8

Tree density in cacao plantations in both the Venezuelan and Brazilian sites is approximately 1000 trees ha^{-1}. Venezuelan biomass data for an average cacao tree was multiplied by 1000 to calculate total biomass per ha.

In both coffee studies (Mexico and Venezuela), shade tree density was about 200 trees ha^{-1}. However, all shade trees in the Mexican study were 15 m high, while half the trees in the Venezuelan study were this height and the other half 30 m. A 15 m shade tree has a mass of about 500 kg. Of this amount, *ca.* 75% is wood, 20% roots, and 5% leaves. We assumed that a 30 m tree has the same leaf biomass, the same percentage of roots, and twice the wood biomass of a 15 m tree. These assumptions were used to calculate the biomass for the shade tree compartment of the two sites. A value of half of that for the Mexican site was used for the traditional system estimate.

The Venezuelan cacao site had 325 shade trees ha^{-1} and the Brazilian site, 25. In the Venezuelan site half the trees were 15 m high and the other half 30 m; all shade trees in the Brazilian site were 30 m high. The same value for a 15 and 30 m tree used in the coffee section was used to calculate tree biomass for the cacao.

Fluxes of nitrogen into and out of these systems are presented in Table 3. Nitrogen fixed by shade trees was estimated for coffee plantations using data from Mexican sites. For cacao plantations, Brazilian data indicated an Erythrina nodule density of *ca.* 2.0 g dry nodules m^{-2}. We used this and an average nodule fixation rate from the Mexican sites of 4.5×10^{-6} moles N-fixed $g^{-1} h^{-1}$ for another species of Erythrina, together with 5 and 8 month activity periods for the nodules, to calculate fixation by this legume shade tree for the cacao plantation. For N-fixation by free-living bacteria, we used data from the Mexican sites for coffee; no data is available for cacao plantations.

We estimate that between 5 and 15 kg N $ha^{-1} yr^{-1}$ are added to both systems *via* wet precipitation. We consider dry fallout inputs to be insignificant since humidity in both systems is quite high and precipitation frequent.

Fertilizer inputs were estimated from knowledge of current management practices. Data for this parameter together with harvest data is probably the most reliable and precise.

Table 2. Nitrogen distrubution in the biomass of shaded coffee and cacao plantations in Latin America

System and component	Leaves			Wood			Roots			Total			
	Biomass (kg tree⁻¹)	%N	N (kg N tree⁻¹)	Biomass (kg tree⁻¹)	%N	N (kg N tree⁻¹)	Biomass (kg tree⁻¹)	%N	N (kg N tree⁻¹)	N (kg N tree⁻¹)	Trees (trees ha⁻¹)	N (kg N ha⁻¹)	
Coffee system													
Coffee trees	0.20	2.0	0.004	0.30	0.5	0.002	0.40	1.8	0.007	0.013	1200–6000[7]	16.–78.	
Shade trees (500 kg)	20.0	3.2[1]	0.64	380.	0.5[3]	1.90	100.	1.8[2]	1.80	4.34	100–200[4]	434.–868.	
Shade trees (1000 kg)	20.0	3.2[1]	0.64	780.	0.5[3]	3.90	200.	1.8[2]	3.60	8.14	100–0[4]	814.–0.	
Weeds	na	na	na	na	na	na	na	na	na	na	na	0.25–0.38[5]	
Total												1260.–946.	
Cacao system													
Cacao trees	0.42	1.8	0.008	4.5	1.3	0.58	2.00	1.2	0.24	0.090	1000	90.	
Shade trees (500 kg)[6]										4.34	0–160	0.–694.	
Shade trees (1000 kg)[6]										8.14	25–160	204.–1300.	
Total												294.–2080.	

[1] *Inga* spp.
[2] Based on coffee roots.
[3] Estimated.
[4] Mexican plantations = 200 trees of 500 kg, Venezuelan = 100 trees of 500 kg + 100 trees of 1000 kg.
[5] 10–15 kg weeds ha⁻¹ at 2.5% N.
[6] N-pool per tree same as for coffee system; Brazilian plantations = 25 trees of 1000 kg, Venezuelan = 160 trees of 500 kg + 160 trees of 1000 kg. Neither site has significant weed biomass.
[7] Lower value represents traditional cultivation regime.

Table 3. Nitrogen fluxes in coffee and cacao systems of Latin America (kg N ha^{-1} yr^{-1})

Process	Coffee	Cacao
Inputs		
N-fixation		
by Rhizobia	1.–50.	9.–15.
by free-living bacteria	0.5+	na
Atmospheric deposition		
wet	5.–15.	5.–15.
dry	0.0	0.0
Fertilizer		
commercial	50.–300.	30.–60.
manure	0.0	0.0
Total in	55.–400.	44.–90.
Outputs		
Harvest	30.–60.	20.–40.
NH$_3$-volatilization	na	na
Leaching	na	na
Denitrification	na	na
Erosion	na	na
Burning	na	na
Grazing	na	na

na, data not available.

The only nitrogen output known well is that lost in the harvest. All other outputs are essentially unknown. However, given the well-drained nature of the soils on which coffee and cacao are normally grown, losses *via* denitrification are probably small. Leaching losses, for the same reason, may be a major output. Losses through volatilization could also be significant given the amount of fertilizer applied in some plantations. Erosion losses are not generally thought to be high, except in the case of coffee grown on steep slopes without shade or in the early stages of plantation establishment when initial shade cover may be small. However, well-managed plantations, particularly those with shade, probably suffer little nitrogen loss through erosion.

Research priorities

Table 4 presents our view of the relative importance of studies concerning the various N-budget parameters in coffee and cacao cropping systems. In general, losses need still to be adequately quantified. This data is essential so that management techniques can be developed to reduce nitrogen losses from the systems. The reduction of such losses are highly desirable from both an agronomic and economic point of view, as well as from an environmental health standpoint. Given the large amounts of fertilizer that are applied to some coffee plantations in particular, sizeable leaching losses, primarily in the form of nitrate, could lead to contamination of groundwater and streams and rivers near the plantations. The chance of such contamination is increased given the porous nature of most of the soils on which these two crops are planted.

Similarly, coffee pulp that is sometimes burned rather than returned to the plantation represents a further loss of nutrients to the system and an additional source of contamination. Coffee processing plants commonly dump coffee wash-water that contains high concentrations of carbohydrates into adjacent streams, thereby increasing the biological oxygen demand of these aquatic systems.

Table 4. Priorities for N-cycle research in coffee and cacao systems of Latin America

Process	Priority	
	Coffee	Cacao
Inputs		
N-fixation		
by Rhizobia	+ + +	+ + +
by algae	+	+
in rhizosphere	O	O
by free-living bacteria	+	+
Atmospheric deposition		
wet	+ +	+ +
dry	O	O
Fertilizer		
commercial	O	O
manure	O	O
Outputs		
Harvest	+ +	+ +
Pruning export (firewood)	+	O
NH_3-volatilization	+	+
Leaching	+ + +	+ + +
Denitrification	+ +	+ +
Erosion	+ +	+ +
Burning	O	O
Grazing	O	O
Within system		
Humification*	+ +	+ +
Mineralization*	+ +	+ +
Mycorrhizal associations (shade trees)	+ +	+
Genetic selection (shade trees)	+	+
Animal import/export of N	O	+
N-dynamics during plantation establishment	+ +	+ +
Weeds N-cycling	+	O

* Including studies covering the soil fauna and micro-organisms responsible for the processes.
+ + + = Highest priority.
O = not applicable.

Nitrogen fixation by leguminous shade trees in both coffee and cacao systems is another area that deserves high priority.

These research priorities (Table 3) represent the basic internal processes and inputs and outputs about which little is known at present. Managed crops, such as coffee and cacao, present an ideal opportunity for the efficient use of nitrogen resources. However, until more information is available, particularly for the important nitrogen outputs, efficient nitrogen management will be unlikely.

Plant and Soil 67, 409–413 (1982).   0032-079X/82/0673-0409$00.75.   SU-40
© 1982 *Martinus Nijhoff/Dr W. Junk Publishers, The Hague.*

Report of the work group on savannas and shrublands

L. R. YATES (chairman), P. W. RUNDEL (rapporteur), R. SYLVESTER-BRADLEY and R. E. CISTERNAS

Introduction

Savannas and shrublands of Latin America and the Caribbean include a great diversity of ecosystems. For the purpose of this review, we have artificially divided these communities into three major divisions. The first are the true tropical savannas, which generally have precipitation in excess of 900 mm yr^{-1}. These include the llanos of Venezuela and the cerrado of Brazil. The other two groups can be classified as semi-arid or arid ecosystems – dry savannas and shrublands. The former are primarily legume-dominated savannas occurring in regions with less than 500 mm yr^{-1} precipitation. The shrublands are a heterogeneous group including matorral, caatinga, warm desert scrub and austral-patagonian scrub.

Tropical savannas

The tropical savannas of Latin America cover a total of 300 million ha. These savannas include the cerrado of Brazil (183 million ha.), the llanos of Colombia, Venezuela, Guyana and Suriname, and the savannas of Roraima and Macapé in Brazil. These vegetation types have been defined by CIAT using a total wet season potential evapotranspiration (TWPE), a factor considered limiting for cattle production. The TWPE range for tropical savannas is 910–1060 mm. Subdivisions within this classification separate the Brazilian cerrado from the other tropical savannas on the basis of its cooler wet season. Thus the Brazilian cerrado is termed isohyperthermic ($< 23.5°C$) and the others are hyperthermic ($> 23.5°C$).

Soil fertility and soil drainage vary greatly in tropical savanna regions. Four subdivisions of the Brazilian cerrado can be separated on a gradient of increasing soil fertility – campo limpio, campo cerrado, cerrado, and cerradão. Other areas of the cerrado, such as Partoral in Mato Grosso, are characterized by poorly-drained soils. Land use patterns of these poorly-drained soils are very different from those of well-drained regions. The former are very important for grazing during the dry season.

There have been extensive studies of the ecological relationships of the Brazilian cerrado, but intensive ecosystem studies are lacking. With careful literature searching it would be possible to characterize the nutrient pool sizes for these ecosystems. Data on nutrient fluxes are generally lacking, however. Agricultural studies suggest that phosphorus is generally the primary limiting nutrient for plant productivity.

Legumes are very important elements of the vegetation of the Brazilian cerrado, suggesting the symbiotic fixation may be important. The Triangulo mineiro, for example, contains 44 genera and 107 species of legumes[1]. Estimates of total soil nitrogen levels range from about 1200–3000 kg N ha^{-1}, the great majority in organic forms. The above-ground biomass of cerrado ecosystems is quite variable, with increasing amounts apparently strongly correlated to soil fertility. Goodland and Ferri[1] list the following comparative basal areas of woody vegetation for cerrado communities:

campo sucio	25,000 cm^2 ha^{-1}
campo cerrado	75,000
cerrado	175,000
cerradão	330,000

These values should be strongly correlated with biomass. Total levels of nitrogen in above-ground vegetation should also be related.

Inputs of nitrogen to tropical savannas are poorly known. A large number of the native legumes have been shown to be nodulated in the field. Herbaceous species are more heavily nodulated in sandy soils than in clay soils. This information suggests that symbiotic nitrogen fixation may be a significant input. Pereira (unpublished data) has estimated a value of 20 kg ha^{-1} yr^{-1} for a cerrado site. Other forms of nitrogen fixation may also be important. Fixation associated with grass rhizosphere microorganisms undoubtedly occurs. Blue-green algal mats are widely present in the cerrado limpio, while epiphytic N-fixers are present in the cerradão. Termites are very abundant and may also contribute to nitrogen fixation. All factors considered then, N-fixation is clearly much more important in tropical savannas than in tropical forests. Given the dominance of woody legumes in most tropical savannas, it would be interesting to consider the comparative nitrogen balance of some notable savanna areas without legumes.

Precipitation inputs of nitrogen in tropical savannas are very poorly studied. These should be significant, however, because of the high precipitation. Dry atmospheric inputs would be expected to be very low.

Nitrogen outputs are also poorly studied. Almost no data are available on denitrification and ammonia volatilization. Leaching losses would be expected to be relatively high, while erosion should be minimal. Seasonal fluxes of mineral nitrogen are, likewise, a subject with little information. Generally, grassland fluxes are low, with variable seasonal peaks of mineralization related to soil wetting and drying and the subsequent effects on plant growth, microbial activity, nitrification rates, leaching and upward capillary movement of soil water. Experiments on such fluxes at the end of the wet season have shown more NH_4^+ release than NO_3^-. Since soils are frequently highly acid, nitrification may be limited.

Allelopathic factors in soils may affect many aspects of microbial activity as well.

Internal shunts in the cycles of nitrogen are important areas for new research. Very little is known, for example, about the long-term impacts of the frequency and intensity of savanna fires. These must certainly have a very strong impact on nitrogen flux rates. Animal impacts also need investigation. Armadillo activity, termite mounds and ant nests are nearly ubiquitous in many tropical savannas. The current carrying capacity for cattle in unmanaged savanna ecosystems is about 1 animal per 5 ha. This level of grazing would mean a grazing output from the system of about 1 kg N ha^{-1}. Research on managed pastures has suggested that carrying capacities ten times this level may be possible.

Arid and semi-arid savannas

Open savannas dominated by woody legumes cover several hundred million hectares of arid and semi-arid land in Latin America. One genus, Prosopis (mesquite or algarrobal) provides the major part of this coverage. It is particularly important in the Sonoran and Chihuahuan Deserts and adjacent areas in North America and the monte region of West Central Argentina. Prosopis is also important, however, in arid coastal savannas of northern Colombia and Venezuela and in arid regions of Peru, Equador and Chile. Other arid zone legumes such as Acacia and Cercidium may also dominate large savanna areas.

Typical levels of precipitation in Prosopis savannas are from 200–350 mm yr^{-1}, but the range overall is from 0 to 500 mm yr^{-1} or more. The lowest precipitation regimes typically occur in winter rainfall regimes in the western Sonoran Desert and in Chile. In these areas Prosopis is commonly restricted to habitats where stable groundwater is available. Outside of these winter rainfall areas, Prosopis is not restricted to ground water supplies. While Acacia savannas are most characteristic of arid regions of Africa and Australia, the 'espinal' of central Chile dominated by *A. caven* covers a significant area with main annual precipitation of about 400 mm.

The above-ground biomass of arid zone savannas ranges from 5,000–15,000 kg ha^{-1}, or even more in managed stands. Almost all of the biomass is comprised of woody legumes. The root-shoot ratios of savanna legumes have not been measured, but they are probably in the range of 0.5–1.0. Productivity rates of above-ground tissues are generally very high, a condition almost certainly related to a freedom from nitrogen-limitation in nodulated species. A stand of *Prosopis glandulosa* in the Sonoran

Desert, described as a case study below in this report, had $3700 \, kg \, ha^{-1} \, yr^{-1}$ above-ground production from $13,000 \, kg \, ha^{-1}$ of biomass in an area with only $70 \, mm \, yr^{-1}$ precipitation.

Soil nitrogen pools in Prosopis woodlands vary from $600-4000 \, kg \, ha^{-1}$. Nitrogen levels under the canopy of Prosopis, however, reach 2–3 times this upper limit. Large parts of this total nitrogen (up to 25% or more) may occur as inorganic nitrates. Ammonia levels are low.

The above-ground pool of nitrogen in arid savannas is directly related to biomass. Leaf tissues are extremely high in nitrogen and provide an important potential fodder for grazing animals. Prosopis savannas in the monte and Sonoran Deserts have a range of $120-200 \, kg \, N \, ha^{-1}$. Since production is very high, there is a relatively low biomass accumulation ratio (biomass/net production) for nitrogen in these stands.

Symbiotic fixation of nitrogen appears to provide the major input of nitrogen to arid savannas. Suggested levels of fixation are in the range of $20-30 \, kg \, ha^{-1} \, yr^{-1}$. Other forms of fixation have not been studied, but these appear to be relatively unimportant. Since precipitation is low, nitrogen inputs through this route are generally very small – certainly less than $1-2 \, kg \, ha^{-1} \, yr^{-1}$. Dry inputs of nitrogen through atmospheric dust may be more significant, but no data are available. The potential importance of nitrogen inputs through deposition of fecal matter and urea or uric acid should also be carefully investigated. The shade of savanna tree canopies provide roots for insectivorous birds and resting sites for grazing animals.

Outputs of nitrogen from arid savannas are not well studied. While leaching and denitrification appear to be relatively low, particularly in regions with very low precipitation, erosional losses at irregular intervals may be considerable. Ammonia volatilization during litter decomposition could be considerable.

Desert shrublands typically have an above-ground biomass which is very small in comparison with other shrublands and savannas. Typical levels are $2000-4000 \, kg \, ha^{-1}$ for Larrea communities in North America and in Argentina, while most other communities are somewhat lower. Productivity rates are commonly $200-900 \, kg \, ha^{-1}$. Although water availability is generally considered to be the most limiting factor for productivity, nitrogen availability may also be very significant. Soil nitrogen levels in Larrea communities fall in the range of only about $450 \, kg \, N \, ha$. Fluxes of nitrogen in warm desert ecosystems are extremely poorly studied. Nitrogen fixation does not appear to be important.

Shrubland (matorral)

The matorral is a common vegetational formation in Central Chile, Northern Mexico and Southern California. The climate of this region is of Mediterranean type with cool moist winters and hot dry summers. Mean temperatures and precipitation vary considerably from one extreme to the other.

In Central Chile the matorral covers approximately 1.0 to 1.5 million hectares, much of which have been turned to agricultural uses. Mean annual temperature is about $14°C$, and mean annual rainfall amounts to $360 \, mm$. Its plant community structure is very similar to other Mediterranean ecosystems of the world. The most abundant plant species are *Lithraea caustica*, *Quillaja saponaria*, *Kageneckia oblonga*, *Trevoa trinervis*, *Colliguaya odorifera*, *Bacharis* sp. and *Cryptocarya alba*. The soils are of alluvial origin and are characterized by low nitrogen content. Much of this soil nitrogen is organic nitrogen tied up and unavailable to the plants.

The importance of the Chilean matorral is not in its agricultural potential, but in the capacity of its vegetation to reduce erosion losses and keep a good water balance in the watershed.

Very few herbaceous legumes species and woody legumes have been described for the Chilean matorral. It is suspected that *Trevos trinervis*, which shows very high nitrogen concentration in its tissue, could fix nitrogen through rhizobia activity. It is postulated (R. Cisternes and L. R. Yates) that nitrogen fixation by free-living bacteria and rainfall are the main soil nitrogen sources. Denitrification will be the most important mechanism to release nitrogen to the soil in the system.

The main source of organic and soluble nitrogen to the soil is litterfall. This amounts to $1.2 \, kg \, m^{-2} \, yr^{-1}$ and 60% of it takes place during the dry months. Decomposition rates are very low ($k = 2.7$) and

Table 1. Case studies of nitrogen cycles in savanna and arid shrub ecosystems of Latin America. Many of these data are tentative

	Cerrado (Brazil)	Prosopis woodland (Sonoran Desert)	Matorral (Chile)
Community type	Tropical savanna	Desert legume savanna	Evergreen shrubland
Precipitation	1500	70	350
Coverage (%)			
Woody	25%	30	55–80
Herbaceous	75%	+	30–80
Biomass (kg ha^{-1})			
Woody	75,000 (m^2/ha)	13,000	~10,000
Herbaceous		+	very low
Productivity (kg ha^{-1} yr^{-1})			
Woody		3700	~1000–1500
Herbaceous		+	very low
Nitrogen pools (kg ha^{-1})			
Soil			
Total	2000–3000 (0–20 cm)	4000	3000 (0–10 cm)
Inorganic	40–120	1150	60–100
A–G Biomass			
Woody	Biomass × 1.5	(7)	60–100
Herbaceous	–	–	0.5–1.0
Nitrogen inputs (kg ha^{-1} yr^{-1})			
Precipitation	5	very small	4–5 (8)
Dry atmospheric			up to 2
N-fixation			
Symbiotic	25	20–30?	<1–2
Free-living	4	small	<1
Nitrogen outputs (kg ha^{-1} yr^{-1})			
Leaching		very small	
Denitrification		<1	<1
Ammonia volatilization			
Erosion		important?	small
Harvesting	1.0 (cattle)		1.6 (fire wood)

show a seasonal trend, with higher rates during spring and fall. It seems very important to establish the influence of leaching, grazing and harvesting in the nitrogen cycle stability of the system. Data on nitrogen cycling in these savanna and arid shrub ecosystems are summarized on Table 1.

Research priorities
1. Soils

The presently available nitrogen data based on the uppermost levels of soil are good, but there is too little information on inorganic nitrogen. Much more data on phosphorus, cation and micronutrient interactions with productivity are also important, particularly for tropical savannas.

2. Biomass

More baseline studies of biomass and productivity in representative savanna sites are badly needed. Below-ground data are almost totally lacking.

3. Fluxes

a) Inputs   The quantitative importance of symbiotic fixation in legume-dominated savannas needs careful documentation. The biological and physical factors limiting fixation on both seasonal and annual bases should be investigated. The importance of dry deposition of atmospheric nutrients in arid regions should also be studied.

b) Outputs   Very few accurate data are available on any outputs. The relative importance of individual outputs varies considerably with precipitation.

c) Shunts   Fire-cycling of nutrients, decomposition and mineralization, and herbivory should be carefully considered as ecosystem processes.

d) Standardization of methodology   The current variation in sampling and analysis techniques makes comparative ecosystem studies more difficult.

4. Geographical representation of data

Considering the very extensive range of savanna ecosystems, the available data base for understanding nitrogen cycling processes is very small. While there is some extensive data for the Brazilian cerrado, there is a clear lack of intensive ecosystem studies in both the cerrado and llanos.

5. Stability of nitrogen cycles

Without better knowledge of the interactive forces of environmental and nutritional factors which limit productivity, it is very difficult to predict the stability of nitrogen cycles or the agronomic capacity of savannas and shrublands. This subject is of critical importance.

Reference

1   Goodland R and Ferri M G 1979 Ecologia do Cerrado. Series Reconquista do Brasil. Univ. São Paulo 52, 193 p.

Plant and Soil 67, 415–420 (1982). 0032-079X/82/0673-0415$00.90.
© 1982 *Martinus Nijhoff/Dr W. Junk Publishers, The Hague*.

Report of the work group on Latin American forests

B. L. BENTLEY (rapporteur), R. HERRERA (chairman), J. T. ARNASON, J. S. MOLINA BUCK,
R. CASTILLEJA, L. E. GARCIA GARCIA, C. F. JORDAN, C. E. RUSSELL, E. SALATI and
E. SANHUEZA

Introduction

Given the variability of tropical forests, it would be presumptious to try to establish a nitrogen budget for them. For this reason we have chosen to analyze the available data across the soil and climate ranges on which tropical forests are found in Latin America. Then, we take the data given for the Amazon elsewhere in this volume and on the basis of soil types try to arrive at a general nitrogen balance. These estimates are then cross-checked with better-known continental atmospheric parameters for nitrogen.

General nitrogen budgets of tropical forest systems

Major sources and sinks of nitrogen in a variety of tropical forests in Latin America (Table 1) are presented in Table 2. These data are initial approximations based on data available at the time that the table was assembled, but nonetheless could be useful for broad-scale general comparisons among systems.

The most striking pattern that emerges from these data is that despite differences in soils, rainfall, and vegetation cover among the different sites, values for biomass, soil carbon, and soil nitrogen show relatively small variations among all undisturbed sites. In contrast, values for soil P, K, and pH are highly variable. Although the significance of the variation is unclear since within-site variability estimates are not presently available, much of this variation can be reduced by dividing the forests into 'oligotrophic' and 'eutrophic' nutrient status classes (Table 2). When this is done, there apparently remains very little overlap of values for all nutrient factors, including carbon and nitrogen.

Nitrogen cycling in the Amazon Basin

On the basis of these and other measurements, we have estimated nitrogen input and output values for the entire Amazon Basin; these estimates are presented in Table 3.

If the NH_4^+ in wet deposition is derived from NH_3 volatilized from soil and foliage, then the NH_3 volatilization can be considered an internal transfer, and denitrification losses would equal total nondeposition N-inputs less hydrologic outputs. This means that *ca.* 14 kg N ha^{-1} yr^{-1} is lost via denitrification in the Amazon Basin (Table 3). This estimate is supported by the facts that 1) the ratio of N_2O plus N_2 loss to NH_4 loss for the Amazon Basin (2.3:1) is very close to that for continental areas in general (2.0:1)[1], and 2) the ratio of biological fixation to denitrification (1.4:1) also matches that estimated for global continental areas (1.4:1)[1]. A schematic nitrogen budget for the Amazon Basin is shown in Fig. 1.

Research priorities

Adequate information is lacking for a number of important nitrogen cycle processes in tropical forest systems of Latin America. We discuss below (in no particular order) those areas that deserve special attention.

Soils

Because methods for soil nutrient analyses are well standardized, data from different studies are generally comparable. Soil data should be collected for any given area under study, and are particularly important in programs investigating other aspects of nitrogen cycling.

Table 1. Selected characteristics of representative tropical forests in Latin America

Feature	Manaus (Brazil)	San Carlos (Venezuela) 1	San Carlos (Venezuela) 2	Indian Church (Belize)	Tabasco (Mexico)	Yurimagnas (Peru)	Pedrera Colombia	Range	Remarks Eutrophic	Remarks Oligotrophic
Climate										
Mean annual temperature	26	26		25	25	26	28	25–28		
Precipitation (mm yr^{-1})	2000	3600		1500	4000	2100	4170	1500–4000		
Rainy season (days yr^{-1})	265	365		265	365	305	183	183–365		
Number of dry months								0–6		
Natural vegetation	dense semi-deciduous forest	evergreen rainforest	caatinga	semi-deciduous, seasonally-dry forest	evergreen rainforest	evergreen rainforest	evergreen rainforest	dry-semi-deciduous to evergreen rainforest		
Soil (to 30 cm)										
C (%)	ca. 1.0	2.8	3.0	7.0	2.9	1.5	na	1–7	3–7	1–2.8
N$_{total}$ (%)	ca. 0.05	0.3	0.2	0.35	0.3	0.2	na	0.02–0.35	0.3–0.35	0.02–0.3
(kg ha^{-1})	na	4000	1000	na	na	na	na	1000–4000	–	–
P (available) (mg kg^{-1})	2.0	2.0	na	15	5	1	na	1–15	5–15	1–2
K (meq 100 g^{-1})	0.08	0.003	na	1.85	1.2	0.10	na	0.003–1.2	1.2–1.9	0.003–0.1
pH	4.6	4.0	4.3	7.2	5.5	4.0	4.3	4.0–7.2	5.5–7.2	4.0–4.6
type	oxisol	oxisol	spodosol	pendoll (mollisol)	ultisol	palendult	na			
Biomass										
total (10^3 kg ha^{-1})	700	500	400	ca. 150	600	500	400	150–700		
N (kg N ha^{-1})	2000	1770	1170	550	na	1600	na	550–2000		

na, not available

Table 2. N-fluxes in selected tropical forests described in Table 1 (kg N ha^{-1} yr^{-1})

Process	Manaus	San Carlos 1	Indian Church	San Carlos 2	Range	Comments
Inputs						
N-fixation						
Rhizobia	200.	na	na	na	na	Varzea forest
Algae	na	na	na	na	na	
Bacteria						
in rhizosphere	15–30	15.	na	30.	15–30	
free-living	1–5	1.	na	5.	1–5.	
Bulk precipitation	0.2	11.	na	12.	0.2–12	predominantly NH$_4^+$
Total in	na	na	na	na	16–235	
Outputs						
NH$_3$-volatilization	na	na	na	na	na	possibly low or patchy (acid soils)
Leaching	0.024	14.	0.	9.	0–14	predominantly NH$_4^+$–N
Denitrification	na	3.	na	na	0–3	range is crude estimate
Erosion	0.8–1.2	0.	0.	0.	0–1.2	Manaus based on Amazon R. load
Burning	na	0.	0.	0.	0.	
Total out	4–6	17.	na	na	0–17	

na, not available

Table 3. Local and regional nitrogen fluxes in the Amazon Basin. Regional fluxes are based on a total area of 6×10^8 ha

Feature	Local (kg N ha^{-1} yr^{-1})	Regional (10^9 kg N)
N-inputs (kg N ha^{-1} yr^{-1})		
N-fixation.		
forests on oxisols (50% of area)	2	
forests on ultisols (45% of area)	20	
Varzea forest (5% of area)	200	
Total (regional average)	20	12.
Bulk precipitation (as NH$_4^+$)	6	3.6
Total in	26	16.
N losses (kg N ha^{-1} yr^{-1})		
Hydrologic*	6	3.6
NH$_3$-volatilization**	6	3.6
Denitrification***	14	8.4
Total out	26	16.

* NH$_4^+$ and organic matter discharged by rivers.
** assumes that volatilization is principal source of bulk-precipitation-N.
*** by difference: N-fixation less hydrologic losses.

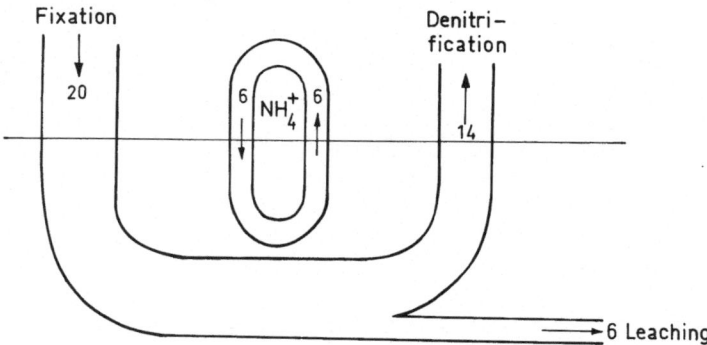

Fig. 1.   Schematic nitrogen budget for the Amazon Basin vased on data in Table 3.

Vegetation and biomass

Much more emphasis should be given to studies of the below-ground aspects of nitrogen in forested ecosystems. Information is needed for both biomass and N-pool sizes as well as for N-fluxes through these compartments and regulating mechanisms.

Climate

More long-term climatic data need to be gathered for most areas. The data we have at present indicate that year-to-year fluctuations can be very great. There is also a need to standardize methodology and the format for reporting this data so that comparisons can be made among different ecosystems. In addition to the number of hours of sunshine, total solar radiation should be included in the data sets.

Nitrogen fluxes

Much greater effort must be invested in studies that attempt to elucidate patterns of fluxes in ecosystems. Several specific areas require attention. First, since variation in time and space is likely to be a very important aspect of fluxes of living systems, standard procedures for adequate sampling must be established. These should include protocols for establishing minimum sample size, inclusions of variance estimates appropriate for the process under study, and a clarification of the effects of patchiness in both time and space. Second, standard methods must be developed for accurately measuring denitrification. Ideally, this should include methods for measuring all forms of gaseous nitrogen denitrified. Third, more complete-watershed studies are necessary in order to establish accurate measures of leaching. Because protocols for water analyses are relatively well-established, the limiting factor in estimates of the magnitude of leaching can be determined from accurate water budgets. In this connection, standard methods for water sample collection, storage, and analysis should be tested and established.

Geographical representation

Tropical forests are diverse and a greater number of different types should be analyzed in detail. The forests most poorly studied currently include montane rainforests, Central American and Mexican forests, and aquatic systems within all forested areas. These studies should be designed so that the data collected can be compared with data from other forest types and with data from various agroecosystems studies.

Mechanisms

Many of the mechanisms that drive the nitrogen cycle are very poorly understood. In particular, mineralization, NH_3 emission from the forest canopy, and potential N_2O uptake need attention. In addition, the general ecology of the microorganisms involved in nitrogen transformations must be studied. These studies should include not only population estimates, but also studies of those factors that control the rates of microbial activities.

Interactions

A greater emphasis also needs to be placed on understanding the interactions of nitrogen with other components of the ecosystems. For example, what are the effects of deficiencies in other elements on nitrogen availability? Will changes in organic matter or pH change nitrogen pools or fluxes? Furthermore, a recognition of the potential effects of insects and larger animals as possible 'bottlenecks,' concentrators, or processors in nutrient flow should be incorporated into ecosystem studies.

Land transformation

Although this report mainly concerns the N-cycle in undisturbed forests, we think that it is of the utmost importance to determine the rate at which forested land is being transformed to other uses. Virtually all aspects of the nitrogen cycle in forests are affected by such transformations, and studies of successional and agricultural systems are needed to assess short and long-term impacts of these changes.

Possible areas for international collaboration

Research areas for which international collaboration is particularly desirable include:

1) identification of the chemical variability, especially of nitrogen compounds, in the large hydrographic basins of the region such as the Amazon, Orinoco, and Paraná Rivers;

2) comparisons of the chemistry of rainwater collected from a series of weather stations in diverse climatic regions;

3) establishment of complete-watershed studies with identical methodologies in distinct ecosystems within the major geomorphological/geochemical areas of the region.

In addition, we feel that collaboration among the various analytical laboratories in the region should be encouraged. An example of the usefulness of such collaboration is the availability of isotope facilities (including ^{15}N production and analyses) at CENA in Piracicaba, Brazil, to cooperating research programs.

Finally, the exchange of scientists and students among research institutions in Latin America and Caribbean countries should also be strongly encouraged.

Reference

1   Söderlund R and Svensson B H 1976 The global nitrogen cycle. *In* Svensson B H and Söderlund R (Eds). Nitrogen, Phosphorus and Sulfur-Global Cycles. SCOPE Report 7. Ecol. Bull. Stockholm 22, 23–73.

Plant and Soil 67, 421–424 (1982).   0032-079X/82/0673-0421$00.60.   SU-42
© 1982 *Martinus Nijhoff/Dr W. Junk Publishers, The Hague.*

Report of the work group on wetlands (including rice) and aquatic systems

K. R. REDDY (rapporteur), V. MONTECINO (chairman), A. J. HOLDING, U. IRMLER, P. S. C. RAO, I. VERGARA and H. W. WILSON

Introduction

General information on the main types of freshwater ecosystems in the region is provided in the following table.

Ecosystem	Area (km^2)	Total N ($\mu g \cdot l^{-1}$)	
		Range	Mean
Standing waters			
Lakes	$62 \cdot 10^3$*	5–2210*	467
Running waters			
Rivers	$81 \cdot 10^5$**		
Flood plains	–	40–2860***	
Rice field			
Upland	$54 \cdot 10^3$		
Irrigated	$21 \cdot 10^3$		

* Greater than 50 km^2.
** Basin area (from FAO: The Inland Waters of Latin America (1979)).
*** Flood plains of white and black waters of Amazonia; data from ref.[1,2,3,4,5].

Freshwater systems

The climatic, topographical, agricultural and urban factors on the continent are associated with a wide range of trophic states in the freshwaters. These states range from very pure oligotrophic waters, mainly in the Andean region, to highly productive waters at high and low altitudes. These ecosystems occur in three main geographical regions:

i) The warm humid region, examples: rivers, and associated flood plains in Colombia and northern parts of Brazil and Peru.

ii) Semi-arid zones, where there are a few natural freshwaters and man made reservoirs are the dominant freshwaters.

iii) Cool per-humid regions typical of the southern parts of Chile and Argentina. These waters are primarily oligotrophic.

Collection of data

Since the end of the meeting a form was prepared and circulated (May–June 1981) among limmologists in Latin America and the Caribbean. 83 forms were sent to different institutions in the following countries: Argentina, Barbados, Bolivia, Brazil, Chile, Colombia, Costa Rica, Ecuador, El Salvador, Germany, Guatemala, Honduras, México, Nicaragua, Panamá, Perú, Trinidad Tobago, Uruguay, U.S.A., and Venezuela.

Sixteen answers were received, 10 of which had no nitrogen information available. Information was received from: Sn. Roque and Salto Grande Reservoir (Argentina): Totoras, Herradura and Gonzales lagoons, Argentina; Laguna El Plateado, Aculeo (Chile); Broa Reservoir and Amazonas (Brazil).

This information was in most of the cases restricted to total concentrations of N in the epilimnion. From the data collected it is evident that it would be of first priority to develop in the region a limnological survey of N budgets and fluxes considering the complete watersheds.

Research priorities

Nitrogen and phosphorus availability is a major factor influencing primary productivity in most lakes. For the global N cycle the aquatic phase is very important. Two priorities stand out:

(A) To quantify the major inputs and outputs for measuring N budgets of representative sites for the major types of ecosystems. At least, two sites for each type of ecosystem in the three geographical regions of the continent should be investigated. These inputs and outputs should be measured by standardized methodology and established techniques for measuring the sizes of the components and fluxes across them. The input and output processes to be investigated are listed below (*** = highest priority):

Inputs: *** (i)   stream flow
          *** (ii)   runoff
          ** (iii) groundwater
          ** (iv)   precipitation
          ** (v)   dry deposition
          *** (vi) N_2-fixation
          ** (vii) organic N mineralization in the sediments and upward flux of N.

Outputs: ** (i)   outflow
          *   (ii)   groundwater
          *** (iii) sedimentation
          *** (iv)   ammonia volatilization
          *** (v)   denitrification
          *   (vi)   nitrate reduction to ammonia and organic N.

(B) To verify and use the existing limnological models as a predictive tool to estimate N budget.

International cooperation

UNEP/MAB/OAS should be encouraged to coordinate this research in Latin America. This is essential since no national scientists have manpower or sites to cover comprehensive ranges of sites. International cooperation is also necessary for standardizing methodologies and organizing workshops to synthesize the data.

Rice production ecosystems

The total area under rice cultivation in South America was reported to be 5.4 million ha under upland conditions, whith 2.1 million ha under lowland or flooded conditions, and total annual production of about 13.5 million tons of grain. Average yields under upland conditions were reported to be about 1.1 t ha^{-1}, while 3.6 t ha^{-1} were reported for lowland conditions. In this report nitrogen cycling under lowland rice will be discussed.

Rice production in Latin America has received considerable attention in recent years, with significant increases in areas under cultivation. However, very little or no information is available on the N inputs or outputs for these ecosystems to calculate N budgets. In this report an attempt will be made to estimate the N inputs for Latin American rice production systems, based on the literature values reported for The southern United States. The estimated fluxes are shown in Table 1. These values should be used with some caution, until additional data are obtained for these ecosystems.

Table 1. Rice ecosystems of Latin America. No research data are available for Latin American countries to calculate the N budget for rice ecosystem

Lowland Rice
Rates of influx and outflux of N (in kg N ha^{-1} growing season^{-1})

Influx	kg N ha^{-1}	Outflux	kg N ha^{-1}
Fixation		Harvest	70–120
by algae	10– 15	NH$_3$-volatilization	15– 75
by bacteria		Leaching and interflow	1– 10
in rhizosphere	5– 15	Denitrification	20–100
or free living	–	Immobilization	10– 30
Atmospheric deposition			
wet	2– 5	Total out	116–335
dry	–		
Crop residues	15– 40		
Fertilizer	50–100		
Soil N mineralized	70–150		
Total in	159–325		

Research priorities

Rice N-cycle research priorities include:

(1) To quantify the input/output processes using standard methodologies. These processes include (*** = highest priority):

Inputs:	***	(i)	fertilizer application rates
	***	(ii)	crop residue application rates and decomposition rates
	***	(iv)	biological N$_2$-fixation at the floodwater-soil interface and root rhizosphere
	**	(v)	root decomposition and N release and
	*	(vi)	atmospheric deposition.

Outputs:	***	(i)	fertilizer N recovery by rice in grain, straw and roots
	***	(ii)	gaseous losses *via* (a) ammonia volatilization and (b) nitrification-denitrification
	***	(iii)	biological immobilization
	*	(v)	leaching, interflow and surface run off.

(2) Mass balance studies using ^{15}N in selected locations. These studies are necessary to follow the fate of applied fertilizer N in the soil-plant system. Studies should be conducted during at least 3–4 years to follow the release of residual fertilizer N.

(3) Nitrogen input/output processes should also be studied during fallow periods of rice fields. This is important for rice ecosystems because of drained conditions that exist during fallow periods which can influence soil processes. For example, when a flooded soil is drained, nitrate-N can accumulate in the system, and upon reflooding nitrate-N can be lost through denitrification.

(4) To develop simulation models to predict the components of the N cycle and the fluxes between them. These models can be useful in fertilizer and water management programmes for rice ecosystems.

References

1   Howard Williams C and Junk W J 1977 The chemical composition of Central Amazonian aquatic macrophytes with special reference to their role in the ecosystem. Arch. Hydrobiol. 79, 446–464.

2   Irmler U 1979 Considerations on the structure and function of the Central Amazonian inundation forest ecosystem with particular emphasis on selected soil animals. Oecologia 43, 1–18.

3   Rai U and Hill G 1981 Physical and chemical studies of Lago Tupí; a Central Amazonian black water 'Ria Lake'. Int. Rev. ges. Hydrobiol. 66, 37–82.

4   Schmidt G 1972 Chemical properties of some waters in the rainforest region of Central Amazonia along the new road Manaus-Caracari. Amazonia 3, 175–185.

5   Schmidt G 1972 Seasonal changes in water chemistry of a tropical lake (Lago de Castanho, Amazonia, South America). Ver. Internat. ver. Limnol. 18, 613–621.

List of participants

N. Ahmad, Department of Soil Science, University of the West Indies, St. Augustine, Trinidad, West Indies

J. Aranguren C., Departamento de Biología y Química, Instituto Universitario Pedagógico, Caracas, Venezuela

J. T. Arnason, Department of Biology, University of Ottawa, Ottawa, Canada

J. Arrivets, Mission IRAT, B.P. 853, Tananarive, Madagascar

B. L. Bentley, Department of Ecology and Evolutionary Biology, SUNY-Stony Brook, NY 11794, USA

E. Bornemisza, Faculdad de Agronomía, Universidad de Costa Rica, Ciudad Universitaria, Costa Rica

R. Sylvester-Bradley, CIAT, Apartado Aéreo 6713, Cali, Colombia

G. Castilleja, Instituto Nat. de Investigaciones Sobre Recursos Bióticos, Apto. 63, Xalapa, Ver., México

R. E. Cisternas, Laboratorio de Ecología, Instituto de Ciencias Biológicas, Pontificia Universidad Católica de Chile, Casilla 114-D, Santiago, Chile

H. Corredor Triana, Departamento de Botánica, Universidad Nacional de Colombia, Bogotá, Colombia

J. Döbereiner, EMBRAPA, Km 47, Seropedica, 23460 Rio de Janeiro, Brazil

G. Escalante R., Centro de Ecología, Instituto de Investigaciones Científicas, Apartado 1827, Caracas 1010A, Venezuela

A. A. Franco, Department of Land, Air and Water Resources, University of California, Davis, CA 95616, USA

L. R. Frederick, Agency for International Development, United States International Development Cooperation Agency, Washington, DC 20523, USA

J. R. Jardim Freire, Faculdade Agronomia, Universidad Federal do Rio Grande do Sul, 90 000 Porto Alegre, RS, Brazil

L. Frioni, Facultad de Agronomía y Veterinaria, Universidad Nacional de Rio Cuarto, 5800 Rio Cuarto, Cordoba, Argentina

P. Fuentes Goda, Bouchard 454–6° piso, Asociación Amigos del Suelo, 1106 Buenos Aires, Argentina

L. E. García García, Avenida Lauro Guerrere 14–15 y Venezuela, Loja, Ecuador

S. R. Gliessman, Environmental Studies, University of California, Santa Cruz, CA 95064, USA

J. Gómez Carrión, Universidad Nacional Mayor de San Marcos, Museo de Historia Natural 'Javier Prado', Av. Arenales 1256 – Apartado 1109, Lima, Peru

C. Gonçalves de Medeiros, Rua Serinhaém 136, Apt. 101, CEP, 50.000, Boa-Viagem, Recife, Pernambuco, Brazil

P. Graham, CIAT, Apartado Aéreo 6713, Cali, Colombia

D. J. Greenwood, National Vegetable Research Station, Wellsbourne, Warwick, CV35 9EF, UK

J. Halliday, NifTAL Project, College of Tropical Agriculture and Human Resources, P.O. Box '0', Paia, HI 96779, USA

G. Hartshorn, Tropical Science Centre, Apartado 8–3870, San José, Costa Rica

R. Herrera, Centro de Ecología, Instituto Venezolano de Investigaciones Científicas, Apartado 1827, Caracas 1010A, Venezuela

A. J. Holding, Department of Agricultural and Food Bacteriology, The Queen's University of Belfast, Newforge Lane, Belfast BT9 5PX, Northern Ireland, UK

U. Irmler, Zoologisches Institut der Universität, Olshaussenstrasse 40–60, D-2300 Kiel, FRG

Y. Z. Ishac, Department of Agricultural Microbiology, Faculty of Agriculture, Ain-Shams University, Shobra, Cairo, Egypt

426

C. F. Jordan, Institute of Ecology, University of Georgia, Athens, GA 30602, USA

J. D. H. Lambert, Department of Biology, Carleton University, Ottawa, Canada K1S 5B6

M. A. Lazzari, Laboratorio de Humus y Biodinámica del Suelo, Departamento de Ciencias Agrarias, Universidad Nacional del Sur, 8000 Bahia Blanca, Argentina

S. N. Levine, Apartado 2132, Las Delicias, Maracay, Venezuela

P. L. Libardi, Centro de Energia Nuclear na Agricultura, Caixa Postal 96, 13400 Piracicaba, S.P., Brazil

L. Longeri S., Departamento de Microbiologia, Universidad de Concepción, Casilla 2407, Concepción, Chile

M. C. J. de J. Martínez Hernández, Colegio de Postgraduados, Centro de Edafología, Chapingo, Mexico

H. H. Mendoza-Placios, Apartado Postal 6331, Guayaquil, Ecuador

V. Montecino B., Departamento Biológia, Faculdad de Ciencias, Universidad de Chile, Casilla 653, Santiago, Chile

J. Pichott Arzusa, Laboratorio de Suelos, Carrera 30 No 48–51, Bogotá, Colombia

J. Pinna Cabrejos, Instituto Central de Investigaciones Azucareras (ICIA), Carretera S/N Casas Grande, Apartado 1071, Trujillo, Peru

P. S. C. Rao, Soil Science Department, Institute of Food and Agricultural Sciences, University of Florida, Gainesville, FL 32611, USA

K. R. Reddy, Agricultural Research and Education Center, Institute of Food and Agricultural Sciences, University of Florida, P.O. Box 909, Sanford, FL 32771, USA

G. P. Robertson, SCOPE/UNEP International Nitrogen Unit, Royal Swedish Academy of Sciences, Box 50005, S-104 05 Stockholm, Sweden

J. P. Roskoski, Instituto Nacional de Investigaciones sobre Recursos Bióticos, Apartado Postal 63, Xalapa, Veracruz, México

T. Rosswall, SCOPE/UNEP International Nitrogen Unit, Royal Swedish Academy of Sciences, Box 50005, S-104 05 Stockholm, Sweden

P. W. Rundel, Department of Ecology and Evolutionary Biology, School of Biological Sciences, University of California, Irvine, CA 92717, USA

A. P. Ruschel, Centro de Energia Nuclear na Agricultura, Caixa Postal 96, 13400 Piracicaba, S.P., Brazil

C. E. Russell, Institute of Ecology, University of Georgia, Athens, GA 30602, USA

E. Salati, Instituto Nacional de Pesquisas da Amazona, Estrado do Aleixo 1756, Manaus, Amazonas, Brazil

P. A. Sanchez, Tropical Soils Program, North Carolina State University, Raleigh, NC 27650, USA

E. Sanhueza, Centro de Ingeniería y Computación, IVIC, Apartado 1827, Caracas 1010-A, Venezuela

M. P. Salema, Department of Soil Science and Plant Nutrition, University of Western Australia, Nedlands, W.A. 6009, Australia

M. B. M. Santana, Centro de Pesquisas do Cacau, Km 26, Rodovia Ilhéus-Itabuna, Caixa Postal 7, Itabuna, BA, Brazil

D. Sanvincenti, Programa de los Naciones Unidas para el Medio Ambiente, Oficina Regional para América Latina, Presidente Masaryk No. 29, México 5, D.F., México

E. C. Schroder, Departamento de Agronomía y Suelos, Universidad de Puerto Rico, Mayaguez, PR 00708, USA

F. R. Tamas, Academia de Ciencias de la República Dominicana, P.O. Box 1365, Santo Domingo, República Dominicana

S. Valdivia Vega, Instituto Central de Investigaciones Azucareras, Carretera S/N, Casa Grande, Apartado Postal 1071, Trujillo, Perú

R. Varela G., Instituto Colombiano Agropecuario ICA, Apartado Aéreo 233, Palmira, Colombia

I. Vergara Favi, Departamento de Química Inorgánica y Analítica, Facultad de Ciencias Químicas y Farmacológicas, Universidad de Chile, Olivos 1007, Casilla 233, Santiago, Chile

M. N. Versteeg, Proyecto FAPROCAF, Apartado 1319, Arequipa, Perú

R. L. Victoria, Centro de Energia Nuclear na Agricultura, Caixa Postal 96, 13 400 Piracicaba, S.P., Brazil

S. Whitney, CIAT, Apartado Aéreo, 6713, Cali, Colombia

H. W. Wilson, la Florissante Garden, D'Abadie, Trinidad and Tobago

L. R. Yates, Laboratorio de Ecología, Instituto de Ciencias Biológicas, Pontificia Universidad Católica de Chile, Casilla 114-D, Santiago, Chile

J. M. Zapater R., Departamento Suelos, Universidad Nacional Agraria, Apartado 456, La Molina, Perú

Subject index

This index lists the key words given at the head of the papers. The figures refer to the first pages of papers, not to the pages on which the subjects are mentioned.